FILTERING COMPLEX TURBULENT SYSTEMS

FILTERING COMPLEX TURBULENT SYSTEMS

Many natural phenomena ranging from climate through to biology are described by complex dynamical systems. Getting information about these phenomena involves filtering noisy data and making predictions based on incomplete information, and often we need to do this in real time, e.g. for weather forecasting or pollution control. All this is further complicated by the sheer number of parameters involved, leading to further problems associated with the "curse of dimensionality" and the "curse of small ensemble size".

The authors develop, for the first time in book form, a systematic perspective on all these issues from the standpoint of applied mathematics. Their approach follows several strands:

- blending classical stability analysis of partial differential equations and their finite difference approximations;
- extending classical Kalman filters and applying them to stochastic models of turbulence to deal with large model errors;
- developing test suites of statistically exactly solvable models and new SPEKF algorithms for filtering slow–fast systems, moist convection, turbulent tracers, and geophysical turbulent systems.

The book contains enough background material from filtering, turbulence theory, and numerical analysis to make the presentation self-contained, and is suitable for graduate courses as well as for researchers in a range of disciplines across science and engineering where applied mathematics is required to enlighten observations and models.

ANDREW J. MAJDA is the Morse Professor of Arts and Sciences at the Courant Institute of New York University.

JOHN HARLIM is an Assistant Professor in the Department of Mathematics at North Carolina State University.

FILTERING COMPLEX TURBULENT SYSTEMS

ANDREW J. MAJDA
New York University

JOHN HARLIM
North Carolina State University

CAMBRIDGE UNIVERSITY PRESS

Cambridge, New York, Melbourne, Madrid, Cape Town,
Singapore, São Paulo, Delhi, Mexico City

Cambridge University Press
The Edinburgh Building, Cambridge CB2 8RU, UK

Published in the United States of America by Cambridge University Press, New York

www.cambridge.org
Information on this title: www.cambridge.org/9781107016668

First published 2012

Printed in the United Kingdom at the University Press, Cambridge

A catalogue record for this publication is available from the British Library

ISBN 978-1-107-01666-8 Hardback

Contents

Preface

This book is an outgrowth of lectures by both authors in the graduate course of the first author at the Courant Institute during spring 2008 and 2010 on the topic of filtering turbulent dynamical systems as well as lectures by the second author at the North Carolina State University in a graduate course in fall 2009. The material is based on the authors' joint research as well as collaborations with Marcus Grote and Boris Gershgorin; the authors thank these colleagues for their explicit and implicit contributions to this material. Chapter 1 presents a detailed overview and summary of the viewpoint and material in the book. This book is designed for applied mathematicians, scientists and engineers, ranging from first- and second-year graduate students to senior researchers interested in filtering large-dimensional complex nonlinear systems.

The first author acknowledges the generous support of DARPA through Ben Mann and ONR through Reza Malek-Madani which funded the research on these topics and helped make this book a reality.

1

Introduction and overview: Mathematical strategies for filtering turbulent systems

Filtering is the process of obtaining the best statistical estimate of a natural system from partial observations of the true signal from nature. In many contemporary applications in science and engineering, real-time filtering of a turbulent signal from nature involving many degrees of freedom is needed to make accurate predictions of the future state. This is obviously a problem with significant practical impact. Important contemporary examples involve the real-time filtering and prediction of weather and climate as well as the spread of hazardous plumes or pollutants. Thus, an important emerging scientific issue is the real-time filtering through observations of noisy signals for turbulent nonlinear dynamical systems as well as the statistical accuracy of spatio-temporal discretizations for filtering such systems. From the practical standpoint, the demand for operationally practical filtering methods escalates as the model resolution is significantly increased. In the coupled atmosphere–ocean system, the current practical models for prediction of both weather and climate involve general circulation models where the physical equations for these extremely complex flows are discretized in space and time and the effects of unresolved processes are parametrized according to various recipes; the result of this process involves a model for the prediction of weather and climate from partial observations of an extremely unstable, chaotic dynamical system with several billion degrees of freedom. These problems typically have many spatio-temporal scales, rough turbulent energy spectra in the solutions near the mesh scale, and a very large-dimensional state space, yet real-time predictions are needed.

Particle filtering of low-dimensional dynamical systems is an established discipline (Bain and Crisan, 2009). When the system is low dimensional or when it has a low-dimensional attractor, Monte Carlo approaches such as the particle filter (Chorin and Krause, 2004) with its various up-to-date resampling strategies (Del Moral, 1996; Del Moral and Jacod, 2001; Rossi and Vila, 2006) provide better estimates in the presence of strong nonlinearity and highly non-Gaussian distributions. However, with the above practical computational constraint in mind, these accurate nonlinear particle filtering strategies are not feasible since sampling a high-dimensional variable is computationally impossible for the foreseeable future. Recent mathematical theory strongly supports this curse of dimensionality for particle filters (Bengtsson *et al.*, 2008; Bickel *et al.*, 2008). Nevertheless some progress in developing particle filtering with small ensemble size for non-Gaussian turbulent dynamical systems is discussed in Chapter 15. These approaches, including the new maximum entropy particle filter (MEPF) due to the authors, all make judicious use of partial marginal distributions to avoid particle collapse. In the second direction, Bayesian hierarchical modeling (Berliner *et al.*, 2003) and reduced-order filtering strategies (Miller *et al.*, 1999; Ghil and Malanotte-Rizolli, 1991; Todling and Ghil, 1994; Anderson, 2001, 2003; Chorin and Krause, 2004; Farrell and Ioannou, 2001, 2005; Ott *et al.*, 2004; Hunt *et al.*, 2007; Harlim and Hunt, 2007b) based on the Kalman filter (Anderson and Moore, 1979; Chui and Chen, 1999; Kaipio and Somersalo, 2005) have been developed with some success in these extremely complex high-dimensional nonlinear systems. There is an inherently difficult practical issue of small ensemble size in filtering statistical solutions of these complex problems due to the large computational overload in generating individual ensemble members through the forward dynamical operator (Haven *et al.*, 2005). Numerous ensemble-based Kalman filters (Evensen, 2003; Bishop *et al.*, 2001; Anderson, 2001; Szunyogh *et al.*, 2005; Hunt *et al.*, 2007) show promising results in addressing this issue for synoptic-scale mid-latitude weather dynamics by imposing suitable spatial localization on the covariance updates; however, all these methods are very sensitive to model resolution, observation frequency and the nature of the turbulent signals when a practical limited ensemble size (typically less than 100) is used. They are also less skillful for more complex phenomena like gravity waves coupled with condensational heating from clouds which are important for the tropics and severe local weather.

Here is a list of fundamental new difficulties in the real-time filtering of turbulent signals that need to be addressed as mentioned briefly above.

1(a) **Turbulent dynamical systems to generate the true signal.** The true signal from nature arises from a turbulent nonlinear dynamical system with extremely complex noisy spatio-temporal signals which have significant amplitude over many spatial scales.

1(b) **Model errors.** A major difficulty in accurate filtering of noisy turbulent signals with many degrees of freedom is model error; the fact that the true signal from nature is processed for filtering and prediction through an imperfect model where by practical necessity, important physical processes are parametrized due to inadequate numerical resolution or incomplete physical understanding. The model errors of inadequate resolution often lead to rough turbulent energy spectra for the truth signal to be filtered on the order of the mesh scale for the dynamical system model used for filtering.

1(c) **Curse of ensemble size.** For forward models for filtering, the state space dimension is typically large, of order 10^4–10^8, for these turbulent dynamical systems, so generating an ensemble size with such a direct approach of order 50–100 members is typically all that is available for real-time filtering.

1(d) **Sparse, noisy, spatio-temporal observations for only a partial subset of the variables.** In systems with multiple spatio-temporal scales, the sparse observations of the truth signal might automatically couple many spatial scales, as shown below in Chapter 7 or in Harlim and Majda (2008b), while the observation of a partial subset of variables might mix together temporal slow and fast components of the system (Gershgorin and Majda, 2008, 2010) as discussed in Chapter 10. For example, observations of pressure or temperature in the atmosphere mix slow vortical and fast gravity wave processes.

This book is an introduction to filtering with an emphasis on the central new issues in 1(a)–(d) for filtering turbulent dynamical systems through the "modus operandi" of the modern applied mathematics paradigm (Majda, 2000a) where rigorous mathematical theory, asymptotic and qualitative models, and novel numerical algorithms are all blended together interactively to give insight into central "cutting edge" practical science problems. In the last several years, the authors have utilized the synergy of modern applied mathematics to address the following:

2(a) How to develop simple off-line mathematical test criteria as guidelines for filtering extremely stiff multiple space–time scale problems that often arise in filtering turbulent signals through plentiful and sparse observations? (Majda and Grote, 2007; Castronovo *et al.*, 2008; Grote and Majda, 2006; Harlim and Majda, 2008b)
2(b) For turbulent signals from nature with many scales, even with mesh refinement, the model has inaccuracies from parametrization, under-resolution, etc. Can judicious model errors help filtering and simultaneously overcome the curse of dimensionality? (Castronovo *et al.*, 2008; Harlim and Majda, 2008a,b, 2010a)
2(c) Can new computational strategies based on stochastic parametrization algorithms be developed to overcome the curse of dimensionality, to reduce model error and improve the filtering as well as the prediction skill? (Gershgorin *et al.*, 2010a,b; Harlim and Majda, 2010b)
2(d) Can exactly solvable models be developed to elucidate the central issues in 1(d) for turbulent signals, to develop unambiguous insight into model errors and to lead to efficient new computational algorithms? (Gershgorin and Majda, 2008, 2010)

The main goals of this book are the following: first, to introduce the reader to filtering from this viewpoint in an elementary fashion where no prior background on these topics is assumed (Chapters 2–4); secondly, to describe in detail the recent and ongoing developments, emphasizing the remarkable new mathematical and physical phenomena that emerge from the modern applied mathematics modus operandi applied to filtering turbulent dynamical systems. Next, in this introductory chapter, we provide an overview of turbulent dynamical systems and basic filtering followed by an overview of the basic applied mathematics motivation which leads to the new developments and viewpoint emphasized in this book.

1.1 Turbulent dynamical systems and basic filtering

The large-dimensional turbulent dynamical systems which define the true signal from nature to be filtered in the class of problems studied here have a fundamentally different statistical character than in more familiar low-dimensional chaotic dynamical systems. The most well-known low-dimensional chaotic dynamical system is Lorenz's famous three-equation model (Lorenz, 1963) which is weakly mixing with one unstable direction on an attractor with high symmetry. In contrast, realistic turbulent dynamical systems have a large phase space dimension, a large dimensional unstable manifold on the attractor, and are strongly mixing with exponential decay of correlations. The simplest prototype example of a turbulent dynamical system is also due to Lorenz and is called the L-96 model (Lorenz, 1996; Lorenz and Emanuel, 1998). It is widely used as a test model for algorithms for prediction, filtering and low-frequency climate response (Majda *et al.*, 2005; Majda and Wang, 2006). The L-96 model is a discrete periodic model given by the following system

$$\frac{du_j}{dt} = (u_{j+1} - u_{j-2})u_{j-1} - u_j + F, \quad j = 0, \ldots, J-1, \tag{1.1}$$

with $J = 40$ and with F the forcing parameter. The model is designed to mimic baroclinic turbulence in the mid-latitude atmosphere with the effects of energy-conserving nonlinear advection and dissipation represented by the first two terms in (1.1). For sufficiently strong forcing values such as $F = 6, 8, 16$, the L-96 model is a prototype turbulent dynamical system which exhibits features of weakly chaotic turbulence ($F = 6$), strongly chaotic turbulence ($F = 8$), and strong turbulence ($F = 16$) (Majda *et al.*, 2005). In order to quantify and compare the different types of turbulent chaotic dynamics in the L-96 model as F is varied, it is convenient to rescale the system to have unit energy for statistical fluctuations around the constant mean statistical state, $\bar{u}$ (Majda *et al.*, 2005); thus, the transformation $u_j = \bar{u} + E_p^{1/2}\tilde{u}_j, t = \tilde{t}E_p^{-1/2}$ is utilized where E_p represents the energy fluctuations (Majda *et al.*, 2005). After this normalization, the mean state becomes zero and the energy fluctuations are unity for all values of F. The dynamical equation in terms of the new variables, $\tilde{u}_j$, becomes

$$\frac{d\tilde{u}_j}{d\tilde{t}} = (\tilde{u}_{j+1} - \tilde{u}_{j-2})\tilde{u}_{j-1} + E_p^{-1/2}((\tilde{u}_{j+1} - \tilde{u}_{j-2})\bar{u} - \tilde{u}_j) + E_p^{-1}(F - \bar{u}). \tag{1.2}$$

Table 1.1 lists, in the non-dimensional coordinates, the leading Lyapunov exponent, λ_1, the dimension of the unstable manifold, N^+, the sum of the positive Lyapunov exponents (the KS entropy) and the correlation time, T_{corr}, of any $\tilde{u}_j$ variable with itself as F is varied

Table 1.1 Dynamical properties of the L-96 model for regimes with $F = 6, 8, 16$. λ_1 denotes the largest Lyapunov exponent, N^+ denotes the dimension of the expanding subspace of the attractor, KS denotes the Kolmogorov–Sinai entropy and T_{corr} denotes the decorrelation time of the energy-rescaled time correlation function.

	F	λ_1	N^+	KS	T_{corr}
Weakly chaotic	6	1.02	12	5.547	8.23
Strongly chaotic	8	1.74	13	10.94	6.704
Fully turbulent	16	3.945	16	27.94	5.594

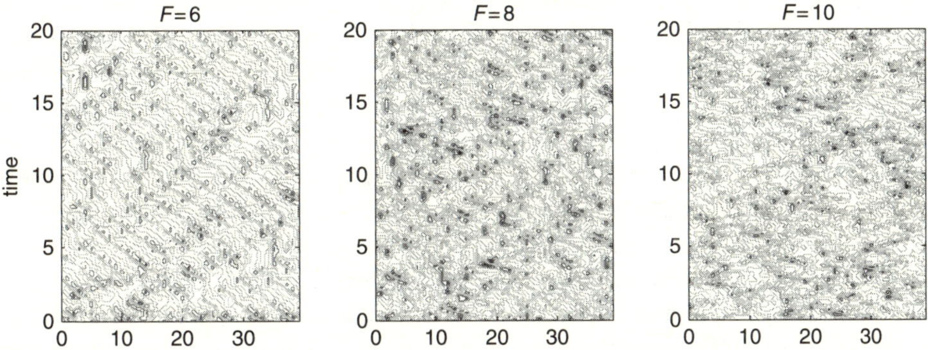

Figure 1.1 Space–time diagrams of numerical solutions of the L-96 model for the weakly chaotic ($F = 6$), strongly chaotic ($F = 8$) and fully turbulent ($F = 16$) regimes.

through $F = 6, 8, 16$. Note that λ_1, N^+ and KS increase significantly as F increases while T_{corr} decreases in these non-dimensional units; furthermore, the weakly turbulent case with $F = 6$ already has a 12-dimensional unstable manifold in the 40-dimensional phase space. Snapshots of the time series for (1.1) with $F = 6, 8, 16$, as depicted in Fig. 1.1, qualitatively confirm the above quantitative intuition with weakly turbulent patterns for $F = 6$, strongly chaotic wave turbulence for $F = 8$, and fully developed wave turbulence for $F = 16$. It is worth remarking here that smaller values of F around $F = 4$ exhibit the more familiar low-dimensional weakly chaotic behavior associated with the transition to turbulence.

In regimes to realistically mimic properties of nature, virtually all atmosphere, ocean and climate models with sufficiently high resolution are turbulent dynamical systems with features as described above. The simplest paradigm model of this type is the two-layer quasi-geostrophic (QG) model in doubly periodic geometry that is externally forced by a mean vertical shear (Smith *et al.*, 2002), which has baroclinic instability (Salmon, 1998); the properties of the turbulent cascade have been extensively discussed in this setting, e.g. see Salmon (1998) and citations in Smith *et al.* (2002). The governing equations for the two-layer QG model with a flat bottom, rigid lid and equal-depth layers H can be written as

$$\begin{aligned}
\frac{\partial q_1}{\partial t} + J(\psi_1, q_1) + U\frac{\partial q_1}{\partial x} + (\beta + k_d^2 U)\frac{\partial \psi_1}{\partial x} + \nu\nabla^8 q_1 = 0,\\
\frac{\partial q_2}{\partial t} + J(\psi_2, q_2) - U\frac{\partial q_2}{\partial x} + (\beta - k_d^2 U)\frac{\partial \psi_2}{\partial x} + \kappa\nabla^2\psi_2 + \nu\nabla^8 q_2 = 0,
\end{aligned} \tag{1.3}$$

where subscript 1 denotes the top layer and 2 the bottom layer; ψ is the perturbed stream function; $J(\psi, q) = \psi_x q_y - \psi_y q_x$ is the Jacobian term representing nonlinear advection; U is the zonal mean shear; β is the meridional gradient of the Coriolis parameter; q is the perturbed quasi-geostropic potential vorticity, defined as follows

$$q_i = \beta y + \nabla^2\psi_i + \frac{k_d^2}{2}(\psi_{3-i} - \psi_i), \quad i = 1, 2, \tag{1.4}$$

where $k_d = \sqrt{8}/L_d$ is the wavenumber corresponding to the Rossby radius L_d; κ is the Ekman bottom drag coefficient; and ν is the hyperviscosity coefficient. Note that Eqns (1.3) are the prognostic equations for perturbations around a uniform shear with stream function $\Psi_1 = -Uy$, $\Psi_2 = Uy$ as the background state, and the hyperviscosity term, $\nu\nabla^8 q$, is added to filter out the energy buildup on the smaller scales.

This is the simplest climate model for the poleward transport of heat in the atmosphere or ocean and with a modest resolution of $128 \times 128 \times 2$ grid points has a phase space of more than 30,000 variables. Again for modeling the atmosphere and ocean, this model in the appropriate parameter regimes is a strongly turbulent dynamical system with strong cascades of energy (Salmon, 1998; Smith *et al.*, 2002; Kleeman and Majda, 2005); it has been utilized recently as a test model for algorithms for filtering sparsely observed turbulent signals in the atmosphere and ocean (Harlim and Majda, 2010b).

1.1.1 Basic filtering

We assume that observations are made at uniform discrete times, $m\Delta t$, with $m = 1, 2, 3, \ldots$ For example, in global weather prediction models, the observations are given as inputs in the model every six hours and for large-dimensional turbulent dynamical systems, it is a challenge to implement continuous observations, practically. As depicted in Fig. 1.2, filtering is a two-step process involving statistical prediction of a probability distribution for the state variable u through a forward operator on the time interval between observations followed by an analysis step at the next observation time which corrects this probability distribution on the basis of the statistical input of noisy observations of the system. In the present applications, the forward operator is a large-dimensional dynamical system perhaps with noise written in the Itô sense as

$$\frac{du}{dt} = F(u,t) + \sigma(u,t)\dot{W}(t) \tag{1.5}$$

for $u \in \mathbb{R}^N$, where σ is an $N \times K$ noise matrix and $\dot{W} \in \mathbb{R}^K$ is K-dimensional white noise. The Fokker–Planck equation for the probability density, $p(u,t)$, associated with (1.5) is

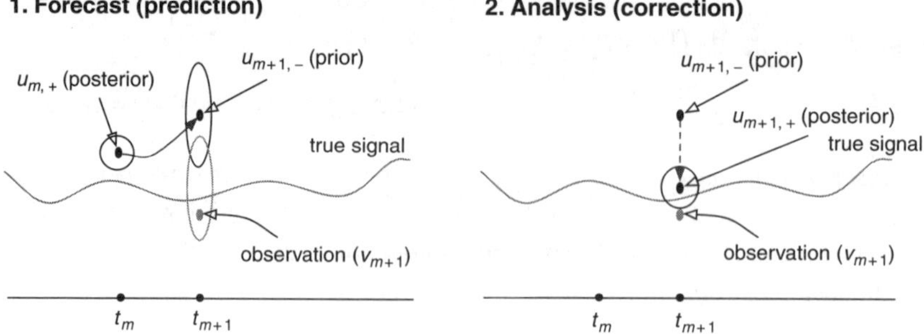

Figure 1.2 Filtering: Two-step predictor–corrector method.

$$p_t = -\nabla_u \cdot (F(u,t)p) + \frac{1}{2}\nabla_u \cdot \nabla_u(Qp) \equiv L_{\mathrm{FP}}p \tag{1.6}$$
$$p_t|_{t=t_0} = p_0(u)$$

with $Q(t) = \sigma\sigma^T$. For simplicity in exposition, here and throughout the remainder of the book we assume M linear observations, $\vec{v}_m \in \mathbb{R}^M$, of the true signal from nature given by

$$\vec{v}_m = Gu(m\Delta t) + \vec{\sigma}_m^o, \ m = 1, 2, \ldots \tag{1.7}$$

where G maps $\mathbb{R}^N$ into $\mathbb{R}^M$ while the observational noise, $\vec{\sigma}_m^o \in \mathbb{R}^M$, is assumed to be a zero-mean Gaussian random variable with $M \times M$ covariance matrix,

$$R^o = \langle \vec{\sigma}_m^o \otimes (\vec{\sigma}_m^o)^T \rangle. \tag{1.8}$$

Gaussian random variables are uniquely determined by their mean and covariance; here and below, we utilize the standard notation $\mathcal{N}(\vec{X}, R)$ to denote a vector Gaussian random variable with mean $\vec{X}$ and covariance matrix R. With these preliminaries, we describe the two-step filtering algorithm with the dynamics in (1.5), (1.6) and the noisy observations in (1.7), (1.8). Start at time step $m\Delta t$ with a posterior probability distribution, $p_{m,+}(u)$, which takes into account the observations in (1.7) at time $m\Delta t$. Calculate a prediction or forecast probability distribution, $p_{m+1,-}(u)$, by using (1.6), in other words, let p be the solution of the Fokker–Planck equation,

$$p_t = L_{\mathrm{FP}}p, \quad m\Delta t < t \le (m+1)\Delta t \tag{1.9}$$
$$p|_{t=m\Delta t} = p_{m,+}(u).$$

Define $p_{m+1,-}(u)$, the prior probability distribution before taking observations at time $m+1$ into account, by

$$p_{m+1,-}(u) \equiv p(u, (m+1)\Delta t) \tag{1.10}$$

with p determined by the forward dynamics in (1.9). Next, the analysis step at time $(m+1)$ Δt which corrects this forecast and takes the observations into account is implemented by using Bayes' theorem

$$\begin{aligned} p_{m+1,+}(u)p(v_{m+1}) &= p_{m+1}(u|v_{m+1})p(v_{m+1}) \\ &= p_{m+1}(u,v) = p_{m+1}(v_{m+1}|u)p_{m+1,-}(u). \end{aligned} \tag{1.11}$$

With Bayes' formula in (1.11), we calculate the posterior distribution

$$p_{m+1,+}(u) \equiv p_{m+1}(u|v_{m+1}) = \frac{p_{m+1}(v_{m+1}|u)p_{m+1,-}(u)}{\int p_{m+1}(v_{m+1}|u)p_{m+1,-}(u)du}. \tag{1.12}$$

The two steps described in (1.9), (1.10), (1.12) define the basic nonlinear filtering algorithm which forms the theoretical basis for practical design of algorithms for filtering turbulent dynamical systems (Jazwinski, 1970; Bain and Crisan, 2009). While this is conceptually clear, practical implementation of (1.9), (1.10), (1.12), directly in turbulent dynamical systems, is impossible due to large state space, $N \gg 1$, as well as the fundamental difficulties elucidated in 1(a)–(d) in the introduction.

The most important and famous example of filtering is the Kalman filter where the analysis step in (1.5) is associated with linear dynamics which can be integrated between observation time steps $m\Delta t$ and $(m+1)\Delta t$ to yield the forward operator

$$u_{m+1} = Fu_m + \bar{f}_{m+1} + \sigma_{m+1}. \tag{1.13}$$

Here F is the $N \times N$ system operator matrix and σ_m is the system noise assumed to be zero-mean and Gaussian with $N \times N$ covariance matrix

$$R = \langle \sigma_m \otimes \sigma_m^T \rangle, \; \forall m, \tag{1.14}$$

while $\bar{f}_m$ is a deterministic forcing. Next, we present the simplified Kalman filter equations for the linear case. First assume the initial probability density $p_0(u)$ is Gaussian, i.e. $p_0(u) = \mathcal{N}(\bar{u}_0, R_o)$ and assume by recursion that the posterior probability distribution, $p_{m,+}(u) = \mathcal{N}(\bar{u}_{m,+}, R_{m,+})$, is also Gaussian. By using the linear dynamics in (1.13), the forecast or prediction distribution at time $(m+1)\Delta t$ is also Gaussian,

$$\begin{aligned} p_{m+1,-}(u) &= \mathcal{N}(\bar{u}_{m+1,-}, R_{m+1,-}) \\ \bar{u}_{m+1,-} &= F\bar{u}_{m,+} + \bar{f}_{m+1} \\ R_{m+1,-} &= FR_{m,+}F^T + R. \end{aligned} \tag{1.15}$$

With the assumptions in (1.7), (1.8) and (1.13), (1.15), the analysis step in (1.12) becomes an explicit regression procedure for Gaussian random variables (Chui and Chen, 1999; Anderson and Moore, 1979) so that the posterior distribution, $p_{m+1,+}(u)$, is also Gaussian yielding the **Kalman filter**

$$\begin{aligned} p_{m+1,+}(u) &= \mathcal{N}(\bar{u}_{m+1,+}, R_{m+1,+}) \\ \bar{u}_{m+1,+} &= (\mathcal{I} - K_{m+1}G)\bar{u}_{m+1,-} + K_{m+1}v_{m+1} \\ R_{m+1,+} &= (\mathcal{I} - K_{m+1}G)R_{m+1,-} \\ K_{m+1} &= R_{m+1,-}G^T(GR_{m+1,-}G^T + R^o)^{-1}. \end{aligned} \tag{1.16}$$

The $N \times M$ matrix, K_{m+1}, is the Kalman gain matrix. Note that the posterior mean after processing the observations is a weighted sum of the forecast and analysis contributions through the Kalman gain matrix and also that the observations reduce the covariance, $R_{m+1,+} \leq R_{m+1,-}$. In this Gaussian case with linear observations, the analysis step going from (1.15) to (1.16) is a standard linear least-squares regression. An excellent treatment of this can be found in chapter 3 of Kaipio and Somersalo (2005). There is a huge literature on Kalman filtering; two excellent basic texts are Chui and Chen (1999) and Anderson and Moore (1979) where more details and references can be found. Our intention in the introductory parts in this book in Chapters 2 and 3 is not to repeat the well-known material in (1.15), (1.16) in detail; instead we introduce this elementary material in a fashion to set the stage for the mathematical guidelines developed in Part II (Chapters 5–8) and the applications to filtering turbulent nonlinear dynamical systems presented in Part III (Chapters 9–15).

Naively, the reader might expect that everything is known about filtering linear systems; however, when the linear system is high dimensional, i.e. $N \gg 1$, the same issues elucidated in 1(a)–(d) occur for linear systems in a more transparent fashion. This is the viewpoint emphasized and developed in Part II of the book (Chapters 5–8) which is motivated next. For linear systems without model errors, the recursive Kalman filter is an optimal estimator but the recursive nonlinear filter in (1.7)–(1.12) may not be an optimal estimator for the nonlinear stochastic dynamical system without model error in (1.5).

1.2 Mathematical guidelines for filtering turbulent dynamical systems

How can useful mathematical guidelines be developed in order to elucidate and ameliorate the central new issues in 1(a)–(d) from the introduction for turbulent dynamical systems? This is the topic of this section. Of course, to be useful, such mathematical guidelines have to be general yet still involve simplified models with analytical tractability. Such criteria have been developed recently by Majda and Grote (2007); Castronovo *et al.* (2008) and Harlim and Majda (2008b) through the modern applied mathematics paradigm and the goal here is to outline this development and discuss some of the remarkable phenomena which occur. The starting point for this development for filtering turbulent dynamical systems involves the symbiotic interaction of three different disciplines in applied mathematics/physics, as depicted in Fig. 1.3: stochastic modeling of turbulent signals, numerical analysis of PDEs and classical filtering theory outlined in (1.13)–(1.16) of Section 1.1. Here is the motivation from the three legs of the triangle.

First, the simplest stochastic models for modeling turbulent fluctuations consist of replacing the nonlinear interaction at these modes by additional dissipation and white noise

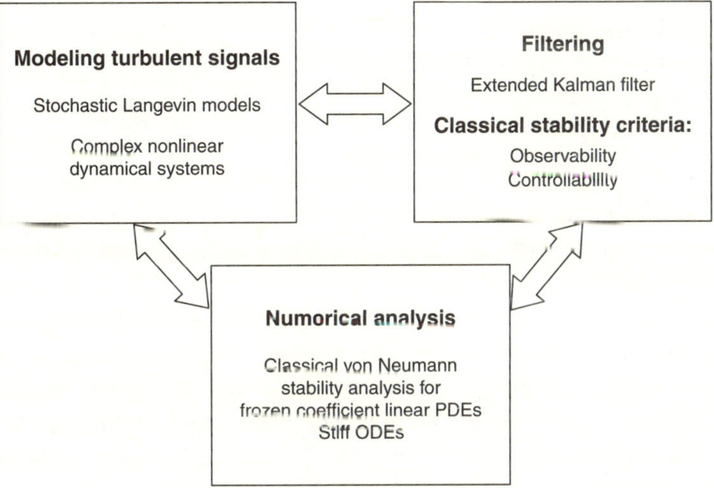

Figure 1.3 Modern applied mathematics paradigm for filtering.

forcing to mimic rapid energy transfer (Salmon, 1998; Majda *et al.*, 1999, 2003; Majda and Timofeyev, 2004; Delsole, 2004; Majda *et al.*, 2008). Conceptually, we view this stochastic model for a given turbulent (Fourier) mode as given by the linear Langevin SDE or Ornstein–Uhlenbeck process for the complex scalar

$$du(t) = \lambda u(t)dt + \sigma dW(t), \quad (1.17)$$
$$\lambda = -\gamma + i\omega, \gamma > 0,$$

with $W(t)$ a complex Wiener process, and σ its noise strength. Of course, the amplitude and strength of these coefficients, γ, σ, vary widely for different Fourier modes and depend empirically on the nonlinear nature of the turbulent cascade, the energy spectrum, etc. These simplest turbulence models are developed in detail in Chapter 5 and an important extension with intermittent instability at large scales is developed in Chapter 8. Quantitative illustrations of this modeling process for the L-96 model in (1.1) in a variety of regimes and the two-layer model in (1.3) are developed in Part III in Chapters 12 and 13, together with cheap stochastic filters with judicious model errors based on these linear stochastic models.

Secondly, the most successful mathematical guideline for numerical methods for deterministic nonlinear systems of PDEs is von Neumann stability analysis (Richtmeyer and Morton, 1967): The nonlinear problem is linearized at a constant background state, and Fourier analysis is utilized for this constant-coefficient PDE, resulting in discrete approximations for a complex scalar test model for each Fourier mode,

$$\frac{du(t)}{dt} = \lambda u(t), \lambda = -\gamma + i\omega, \gamma > 0. \quad (1.18)$$

All the classical mathematical phenomena such as, for example, the CFL stability condition on the time step Δt and spatial mesh h, $|c|\Delta t/h < 1$, for various explicit schemes for the advection equation $u_t + cu_x = -du$, occur because, at high spatial wavenumbers, the scalar test problem in (1.18) is a stiff ODE, i.e.

$$|\lambda| \gg 1. \quad (1.19)$$

For completeness, Chapter 4 provides a brief introduction to this analysis.

The third leg of the triangle involves classical linear Kalman filtering as outlined in (1.13)–(1.16). In conventional mathematical theory for filtering linear systems, one checks algebraic observability and controllability conditions (Chui and Chen, 1999; Anderson and Moore, 1979) and is automatically guaranteed asymptotic stability for the filter; this theory applies for a fixed state dimension and is a very useful mathematical guideline for linear systems that are not stiff in low-dimensional state space. Grote and Majda (2006) developed striking examples involving unstable differencing of the stochastic heat equation where the state space dimension is $N = 42$ with 10 unstable modes where the classical observability (Cohn and Dee, 1988) and controllability conditions were satisfied yet the filter covariance matrix had condition number 10^{13} so there is no practical filtering skill!

This suggested that there were new phenomena in filtering turbulent signals from linear stochastic systems which are suitably stiff with multiple spatio-temporal scales.

Chapter 2 provides an elementary self-contained introduction to filtering the complex scalar test problem in (1.17) and in Section 2.3 describes the new phenomena that can occur in stiff regimes with model error as a prototype for the developments in Part II. In Chapters 6 and 7, the analogue of von Neumann stability analysis for filtering turbulent dynamical systems is developed. The new phenomena that occur and the robust mathematical guidelines that emerge are studied for plentiful observations, where the number of observations equals the number of mesh points in Chapter 6, and for the practically important and subtle case of sparse regular observations in Chapter 7.

Clearly, successful guidelines for filtering turbulent dynamical systems need to depend on many interacting features for these turbulent signals with complex multiple spatio-temporal structure:

- 3(a) The specific underlying dynamics.
- 3(b) The energy spectrum at spatial mesh scales of the observed system and the system noise, i.e. decorrelation time, on these scales.
- 3(c) The number of observations and the strength of the observational noise.
- 3(d) The time-scale between observations relative to 3(a), (b).

Examples of what are two typical practical computational issues for the filtering in the above context to avoid the "curse of ensemble size" from (1.1) are the following:

- 4(a) When is it possible to use for filtering the standard explicit scheme solver of the original dynamic equations by violating the CFL stability condition with a large time step equal (proportional) to the observation time to increase ensemble size yet retain statistical accuracy?
- 4(b) When is it possible to use for filtering a standard implicit scheme solver for the original dynamic equations by using a large time step equal to the observation time to increase ensemble size yet retain statistical accuracy?

Clearly resolving the practical issues in 4(a),(b) involves the understanding of 3(a)–(d) in a given context and this is emphasized in Part II of the book. In particular, the role of model error in filtering stochastic systems with intermittent large-scale instability is emphasized in Chapters 8 and 13.

1.3 Filtering turbulent dynamical systems

Part III of the book (Chapters 9–15) is devoted to contemporary strategies for filtering turbulent dynamical systems as described earlier in Section 1.1 and coping with the difficult issues in 1(a)–(d) mentioned earlier. In Chapter 9, contemporary strategies for filtering nonlinear dynamical systems with a perfect model are surveyed and their relative skill and merits are discussed for the three-mode, Lorenz-63 model. The application of the finite ensemble filters from Chapter 9 to high-dimensional turbulent dynamical system such as the L-96 model and the two-layer QG model are described in Chapter 11.

Given all the complexity in filtering turbulent signals described in 1(a)–(d), an important topic is to develop nonlinear models with exactly solvable nonlinear statistics which provide unambiguous benchmarks for these various issues in 1(a)–(d) for more general turbulent dynamical systems. Of course, it is a challenge for applied mathematicians to develop such types of test models which are simple enough for tractable mathematical analysis yet capture key features of complex physical processes which they try to mimic. Once such exactly solvable test models have been developed, all of the issues regarding model error in 1(b) as well as new nonlinear algorithms for filtering can be studied in an unambiguous fashion. Many important problems in science and engineering have turbulent signals with multiple time-scales, i.e. slow–fast systems. The development of such a test model for prototype slow–fast systems is the topic of Chapter 10.

With the mathematical guidelines for filtering turbulent dynamical systems based on linear stochastic models with multiple spatio-temporal scales as discussed in Part II, we discuss the real-time filtering of nonlinear turbulent dynamical systems by such an approach in Chapter 12. The mathematical guidelines and phenomena in Part II suggest the possibility that there might be cheap skillful alternative filter algorithms which can cope with the issues in 1(c),(d) where suitable linear stochastic models with judicious model error are utilized to filter true signals from nonlinear turbulent dynamical systems like the two models discussed in Section 1.1 above. We discuss this radical approach to filtering turbulent signals in Chapter 12. In Chapter 13, we show how exactly solvable test models can be utilized to develop new algorithms for filtering turbulent signals which correct the model errors for the linear stochastic models developed in Chapter 12 by updating the damping and forcing "on the fly" through a stochastic parametrized extended Kalman filter (SPEKF) algorithm (Gershgorin *et al.*, 2010b,a; Harlim and Majda, 2010b). Chapter 14 is devoted to the development and filtering of exactly solvable test models for turbulent diffusion. In particular, we emphasize the recovery of detailed turbulent statistics including the energy spectrum and non-Gaussian probability distributions from sparse space–time observations through a generalization of the specific nonlinear extended Kalman filter (NEKF) which we introduced earlier in Chapter 10. Finally, in Chapter 15, we describe the search for efficient skillful particle filters for high-dimensional turbulent dynamical systems; this requires successful particle filtering with small ensemble sizes in non-Gaussian statistical settings. In this context, trying to filter the L-63 model with temporally sparse partial observations with small noise with only 3–10 particles is a challenging toy problem. We introduce a new maximum entropy particle filter (MEPF) with exceptional skill on this toy model and discuss the strengths and limitations of current particle filters with small ensemble size for the L-96 model.

Part I

Fundamentals

2

Filtering a stochastic complex scalar: The prototype test problem

As discussed in the introductory chapter, the scientific issue in real-time prediction problems is to provide a statistical estimate of a true state given that the nature of the physical process is chaotic and given the fact that the measurements (observations) are inaccurate or sometimes even unavailable. Essentially, at every time when observations become available, one chooses the best estimate of the true state by accounting for the prior forecasts and these observations. There are two challenges for improving the real-time prediction of a turbulent signal with multiple scales: the first is to improve the model which suffers from model error since we don't yet understand the underlying physical processes. Even if we do, we cannot realize these processes at every temporal or spatial scale with our limited computing power. Thus, the second challenge is to provide efficient and accurate strategies that meet this practical constraint.

In this chapter, we derive the one-dimensional Kalman filter formula which is a specific analytical solution of the Bayesian update in a simplified setting. As mentioned in the introductory chapter, this is an important test problem for filtering multi-scale turbulent systems and this point of view is emphasized here. We show the numerical results of filtering a one-dimensional complex Ornstein–Uhlenbeck process. We then discuss the conditions for filtering stability and compute the asymptotic off-line variables in closed form for the one-dimensional Kalman filter. In Section 2.3, we discuss the model error due to finite difference approximations in a stiff regime as an important but elementary prototype problem for future developments. Finally, we close this chapter with one judicious strategy for dealing with model errors through an information criterion and illustrate how it improves filter performance with model errors in the stiff regime.

2.1 Kalman filter: One-dimensional complex variable

As discussed in Chapter 1, discrete-time filtering theory is a two-step predictor–corrector method that treats the model as a black box for the forecast (or prediction) step while the analysis (or correction) step is basically a Bayesian update that accounts for both the prior forecasts and the observations.

Let $u_m \in \mathbb{C}$ be a complex random variable whose dynamics is given by the following discrete process:

$$u_{m+1} = Fu_m + \sigma_{m+1},$$

where $F \in \mathbb{C}$ is a complex number and $\sigma_{m+1} \in \mathbb{C}$ is a complex Gaussian system noise

$$\sigma_{m+1} \equiv \frac{\sigma_{1,m+1} + \mathrm{i}\sigma_{2,m+1}}{\sqrt{2}},$$

where each component $\sigma_{j,m+1}$ is a real unbiased Gaussian noise with variance $\langle \sigma_{j,m+1}^2 \rangle$ with the property $\langle \sigma_{i,m+1}\sigma_{j,m+1} \rangle = 0$ when $i \neq j$ so that different noises are independent. The notation $\langle \cdot \rangle$ denotes expectation. Thus, the complex Gaussian noise σ_{m+1} has zero mean and variance

$$r = \langle \sigma_{m+1}\sigma_{m+1}^* \rangle = \frac{1}{2}\sum_{j=1}^{2} \langle \sigma_{j,m+1}^2 \rangle.$$

The goal of the Kalman filter is to estimate the unknown true state u_{m+1}, given noisy observations

$$v_{m+1} = gu_{m+1} + \sigma_{m+1}^o, \tag{2.1}$$

where $g \in \mathbb{R}$ is a linear observation operator and $\sigma_m^o \in \mathbb{C}$ is an unbiased Gaussian noise with variance

$$r^o \equiv \langle \sigma_m^o (\sigma_m^o)^* \rangle.$$

The Kalman filter is the optimal (in the least-squares sense) solution found by assuming that the model and the observation operator that relates the model state with the observation variables are both linear and both the observation and prior forecast error uncertainties are Gaussian, unbiased and uncorrelated. In particular, the observation error distribution of v at time t_{m+1} is a Gaussian conditional distribution

$$p(v_{m+1}|u_{m+1}) \sim \mathcal{N}(gu_{m+1}, r^o),$$

which depends on the true state u_{m+1} through Eqn (2.1). We often refer to $p(v_{m+1}|u_{m+1})$ as the likelihood of estimating u_{m+1} given observation v_{m+1}.

Suppose that the filter model is perfectly specified, i.e. exactly the model that governs the dynamics of the true signal (this situation is referred to as the perfect model experiment, the twin experiment, or the true filter by Castronovo *et al.* (2008)). An estimate of the true state prior to knowledge of the observation at time t_{m+1}, the prior state (sometimes also called the forecast or background), $u_{m+1|m} \in \mathbb{C}$, is given by

$$u_{m+1|m} = Fu_{m|m} + \sigma_{m+1}, \tag{2.2}$$

where $u_{m|m} \in \mathbb{C}$ is the posterior state (an estimate of the true state after the observation at time m has been considered, sometimes also called the analysis state). These notations are defined as estimates of the true signal $u_m \in \mathbb{C}$. The notation utilized here and throughout the remainder of the book differs from that utilized in (1.10), (1.15), (1.16) in Chapter 1 but should not confuse the reader; $u_{m+1|m}$ corresponds to the prior distribution $p_{m+1,-}(u)$ after the the forecast in (1.9) while $u_{m+1|m+1}$ corresponds to the posterior distribution $p_{m+1,+}(u)$ after the analysis step in (1.12).

From the probabilistic point of view, we can represent this prior estimate with a probability density (distribution), $p(u_{m+1})$. This prior distribution accounts only for the earlier observations up to time t_m:

$$p(u_{m+1}) \sim \mathcal{N}(\bar{u}_{m+1|m}, r_{m+1|m}),$$

where

$$\bar{u}_{m+1|m} \equiv \langle u_{m+1|m} \rangle \tag{2.3}$$

and

$$r_{m+1|m} \equiv \langle e_{m+1|m} e^*_{m+1|m} \rangle \equiv \langle (u_{m+1} - \bar{u}_{m+1|m})(u_{m+1} - \bar{u}_{m+1|m})^* \rangle$$

are, consecutively, the prior mean state and the prior error covariance, computed from (2.2) respectively via

$$\bar{u}_{m+1|m} = F \bar{u}_{m|m}, \tag{2.4}$$

$$r_{m+1|m} = F r_{m|m} F^* + r, \tag{2.5}$$

where

$$r_{m|m} = \langle e_{m|m} e^*_{m|m} \rangle = \langle (u_m - \bar{u}_{m|m})(u_m - \bar{u}_{m|m})^* \rangle \tag{2.6}$$

is the posterior error covariance at time m to be computed through the Kalman filter formula (to be described later). Next, we present an informal derivation of this formula. See Chapter 3 for the generalization to more complex systems.

In each assimilation step, we are interested in obtaining an estimate of the true state after using knowledge of the observations at that time. This estimate for the mean is given in the probabilistic sense by the Bayesian update through maximizing the following conditional density

$$p(u_{m+1} | v_{m+1}) \sim p(u_{m+1}) p(v_{m+1} | u_{m+1}) = e^{-\frac{1}{2} J(u_{m+1})}, \tag{2.7}$$

which is equivalent to minimizing

$$J(u) = \frac{(u - \bar{u}_{m+1|m})^*(u - \bar{u}_{m+1|m})}{r_{m+1|m}} + \frac{(v_{m+1} - gu)^*(v_{m+1} - gu)}{r^o}.$$

The value of u at which $J(u)$ attains its minimum is the estimate for the mean and is given by

$$\bar{u}_{m+1|m+1} = \bar{u}_{m+1|m} + K_{m+1}(v_{m+1} - g\bar{u}_{m+1|m}), \tag{2.8}$$

where

$$K_{m+1} = \frac{g r_{m+1|m}}{r^o + g^2 r_{m+1|m}} \tag{2.9}$$

is the Kalman gain. Note that this is a real number restricted by $0 \leq K_{m+1}g \leq 1$ since every component in the right-hand side of (2.9) is non-negative and (2.9) is derived by assuming that the denominator on the right-hand side is nonzero. The filter fully weighs to the model or prior forecast when $K_{m+1}g = 0$ (see (2.8)) and fully weighs to the observation when $K_{m+1}g = 1$.

The posterior error covariance is obtained by taking the expectation of the square of the following posterior difference

$$\begin{aligned}
u_{m+1} - \bar{u}_{m+1|m+1} &= u_{m+1} - \bar{u}_{m+1|m} \\
&\quad - K_{m+1}(v_{m+1} - g u_{m+1} - g(\bar{u}_{m+1|m} - u_{m+1})) \\
e_{m+1|m+1} &= (1 - K_{m+1}g)e_{m+1|m} - K_{m+1}\sigma^o_{m+1}.
\end{aligned}$$

Next, use (2.9) and explicitly calculate,

$$\begin{aligned}
\langle e_{m+1|m+1} e^*_{m+1|m+1} \rangle &= (1 - K_{m+1}g)\langle e_{m+1|m} e^*_{m+1|m} \rangle (1 - K^*_{m+1}g) \\
&\quad + K_{m+1} r^o K^*_{m+1} \\
\langle e_{m+1|m+1} e^*_{m+1|m+1} \rangle &= (1 - K_{m+1}g)\langle e_{m+1|m} e^*_{m+1|m} \rangle \\
r_{m+1|m+1} &= (1 - K_{m+1}g) r_{m+1|m}
\end{aligned} \tag{2.10}$$

assuming that

$$\langle e_{m+1|m} e^*_{m+1|m+1} \rangle = \langle e_{m+1|m+1} e^*_{m+1|m} \rangle = 0.$$

We refer to Eqns (2.4)–(2.10) as the Kalman filter formula where the forecast steps consist of Eqns (2.4) and (2.5) while the correction steps consist of Eqns (2.8)–(2.10).

2.1.1 Numerical simulation on a scalar complex Ornstein–Uhlenbeck process

We consider a complex linear stochastic differential equation

$$du(t) = (-\gamma + i\omega)u(t)dt + \sigma dW(t), \tag{2.11}$$

where $\gamma, \sigma > 0$ and ω are all real numbers and

$$dW(t) \equiv \frac{dW_1(t) + i dW_2(t)}{\sqrt{2}} \tag{2.12}$$

is a complex Gaussian white noise where each component satisfies

$$dW_j(t) \equiv \dot{W}_j(t)dt, \quad j = 1, 2, \tag{2.13}$$

that is, white noise is a "derivative" of the Wiener process $W_j(t)$ and it satisfies the following properties (see Gardiner, 1997):

$$\begin{aligned}
\langle \dot{W}_j(t) \rangle &= 0 \\
\langle \dot{W}_j(t) \dot{W}_j(s) \rangle &= \delta(t - s) \\
\langle \dot{W}_i(t) \dot{W}_j(s) \rangle &= 0 \quad \text{for } i \neq j
\end{aligned} \tag{2.14}$$

where $\delta(x)$ is the Dirac delta function, that is, $\delta(0) = \infty$ and zero otherwise and $\int_{\mathbb{R}} \delta(x)dx = 1$.

The exact solution of (2.11) is

$$u(t) = e^{(-\gamma+i\omega)t}u(0) + \sigma \int_0^t e^{(-\gamma+i\omega)(t-s)}dW(s). \tag{2.15}$$

At any time t, $u(t)$ is a random Gaussian variable with mean and variance given by

$$\bar{u}(t) \equiv \langle u(t) \rangle = e^{(-\gamma+i\omega)t}\langle u(0) \rangle \tag{2.16}$$

$$\mathrm{Var}[u(t)] \equiv \langle (u(t) - \bar{u}(t))(u(t) - \bar{u}(t))^* \rangle = \frac{\sigma^2}{2\gamma}(1 - e^{-2\gamma t}). \tag{2.17}$$

For simplicity in exposition, we have assumed zero variance in the initial data in (2.15) and (2.17). We present the calculation for the variance to give some intuition,

$$\begin{aligned}
\mathrm{Var}[u(t)] &= \left\langle \sigma \int_0^t e^{(-\gamma+i\omega)(t-t')}dW(t') \left(\sigma \int_0^t e^{(-\gamma+i\omega)(t-s')}dW(s') \right)^* \right\rangle \\
&= \sum_{j=1,2} \frac{\sigma^2}{2} e^{-2\gamma t} \int_0^t \int_0^t e^{\gamma(t'+s')-i\omega(t'-s')} \langle dW_j(t')dW_j(s') \rangle \\
&= \sum_{j=1,2} \frac{\sigma^2}{2} e^{-2\gamma t} \int_0^t \int_0^t e^{\gamma(t'+s')-i\omega(t'-s')} \langle \dot{W}_j(t')\dot{W}_j(s') \rangle dt'ds' \\
&= \sum_{j=1,2} \frac{\sigma^2}{2} e^{-2\gamma t} \int_0^t \int_0^t e^{\gamma(t'+s')-i\omega(t'-s')} \delta(t' - s')dt'ds' \\
&= \sigma^2 e^{-2\gamma t} \int_0^t e^{2\gamma s'}ds' \\
&= \frac{\sigma^2}{2\gamma}(1 - e^{-2\gamma t}).
\end{aligned}$$

In this derivation, we used the definition (2.12), (2.13), and we applied properties (2.14). In the last line of the double integral, we use the following property

$$\int_0^t f(x)\delta(x - s)dx = \begin{cases} f(s), & \text{if } 0 < s < t \\ 0, & \text{if } s = t. \end{cases} \tag{2.18}$$

Alternatively, one can verify the variance above (2.17) using the Itô isometry (Gardiner, 1997). As $t \to \infty$, every solution converges to the statistical attractor which is a Gaussian invariant measure with zero mean and variance

$$E = \frac{\sigma^2}{2\gamma}. \tag{2.19}$$

This statistical attractor is called the climatological mean state in geophysical applications and E in (2.19) is the climatological variance.

To be consistent with the earlier notation, we define a homogeneous observation time interval $\Delta t = t_{m+1} - t_m$ (in general, Δt can depend on time too):

$$F = e^{(-\gamma+\mathrm{i}\omega)\Delta t},$$
$$r = \frac{\sigma^2}{2\gamma}(1 - e^{-2\gamma\Delta t}). \tag{2.20}$$

We address the filtering skill in this problem in a noisy, moderately stiff regime of parameters mimicking moderately large spatial wavenumbers in a turbulent signal. We use the values

$$\gamma = \frac{1}{2}, \omega = 10, \sigma = 1$$

both in generating the truth signal and in the filter model. The climatological variance with these parameters is one and we use an observational noise variance $r^o = 0.25$ with sparse observation time $\Delta t = 2$, near the decorrelation time of (2.11). The observations are simulated by adding a sampled Gaussian noise σ_m^o as in (2.1) with $g \equiv 1$ and noise variance r^o.

In this numerical experiment and the others reported below, we run the assimilation cycle for 400 times. The top panel in Fig. 2.1 shows that the filtered solution with the perfect model (solid) tracks the true signal (dashes) pretty well. The RMS error (the difference between the posterior state and the true signal) is smaller than the observation error $\sqrt{r^o}$ as shown in the bottom panel in Fig. 2.1. The temporal RMS error for this experiment is 0.43, which is still smaller than the observation error $\sqrt{r^o} = 0.5$ in this noisy environment with sparse observation time.

2.2 Filtering stability

For the scalar filter described earlier, if the pair of dynamical systems (2.1), (2.2) is observable (i.e. $g \neq 0$) and controllable ($r \neq 0$), the filter is stable for any values of $|F|$. Thus, we claim there exists a unique positive asymptotic value r_∞ for the covariance that is a fixed point for the recursive formula

$$r_\infty = \Phi(\Psi(r_\infty, r), r^o), \tag{2.21}$$

where

$$\Phi(r_{m+1|m}, r^o) \equiv r_{m+1|m+1} = (1 - r_{m+1|m}g^2(g^2 r_{m+1|m} + r^o)^{-1})r_{m+1|m},$$
$$\Psi(r_{m|m}, r) \equiv |F|^2 r_{m|m} + r.$$

Note that Φ and Ψ are nothing but the two covariance updates from the observations (2.10) and the dynamics (2.5), respectively. We will also show that when $|F| > 1$ the controllability condition is not necessary for filter stability so we can have $r = 0$, but when $|F| = 1$ controllability is necessary.

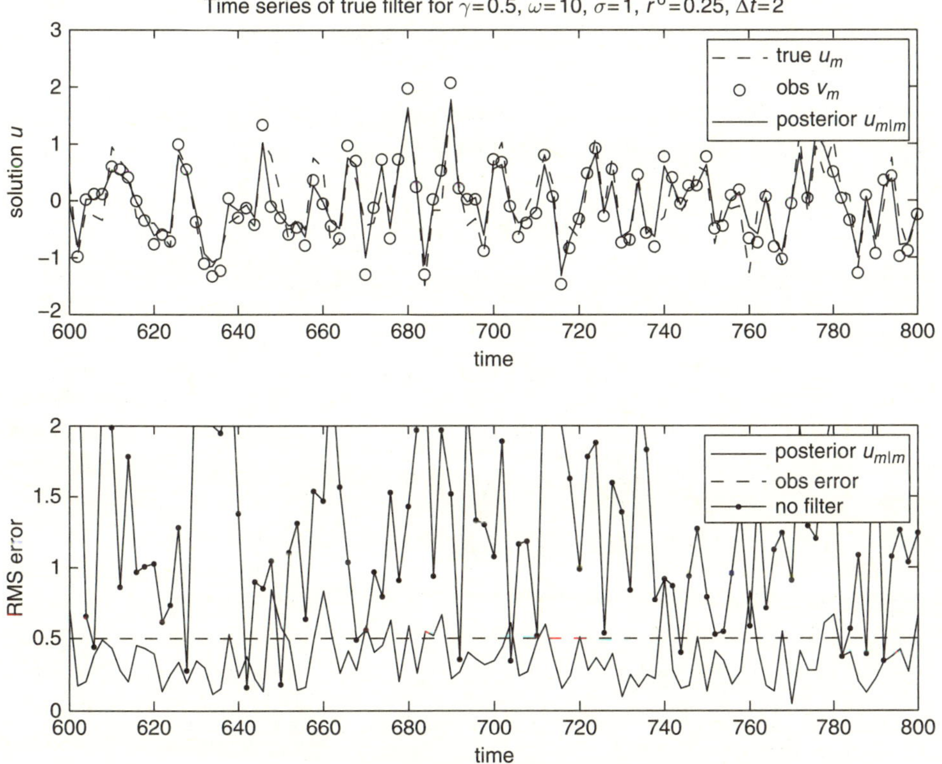

Figure 2.1 True filter (perfect model simulation). The first row shows the real part of the solutions as functions of time: true signal (dashes), filtered solution (solid), observations (circle). The second row shows the RMS difference between the mean posterior state and the true signal (solid) as a function of time.

To justify this, we start by rewriting Eqn (2.21) as follows

$$
\begin{aligned}
r_\infty &= |F|^2 r_\infty + r - \frac{g^2(|F|^2 r_\infty + r)^2}{r^o + g^2(|F|^2 r_\infty + r)} \\
&= \frac{r^o(|F|^2 r_\infty + r)}{r^o + g^2(|F|^2 r_\infty + r)} \\
&= r^o \left(g^2 + \frac{r^o}{|F|^2 r_\infty + r} \right)^{-1}.
\end{aligned} \tag{2.22}
$$

Now let us evaluate (2.9) at its asymptotic value, i.e.

$$
\begin{aligned}
K_\infty &= g \frac{(|F|^2 r_\infty + r)}{r^o + g^2(|F|^2 r_\infty + r)} \\
&= g \left(g^2 + \frac{r^o}{|F|^2 r_\infty + r} \right)^{-1}.
\end{aligned} \tag{2.23}
$$

Combining (2.22) and (2.23), we have

$$r_\infty = \frac{r^o}{g} K_\infty, \tag{2.24}$$

which requires observability, i.e. $g \neq 0$, otherwise r_∞ is potentially unbounded. Equation (2.22) can be rewritten as a quadratic function of r_∞:

$$r_\infty^2 + \left(\frac{r}{|F|^2} + \frac{r^o}{g^2|F|^2} - \frac{r^o}{g^2}\right) r_\infty - \frac{r^o r}{g^2|F|^2} = 0,$$

and the positive solution is

$$r_\infty = \frac{1}{2}\left(\frac{r^o}{g^2} - \frac{r}{|F|^2} - \frac{r^o}{g^2|F|^2} + \left[\left(\frac{r}{|F|^2} + \frac{r^o}{g^2|F|^2} - \frac{r^o}{g^2}\right)^2 + 4\frac{r^o r}{g^2|F|^2}\right]^{1/2}\right). \tag{2.25}$$

We define rescaled variables

$$x = \frac{r^o}{g^2}, \quad y = \frac{r}{|F|^2}, \quad z = \frac{r^o}{g^2|F|^2}, \quad \tilde{y} = \frac{y}{x}, \quad \tilde{z} = \frac{z}{x} = \frac{1}{|F|^2},$$

so that we have the explicit formula for the **asymptotic filter covariance**

$$r_\infty = \frac{x}{2}\left(1 - \tilde{y} - \tilde{z} + \left[(1 - \tilde{y} - \tilde{z})^2 + 4\tilde{y}\right]^{1/2}\right). \tag{2.26}$$

Combining (2.24) and (2.26), we obtain the **explicit asymptotic Kalman gain**

$$\begin{aligned} K_\infty(F, r^o, r) &= K_\infty(\tilde{y}, \tilde{z}) \\ &= \frac{1}{2g}\left(1 - \tilde{y} - \tilde{z} + \left[(1 - \tilde{y} - \tilde{z})^2 + 4\tilde{y}\right]^{1/2}\right). \end{aligned} \tag{2.27}$$

Note that it is easy for the reader to check that this explicit asymptotic Kalman gain satisfies:

(A) $K_\infty(\tilde{y}, \tilde{z})g$ is monotonically increasing in $\tilde{y}$ for fixed $\tilde{z}$.
(B) $K_\infty(\tilde{y}, \tilde{z})g \to 1$ as $\tilde{y} \to \infty$ for fixed $\tilde{z}$.
(C) $K_\infty(0, \tilde{z})g = 0$, when $\tilde{z} \geq 1$ or $|F|^2 \leq 1$.
(D) $K_\infty(0, \tilde{z})g = 1 - \tilde{z}$, when $0 < \tilde{z} < 1$ or $|F|^2 > 1$ since $\tilde{z} = |F|^{-2}$.

In particular, conditions (A), (B), (C), (D) guarantee $0 \leq K_\infty g \leq 1$. Notice that conditions (C) and (D) occur when controllability is violated ($r = 0$).

Furthermore, (2.8) can also be rewritten as

$$\bar{u}_{m+1|m+1} = F(1 - K_{m+1}g)\bar{u}_{m|m} + K_{m+1}v_{m+1},$$

and as $m \to \infty$, the stability of this dynamical system for the mean requires

$$|F(1 - K_\infty g)| < 1. \tag{2.28}$$

Table 2.1 Off-line asymptotic variables for finite difference approximations.

	$\lvert F_h \rvert$	r_h	$K_{h,\infty}$	$\lvert F(1 - K_{h,\infty}g)\rvert$	RMS
True filter	0.3679	0.8647	0.7809	0.0806	0.42
Forward Euler	20	2	0.9975	0.0490	0.49
Backward Euler	0.0498	0.005	0.0195	0.0488	0.95
Trapezoidal	0.9756	0.0196	0.2279	0.7533	0.82
Forward Euler with information	20	0	0.9975	0.05	0.49
Backward Euler with information	0.0498	0.8906	0.7809	0.0109	0.43
Trapezoidal with information	0.9756	0.7053	0.7809	0.2137	0.48

When $|F| < 1$, the stability condition (2.28) is obviously satisfied since $0 \leq K_\infty g \leq 1$. When $|F| > 1$, condition (D) guarantees that $|F(1 - K_\infty g)| \leq |F||1 - K_\infty g| = |F|^{-1} < 1$ hence it is clear as well that controllability is not necessary here, i.e. r can be zero for $|F| > 1$. When $|F| = 1$, controllability, $r > 0$, is a necessary condition for filter stability otherwise condition (C) yields $|F(1 - K_\infty g)| = 1$ which implies that asymptotic convergence can never be achieved.

At the top of Table 2.1, we show the off-line variables for the numerical experiment shown earlier in Section 2.1.1. For this case, we see that the filter is stable for $|F| < 1$ and $r > 0$.

2.3 Model error

Suppose now that the filter model is different from the dynamical model of the true signal, that is, instead of Eqn (2.2) the filter model is given by

$$u_{h,m+1|m} = F_h u_{h,m|m} + \sigma_{h,m+1}, \tag{2.29}$$

where $F_h \neq F$ and $\sigma_{h,m+1}$ is a complex Gaussian random variable $\mathcal{N}(0, r_h)$ with $r_h \neq r$. While the Kalman gain and variance provide information on the dynamics of the filtered state variable u_h, we are also interested in the statistics of the difference between the prior state $u_{h,m|m-1}$ and the truth u_m. Particularly we are interested in finding the mean model error and model error covariance.

2.3.1 Mean model error

The most interesting error statistics for the complex scalar test problem are given by

$$y_m \equiv \langle u_m - u_{h,m|m-1} \rangle \tag{2.30}$$

or the mean deviation of the filtered solution from the truth signal. Substituting (2.2) and (2.8), we obtain the recursive form of the mean model error

$$
\begin{aligned}
y_{m+1} &= \langle u_{m+1} - u_{h,m+1|m} \rangle \\
&= \langle e^{(-\gamma+i\omega)\Delta t} u_m + \sigma_{m+1} \rangle \\
&\quad - \langle F_h (u_{h,m|m-1} + K_{h,m} v_m - K_{h,m} g u_{m|m-1}) \rangle \\
&= e^{(-\gamma+i\omega)\Delta t} \langle u_m \rangle - F_h (1 - K_{h,m} g) \langle u_{h,m|m-1} \rangle - F_h K_{h,m} \langle v_m \rangle \\
&= e^{(-\gamma+i\omega)\Delta t} \langle u_m \rangle + F_h (1 - K_{h,m} g) y_m \\
&\quad - F_h (1 - K_{h,m} g) \langle u_m \rangle - F_h K_{h,m} \langle v_m \rangle \\
&= (e^{(-\gamma+i\omega)\Delta t} - F_h) \langle u_m \rangle + F_h (1 - K_{h,m} g)\, y_m.
\end{aligned} \tag{2.31}
$$

In (2.31) the first term represents the effect of the discretization error in the dynamics towards the mean of the true signal $\langle u_m \rangle$. From this dependence, it is clear that this mean model error is an average model error obtained by averaging over all ensemble trajectories drawn from the initial statistical state for both the true dynamics and the filter as well as an average over all realizations of the random noise in the observations. The second term measures the effect of the filter evolution operator and of its Kalman gain, $F_h(1 - K_{h,m} g)$. Note that the first term vanishes if there is no model error and trivially yields the fact that the Kalman filter is unbiased on the perfect model provided there is no bias at time $t_m = 0$.

2.3.2 Model error covariance

Consider the following augmented system

$$
\begin{bmatrix} u_{m+1} \\ u_{h,m+1|m} \end{bmatrix} = \begin{bmatrix} e^{(-\gamma+i\omega)\Delta t} & 0 \\ F_h K_{h,m} g & F_h (1 - K_{h,m} g) \end{bmatrix} \begin{bmatrix} u_m \\ u_{h,m|m-1} \end{bmatrix} + \begin{bmatrix} \sigma_{m+1} \\ F_h K_{h,m} \sigma^o_{m+1} + \sigma_{h,m+1} \end{bmatrix}.
$$

Following the notation in Anderson and Moore (1979), we define $\mathcal{X}_m$ as the concatenation of u_m and $u_{h,m|m-1}$, or

$$
\mathcal{X}_m = \begin{bmatrix} u_m \\ u_{h,m|m-1} \end{bmatrix}
$$

so that

$$
\mathcal{X}_{m+1} = \mathcal{F}_m \mathcal{X}_m + \mathcal{W}_m \tag{2.32}
$$

where

$$
\mathcal{F}_m = \begin{bmatrix} e^{(-\gamma+i\omega)\Delta t} & 0 \\ F_h K_{h,m} g & F_h (1 - K_{h,m} g) \end{bmatrix}
$$

and

$$
\mathcal{W}_m = \begin{bmatrix} \sigma_{m+1} \\ F_h K_{h,m} \sigma^o_{m+1} + \sigma_{h,m+1} \end{bmatrix}.
$$

Note that in this case we have

$$\langle \mathcal{X}_{m+1} \rangle = \mathcal{F}_m \langle \mathcal{X}_m \rangle \tag{2.33}$$

and

$$\langle | \, u_{m+1} - u_{h,m+1|m} \, |^2 \rangle = \begin{bmatrix} 1 & -1 \end{bmatrix} \langle \mathcal{X}_{m+1} \mathcal{X}_{m+1}^* \rangle \begin{bmatrix} 1 \\ -1 \end{bmatrix}. \tag{2.34}$$

The evolution of $\langle \mathcal{X}_{m+1} \mathcal{X}_{m+1}^* \rangle$ is given by

$$\begin{aligned} \langle \mathcal{X}_{m+1} \mathcal{X}_{m+1}^* \rangle &= \mathcal{F}_h \langle \mathcal{X}_m \mathcal{X}_m^* \rangle \mathcal{F}_h^* + \langle \mathcal{W}_m \mathcal{W}_m^T \rangle \\ &= \mathcal{F}_h \langle \mathcal{X}_m \mathcal{X}_m^* \rangle \mathcal{F}_h^* + \begin{pmatrix} r & 0 \\ 0 & F_h K_{h,m} r^o K_{h,m}^* F_h^* + r_h \end{pmatrix}. \end{aligned} \tag{2.35}$$

The model error covariance is then defined as

$$\begin{aligned} \mathrm{Cov}\left[\mathcal{X}_{m+1} \mathcal{X}_{m+1}^* \right] = \langle | \, u_{m+1} - u_{h,m+1|m} \, |^2 \rangle \\ - \begin{bmatrix} 1 & -1 \end{bmatrix} \langle \mathcal{X}_{m+1} \rangle \langle \mathcal{X}_{m+1}^* \rangle \begin{bmatrix} 1 \\ -1 \end{bmatrix}. \end{aligned} \tag{2.36}$$

Using Eqns (2.31) and (2.35) we obtain a recursive formula for the covariance of $\mathcal{X}_m$. This gives us an off-line algorithm for computing the evolution of the mean model error, mean square model error and model error covariance:

1. Start with $\langle u_0 \rangle = u_0$ and with any positive definite $\langle \mathcal{X}_0 \mathcal{X}_0^* \rangle$. (Note this define also $\mathrm{Cov}\left[\mathcal{X}_0 \mathcal{X}_0^*\right]$.)
2. Use Eqns (2.33) and (2.35) to evolve the mean and mean square error.
3. Use Eqn (2.36)) to obtain $\mathrm{Cov}\left[\mathcal{X}_m \mathcal{X}_m^*\right]$.
4. Obtain the mean model error using (2.31).

2.3.3 Example: Model error through finite difference approximation

In Section 2.1.1, we considered a perfect filter by letting the filter model be the exact solution of (2.11). Now, we consider the model error through finite difference approximations: the forward Euler, backward Euler and trapezoidal methods. For forward Euler, we substitute

$$\frac{u(t + \Delta t) - u(t)}{\Delta t}$$

for $du(t)/dt$ in (2.11) and recalling that $\sigma \dot{W}(t) \sim \sigma \Delta t^{-\frac{1}{2}} \mathcal{N}(0, 1)$, with $\mathcal{N}(0, 1)$ a Gaussian random variable with mean zero and variance one, we obtain formally

$$u(t + \Delta t) = (1 - \gamma \Delta t + \mathrm{i}\omega \Delta t) u(t) + \sigma \Delta t^{\frac{1}{2}} \mathcal{N}(0, 1).$$

Thus, the model operator is $F_h = 1 - \gamma \Delta t + \mathrm{i}\omega \Delta t$ in (2.29) and the system noise variance is given by $r_h = \sigma^2 \Delta t$. For backward Euler, we have

$$u(t + \Delta t) - u(t) = (-\gamma \Delta t + \mathrm{i}\omega \Delta t) u(t + \Delta t) + \sigma \Delta t^{\frac{1}{2}} \mathcal{N}(0, 1)$$

and solving for $u(t+\Delta t)$ we obtain

$$u(t+\Delta t) = (1+\gamma\Delta t - \mathrm{i}\omega\Delta t)^{-1}u(t) + (1+\gamma\Delta t - \mathrm{i}\omega\Delta t)^{-1}\sigma\Delta t^{\frac{1}{2}}\mathcal{N}(0,1)$$

with system variance $r_h = \mid 1+\gamma\Delta t - \mathrm{i}\omega\Delta t \mid^{-2} \sigma^2\Delta t$. For the trapezoidal method, we have

$$u(t+\Delta t) - u(t) = \frac{-\gamma\Delta t + \mathrm{i}\omega\Delta t}{2}(u(t+\Delta t)+u(t)) + \sigma\Delta t^{\frac{1}{2}}\mathcal{N}(0,1)$$

which results in

$$u(t+\Delta t) = \frac{1+\dfrac{-\gamma\Delta t + \mathrm{i}\omega\Delta t}{2}}{1-\dfrac{-\gamma\Delta t + \mathrm{i}\omega\Delta t}{2}}u(t) + \left(1-\frac{-\gamma\Delta t + \mathrm{i}\omega\Delta t}{2}\right)^{-1}\sigma\Delta t^{\frac{1}{2}}\mathcal{N}(0,1)$$

with variance

$$r_h = \mid 1 - \frac{-\gamma\Delta t + \mathrm{i}\omega\Delta t}{2} \mid^{-2} \sigma^2\Delta t.$$

In the first column of Fig. 2.2, we show the numerical simulations with these three finite difference approximations utilizing parameters as in the earlier perfect model example. The strongly unstable forward Euler simply trusts the observations (see Table 2.1 where $K_{h,\infty} \approx 1$) here with RMS error 0.49. In the backward Euler, the filtering failure (with RMS error 0.95) is due to the fact that the amplitude of F_h is very small ($\approx 10^{-2}$) whereas the Kalman gain (≈ 0) weighs the filtered solution almost fully to the dynamics with model error; hence the solution is nearly zero when the system noise variance is small too, as occurs here. The trapezoidal method is not too accurate as well (with RMS error 0.82) since the system noise is very small and the asymptotic stability factor for the mean $|F_h(1-K_{h,\infty}G)|$ in Table 2.1 is near one. These last two unskillful filtered solutions are sometimes called filter divergence in the data assimilation community where the filtered solution with model error has a nice looking but totally inaccurate behavior with too little variance. These are all novel features of filtering stiff noisy turbulent signals. While these methods are both observable and controllable, their noise level is so small that they violate practical controllability. How can skill be restored in the filtering of turbulent signals in this stiff regime with model errors with methods which might still retain significant computational advantages? An explicit information-theoretic criterion is discussed next to avoid the filter divergence discussed above.

2.3.4 Information criteria for filtering with model error

The relative entropy

$$\mathcal{P}(p, p_h) = \int p\log\left(\frac{p}{p_h}\right)$$

measures the lack of information in the probability measure p_h compared to p (Majda *et al.*, 2005; Majda and Wang, 2006). In the present application, p is the asymptotic filtering

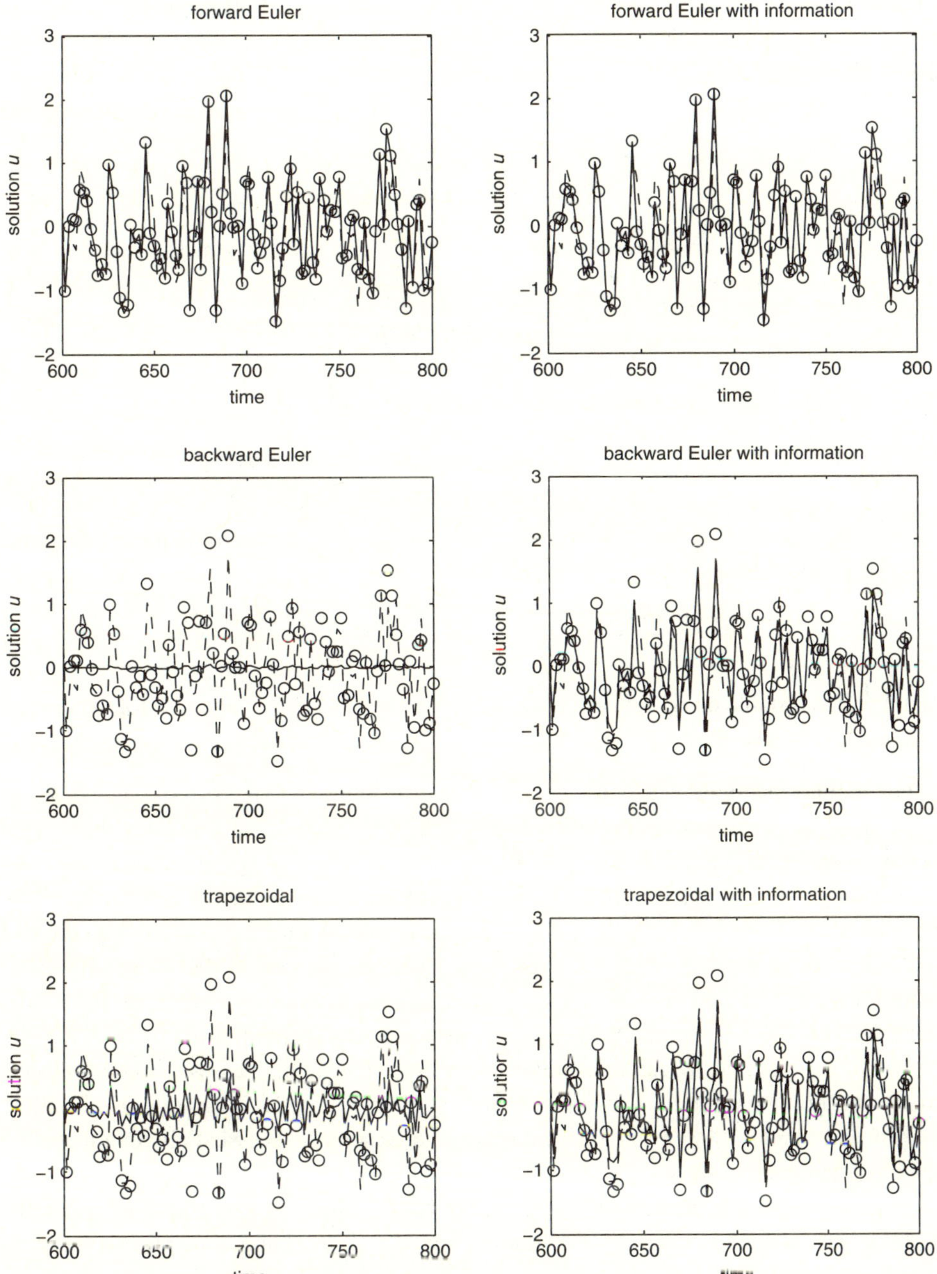

Figure 2.2 Finite difference schemes: The first column shows solutions as functions of time with finite difference noise whereas the second column shows solutions with information criteria. In each panel, we show the true signal (dashes), the filtered solution (solid) and the observations (circle).

limit mean zero Gaussian measure with variance r_∞, while p_h is the asymptotic limit mean zero Gaussian measure with variance $r_{h,\infty}$ for the fixed model F_h, r^o (see Castronovo *et al.*, 2008; Majda and Grote, 2007). Specifically, the approximate filter asymptotic variables are

$$\begin{aligned} r_{h,\infty} &= \frac{r^o}{g} K_{h,\infty}, \\ K_{h,\infty} &= K_\infty(F_h, r^o, r_h). \end{aligned}$$

Clearly, the best choice of the system noise $r^*_{h,m}$ is the one for which p has the least additional information, i.e.

$$\mathcal{P}(p, p_h(r^*_h)) = \min_{0 \le r_h < \infty} \mathcal{P}(p, p_h(r_h)).$$

Thus, this information criterion determines r^*_h uniquely by

(A) For the stable case $|F_h| \le 1$, r^*_h is uniquely determined by solving

$$K_\infty(F, r^o, r) = K_\infty(F_h, r^o, r^*_h).$$

That is,

$$r^*_h = \frac{r^o K_\infty(F, r^o, r)(1 - |F_h|^2(1 - K_\infty(F, r^o, r)g))}{(1 - K_\infty(F, r^o, r)g)g}.$$

(B) For the unstable case $|F_h| > 1$, if $K_\infty(F, r^o, r) \le 1 - |F_h|^{-2}$, use $r^*_h = 0$. Otherwise, use (A) to determine r^*_h.

For implicit methods with strong stability in stiff regimes like the backward Euler or trapezoidal methods, this information-theory criterion automatically inflates the system noise variance and guarantees $K_{h,\infty} = K_\infty$ where K_∞ is the asymptotic Kalman gain factor for the perfect model. The filtered solutions with model error and the new system noise are shown on the right-hand side of Fig. 2.2 and significant filtering skill through the two implicit methods has been restored. Note from Table 2.1 that the RMS errors with backward Euler, 0.43, and the information criteria are as good as filtering with the perfect model with the same RMS, 0.43; the RMS error for the trapezoidal methods also improves to 0.48. Note that the system noise has been inflated to a large value in both cases through the information criterion and practical controllability has been restored (Castronovo *et al.*, 2008; Harlim and Majda, 2008a). The forward Euler filtered solution is essentially unaffected by the information criterion.

To gain insight into the behavior of the mean model error it is interesting to replace the Kalman gain at each iteration by the limiting one, $K_{h,\infty}$, to obtain an explicit solution of (2.31),

$$\begin{aligned} y_{m+1} = {} & (F_h(1 - K_{h,\infty}g))^{m+1} y_0 + (\mathrm{e}^{(-\gamma+\mathrm{i}\omega)\Delta t} - F_h)\langle u_0 \rangle \\ & \times \sum_{n=0}^{m} \mathrm{e}^{-n(\gamma+\mathrm{i}\omega)\Delta t} (F_h(1 - K_{h,\infty}g))^{m-n}. \end{aligned}$$

For stable schemes the rate at which y_m decays depends strongly on the limiting Kalman gain.

When the information criterion is applied for an unstable approximate filter with $y_0 = 0$, the mean model error is readily estimated by

$$|y_m| \leq |e^{(-\gamma\Delta t + i\omega\Delta t)} - F_h||\langle u_0 \rangle| \sum_{n=0}^{m} |F_h|^{-(m-n)} e^{-\gamma n \Delta t}. \tag{2.37}$$

In spite of the potentially large pre-constant in (2.37), the model error decays quickly due to the rapidly convergent sum as $m \to \infty$. This helps to explain why stable accurate filtering can be achieved with unstable difference approximations.

3

The Kalman filter for vector systems: Reduced filters and a three-dimensional toy model

In this chapter, we generalize the Kalman filter formula and the filter stability for higher-dimensional systems. We then demonstrate the practical meaning of satisfying filter stability with a three-dimensional toy model. In particular, we would like to check whether accurate filtered solutions can be produced when filter stability conditions are satisfied. This simple question turns out to have non-trivial answers as shown by our numerical experiments. Subsequently, we discuss a reduced filtering strategy for large systems with stable and unstable subspaces. We close this chapter with discussions of the stability criteria for such a reduced filtering strategy and we provide the covariance bound for the unstable subspace when filter stability conditions are met.

3.1 The classical N-dimensional Kalman filter

As we discussed in Section 2.1, the real-time prediction problem is basically to obtain the posterior distribution $p(\vec{u}|\vec{v}_{m+1})$ through the Bayesian formula assuming a knowledge of the prior distribution $p(\vec{u})$ of a true signal $\vec{u}_{m+1} \in \mathbb{R}^N$ and the observed value $\vec{v}_{m+1} \in \mathbb{R}^M$. Consider the following linear observation model

$$\vec{v}_{m+1} = G\vec{u}_{m+1} + \vec{\sigma}^o_{m+1}, \tag{3.1}$$

where the matrix $G \in \mathbb{R}^{M\times N}$ and $\vec{\sigma}^o_m = \{\sigma^o_{j,m}\}$ is an M-dimensional Gaussian white noise vector with zero mean and covariance

$$R^o = \langle \vec{\sigma}^o_m \otimes (\vec{\sigma}^o_m)^T \rangle = \{\langle \sigma^o_{i,m} (\sigma^o_{j,m})^T \rangle\} = \{\delta(i-j) r^o\}, \tag{3.2}$$

where $\delta(i-j) = 1$ if $i = j$ and zero otherwise. The observation error covariance (3.2) is diagonal assuming that the observation noises of two distinct locations are uncorrelated and each location has an observation noise variance $r^o \in \mathbb{R}$. Assume also that the filter model is given as follows

$$\vec{u}_{m+1|m} = F\vec{u}_{m|m} + \vec{\sigma}_{m+1}, \tag{3.3}$$

where $\vec{u}_{m|m}$ denotes the posterior state at time t_m as an estimate of the true signal $\vec{u}_m$ and $\vec{u}_{m+1|m}$ is the prior state at time t_{m+1} which is propagated forward with a linear dynamical operator $F \in \mathbb{R}^{N\times N}$ and a Gaussian noise N-dimensional vector $\vec{\sigma}_m$ with zero mean and covariance

$$R = \langle \vec{\sigma}_m \otimes \vec{\sigma}_m^T \rangle.$$

Note also that the system noise $\vec{\sigma}_m$ is uncorrelated in time, i.e.

$$\langle \vec{\sigma}_i \otimes \vec{\sigma}_j^T \rangle = \delta(i-j)R.$$

With all these assumptions, the posterior mean state $\bar{\vec{u}}_{m+1|m+1}$ and the posterior error covariance

$$R_{m+1|m+1} = \langle (\vec{u}_{m+1} - \bar{\vec{u}}_{m+1|m+1}) \otimes (\vec{u}_{m+1} - \bar{\vec{u}}_{m+1|m+1})^T \rangle, \tag{3.4}$$

can be obtained through maximizing the conditional density $p(\vec{u}|\vec{v}_{m+1})$, generalizing the result for a complex scalar in Chapter 2. For Gaussian prior and observation distributions, this is equivalent to minimizing the following cost function

$$\begin{aligned} J(\vec{u}) &= (\vec{u} - \bar{\vec{u}}_{m+1|m})^T R_{m+1|m}^{-1} (\vec{u} - \bar{\vec{u}}_{m+1|m}) \\ &\quad + (\vec{v}_{m+1} - G\vec{u})^T (R^o)^{-1} (\vec{v}_{m+1} - G\vec{u}), \end{aligned}$$

where the first term corresponds to the exponent of the prior density $p(\vec{u})$ with mean $\bar{\vec{u}}_{m+1|m}$ obtained by taking the mean of (3.3)

$$\bar{\vec{u}}_{m+1|m} = F\bar{\vec{u}}_{m|m}, \tag{3.5}$$

and covariance, defined as

$$R_{m+1|m} \equiv \langle (\vec{u}_{m+1} - \bar{\vec{u}}_{m+1|m}) \otimes (\vec{u}_{m+1} - \bar{\vec{u}}_{m+1|m})^T \rangle = F R_{m|m} F^T + R,$$

while the second term arises from the observations in (3.1) and (3.2).

The minimum is given by

$$\bar{\vec{u}}_{m+1|m+1} = \bar{\vec{u}}_{m+1|m} + K_{m+1}(\vec{v}_{m+1} - G\bar{\vec{u}}_{m+1|m}),$$

where

$$\begin{aligned} K_{m+1} &= (R_{m+1|m}^{-1} + G^T (R^o)^{-1} G)^{-1} G^T (R^o)^{-1} \\ &= R_{m+1|m} G^T (G R_{m+1|m} G^T + R^o)^{-1}, \end{aligned} \tag{3.6}$$

which is also referred to as the Kalman gain matrix. The second identity in (3.6) can be verified by multiplying each left side by $(R_{m+1|m}^{-1} + G^T (R^o)^{-1} G)$ and each right-hand side by $(G R_{m+1|m} G^T + R^o)$, i.e.

$$\begin{aligned} G^T (R^o)^{-1} (G R_{m+1|m} G^T + R^o) &= (R_{m+1|m}^{-1} + G^T (R^o)^{-1} G) R_{m+1|m} G^T \\ &= G^T + G^T (R^o)^{-1} G R_{m+1|m} G^T \end{aligned}$$

We leave it as an exercise for the reader to obtain the following posterior covariance matrix

$$R_{m+1|m+1} = (\mathcal{I} - K_{m+1} G) R_{m+1|m}.$$

Hint: Follow the derivation for the scalar case discussed in Section 2.1.

3.2 Filter stability

Before we state the stability condition, let us review the controllability and observability. In the control theory literature (e.g. Anderson and Moore, 1979; Chui and Chen, 1999), controllability of a dynamical system can be loosely understood as the ability to steer the system such that its solution at later time reaches a desirable configuration state. In particular, for a linear system

$$\frac{d\vec{u}}{dt} = F\vec{u} + V\vec{\eta},$$

the control $\vec{\eta}$ exists (or the system is **controllable**) if the following algebraic condition is satisfied,

$$\text{Rank}[V, FV, F^2V, \ldots, F^{N-1}V] = N. \tag{3.7}$$

When the control is a stochastic white noise as in (3.3), we can write

$$\vec{\sigma}_{m+1} = V\vec{\eta}_{m+1}$$

where $\vec{\eta}_{m+1}$ is a Gaussian white noise with mean zero and unit variance, i.e.

$$\langle \vec{\eta}_{m+1} \otimes \vec{\eta}^T_{m+1} \rangle = \mathcal{I},$$

and thus

$$R \equiv \langle \vec{\sigma}_{m+1} \otimes \vec{\sigma}^T_{m+1} \rangle = VV^T.$$

Suppose we are given an observation to the following linear system,

$$\frac{d\vec{u}}{dt} = F\vec{u}, \tag{3.8}$$

$$\vec{v} = G\vec{u}, \tag{3.9}$$

where $G \in \mathbb{R}^{M \times N}$ and $M < N$. The pair of F and G in (3.8), (3.9) is **observable** if and only if observations $\vec{v}$ at time interval $[0, t]$ allow us to reconstruct the initial condition, $\vec{u}_0$. It turns out that system (3.8), (3.9) is observable if and only if the dual,

$$\frac{d\vec{z}}{dt} = F^T\vec{z} + G^T\vec{\eta}, \tag{3.10}$$

is controllable (Lee and Markus, 1967), i.e. the following algebraic condition holds,

$$\text{Rank}[G^T, F^TG^T, (F^T)^2G^T, \ldots, (F^T)^{N-1}G^T] = N. \tag{3.11}$$

Note that the algebraic conditions are identical for a discrete-time system as in (3.1), (3.3).

Stability condition

Consider the pair of observation model (3.1) and dynamical model (3.3). When $\|F\| < 1$ (for any appropriate norm $\|\cdot\|$), the filter is guaranteed to be stable. Secondly, a necessary and sufficient condition for filtering stability is observability regardless of any value $\|F\|$.

Kalman and Bucy (Kalman and Bucy, 1961; Jazwinski, 1970; Anderson and Moore, 1979; Chui and Chen, 1999) showed that if the filter dynamical model is also controllable, then it is possible to obtain an optimal weight K_m at every time step. This optimal K_m is called the Kalman gain matrix and the resulting filter is called the Kalman filter. By filter stability, we mean there is a unique asymptotic covariance R_∞ such that

$$R_\infty = \lim_{t_m \to \infty} R_{m|m}.$$

This also means that the Kalman gain matrix K_m converges to an asymptotic limit K_∞ as $t_m \to \infty$. A necessary condition for filter stability is

$$\|F(\mathcal{I} - K_\infty G)\| < 1.$$

It is not our intention to develop these results here but refer the interested reader to well-known texts (Anderson and Moore, 1979; Chui and Chen, 1999). Instead, here, we illustrate this theory on an instructive toy model and perform numerical experiments illustrating the significance of these criteria both with and without model error.

3.3 Example: A three-dimensional toy model with a single observation

Let the truth signal $\vec{u} \in \mathbb{R}^3$ be generated by the analytical solution of

$$d\vec{u} = -\Gamma\vec{u}dt + \Sigma d\vec{W}(t) \tag{3.12}$$

where Γ and Σ are diagonal matrices with diagonal components $\{\gamma_1, \gamma_2, \gamma_3\}$ and $\{\sigma_1, \sigma_2, \sigma_3\}$, respectively. In (3.12), each component of the vector $\vec{W}(t) \in \mathbb{R}^3$ is a Wiener process whose "derivative" is a Gaussian random variable with zero mean and unit variance . Thus, each component of (3.12) is a real-variable Ornstein–Uhlenbeck process (see Section 2.1.1).

3.3.1 Observability and controllability criteria

Since (3.12) is a decoupled system, the discrete dynamical operator F is basically a diagonal matrix with diagonal components

$$F_i = e^{-\gamma_i \Delta t}, \quad i = 1, 2, 3,$$

where Δt denotes the observation time. In our experiment, we consider the following observation operator

$$G = [g_1, g_2, g_3]. \tag{3.13}$$

Thus, the observability condition (3.11) is satisfied whenever

$$\text{Rank}(\mathcal{O}) = \text{Rank} \begin{bmatrix} g_1 & F_1 g_1 & F_1^2 g_1 \\ g_2 & F_2 g_2 & F_2^2 g_2 \\ g_3 & F_3 g_3 & F_3^2 g_3 \end{bmatrix} = 3. \tag{3.14}$$

The matrix $\mathcal{O}$ is full rank if its determinant is nonzero which then yields the following neccessary and sufficient conditions for observability.

$$\det(\mathcal{O}) \neq 0,$$
$$g_j \neq 0, \text{ for every } j, \text{ and}$$
$$F_1 \neq F_2 \neq F_3.$$

The time discretized solution of (3.12) has a time-independent noise variance $R \in \mathbb{R}^{3\times 3}$, which is a diagonal matrix with diagonal components as in Section 2.1.1 with

$$R_{i,i} = \frac{\sigma_i^2}{2\gamma_i}(1 - e^{-2\gamma_i \Delta t}), \quad i = 1, 2, 3.$$

From (3.7) , the system is controllable if $\text{Rank}(V) = 3$, which requires $R_{i,i} \neq 0$ for all i.

3.3.2 Numerical simulations

The main goal of the following numerical simulations is to check the filter stability and its performance when the pair of model (3.12) and observation (3.1) with operator (3.13) nearly violates either the observability or controllability criteria. Secondly, we are also interested to understand the effect of model errors through numerical discretizations such as forward Euler and backward Euler.

To check the role of the observable condition, we test four observation operators:

$$\text{Fully observable: } G = [1, 1, 1],$$
$$\text{First mode weakly observable: } G1 = [10^{-2}, 1, 1],$$
$$\text{Second mode weakly observable: } G2 = [1, 10^{-2}, 1],$$
$$\text{Third mode weakly observable: } G3 = [1, 1, 10^{-2}].$$

We fix the damping coefficients $\gamma_1 = 1/2, \gamma_2 = 1, \gamma_3 = 5$; these damping strengths and observation time $\Delta t = 1/2$ yield amplification factors $|F_1| = 0.78, |F_2| = 0.61$ and $|F_3| = 0.08$. We also check a fully unobservable system by letting $|F_1| = |F_2| = |F_3| = 0.78$.

In our numerical simulations, we specify a uniform climatological energy spectrum

$$E_i = \frac{\sigma_i^2}{2\gamma_i} = 1, \tag{3.15}$$

for all $k = 1, 2, 3$ for this fully controllable system. To tamper with a practical controllability condition, we also consider a weak energy $E_2 = 0.01$ and let the remaining modes' energy be unity as in (3.15). The Kalman filter is implemented with a fixed observation time $\Delta t = 0.5$ which is longer than the damping time of the third component $1/\gamma_3 = 0.2$, but shorter than the damping times of the first two components $1/\gamma_1 = 2$ and $1/\gamma_2 = 1$, respectively.

True filter: Simulations with no model error

We check four cases here: fully controllable and weakly observable (Fig 3.1), weakly controllable and weakly observable (Fig 3.2), fully controllable and fully unobservable (Fig 3.3), and weakly controllable and fully unobservable (Fig 3.4). In each figure we show the RMS errors (averaged over 1000 assimilation cycles), the condition number of the asymptotic posterior error covariance matrix R_∞, the largest eigenvalue of the asymptotic posterior error covariance matrix R_∞, and the largest eigenvalue of the asymptotic model error covariance matrix $\text{cov}(\mathcal{X}_\infty \mathcal{X}_\infty^T)$ (see the detailed derivation in Section 2.3.2), all plotted as functions of the observation error variance r^o.

From these simulations, we learn that the filter performance of the weakly controllable system is slightly better than the fully controllable system (compare Fig. 3.1 and Fig. 3.2 or Fig. 3.3 and Fig. 3.4). When controllability is fully satisfied, the filter performance with weak observability due to either one of $G1$, $G2$ or $G3$ does not differ too much from each other. An interesting fact is that when controllability is weakened on the

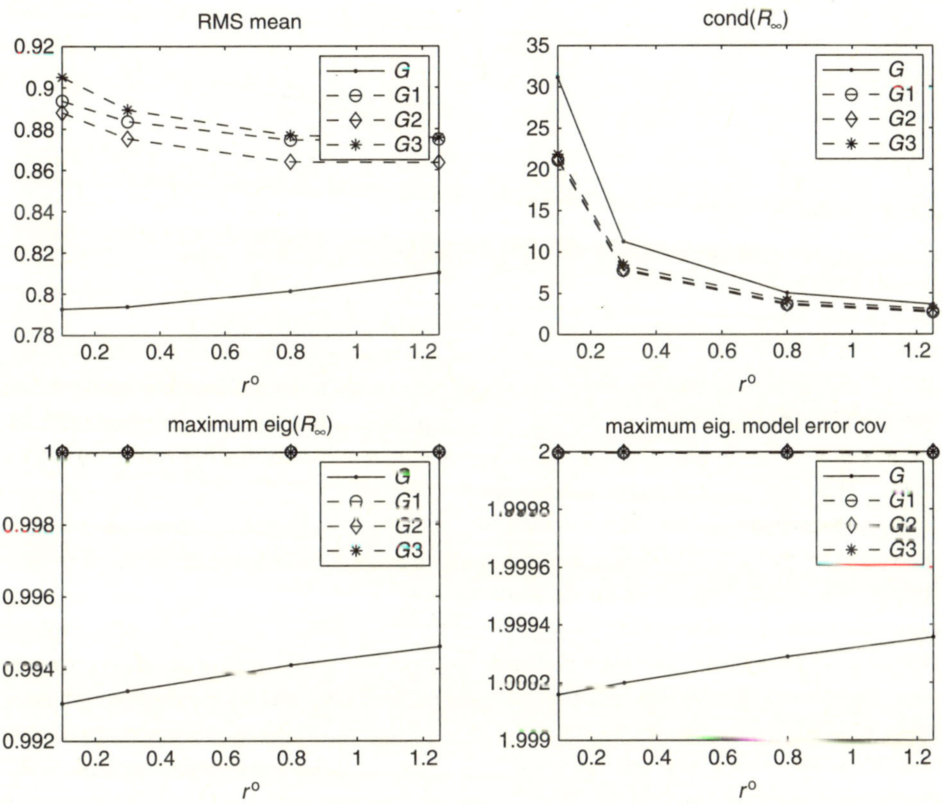

Figure 3.1 True filter: fully controllable.

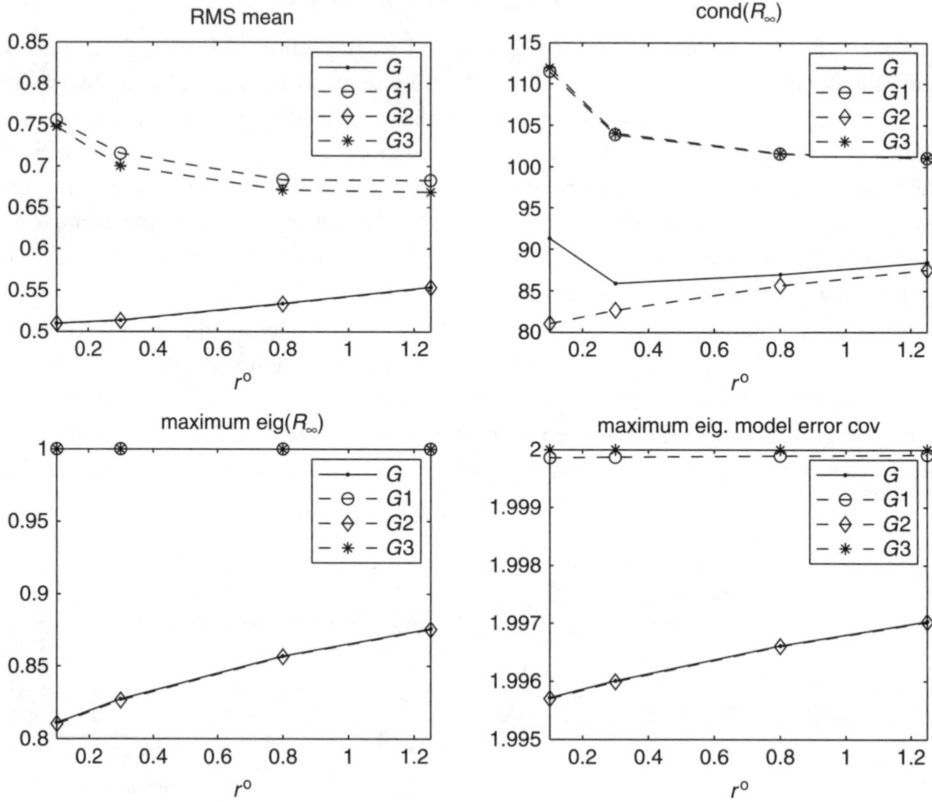

Figure 3.2 True filter: weakly controllable due to $E_2 = 0.01$.

second mode, the filter performances with full observability G and with weakly observable $G2$ are indistinguishable. Here, the weak controllability on the second mode suggests that the filter model for this mode is nearly deterministic and therefore the availability of the observation on this mode becomes irrelevant since the filter weighs more toward the model. The condition number of the asymptotic posterior covariance R_∞ can be very misleading in determining the filter performance as shown in Fig. 3.2 where the condition number is on the order 10^2 when controllability is nearly violated but the actual filter performance is slightly better than the fully controllable system (Fig. 3.1); it is more relevant to check the maximum eigenvalues of R_∞. When the observability condition is fully violated (Fig. 3.3 and Fig. 3.4), the filter performance deteriorates only slightly. These simulations also suggest that the results above are quite robust for variations of observation noise variance r^o.

One can also check that when a model error is introduced with implicit scheme backward Euler, the filter performance and its off-line variables are very similar to the results

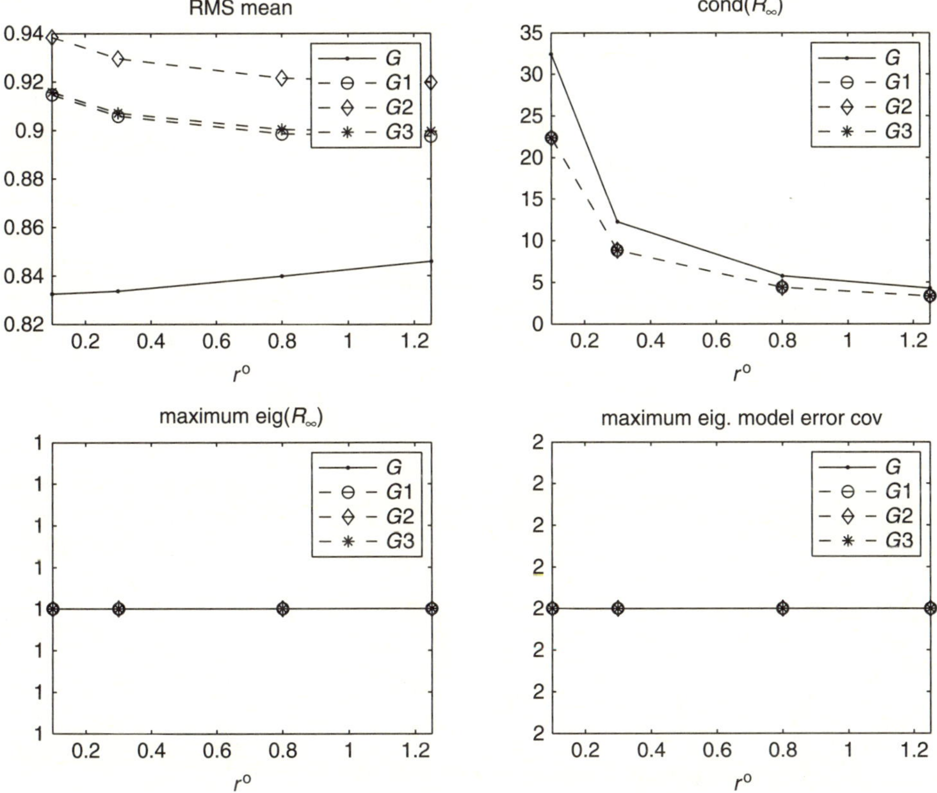

Figure 3.3 True filter: fully controllable and fully unobservable due to $F_1 = F_2 = F_3$.

in this section. This numerical experiment shows that the observability and controllability conditions are not essential when every mode is strongly stable, i.e. $|F_i| < 1$ for every i. However, we will encounter in Chapter 7 that these conditions play different roles when the filter model is wavelike with small uniform damping (the damped stochastically forced advection–diffusion equation is introduced in Chapter 5). In the situation when the mean state does not decay to zero as used here but decays to a known deterministic time-dependent periodic forcing with resonant frequency (Harlim and Majda, 2008b), the filter suffers from strong divergence. On the other hand, for a decaying mean state as in the present setting, there is significant filter skill even though observability is violated (see Chapter 7).

Forward Euler discretization and model error

In this section, we consider an imperfect filter model through forward Euler in solving differential equation (3.12). Specifically, for this time step $\Delta t = 0.5$, the amplitudes of the

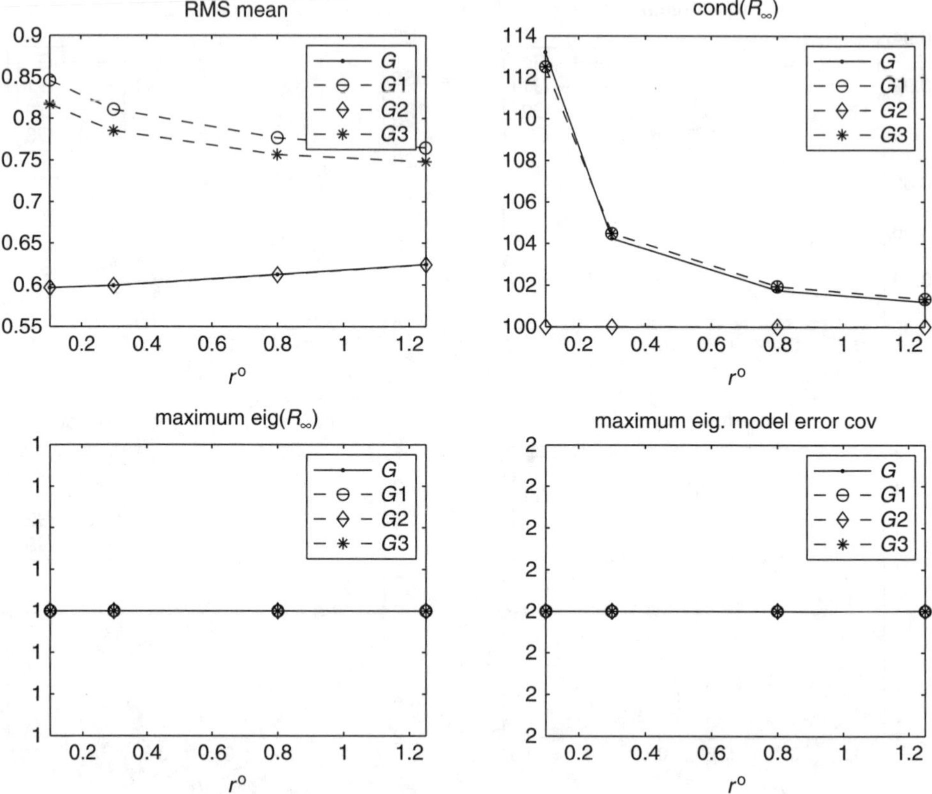

Figure 3.4 True filter: weakly controllable and fully unobservable due to $F_1 = F_2 = F_3$.

dynamical operator are stable for the first two components $|F_1| = 0.75$, $|F_2| = 0.5$ and unstable for the third component $|F_3| = 1.5$. Note that the true signal is generated by the analytical solution of the original model in (3.12), *not* through forward Euler.

The numerical results show that when the observability condition is weakened in the stable modes (e.g. with $G1$ or $G2$), the filter is still stable and the performance skill declines slightly (see Fig. 3.5). However when the unstable mode is also weakly observed ($G3$), the filter becomes divergent. This filtering divergence is detectable by looking at the maximum eigenvalue of both the asymptotic posterior error covariance and the asymptotic model error covariance, which are on the order of 10^4 (see Fig. 3.6). These results also hold even when the controllability is weakened (not shown).

This example illustrates the role of observability and controllability in guaranteeing filter stability and skill in the unstable mode. This numerical example also illustrates that the formal mathematical observability condition for filtering stability for an unstable mode can be satisfied, yet from a practical point of view the filter is useless. In other words, the filter stability does not give us any information about how precise the filtered solutions

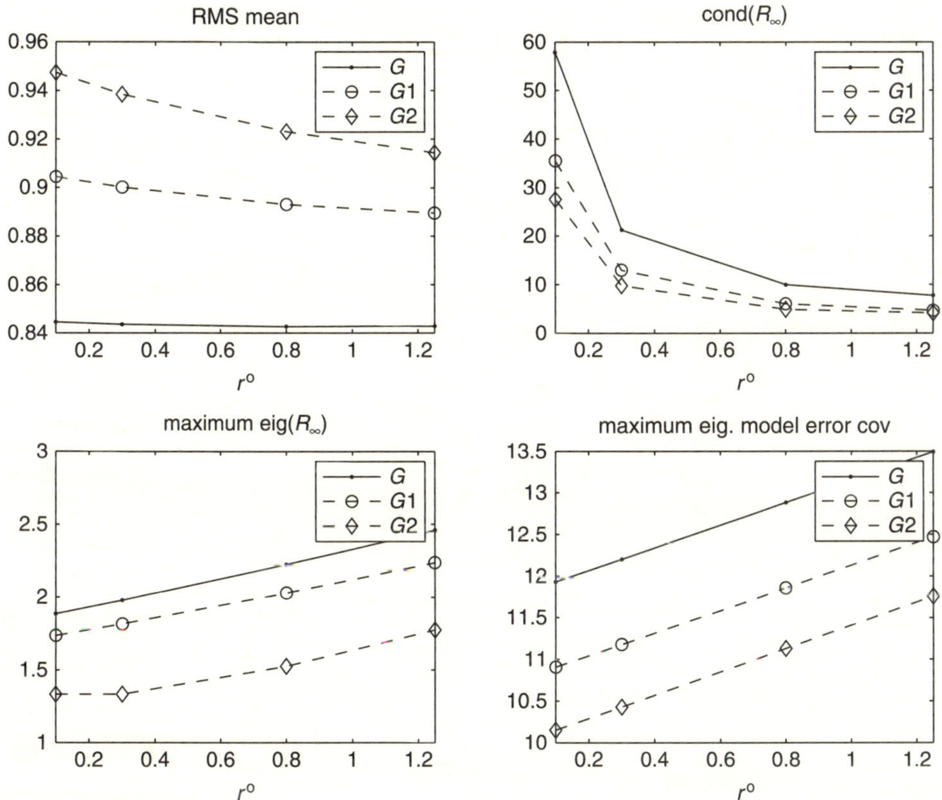

Figure 3.5 Forward Euler: fully controllable, fully observable G and weakly observable on the stable modes $G1$, $G2$.

are compared to the true signals; also recall the example for the simpler scalar field from Chapter 2 of backward Euler with time-discrete noise where the filter is strongly stable but has no practical skill. Things are even worse here for the unstable mode; in Fig. 3.7 we observe that when the model is observable (although weakly with $G3$), we obtain a stable filtered solution that fluctuates near 73 whereas the true signal fluctuates near 1! Even more dramatic examples of the practical failure of observability for large unstable systems can be found in Majda and Grote (2007).

3.4 Reduced filters for large systems

In many practical physical systems the state space for $\vec{u} \in \mathbb{R}^N$ has a very large dimension, $N \gg 1$. The evaluation of the full covariance matrix for a filter becomes prohibitively expensive under these circumstances, for example, if $N = 10^4$ or 10^6. On the other hand, some systems with large N often have a much smaller subspace, N^+, of unstable directions

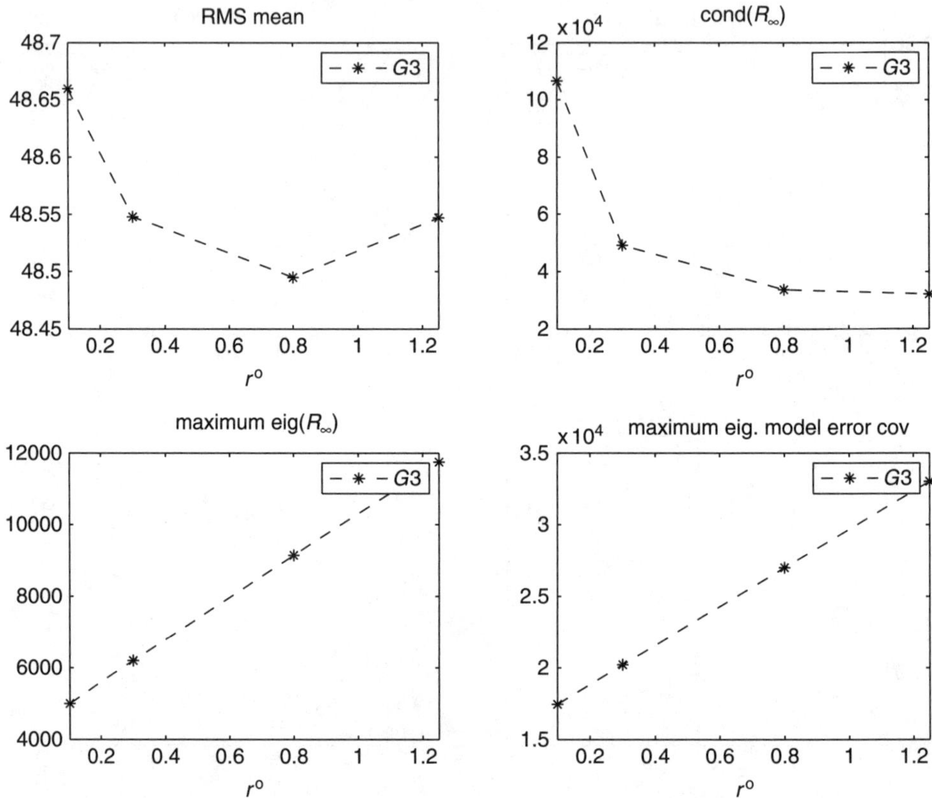

Figure 3.6 Forward Euler: fully controllable and weakly observable on the unstable mode $G3$.

and these directions can contain most of the important practical information to be processed for the filter. This happens, for example, in large-scale models of the mid-latitude atmosphere (Todling and Ghil, 1994; Farrell and Ioannou, 2005; Chorin and Krause, 2004). Thus, it is interesting to develop reduced filters for processing observations and their impact on the unstable modes. Below, we illustrate such a reduced filter in an idealized setting.

Consider a filtering problem for an N-dimensional vector $\vec{u} = (\vec{u}^-, \vec{u}^+)^T$ with $\vec{u}^-, \vec{u}^+$, the subspaces of stable and unstable dynamical directions with dimensions N^-, N^+, respectively, where $N = N^- + N^+$. The linear dynamics that advances the system from the state $\vec{u}_{m|m}$ to $\vec{u}_{m+1|m}$ for $m = 0, 1, \ldots$, with $\vec{u}_{0|0}$ a specified Gaussian distribution, is given by $\vec{u}_{m+1|m} = (\vec{u}^-_{m+1|m}, \vec{u}^+_{m+1|m})$ with

$$\begin{aligned} \vec{u}^-_{m+1|m} &= F_- \vec{u}^-_{m|m} + \vec{\sigma}^-_{m+1}, \\ \vec{u}^+_{m+1|m} &= F_+ \vec{u}^+_{m|m} + \vec{\sigma}^+_{m+1}, \end{aligned} \tag{3.16}$$

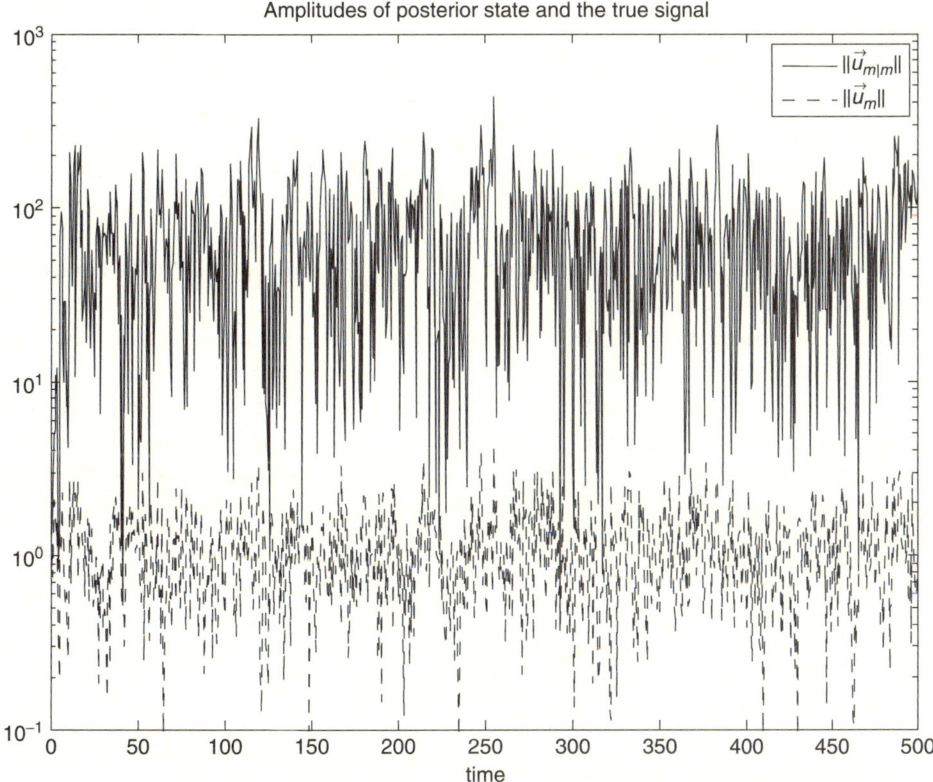

Figure 3.7 Amplitudes of the filtered solution $\|\vec{u}_{m|m}\|$ for $r^o = 0.8$ with forward Euler and the true signal $\|\vec{u}_m\|$, both (in logarithmic scale) plotted as functions of time.

where the stable and unstable dynamics operators F_-, F_+ satisfy

$$\text{The eigenvalues of } F_-(F_+) \text{ have modulus } < 1(> 1).$$

We remark that any general linear system with only stable and unstable modes can be written in this block form through a change of variables. The forcing terms, $\vec{\sigma}^-_{m+1}, \vec{\sigma}^+_{m+1}$, are assumed to be Gaussian white noise vectors with block diagonal covariance

$$R = \begin{pmatrix} R^- & 0 \\ 0 & R^+ \end{pmatrix},$$

independent of each m, with $N^- \times N^-$ and $N^+ \times N^+$ block covariance matrices given respectively by R^-, R^+ with $R^- > 0$, $R^+ > 0$ so controllability is always satisfied ($R > 0$ means $\vec{\zeta}^T R \vec{\zeta} > 0$ for every nonzero column vector $\vec{\zeta}$). The state $\vec{u}_{m+1|m+1}$, as an estimate of the unknown true state $\vec{u}_{m+1}$, is determined by a linear filtering strategy recursively from the prior distribution in $\vec{u}_{m+1|m}$ in (3.16) through a set of observations

$$\vec{v}_{m+1} = G\vec{u}_{m+1} + \vec{\sigma}^o_{m+1}. \tag{3.17}$$

The observation matrix G is $M \times N$, so there are M observations, and $\vec{\sigma}^o_{m+1}$ is Gaussian noise for the observations, which is assumed to be independent for each t_m with $M \times M$ covariance matrix, R^o, and $R^o > 0$.

With the assumption that the initial state, $\vec{u}_{0|0}$, is a Gaussian distribution with mean $\bar{\vec{u}}_{0|0}$ and covariance $R_{0|0} > 0$, the dynamics in (3.16) and the observations in (3.17) through a filtering strategy determine the state $\vec{u}_{m+1|m}, \vec{u}_{m+1|m+1}$ for $t_m, m = 0, 1, 2, \ldots$, which is a Gaussian distribution at each state with mean $\bar{\vec{u}}_{m+1|m}, \bar{\vec{u}}_{m+1|m+1}$ and $N \times N$ covariance matrix $R_{m+1|m}, R_{m+1|m+1}$.

We build a reduced filter for the system in (3.16). Start at time $t = 0$ with a factored covariance matrix $R_{0|0} = (R^-_{0|0}, R^+_{0|0})$. Generate an estimator, $\vec{u}^-_{m+1|m+1} = \vec{u}^-_{m+1|m}$ in the trivial fashion by solving the first equation for the stable directions (3.16) with the initial data and completely ignoring the observations for F_-. Because F_- has all eigenvalues smaller than one in modulus, it is easy to see that $\vec{u}^-_{m+1|m+1}$ is stable. Next, let $\vec{u}^+_{m+1|m+1}$ be the Kalman filter solution for the second unstable dynamical equation in (3.16) with the observations

$$\begin{aligned}\vec{v}^+_{m+1} &\equiv \vec{v}_{m+1} - GP_-\vec{u}_{m+1} = \vec{v}_{m+1} - G^-\vec{u}^-_{m+1}\\ &= GP_+\vec{u}_{m+1} + \vec{\sigma}^o_{m+1},\\ &= G^+\vec{u}^+_{m+1} + \vec{\sigma}^o_{m+1}\end{aligned} \tag{3.18}$$

where P_+ and P_- denote the projection on $\vec{u}^+$ and $\vec{u}^-$, respectively, i.e. $P_+\vec{u} = P_+(\vec{u}^-, \vec{u}^+)^T = (0, \vec{u}^+)^T$ and $P_-\vec{u} = (\vec{u}^-, 0)^T$; in the first and third rows of (3.18), we assume the observation operator has the following block decomposition

$$G = [G^-, G^+], \quad G^- \in \mathbb{R}^{M\times N^-}, \ G^+ \in \mathbb{R}^{M\times N^+}. \tag{3.19}$$

Since the best estimate of the unknown true state $\vec{u}_{m+1}$ is the mean posterior state $\bar{\vec{u}}_{m+1|m+1}$, then the observation $\vec{v}^+_{m+1}$ on the left-hand side of (3.18) is obtained through the following unbiased estimator

$$\vec{v}^+_{m+1} = \vec{v}_{m+1} - G^-\bar{\vec{u}}^-_{m+1|m+1}, \tag{3.20}$$

where $\bar{\vec{u}}^-_{m+1|m+1}$ is the mean of the trivial estimate in the stable direction.

Now, the unstable dynamics in (3.16) together with (3.18), (3.20) define a filtering problem on the lower-dimensional state space, $\vec{u}^+ \in \mathbb{R}^{N_+}$; use the classical Kalman filtering algorithm on this reduced problem in (3.16), (3.18). This is a simple reduced filter which always trusts the dynamics on the stable directions and only filters the unstable ones.

When is this reduced filter stable? Since it is already controllable, we just need to check that the reduced filter in (3.16), (3.18) is observable and satisfies (3.11). In other words, the matrix mapping defined by

$$\vec{u}^+ \mapsto (G^+\vec{u}^+, G^+F_+\vec{u}^+, \ldots, G^+F^L_+\vec{u}^+), \tag{3.21}$$

has full rank for some $L \geq 0$, i.e. there exists $c > 0$ so that

$$\sum_{\ell=0}^{L} \|G^+ F_+^{\ell} \vec{u}^+\|^2 \geq c\|\vec{u}^+\|^2. \tag{3.22}$$

The algebraic test condition in Eqn (3.22) generalizes the observability condition in filtering theory as a requirement only on the unstable modes.

In the present context, this filter provides rigorous justification for the reduced-order Kalman filtering strategies with $N^+ \ll N$ developed by several authors (Todling and Ghil, 1994; Farrell and Ioannou, 2001, 2005) and also provides a precise quantitative requirement for stability of such algorithms through the requirement of observability in (3.21) or (3.22).

It is worthwhile to point out that such a reduced filtering strategy can be expected to have skill only if the ratio between the largest eigenvalue of F_- and the smallest eigenvalue of F_+ is significantly large and these matrices do not have large off-diagonal entries. For example if

$$F_- = \begin{pmatrix} 0.95 & 10^4 \\ 0 & 0.97 \end{pmatrix}$$

and

$$F_+ = 1.05,$$

then obviously, the stable directions need to be dynamically filtered.

3.5 A priori covariance stability for the unstable mode filter given strong observability

Here we consider the situation where the unstable modes are strongly observable, i.e. $L = 0$ in (3.22) so that we have

$$\|G^+\vec{u}^+\|^2 \geq g^2\|\vec{u}^+\|^2,$$

for some scalar $g > 0$ and every $\vec{u}^+ \in \mathbb{R}^{N^+}$ where $(0, \vec{u}^+) = P_+\vec{u}$. Recall that the operator matrix $G^+ \in \mathbb{R}^{M\times N^+}$ is defined as in (3.19).

Recall for positive definite matrices P and Q, $P \leq Q$ means that $\vec{\zeta}^T P\vec{\zeta} < \vec{\zeta}^T Q\vec{\zeta}$ for all column vectors $\vec{\zeta}$. We now give a direct proof of covariance stability for the Kalman filter on the unstable modes alone.

Lemma 3.1 *Consider the map on positive definite covariance matrices*

$$\Phi(P, R) = P - P(G^+)^T(G^+P(G^+)^T + R)^{-1}G^+P,\ P \in \mathbb{R}^{N^+\times N^+}.$$

Then

$$P_1 \leq P_2 \Rightarrow \Phi(P_1, R) \leq \Phi(P_2, R),$$

and

$$0 < R_1 \leq R_2 \Rightarrow \Phi(P, R_1) \leq \Phi(P, R_2).$$

Proof The first statement is a particular case of lemma 6.2 in Chui and Chen (1999) while the second statement immediately follows from lemma 1.3 in Chui and Chen (1999). □

Similarly, we have

Lemma 3.2 *Consider the dynamic map on positive definite covariance matrices*

$$\Psi(P, R) = F_+ P F_+^T + R,\ P \in \mathbb{R}^{N^+ \times N^+}.$$

Then

$$P_1 \leq P_2 \Rightarrow \Psi(P_1, R) \leq \Psi(P_2, R),$$

and

$$0 < R_1 \leq R_2 \Rightarrow \Psi(P, R_1) \leq \Psi(P, R_2).$$

Proof Both statements immediately follow from basic properties of quadratic forms and are left as an exercise for the reader. □

The following identity will be useful below:

Lemma 3.3 *For $\beta > 0$ and an arbitrary matrix G, the following identity holds*

$$G^T (GG^T + \beta \mathcal{I})^{-1} G = G^T G (G^T G + \beta \mathcal{I})^{-1}.$$

Proof Since

$$G(G^T G + \beta \mathcal{I}) = (GG^T + \beta \mathcal{I})G,$$

we immediately conclude that

$$(GG^T + \beta \mathcal{I})^{-1} G = G(G^T G + \beta \mathcal{I})^{-1}.$$

Pre-multiplication by G^T then yields the identity. □

Given the current covariance $R_{m|m} \in \mathbb{R}^{N^+ \times N^+}$, the standard Kalman filtering algorithm for the unstable modes proceeds in two steps as follows:

Time evolution

$$R_{m+1|m} = \Psi(R_{m|m}, R^+).$$

Observation update

$$R_{m+1|m+1} = \Phi(R_{m+1|m}, R^o).$$

We are interested in obtaining an explicit stability bound under the strong observability hypothesis, i.e.

Find c_o so that $R_{m|m} \leq c_o \mathcal{I} \Rightarrow R_{m+1|m+1} \leq c_o \mathcal{I}$ for every m.

We assume that

$$R_{m|m} \leq c_o \mathcal{I},$$

and let the system and observation noise covariance matrices be bounded from above by

$$0 < R^+ < r_+ \mathcal{I}, \quad 0 < R^o < r^o \mathcal{I}.$$

Then by Lemma 3.2, we have

$$\begin{aligned} R_{m+1|m} &= \Psi(R_{m|m}, R^+) \leq \Psi(c_o \mathcal{I}, R^+) \\ &= \Psi(c_o \mathcal{I}, r_+ \mathcal{I}) = c_o F_+ F_+^T + r_+ \mathcal{I} \\ &\leq (c_o f_+^2 + r_+) \mathcal{I}, \end{aligned}$$

where

$$\|(F_+)^T \vec{u}^+\| \leq f_+ \|\vec{u}^+\|, \quad \forall \vec{u}^+.$$

Thus by Lemma 3.1, we have

$$\begin{aligned} R_{m+1|m+1} &= \Phi(R_{m+1|m}, R^o) = \Phi(\Psi(R_{m|m}, R^+), R^o) \\ &\leq \Phi((c_o f_+^2 + r_+)\mathcal{I}, r^o \mathcal{I}) \\ &= (c_o f_+^2 + r_+)\mathcal{I} \\ &\quad -(c_o f_+^2 + r_+)(G^+)^T \left(G^+ (G^+)^T + \frac{r^o}{c_o f_+^2 + r_+} \mathcal{I} \right)^{-1} G^+. \end{aligned}$$

Now, we use Lemma 3.3 to rewrite the last equation as

$$\begin{aligned} R_{m+1|m+1} &\leq (c_o f_+^2 + r_+)\mathcal{I} \\ &\quad -(c_o f_+^2 + r_+)(G^+)^T G^+ \left((G^+)^T G^+ + \frac{r^o}{c_o f_+^2 + r_+} \mathcal{I} \right)^{-1} \\ &= r^o \left((G^+)^T G^+ + \frac{r^o}{c_o f_+^2 + r_+} \mathcal{I} \right)^{-1} \\ &\leq r^o \left(g^2 + \frac{r^o}{c_o f_+^2 + r_+} \right)^{-1} \mathcal{I}. \end{aligned}$$

Therefore, to guarantee the filtering stability bound, we require

$$r^o \left(g^2 + \frac{r^o}{c_o f_+^2 + r_+} \right)^{-1} \leq c_o. \tag{3.23}$$

Now, (3.23) with equality yields a quadratic equation for c_o

$$c_o^2 + \left(\frac{r_+}{f_+^2} + \frac{r^o}{g^2 f_+^2} - \frac{r^o}{g^2} \right) c_o - \frac{r^o r_+}{g^2 f_+^2} = 0,$$

whose positive solution is the desired upper bound,

$$c_o = \frac{1}{2}\left(\frac{r^o}{g^2} - \frac{r_+}{f_+^2} - \frac{r^o}{g^2 f_+^2} + \left[\left(\frac{r_+}{f_+^2} + \frac{r^o}{g^2 f_+^2} - \frac{r^o}{g^2}\right)^2 + 4\frac{r^o r_+}{g^2 f_+^2}\right]^{1/2}\right). \tag{3.24}$$

Note that the upper bound estimate (3.24) is precisely the asymptotic scalar limiting filter from (2.25) in Section 2.2 with $c_o = r_\infty$, $r_+ = r$ and $f_+ = |F|$, where g is the smallest singular value of the matrix G^+, f_+ is the upper bound of the unstable expansion rate F_+^T and positive scalars r_+, r^o are the system and observation noise covariance upper bounds, respectively.

4

Continuous and discrete Fourier series and numerical discretization

In this chapter, we briefly review the basic ideas involved in continuous and discrete Fourier series such as aliasing. After we have the tools which we need from continuous and discrete Fourier series, we will use them to solve differential and difference equations as well as to analyze properties of good difference equations for numerical approximations.

4.1 Continuous and discrete Fourier series

Fourier series are a mathematical tool that represents continuous periodic functions as linear combinations of trigonometric functions $\{e^{i\ell x}\}$. Throughout this chapter, we only consider 2π-periodic functions $f(x) \in L^2(0, 2\pi)$, that is, $f(x) = f(x + 2\pi)$ for all x and $\int_0^{2\pi} |f|^2 < \infty$. Consequently, we can restrict our attention to a domain of $0 \le x \le 2\pi$. The corresponding discrete case has a domain of finite grid points $x_j = jh$, $j = 0, 1, \ldots, 2N$, where $(2N + 1)h = 2\pi$. In this setup, a 2π-periodic discretized function satisfies $f_j = f_{j+(2N+1)}$ for $f_j \equiv f(x_j)$, $j \in \mathbb{Z}$. For functions with different periods, we can always normalize them to be 2π-periodic. To define the discrete equivalent to the integral in the continuous case, we consider a simple example, $f(x) = 1$. A proper normalization for the discrete case chooses h and N such that

$$\int_0^{2\pi} f(x)dx = \sum_{j=0}^{2N} 1h,$$

which explains the choice of $2\pi = (2N + 1)h$. Thus, we can define the discrete complex inner product and the discrete norm, following the respective continuous case (see Table 4.1). One can easily see that the dimension of the discrete vector space is $2N + 1$ since we sum over $2N + 1$ points. On the other hand, we need to prove that the dimension of the continuous vector space is infinite.

Proposition 4.1 *The trigonometric functions* $\{e^{i\ell x} = \cos(\ell x) + i\sin(\ell x), \ell \in \mathbb{Z}\}$ *form a basis for the vector space* $L^2(0, 2\pi)$. *Consequently, this vector space is infinite dimensional.*

Table 4.1 Continuous and discrete equivalent of complex inner product and norm.

	Continuous	Discrete
Complex inner product	$(f, g) = \frac{1}{2\pi}\int_0^{2\pi} f(x)g^*(x)dx$	$(\vec{f}, \vec{g})_h = \frac{h}{2\pi}\sum_{j=0}^{2N} f_j g_j^*$
Norm	$\|f\|^2 = \frac{1}{2\pi}\int_0^{2\pi} \|f(x)\|^2 dx$	$\|\vec{f}\|_h^2 = \frac{h}{2\pi}\sum_{j=0}^{2N} \|f_j\|^2$

Proof To show that they form a basis, we check that they are orthogonal:

$$\left(e^{\mathrm{i}\ell x}, e^{\mathrm{i}mx}\right) = \frac{1}{2\pi}\int_0^{2\pi} e^{\mathrm{i}\ell x} e^{-\mathrm{i}mx} dx = \frac{1}{2\pi}\int_0^{2\pi} e^{\mathrm{i}(\ell-m)x} dx.$$

Note that $\left(e^{\mathrm{i}\ell x}, e^{\mathrm{i}mx}\right) = 1$ when $\ell = m$ and $\left(e^{\mathrm{i}\ell x}, e^{\mathrm{i}mx}\right) = 0$ when $\ell \neq m$ since $\ell - m$ is an integer and $e^{\mathrm{i}x}$ is a 2π-periodic function. Thus, $\{e^{\mathrm{i}\ell x}, \ell \in \mathbb{Z}\}$ form an infinite-dimensional basis for our vector space. □

Now, let us construct a basis for the discrete vector space with dimension $2N+1$. Based on the continuous basis, our first guess would be $\vec{e}^{\,\ell} = (e^{\mathrm{i}\ell x_j})$ where $|\ell| \leq N$ and $x_j = jh$ since we only evaluate our discrete function at lattice points.

Proposition 4.2 $\{\vec{e}^{\,\ell}\}_{\ell=0,\pm1,\ldots,\pm N}$, *form an orthonormal basis for the vector space* $\mathbb{C}^{2N+1}$ *with discrete inner product.*

Proof Let $\ell, m = 0, \pm1, \ldots, \pm N$. The discrete inner product

$$\left(\vec{e}^{\,\ell}, \vec{e}^{\,m}\right)_h = \frac{h}{2\pi}\sum_{j=0}^{2N} e^{\mathrm{i}(\ell-m)jh}$$

is exactly 1 when $\ell = m$ since $(2N+1)h = 2\pi$. When $\ell \neq m$, set $\omega = e^{\mathrm{i}(\ell-m)h}$ such that

$$\left(\vec{e}^{\,\ell}, \vec{e}^{\,m}\right)_h = \frac{h}{2\pi}\sum_{j=0}^{2N}\omega^j = \frac{h}{2\pi}\frac{1-\omega^{2N+1}}{1-\omega} = \frac{h}{2\pi}\frac{1-e^{\mathrm{i}(\ell-m)h(2N+1)}}{1-e^{\mathrm{i}(\ell-m)h}}$$

through geometric series. Notice that the denominator is nonzero but the numerator is always zero since $(2N+1)h = 2\pi$. Therefore, $\vec{e}^{\,\ell}$ is a basis. □

We end this section by stating the main results for the continuous and discrete expansion theory (consult Folland, 1999, for proofs of these facts).

Continuous expansion theory

For the continuous 2π-periodic function $f(x) \in L^2(0, 2\pi)$, if the Fourier coefficients

$$\hat{f}(\ell) = \frac{1}{2\pi}\int_0^{2\pi} f(x)e^{-\mathrm{i}\ell x}dx, \quad \ell = 0, \pm1, \pm2, \ldots$$

satisfy

$$\sum_{\ell=-\infty}^{\infty} |\hat{f}(\ell)| < \infty, \tag{4.1}$$

then

$$f(x) = \sum_{\ell=-\infty}^{\infty} \hat{f}(\ell) e^{\mathrm{i}\ell x}$$

uniformly in x. Note that if we don't have condition (4.1), the convergence is only in the L^2 sense. What is true when condition (4.1) is not satisfied is

$$\|f\|^2 = \sum_{\ell=-\infty}^{\infty} |\hat{f}(\ell)|^2,$$

which is known as Parseval's identity.

Discrete expansion theory

Let $\vec{f} \in \mathbb{C}^{2N+1}$, where $\vec{f} = (f_0, f_1, \ldots, f_{2N})$. Since $\vec{e}^{\ell}$ is an orthonormal basis for the vector space $\mathbb{C}^{2N+1}$, we have

$$\vec{f} = \sum_{\ell=-N}^{N} (\vec{f}, \vec{e}^{\ell})_h \vec{e}^{\ell},$$

where the discrete Fourier coefficients are given by

$$\hat{f}_h(\ell) = (\vec{f}, \vec{e}^{\ell})_h = \frac{h}{2\pi} \sum_{j=0}^{2N} f_j e^{-\mathrm{i}\ell j h}.$$

We can also write $\vec{f}$ component-wise

$$f_j = \sum_{|\ell| \leq N} \hat{f}_h(\ell) e^{\mathrm{i}\ell j h}. \tag{4.2}$$

The natural notion of length (Parseval's identity) follows from Table 4.1.

$$\|\vec{f}\|_h^2 = (\vec{f}, \vec{f})_h = \sum_{|\ell| \leq N} |\hat{f}_h(\ell)|^2.$$

4.2 Aliasing

Aliasing is an artifact of the discretization process. When we try to represent a continuous function by a discrete set of points we lose information. In particular we cannot distinguish high harmonics from their low counterpart (e.g. see Fig. 4.1). Mathematically, we can describe this fact by considering the discrete oscillatory basis $\vec{e}^{\ell} = (e^{\mathrm{i}\ell j h})$ for $|\ell| \leq N$ with the following harmonics

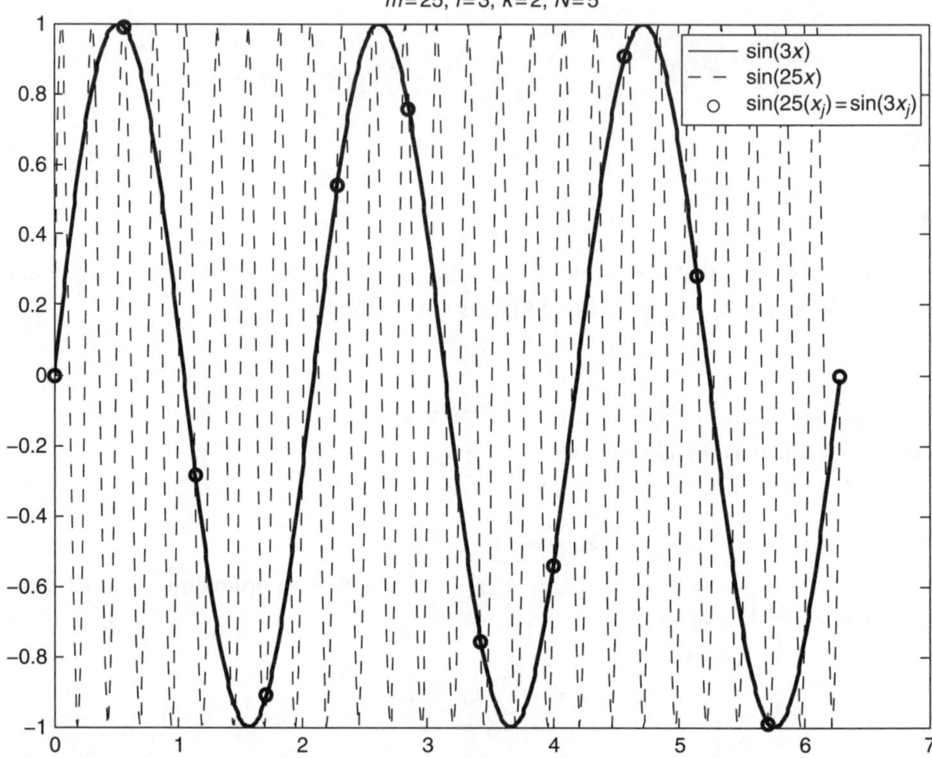

Figure 4.1 Aliasing: identical discrete sample points (circles) for $\sin(3x)$ (solid line) and $\sin(25x)$ (dashes).

$$m = \ell + (2N + 1)k, \quad k = 0, \pm 1, \pm 2, \ldots$$

Since $(2N + 1)h = 2\pi$, it is clear that

$$e^{imjh} = e^{i(\ell+(2N+1)k)jh} = e^{i\ell jh} e^{i2\pi kj} = e^{i\ell jh}.$$

We have shown that when we sample e^{imx} with $2N + 1$ grid points, $x = x_j = jh$, such that $(2N + 1)h = 2\pi$, we cannot distinguish it from its lower harmonics, $e^{i\ell x}$, at the corresponding grid points. In Fig. 4.1, we show an example where aliasing occurs with two sinusoidal functions, one with higher frequency, $m = 25$, and another with lower frequency, $\ell = 3$. Here, both functions are sampled with identically $2N+1 = 11$ uniformly distributed grid points.

Next, we would like to know how the coefficients of the continuous Fourier series, $\hat{f}(\ell)$, $|\ell| \leq N$, compare to the discrete coefficients, $\hat{f}_h(\ell)$. To understand the difference, let us consider

$$f(x) = e^{imx}$$

for any m. No matter what m is, it can be written in the following form

$$m = \ell + (2N + 1)k$$

for some ℓ, $|\ell| \leq N$ and $k \in \mathbb{Z}$. Since $f(x)$ is sufficiently simple, we can calculate $\hat{f}(\ell)$ and $\hat{f}_h(\ell)$. In particular,

$$\hat{f}(\ell) = \frac{1}{2\pi} \int_0^{2\pi} e^{imx} e^{-i\ell x} dx = \begin{cases} 1, & \text{if } \ell = m \\ 0, & \text{else} \end{cases}$$

$$\hat{f}_h(\ell) = \frac{h}{2\pi} \sum_{j=0}^{2N} e^{imjh} e^{-i\ell jh} = 1.$$

The last equality is due to the aliasing $e^{imjh} = e^{i\ell jh}$. If $\ell = m$, both $\hat{f}(\ell) = \hat{f}_h(\ell) = 1$ and we do not have aliasing since $k = 0$, $|\ell| \leq N$. For $\ell \neq m$, there is a difference between $\hat{f}(\ell)$ and $\hat{f}_h(\ell)$, and this difference is the result of aliasing.

For a more general $f(x)$, we can look at it as a superposition of exponentials.

Proposition 4.3 *Consider*

$$f(x) = \sum_{m=-\infty}^{\infty} \hat{f}(m) e^{imx}.$$

For $|\ell| \leq N$, we can write down an expression for $\hat{f}_h(\ell)$ in terms of $\hat{f}(\ell)$,

$$\hat{f}_h(\ell) = \hat{f}(\ell) + \sum_{k \neq 0} \hat{f}(\ell + k(2N + 1)),$$

where the second term on the right-hand side is the aliasing error.

Proof Compute $\vec{f} = (f(0), f(h), \ldots, f(2\pi))$. We want to reorganize the following sum

$$\begin{aligned} f_j &= \sum_{m=-\infty}^{\infty} \hat{f}(m) e^{imjh}, \quad j = 0, 1, \ldots, 2N, \\ &= \ldots + \sum_{\ell=-3N-1}^{-N-1} \hat{f}(\ell) e^{i\ell jh} + \sum_{\ell=-N}^{N} \hat{f}(\ell) e^{i\ell jh} + \sum_{\ell=N+1}^{3N+1} \hat{f}(\ell) e^{i\ell jh} + \ldots \\ &= \ldots + \sum_{|\ell| \leq N} \hat{f}(\ell - (2N + 1)) e^{i(\ell-(2N+1))jh} + \sum_{\ell=-N}^{N} \hat{f}(\ell) e^{i\ell jh} \\ &\quad + \sum_{|\ell| \leq N} \hat{f}(\ell + 2N + 1) e^{i(\ell+2N+1)jh} + \ldots \\ &= \sum_{|\ell| \leq N} \sum_{k=-\infty}^{\infty} \hat{f}(\ell + k(2N + 1)) e^{i(\ell+(2N+1)k)jh} \\ &= \sum_{|\ell| \leq N} \sum_{k=-\infty}^{\infty} \hat{f}(\ell + k(2N + 1)) e^{i\ell jh}. \end{aligned}$$

The last equality uses the fact that $(2N+1)h = 2\pi$. By the discrete expansion theory in (4.2), we have

$$\hat{f}_h(\ell) = \sum_{k=-\infty}^{\infty} \hat{f}(\ell + k(2N+1))$$

and the proof is completed. □

4.3 Differential and difference operators

In Section 4.1, we showed that both the continuous and discrete bases involving oscillations were made up of exponentials. In this section, we are going to see that these exponentials are indeed eigenfunctions of both differential and difference operators, respectively. For the continuous case, let $u(x) \in C^K(0, 2\pi)$ be a 2π-periodic K-times continuously differentiable function. For known constant p_m, we define a differential operator

$$\mathcal{P}\Big(\frac{\partial}{\partial x}\Big)u(x) = \sum_{m=0}^{K} p_m \frac{d^m u(x)}{dx^m}.$$

Proposition 4.4 *Every harmonic* $e^{i\ell x}$ *is a* 2π*-periodic eigenfunction of* $\mathcal{P}(\partial_x)$.

Proof Since

$$\frac{d}{dx}e^{i\ell x} = (i\ell)e^{i\ell x},$$

$$\vdots$$

$$\frac{d^m}{dx^m}e^{i\ell x} = (i\ell)^m e^{i\ell x},$$

we have

$$\mathcal{P}\Big(\frac{\partial}{\partial x}\Big)e^{i\ell x} = \sum_{m=0}^{K} p_m \frac{d^m e^{i\ell x}}{dx^m} = \sum_{m=0}^{K} p_m (i\ell)^m e^{i\ell x}.$$

Thus, we have shown that $e^{i\ell x}$ is an eigenfunction of $\mathcal{P}(\partial_x)$ with eigenvalue

$$\tilde{p}(i\ell) = \sum_{m=0}^{K} p_m (i\ell)^m.$$

□

The equivalent discrete case is for $\vec{u} \in \mathbb{C}^{2N+1}$ with components $u_j = u(x_j)$, $x_j = jh$, where $j \in \mathbb{Z}$ and $(2N+1)h = 2\pi$. Each component satisfies $u_j = u_{j+2N+1}$. We define the difference operator $\vec{G}$ with component

$$(G(u_j))_k = \sum_{m=-K}^{K} a_m u_{k+m},$$

where a_m are known constant.

Proposition 4.5 *Every discrete harmonic $e^{i\ell x_j}$ with $x_j = jh$ is an eigenfunction of the difference operator $\vec{G}$.*

Proof Again the proof follows from the definition of $(G(u_j))_k$. Let $u_j = e^{i\ell jh}$ and

$$(G(e^{i\ell jh}))_k = \sum_{m=-K}^{K} a_m e^{i\ell(k+m)h} = \sum_{m=-K}^{K} a_m e^{i\ell mh} e^{i\ell kh}.$$

Here, the corresponding eigenvalue is

$$\tilde{g}_h(\ell) = \sum_{m=-K}^{K} a_m e^{i\ell mh}, \tag{4.3}$$

which is sometimes called the amplification factor. □

We are now in a position to show how one could get the eigenvalues directly for the linear finite difference operator which is the discrete Laplacian.

Example 4.6 Compute the eigenvalues for

$$(G(u_j))_k = \frac{u_{k+1} + u_{k-1} - 2u_k}{h^2}.$$

For this operator, we have $K = 1, a_{-1} = a_1 = 1/h^2$, and $a_0 = -2/h^2$. From Proposition 4.5, it is clear that $e^{i\ell jh}$ is an eigenvector for the difference operator with eigenvalue

$$\tilde{g}_h(\ell) = \frac{1}{h^2}(e^{-i\ell h} + e^{i\ell h} - 2) = \frac{2}{h^2}(\cos(\ell h) - 1).$$

4.4 Solving initial value problems

In this section, our goal is to understand when we can solve the initial value problem (IVP):

$$\frac{\partial u}{\partial t} = \mathcal{P}\Big(\frac{\partial}{\partial x}\Big)u,$$
$$u(x, 0) = f(x),$$

for $t > 0$ and $u(x + 2\pi, t) = u(x, t)$ and what the general solution is. In particular, we would like to find explicit algebraic criteria which would tell us when we have a solution for the IVP. Some examples of differential operators which arise in applications are the free-space Schrodinger operator, $\mathcal{P}(\partial_x)u = iu_{xx}$, the simple wave equation, $\mathcal{P}(\partial_x)u = Cu_x$, and the heat/diffusion equation, $\mathcal{P}(\partial_x)u = u_{xx}$.

Consider a set of trigonometric function, T_N, defined as follows

$$T_N = \Big\{ \sum_{|\ell| \le N} C_\ell e^{i\ell x}, C_\ell \in \mathbb{C} \Big\}. \tag{4.4}$$

Since we have shown that $e^{i\ell x}$ is an eigenfunction of the differential operator $\mathcal{P}(\partial_x)$, we would not be surprised to see that for functions from T_N, the IVP can always be solved. Formally, this is given by the next proposition.

Proposition 4.7 *For any operator $\mathcal{P}(\partial_x)$, any function $f(x) \in T_N$ and any N, we can always solve the IVP.*

Proof Solve the IVP

$$\begin{aligned} \frac{\partial u^\ell}{\partial t} &= \mathcal{P}\Big(\frac{\partial}{\partial x}\Big) u^\ell, \\ u^\ell(x, 0) &= e^{i\ell x}. \end{aligned} \tag{4.5}$$

We want to use separation of variables so we assume

$$u^\ell(x, t) = a^\ell(t) e^{i\ell x}. \tag{4.6}$$

Recall that since $e^{i\ell x}$ is an eigenfunction of $\mathcal{P}(\partial_x)$, substituting (4.6) into the IVP in (4.5) yields

$$\frac{\partial a^\ell}{\partial t} e^{i\ell x} = a^\ell \tilde{p}(i\ell) e^{i\ell x} \Rightarrow \frac{\partial a^\ell}{\partial t} = \tilde{p}(i\ell) a^\ell, \text{ where, } a^\ell(0) = 1.$$

This equation has the solution $a^\ell(t) = e^{\tilde{p}(i\ell)t}$ and therefore $u^\ell(t) = e^{\tilde{p}(i\ell)t} e^{i\ell x}$ solves (4.5). Now we want to solve the same PDE but with more general initial data, i.e.

$$\begin{aligned} \frac{\partial u}{\partial t} &= \mathcal{P}\Big(\frac{\partial}{\partial x}\Big) u, \\ u(x, 0) &= \sum_{|\ell| \le N} C_\ell e^{i\ell x}. \end{aligned}$$

Since this equation is linear we know that the sum of the solutions is also a solution. From the above we know that we can solve the IVP for initial data which consists of one harmonic. In order to get the solution to match the general initial data we just need to add up all the individual solutions,

$$u(x, t) = \sum_{|\ell| \le N} C_\ell e^{\tilde{p}(i\ell)t} e^{i\ell x}.$$

□

In the discrete setting, consider $\vec{u} = (u_0, u_1, \ldots, u_{2N}) \in \mathbb{C}^{2N+1}$ such that $u_j = u_{j+2N+1}$ and $(2N+1)h = 2\pi$. The corresponding IVP is given as follows

$$\begin{aligned} \vec{u}^{M+1} &= \vec{G}\vec{u}^M, \\ \vec{u}^0 &= \vec{f}, \end{aligned}$$

where superscript M denotes the discrete time step, and the difference operator $\vec{G} : \mathbb{C}^{2N+1} \to \mathbb{C}^{2N+1}$ is defined as follows

$$(G(u_j^M))_k = \sum_{m=-K}^{K} a_m u_{k+m}^M.$$

Based on the Fourier discrete expansion theory, we can write any initial conditions, $\vec{f} \in \mathbb{C}^{2N+1}$, as a linear combination of the exponential basis $\vec{e}^{\ell}$ (see Eqn (4.2)). By superposition, the general solution for the discrete IVP is always given by

$$\vec{u}^M = \sum_{|\ell| \le N} \tilde{g}_h(\ell)^M \hat{f}_h(\ell) \vec{e}^{\ell},$$

where $\tilde{g}_h(\ell)$ is the eigenvalue of the difference operator $\vec{G}$ as defined in (4.3).

4.5 Convergence of the difference operator

Naively one might think that any numerical scheme to discretize in x and t would produce a convergent solution to the IVP. We will show that is not true even in a simple context. To be more explicit, we consider the linear wave equation as an example since it constitutes the simplest prototype model for turbulent systems as we will discuss in the next chapter.

Linear wave equation

$$\begin{aligned} u_t &= C u_x \\ u(x,0) &= f(x). \end{aligned} \tag{4.7}$$

We know that this equation has the solution

$$u(x,t) = f(x + Ct),$$

which is a wave that propagates to the left if $C > 0$ and to the right if $C < 0$.

The fundamental theory in the analysis of finite difference methods for the numerical solutions of partial differential equations is the **Lax equivalence theorem**, which states the following: *Provided that the differential equation $u_t = \mathcal{P}(\partial_x)u$ is stable, the convergence of the difference scheme is guaranteed only when it is stable and consistent.*

The strength of this theorem is that it is quite often easier to check the stability and the consistency relative to directly showing the convergence since the numerical method is defined by recurrence relations while the differential equation involves differentiable functions. To confirm the stability of the differential and difference operators, it suffices to check the following algebraic conditions.

Proposition 4.8 *IVP is stable for a given $\mathcal{P}(\partial_x)$ if and only if*

$$\max_{|\ell| \le \infty, 0 \le t \le T} |e^{\tilde{p}(i\ell)t}|^2 \le C(T).$$

Proposition 4.9 *The difference scheme is stable for a strategy $\Delta t \le S(h)$, where h is the discrete spatial mesh size, if and only if*

$$\max_{|\ell| \le \infty, 0 \le M\Delta t \le T} |\tilde{g}_h(\Delta t, \ell)^M|^2 \le C(T).$$

The proof of these algebraic conditions can be found in many standard numerical PDE textbooks such as Richtmeyer and Morton (1967) and Strikwerda (2004).

In our example, the stability of the PDE is clearly satisfied since the differential operator of the wave equation in (4.7) is bounded from above, $|e^{\tilde{p}(\mathrm{i}\ell)t}|^2 = |e^{C\mathrm{i}\ell}|^2 = 1$. Let us consider the forward Euler time discretization (as described in Chapter 2) as well as the symmetric difference to approximate the spatial derivative,

$$\frac{u_{j+1}^M - u_{j-1}^M}{2h} = u_x + \mathcal{O}(h^2),$$

where $u_j^M \simeq u(jh, M\Delta t)$. This second-order accurate approximation can be easily deduced by substracting the Taylor expansions of u_{j+1}^M and u_{j-1}^M about their mid-point, $x_j = jh$. With these approximations, the numerical estimate of the wave equation is given by the following recurrence relation

$$\begin{aligned} u_j^{M+1} &= u_j^M + \frac{C\Delta t}{2h}(u_{j+1}^M - u_{j-1}^M), \\ u_j^0 &= f_j. \end{aligned} \tag{4.8}$$

We will show that this difference scheme is indeed not stable. The amplification factor of (4.8),

$$\tilde{g}_h(\Delta t, \ell) = 1 + \frac{C\Delta t}{2h}(e^{\mathrm{i}\ell h} - e^{-\mathrm{i}\ell h}) = 1 + \mathrm{i}\frac{C\Delta t}{h}\sin \ell h,$$

satisfies

$$|\tilde{g}_h(\Delta t, \ell)|^2 = 1 + \left(\frac{C\Delta t}{h}\right)^2 \sin^2(\ell h) > 1,$$

when $\ell \ne 0$, $|\ell| \le N$ and for any constant $C\Delta t/h \ne 0$. For a fixed time T where $0 \le M\Delta t \le T$, whenever Δt is small with $\Delta t/h$ constant, we need to increase the time step M, therefore $|\tilde{g}_h(\Delta t, \ell)^M|^2$ keeps growing as M increases, and the algebraic condition in Proposition 4.9 is not satisfied.

Now, let us reduce the accuracy in the spatial derivative approximation by considering a first-order forward difference method,

$$\frac{u_{j+1}^M - u_j^M}{h} = u_x + \mathcal{O}(h).$$

With the forward Euler time discretization, we called the following approximation the upwind difference scheme,

$$u_j^{M+1} = u_j^M + \frac{C\Delta t}{h}(u_{j+1}^M - u_j^M), \qquad (4.9)$$
$$u_j^0 = f_j.$$

Before we check the stability and consistency of this scheme, let us intuitively give a conjecture for this scheme. We know that when $C > 0$ the exact solution propagates to the left and when $C < 0$ it propagates to the right. The difference equation in (4.9) uses the values at the right (grid point $j + 1$) to calculate a value at the left (grid point j). Thus, we expect the difference scheme will work for $C > 0$. On the other hand, for $C < 0$, the difference scheme (4.9) still uses the values at the right side to calculate values at the left side. Since the difference scheme uses information from the wrong side, we do not expect it to work well in this case. When $C < 0$, one needs to consider the backward difference scheme to approximate the spatial derivative since it uses information from the left to calculate the value at the right.

To check the stability, let us denote $\lambda = \Delta t/h$ such that the amplification factor of (4.9) can be written as follows

$$\tilde{g}_h(\Delta t, \ell) = 1 + \lambda C(e^{i\ell h} - 1). \qquad (4.10)$$

The algebraic condition in Proposition 4.9 is satisfied when $|\tilde{g}_h|^2 \leq 1$ for all $|\ell| \leq N$:

$$0 \leq |\tilde{g}_h(\Delta t, \ell)|^2 = 1 - 2C\lambda(1 - C\lambda)(1 - \cos(\ell h)).$$

For $|\tilde{g}_h|^2 \leq 1$, we need

$$2C\lambda(1 - C\lambda)(1 - \cos(\ell h)) \geq 0.$$

Since $(1 - \cos(\ell h)) \geq 0$ for all h, ℓ, the stability holds when $C\lambda(1 - C\lambda) \geq 0$. That is, for $C, \lambda > 0$, $\lambda C \leq 1$. This is the well-known Courant–Friedrichs–Lewy (CFL) criterion for stability.

Consistency is a condition which guarantees that the discrete problem approximates the correct continuous problem. To verify this, let $\tilde{u}_j^M = u(jh, M\Delta t)$ be the exact solution of the wave equation in (4.7), evaluated at grid point jh and time $M\Delta t$. The Taylor expansions about the grid spacing h and discrete time Δt are given as follows

$$\tilde{u}_{j+1}^M = \tilde{u}_j^M + h\tilde{u}_x + h^2\tilde{u}_{xx} + \mathcal{O}(h^3),$$
$$\tilde{u}_j^{M+1} = \tilde{u}_j^M + \Delta t\tilde{u}_t + \mathcal{O}(\Delta t^2).$$

Substituting these expansions into the finite difference scheme in (4.9), we obtain

$$\tilde{u}_t + \mathcal{O}(\Delta t) = C(\tilde{u}_x + h\tilde{u}_{xx} + \mathcal{O}(h^2)).$$

Taking the limit $h, \Delta t \to 0$, we obtain the continuous wave equation in (4.7) and the consistency is satisfied. Therefore, the upwind difference scheme is a convergent method whenever the CFL condition holds.

Part II

Mathematical guidelines for filtering turbulent signals

5

Stochastic models for turbulence

As motivated in Chapter 1, one goal of the present book is to develop an explicit off-line test criterion for stable accurate time filtering of turbulent signals which is akin to the classical frozen linear constant stability test for finite difference schemes for systems of nonlinear partial differential equations as presented in chapter 4 of Richtmeyer and Morton (1967). In applications for complex turbulent spatially extended systems, the actual dynamics is typically turbulent and energetic at the smallest mesh scales but the climatological spectrum of the turbulent modes is known; for example, a mesh truncation of the compressible primitive equations with a fine mesh spacing of 10–50 kilometers still has substantial random and chaotic energy on the smallest 10-kilometer scales due to chaotic motion of clouds, topography and boundary layer turbulence which are not resolved. Similar unresolved features occur in many engineering problems with turbulence. Thus, the first step is the development of an appropriate constant-coefficient stochastic PDE test problem.

The simplest models for representing turbulent fluctuations involve replacing nonlinear interaction by additional linear damping and stochastic white noise forcing in time which incorporates the observed climatological spectrum and decorrelation time for the turbulent field (Majda *et al.*, 2005; Majda and Wang, 2006). Thus, the first step in developing analogous off-line test criteria is to utilize the above approximations. This approach is developed in this chapter and builds on earlier material from Section 2.1.1 and Chapter 4. First, as in standard test criteria for finite difference schemes (Richtmeyer and Morton, 1967), the complex $s \times s$ PDEs are linearized at a constant-coefficient background resulting in the frozen-coefficient PDE, $\vec{u}_t = \mathcal{P}(\partial_x)\vec{u}$. In accordance with the above approximations, additional damping $-\gamma(\partial_x)\vec{u}$ and white noise forcing $\sigma(\vec{x})\dot{\vec{W}}(t)$ are added to the PDE to represent nonlinear interaction with the unresolved turbulent motions resulting in the basic frozen-coefficient test model. For simplicity in notation here we discuss a scalar field in a single space variable but everything generalizes to a matrix system of stochastic PDEs in several space dimensions. In Chapter 12, we develop such stochastic approximations systematically for a family of turbulent nonlinear dynamical systems.

5.1 The stochastic test model for turbulent signals

With the above motivation, we consider solutions of the real-valued scalar stochastically forced PDE

$$\begin{aligned}\frac{\partial u(x,t)}{\partial t} =& \mathcal{P}\left(\frac{\partial}{\partial x}\right)u(x,t) - \gamma\left(\frac{\partial}{\partial x}\right)u(x,t) + \bar{F}(x,t) + \sigma(x)\dot{W}(t), \\ u(x,0) =& u_0(x).\end{aligned} \tag{5.1}$$

Here $\bar{F}(x,t)$ is a known deterministic forcing term, $\sigma(x)\dot{W}(t)$ is a Gaussian statistically stationary spatially correlated scalar random field and $\dot{W}(t)$ is white noise in time while the initial data u_0 is a Gaussian random field with nonzero mean and covariance. As in the usual finite difference linear stability analysis developed in Chapter 4, the problem in (5.1) is non-dimensionalized to a 2π-periodic domain so that continuous and discrete Fourier series can be utilized in analyzing (5.1) and the related discrete approximations.

The operators $\mathcal{P}(\partial_x)$ and $\gamma(\partial_x)$ are defined through unique symbols at a given wavenumber k by

$$\begin{aligned}\mathcal{P}\left(\frac{\partial}{\partial x}\right)\mathrm{e}^{ikx} &= \tilde{p}(\mathrm{i}k)\mathrm{e}^{\mathrm{i}kx}, \\ \gamma\left(\frac{\partial}{\partial x}\right)\mathrm{e}^{ikx} &= \gamma(\mathrm{i}k)\mathrm{e}^{\mathrm{i}kx}.\end{aligned} \tag{5.2}$$

We assume that $\tilde{p}(\mathrm{i}k)$ is wavelike so that

$$\tilde{p}(\mathrm{i}k) = \mathrm{i}\omega_k \tag{5.3}$$

with $-\omega_k$ the real-valued dispersion relation while $\gamma(ik)$ represents both explicit and turbulent dissipative processes so that $\gamma(ik)$ is non-negative with

$$\gamma(ik) > 0 \text{ for all } k \neq 0. \tag{5.4}$$

In geophysical applications, it is natural to have a climatological distribution and as discussed below, (5.3) and (5.4) are needed in order to guarantee this.

5.1.1 The stochastically forced dissipative advection equation

The main example of (5.1) as a prototype in this chapter is given by the stochastically forced dissipative advection equation

$$\frac{\partial u(x,t)}{\partial t} = -c\frac{\partial u(x,t)}{\partial x} - du(x,t) + \mu\frac{\partial^2 u(x,t)}{\partial x^2} + \bar{F}(x,t) + \sigma(x)\dot{W}(t). \tag{5.5}$$

In this example, $\tilde{p}(\mathrm{i}k) = \mathrm{i}\omega_k = -ick$ and the damping symbol $\gamma(ik)$ is given by

$$\gamma(\mathrm{i}k) = d + \mu k^2. \tag{5.6}$$

The slight abuse of notation in (5.5) and (5.6) should not confuse the reader. In (5.6) we require $d \geq 0$ and $\mu \geq 0$, and at least one of these coefficients to be nonzero in order to satisfy (5.4). The case with uniform damping, $d > 0$, but without scale-dependent damping, so that $\mu = 0$, arises often in idealized geophysical problems where d represents radiative damping, Ekman friction or gravity wave absorption (Majda *et al.*, 2005; Majda and

Wang, 2006). In general $\mathcal{P}(\partial_x)$ can be any differential operator which is a combination of odd derivatives to satisfy (5.3) while $\gamma(\partial_x)$ is a suitable combination of even derivatives satisfying (5.4) (see chapter 1 of Majda and Wang (2006) for the precise conditions). The full generality in (5.2) is important for geophysical equations such as the quasi-geostrophic equations where $\tilde{p}(ik)$ is not a polynomial but is given by $\tilde{p}(ik) = \frac{ik}{k^2+F}$ (Majda and Wang, 2006), where F is a non-dimensionalized unit that represents the square of the ratio between the Froude and the Rossby numbers. We discuss a geophysical example of this sort later in this chapter.

As in Chapter 4, we utilize separation of variables to solve (5.1) combined with explicit solution of the complex scalar stochastic equation developed in Section 2.1.1 of Chapter 2. The general solution of (5.1) is defined through Fourier series. The 2π-periodic solution of (5.1) is expanded in Fourier series

$$u(x,t) = \sum_{k=-\infty}^{\infty} \hat{u}_k(t) e^{ikx}, \qquad \hat{u}_{-k} = \hat{u}_k^*, \tag{5.7}$$

where $\hat{u}_k(t)$ for $k > 0$ solves the scalar complex-coefficient stochastic ODEs (Gardiner, 1997),

$$d\hat{u}_k(t) = [\tilde{p}(\mathrm{i}k) - \gamma(\mathrm{i}k)]\, \hat{u}_k(t)\, dt + \hat{F}_k(t)\, dt + \tilde{\sigma}_k\, dW_k(t), \quad \hat{u}_k(0) = \hat{u}_{k,0}. \tag{5.8}$$

The term $\hat{F}_k(t)$ is the Fourier coefficient of the deterministic force $\bar{F}(x,t)$. Here W_k are independent complex Wiener processes for each k and the independent real and imaginary parts have the same variance $t/2$; the coefficients $\hat{u}_{-k}$ for $k > 0$ are defined through the complex conjugate formula $\hat{u}_{-k} = \hat{u}_k^*$ and the constant $k = 0$ Fourier mode is real-valued with a similar single equation with detailed discussion omitted here. Under the natural simplifying assumption that the symbols $\tilde{p}(ik)$ and $\gamma(ik)$ satisfy (5.3) and (5.4), the statistical equilibrium distribution for (5.8) exists provided $\bar{F}(x,t) = 0$ and is a Gaussian with zero mean and variance, E_k, defining the climatological energy spectrum given by

$$E_k = \frac{\tilde{\sigma}_k^2}{2\gamma(\mathrm{i}k)}, \qquad 1 \le k < +\infty. \tag{5.9}$$

Mathematically, one needs to require $\sum E_k < \infty$ to define the stochastic solution of (5.1) correctly with a similar requirement on the Gaussian initial data in $u_0(x)$. However, there is genuine physical interest in situations with an even rougher turbulent spectrum such as white noise where E_k is constant. In these cases we truncate the sum in (5.7) to a large finite sum as in Chapter 4.

While distinct Fourier modes with different magnitudes are uncorrelated, the real part of the temporal correlation function $\langle (\hat{u}_k(t) - \bar{\hat{u}}_k)(\hat{u}_k(t+\tau) - \bar{\hat{u}}_k)^* \rangle = R_k(\tau)$ at a given mode k in the statistical steady state is given by

$$\begin{aligned} \text{Real}\,[R_k(\tau)] &\equiv \text{Real}\,[\langle (\hat{u}_k(t) - \bar{\hat{u}}_k)(\hat{u}_k(t+\tau) - \bar{\hat{u}}_k)^* \rangle] \\ &= E_k \mathrm{e}^{-\gamma(\mathrm{i}k)\tau} \cos(\omega_k \tau). \end{aligned} \tag{5.10}$$

This calculation is based on the exact solution from (2.15) of Chapter 2 and is given in Appendix A in this chapter.

In (5.10), the damping coefficient, $\gamma(ik)$, defines the correlation time, $\gamma(ik)^{-1}$, while $i\omega_k = \tilde{p}(ik)$ defines ω_k, the oscillation frequency at wavenumber k. Clearly, $\gamma(ik)^{-1}$ measures the memory in the signal being filtered. It is easy to see as in Section 2.1.1 of Chapter 2 that the ensemble mean $\bar{\hat{u}}_k = \langle \hat{u}_k \rangle$ of (5.8) satisfies the ODE,

$$d\bar{\hat{u}}_k(t) = [\tilde{p}(ik) - \gamma(ik)]\,\bar{\hat{u}}_k(t)\,dt + \hat{F}_k(t)\,dt, \qquad \bar{\hat{u}}_k(0) = \bar{\hat{u}}_{k,0}. \tag{5.11}$$

The noise in (5.8) and (5.9) represents the turbulent fluctuations on the mesh scale for both unresolved and resolved features of the nonlinear dynamics (see Majda and Wang, 2006; Majda *et al.*, 2005, and references therein) with a given energy spectrum E_k and decorrelation time $\gamma(ik)^{-1}$ at each wavenumber. In this fashion significant features for a turbulent system are incorporated in the constant-coefficient test problem. In practical problems, quite often the nature of this spectrum is known roughly as well as the decorrelation time, expressed through the damping coefficient $\gamma(ik)$ (Majda and Wang, 2006; Majda *et al.*, 2005; Majda and Grote, 2007). One space dimension is not a restriction for any of the results but is utilized to avoid cumbersome notation; the theory also applies in several variables and to systems of equations which are not scalar fields. Examples of physical systems will be discussed in subsequent chapters. It is clearly interesting to generalize these test models for turbulent fields to situations with large-scale instability as in Section 3.4 of Chapter 3 but also having a climatological statistical steady state as in the models here. Such types of models are developed subsequently in Chapter 8.

5.1.2 Calibrating the noise level for a turbulent signal

In applying the test model to turbulent signals from nature, we exploit the facts mentioned earlier from observations and/or laboratory experiments that we know both the energy, E_k, and the correlation time, $\gamma(ik)^{-1}$, at wavenumber k in the turbulent signal. Then the formula in (5.9) determines the noise level with variance at spatial wavenumber k given by

$$\tilde{\sigma}_k = (2\gamma(ik)E_k)^{1/2}. \tag{5.12}$$

Typical turbulent spectra involve power laws, such as

$$E_k = E(k) = E_o|k|^{-\beta}, \quad |k| \geq 1. \tag{5.13}$$

For example, $\beta = 5/3$ is the familiar Kolmogorov spectrum which generates a fractal random field (Majda and Kramer, 1999) while $\beta = 0$ corresponds to a white noise spectrum. In nature, the turbulent spectrum might be white noise in space, reflecting physical processes such as moist convection over a few kilometers, or another steeper power-law spectrum in the upper troposphere reflecting gravity wave activity (Majda, 2000b). An example with barotropic Rossby waves is discussed later in this chapter.

For the stochastic advection–diffusion in (5.5) and a power-law turbulent spectrum in (5.13), the noise level is determined by

$$\tilde{\sigma}_k = E_o^{1/2}|k|^{-\beta/2}(d+\mu|k|^2)^{1/2}. \tag{5.14}$$

Thus if the viscosity is nonzero, $\mu \neq 0$, there is increasing noise at small spatial scales, i.e. $|k| \gg 1$ if $\beta > 2$ and decreasing noise at small scales for $\beta < 2$. These relations can affect the practical controllability in filtering such turbulent signals with multiple scales as we show below in Chapters 6 and 7.

5.2 Turbulent signals for the damped forced advection–diffusion equation

The deterministic forcing, $\bar{F}(x,t)$, in the test model has obvious physical importance. As shown in subsequent chapters, it can also provide stringent tests for filter performance in time. Examples of forcing which we utilize here are forcing with

$$\hat{F}_k(t) = \begin{cases} A\mathrm{e}^{\mathrm{i}\omega_o(k)t}, & \text{if } k \le M \\ 0, & \text{if } k > M \end{cases} \tag{5.15}$$

for prescribed $\omega_o(k)$. The exact solution in general for the mean $\bar{\hat{u}}_k$ from (5.11) in this case is given by

$$\begin{aligned} \bar{\hat{u}}_k(t) &= \bar{\hat{u}}_k(0)\mathrm{e}^{p(\mathrm{i}k)t} + \int_0^t \hat{F}_k(s)\mathrm{e}^{p(\mathrm{i}k)(t-s)}ds \\ &= \bar{\hat{u}}_k(0)\mathrm{e}^{p(\mathrm{i}k)t} + \begin{cases} \frac{A\mathrm{e}^{\mathrm{i}\omega_o(k)t}}{\gamma(\mathrm{i}k)+\mathrm{i}(\omega_o(k)-\omega_k)}(1-\mathrm{e}^{(p(\mathrm{i}k)-\omega_o(k))t}), & \text{if } k \le M \\ 0, & \text{if } k > M, \end{cases} \end{aligned} \tag{5.16}$$

where $p(\mathrm{i}k) = \tilde{p}(\mathrm{i}k) - \gamma(\mathrm{i}k)$. Particularly, stringent test problems are achieved by using *resonant periodic forcing* so that

$$\omega_o(k) = \omega_k \quad \text{for all } k, |k| \le M. \tag{5.17}$$

In this fashion, even in a stochastic linear system, we can generate the development of a localized extreme event within a turbulent field as shown below.

Next, we illustrate the time development of such a localized extreme event in the turbulent field due to resonant periodic forcing. We set $c = 1$ in (5.5) and use the parameters $d = 0$, $\mu = 10^{-2}$ and two different spectra, $E_k = 1$, white noise and $E_k = k^{-5/3}$. There are 61 Fourier modes utilized or equivalently 123 mesh points in generating the signal. Resonant mean forcing with $A = 0.1$ is applied for the first 20 Fourier modes in (5.17). The initial data is chosen at random from the climatological state. The noise level at each wavenumber for the SDE is determined by the formula (5.14) discussed earlier.

In Fig. 5.1, we see that the amplitude of the solutions are large and roughly uniformly distributed when $E_k = 1$. In this equipartition spectrum case, we hardly see any localized

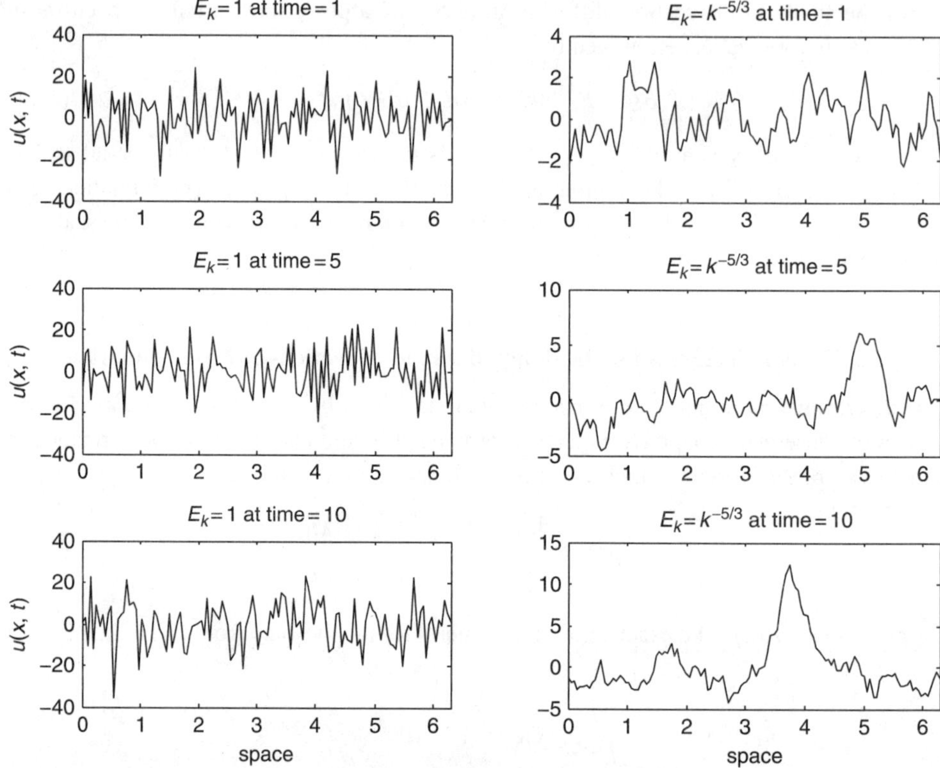

Figure 5.1 Solutions of the stochastically forced advection–diffusion equation with resonant periodic forcing with equipartition spectra (first column) and the smooth $-5/3$ spectra (second column) at time $t = 1$ (first row), $t = 5$ (second row) and $t = 10$ (third row).

extreme event. On the other hand, when $E_k = k^{-5/3}$, there is less noise in the higher-wavenumber modes and we clearly see a distinct localized peak due to the nonzero resonant periodic forcing in the first M modes. This localized extreme event travels eastward in time according to the wave speed $c = 1$ for (5.5). Of course if we increased the amplitude A of the forcing sufficiently to overcome the much stronger noise in the equipartition case, then we would see the emergence of the singular event in that case too.

5.3 Statistics of turbulent solutions in physical space

We introduce the notation, $e_k(x) = e^{2\pi ikx}$, to emphasize the physical space structure. In Section 5.1, we developed Fourier series solutions of the test problem in (5.1) with the form

$$u(x,t) = \sum_{|k|\leq N} \hat{u}_k(t)e_k(x). \tag{5.18}$$

In the remainder of this section, we present the statistical behavior of the random solution $u(x,t)$ of the PDE in (5.1). With (5.11), the ensemble mean of $u(x,t)$ is given by

$$\begin{aligned}\langle u(x,t)\rangle &= \left\langle \sum_{|k|\le N} \hat{u}_k(t)e_k(x)\right\rangle \\ &= \sum_{|k|\le N} \langle \hat{u}_k(t)\rangle e_k(x) \\ &= \sum_{|k|\le N}\left(\hat{u}_k(0)\mathrm{e}^{p(\mathrm{i}k)t} + \int_0^t \hat{F}_k(s)\mathrm{e}^{p(\mathrm{i}k)(t-s)}ds\right)e_k(x), \end{aligned} \tag{5.19}$$

where $p(\mathrm{i}k) = \tilde{p}(\mathrm{i}k) - \gamma(\mathrm{i}k)$. From (5.19) and the solution of (5.8), we have:

$$\hat{u}_k(t) - \langle \hat{u}_k(t)\rangle = \tilde{\sigma}_k \int_0^t \mathrm{e}^{p(\mathrm{i}k)(t-s)}\mathrm{d}W_k(t),$$

where $\mathrm{d}W_k(t) = 2^{-1/2}(\mathrm{d}W_{k,1}(t) + \mathrm{i}\mathrm{d}W_{k,2}(t))$ and each component $W_{k,i}(t)$ is a Wiener process. In Appendix B, for zero initial covariance we compute the general spatio-temporal correlation function

$$\mathcal{R}(x,x',t,t') \equiv \langle [u(x,t) - \langle u(x,t)\rangle][u(x',t') - \langle u(x',t')\rangle]^*\rangle.$$

This yields the covariance

$$\mathrm{Var}\,[u(x,t)] = \mathcal{R}(x,x,t,t) = \sum_{|k|\le N}\frac{\tilde{\sigma}_k^2}{2\gamma(\mathrm{i}k)}\left(1-\mathrm{e}^{-2\gamma(\mathrm{i}k)t}\right), \tag{5.20}$$

and temporal correlation function

$$\begin{aligned}\mathcal{R}(x,x,t,t') &= \sum_{|k|\le N}\frac{\tilde{\sigma}_k^2}{2\gamma(\mathrm{i}k)}\mathrm{e}^{p(\mathrm{i}k)t+p^*(\mathrm{i}k)t'}(\mathrm{e}^{2\gamma(\mathrm{i}k)t}-1) \\ &= \sum_{|k|<N}\frac{\tilde{\sigma}_k^2}{2\gamma(\mathrm{i}k)}\mathrm{e}^{\tilde{p}(\mathrm{i}k)(t-t')}\mathrm{e}^{-\gamma(\mathrm{i}k)(t'-t)}(1-\mathrm{e}^{-2\gamma(\mathrm{i}k)t}).\end{aligned}$$

Fix $t' = t + \tau$ and let $t \to \infty$. Then the temporal correlations converge as $t \uparrow \infty$ to the stationary correlations in physical space of the climatological state,

$$\mathcal{R}(\tau) = \sum_{|k|\le N}\frac{\tilde{\sigma}_k^2}{2\gamma(\mathrm{i}k)}\mathrm{e}^{-\tilde{p}(\mathrm{i}k)\tau}\mathrm{e}^{-\gamma(\mathrm{i}k)\tau}. \tag{5.21}$$

Similarly, the spatial correlation at fixed time in the signal is given by

$$\mathcal{R}(x,x',t,t) = \sum_{|k|\le N}\frac{\tilde{\sigma}_k^2}{2\gamma(\mathrm{i}k)}\left(1-\mathrm{e}^{-2\gamma(\mathrm{i}k)t}\right)e_k(x-x').$$

5.4 Turbulent Rossby waves

Developing the elementary turbulent test model for Rossby waves in geophysical flows is the goal here. Barotropic Rossby waves with phase varying only in the north–south direction (Majda, 2003; Pedlosky, 1979) have the dispersion relation

$$\omega_k = \frac{\beta}{k}$$

and it is also natural to assume uniform damping

$$\gamma(ik) = d > 0$$

representing Ekmann friction. It is known from observations that on scales of order of thousands of kilometers these waves have a k^{-3} energy spectrum (Majda, 2000b) so that

$$E_k = k^{-3}.$$

Below, we confirm the physical space formulas developed in Section 5.3 through Monte Carlo simulation of the Fourier series representation.

On the planetary scale, the mid-latitude beta plane approximates the rotation effect with

$$\beta = \frac{f}{a} = \frac{2\Omega\cos(\theta)}{a},$$

where the natural parameters are given in Table 5.1.

In our model, we consider a periodic domain of length 2π so that the radius is a unit length $a = 1$. Converting the time into days, we find that the parameter β at latitude $\theta = 45°$ is

$$\beta = 2\Omega\cos(45°)\ /\text{sec} = 8.91\ /\text{day}$$

and thus the natural frequency for this model is given by

$$\omega_k = \frac{8.91}{k},$$

so the lowest wavenumber, $k = 1$, has an oscillation period of roughly 17 hours.

In the following numerical experiments, we use parameters as shown in Table 5.2. With the assumed constant damping, the noise levels $\tilde{\sigma}_k^2$ are chosen to be proportional to the energy spectrum of the Rossby waves k^{-3} according to the formula in (5.12). The initial condition $u_k(0)$ is randomly chosen from the climatological Gaussian distribution with mean 0 and variance k^{-3}.

The parameter $d = 1.5$ is chosen such that the decorrelation time is three days which corresponds to realistic weather predictability (Lorenz, 1996). In Fig. 5.2, we compare numerical statistics of the turbulent solutions, generated through a Monte Carlo simulation with an ensemble size of 100, to the analytical statistical solutions in Section 5.3.

Table 5.1 Natural parameters.

Mean radius of the Earth	$a = 6.37 \times 10^6$m
Earth angular speed of rotation	$\Omega = 7.292 \times 10^{-5}$rad/ sec

Table 5.2 Diagnostic parameters for turbulent Rossby waves.

Number of modes	$K = 20$
Time step	$\Delta t = 0.01$
Maximum simulation time	$T = 100$ days
Damping time scale	$d = 1.5$
Natural oscillation frequency	$\omega_k = \frac{8.9}{k}$
Initial conditions	$u_k(0) \sim N(0, k^{-3})$
Ensemble size	$N = 100$
Noise level	$\tilde{\sigma}_k^2 = 2d/k^3$

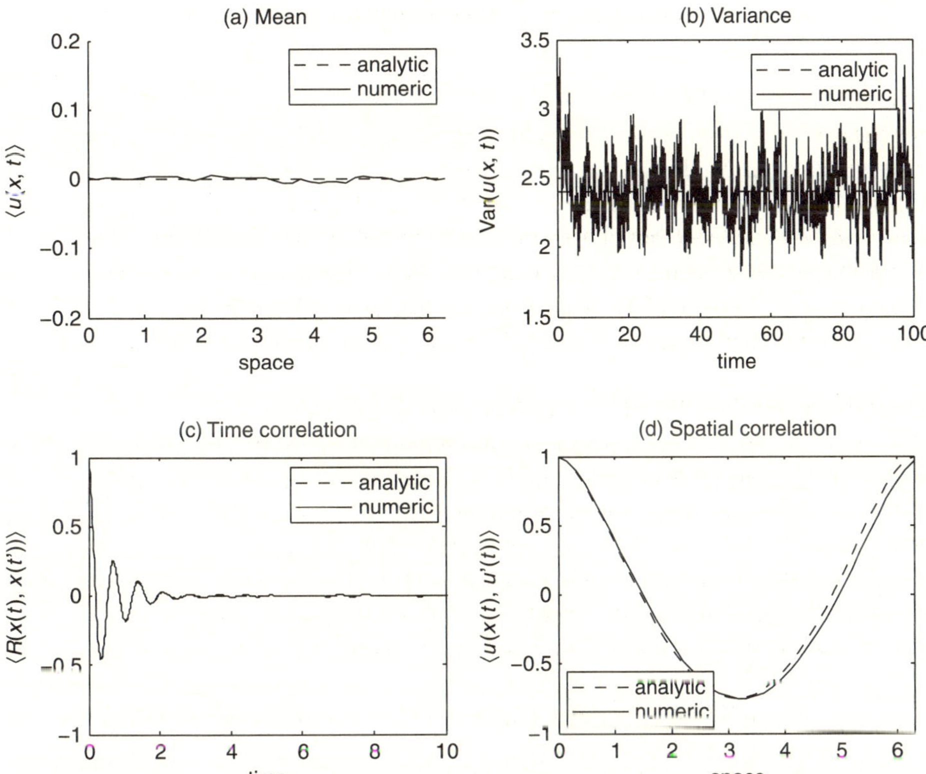

Figure 5.2 Statistics of turbulent Rossby waves: (a) mean of $u(x, t)$ as a function of space, (b) variance as a function of time (in days), (c) temporal correlation function (in days) as a function of time and (d) spatial correlation function as function of space. In each panel, the Monte Carlo simulated statistical quantities are denoted in solid (see text for details) and the analytical formula for the corresponding statistics in Section 5.3 is plotted in dashes.

The Monte Carlo mean is estimated by averaging the solution u ensemble-wise and temporally, the variance is estimated by spatial and ensemble averages, the temporal correlation function at a particular location x is estimated by an ensemble average, and finally the spatial correlation function is estimated by ensemble and temporal averages.

In Fig. 5.2(a), the mean state is plotted as a function of model variables on the interval $[0, 2\pi]$. Here, we expect the mean to be close to zero (that is, $\lim_{t\to\infty}\langle u(x,t)\rangle = 0$). The variance is plotted as a function of time. Figure 5.2(b) shows the fluctuation of the variance around the theoretical variance, $\lim_{t\to\infty} \mathrm{Var}(u(x,t)) = 2.40$. In Fig. 5.2(c), we see the agreement of the theoretical (dashes) ($\mathcal{R}(\tau) = \lim_{t\to\infty}\mathcal{R}(x,x,t,t+\tau)$ as in (5.21)) and computed (solid) temporal correlations, both plotted as functions of time (in days). The last quantity we test is the spatial correlation and compare the theoretical ($\lim_{t\to\infty}\mathcal{R}(x,x',t,t)$) and the computed time series. Once again there is an excellent agreement. We present this example here both to illustrate how to use the test model in a simple physical problem and to confirm the theoretical formulas presented in Section 5.3 in a concrete setting from Monte Carlo simulations.

Appendix A: Temporal correlation function for each Fourier mode

In this appendix, we compute the statistical steady-state temporal correlation function of the kth Fourier mode, $\hat{u}_k(t)$, of the solution of SDE (5.8) which is given by

$$\hat{u}_k(t) = \hat{u}_k(0)\mathrm{e}^{p(\mathrm{i}k)t} + \int_0^t \hat{F}_k(s)\mathrm{e}^{p(\mathrm{i}k)(t-s)}ds + \tilde{\sigma}_k\int_0^t \mathrm{e}^{p(\mathrm{i}k)(t-s)}\mathrm{d}W_k(t), \tag{5.22}$$

with $p(\mathrm{i}k) = \tilde{p}(\mathrm{i}k) - \gamma(\mathrm{i}k)$ and white noise $dW_k = 2^{-1/2}(dW_{k,1} + \mathrm{i}dW_{k,2})$ and each $W_{k,j}$ is a Wiener process with formal properties defined in Section 2.1.1 of Chapter 2.

Note that we assume zero initial covariance for simplicity in exposition. The temporal correlation function is defined as follows

$$\begin{aligned}
\mathcal{R}_k(t,t') &= \langle(\hat{u}_k(t) - \bar{\hat{u}}_k)(\hat{u}_k(t') - \bar{\hat{u}}_k)^*\rangle \\
&= \left\langle \tilde{\sigma}_k\int_0^t \mathrm{e}^{p(\mathrm{i}k)(t-s)}\mathrm{d}W_k(s)\left(\tilde{\sigma}_k\int_0^{t'} \mathrm{e}^{p(\mathrm{i}k)(t'-s')}\mathrm{d}W_k(s')\right)^*\right\rangle \\
&= \tilde{\sigma}_k^2\mathrm{e}^{p(\mathrm{i}k)t+p^*(\mathrm{i}k)t'}\int_0^t\int_0^{t'} \mathrm{e}^{-(p(\mathrm{i}k)s+p^*(\mathrm{i}k)s')}\frac{1}{2}\sum_{i=1}^{2}\langle dW_{k,i}(s)dW_{k,i}(s')\rangle \\
&= \tilde{\sigma}_k^2\mathrm{e}^{p(\mathrm{i}k)t+p^*(\mathrm{i}k)t'}\int_0^t\int_0^{t'} \mathrm{e}^{-(p(\mathrm{i}k)s+p^*(\mathrm{i}k)s')}\delta(s-s')ds\,ds' \\
&= \tilde{\sigma}_k^2\mathrm{e}^{p(\mathrm{i}k)t+p^*(\mathrm{i}k)t'}\int_0^{t'} \mathrm{e}^{-(p(\mathrm{i}k)+p^*(\mathrm{i}k))s'}ds' \\
&= \frac{\tilde{\sigma}_k^2}{2\gamma(\mathrm{i}k)}\mathrm{e}^{p(\mathrm{i}k)t+p^*(\mathrm{i}k)t'}(\mathrm{e}^{2\gamma(\mathrm{i}k)t} - 1) \\
&= \frac{\tilde{\sigma}_k^2}{2\gamma(\mathrm{i}k)}\mathrm{e}^{\gamma(\mathrm{i}k)(t-t')-\tilde{p}(\mathrm{i}k)(t'-t)}(1 - \mathrm{e}^{-2\gamma(\mathrm{i}k)t}).
\end{aligned} \tag{5.23}$$

In deriving this, we assume $\sigma_k \in \mathbb{R}$ and we use the standard properties of white noise and the integral of the delta function as discussed in Section 2.1.1 of Chapter 2. Let $t' = t + \tau$ and let $t \to \infty$, and substitute $E_k = \tilde{\sigma}_k^2/2\gamma(ik)$ as in Eqn (5.9); then we obtain the asymptotic temporal correlation function

$$\mathcal{R}_k(\tau) = E_k \mathrm{e}^{-\gamma(\mathrm{i}k)\tau} \mathrm{e}^{-\tilde{p}(\mathrm{i}k)\tau}. \tag{5.24}$$

Recall that for frequency $\tilde{p}(ik) = \mathrm{i}\omega_k$ as in (5.3), we obtain

$$\mathcal{R}_k(\tau) = E_k \mathrm{e}^{-\gamma(\mathrm{i}k)\tau} \left[\cos(\omega_k \tau) - \mathrm{i}\sin(\omega_k \tau)\right]. \tag{5.25}$$

Appendix B: Spatio-temporal correlation function

Consider $u(x, t)$ as solutions of the stochastically driven PDE (5.1) with $p(ik) = \tilde{p}(ik) - \gamma(ik)$, that is

$$u(x, t) = \sum_{|k| \le N} \hat{u}_k(t) e_k(x), \tag{5.26}$$

where the Fourier coefficients are given in (5.22) of Appendix A. By the general property of white noise, $\langle dW_k \rangle = 0$, the first two terms on the right-hand side of (5.22) determine the ensemble mean. The spatio-temporal correlation function for $u(x, t)$ is given as follows

$$\begin{aligned}
\mathcal{R}(x, x', t, t') &\equiv \langle [u(x, t) - \langle u(x, t) \rangle][u(x', t') - \langle u(x', t') \rangle]^* \rangle \\
&= \left\langle \sum_{|k| \le N} \tilde{\sigma}_k \int_0^t \mathrm{e}^{p(\mathrm{i}k)(t-s)} e_k(x) \mathrm{d}W_k(s) \left(\sum_{|k'| \le N} \tilde{\sigma}_{k'} \int_0^{t'} \mathrm{e}^{p(\mathrm{i}k')(t'-s')} e_{k'}(x') \mathrm{d}W_{k'}(s') \right)^* \right\rangle \\
&= \sum_{|k| \le N} \tilde{\sigma}_k^2 \mathrm{e}^{p(\mathrm{i}k)t + p^*(\mathrm{i}k)t'} \int_0^t \int_0^{t'} \mathrm{e}^{-(p(\mathrm{i}k)s + p^*(\mathrm{i}k)s')} e_k(x - x') \frac{1}{2} \sum_{i=1}^{2} \langle dW_{k,i}(s) dW_{k,i}(s') \rangle \\
&= \sum_{|k| \le N} \tilde{\sigma}_k^2 \mathrm{e}^{p(\mathrm{i}k)t + p^*(\mathrm{i}k)t'} \int_0^t \int_0^{t'} \mathrm{e}^{-(p(\mathrm{i}k)s + p^*(\mathrm{i}k)s')} e_k(x - x') \delta(s - s') ds ds' \\
&= \sum_{|k| \le N} \tilde{\sigma}_k^2 \mathrm{e}^{p(\mathrm{i}k)t + p^*(\mathrm{i}k)t'} \int_0^{t'} \mathrm{e}^{-(p(\mathrm{i}k) + p^*(\mathrm{i}k))s'} e_k(x - x') ds' \\
&= \sum_{|k| \le N} \frac{\tilde{\sigma}_k^2}{2\gamma(\mathrm{i}k)} \mathrm{e}^{p(\mathrm{i}k)t + p^*(\mathrm{i}k)t'} (\mathrm{e}^{2\gamma(\mathrm{i}k)t} - 1) e_k(x - x').
\end{aligned} \tag{5.27}$$

In the third line, we use the fact that $e_k(x) = e^{2\pi \mathrm{i}kx}$ is an orthogonal basis and we use the properties of white noise in Section 2.1.1 of Chapter 2 in addition to the definition of the complex white noise $\mathrm{d}W_k(t) = 2^{-1/2}(\mathrm{d}W_{k,1}(t) + \mathrm{i}\mathrm{d}W_{k,2}(t))$ where each $W_{k,i}(t)$ is a Wiener process. Furthermore, we simply assume that $\tilde{\sigma}_k = \tilde{\sigma}_k^*$.

6

Filtering turbulent signals: Plentiful observations

The difficulties in filtering turbulent complex systems are largely due to our incomplete understanding of the dynamical system that underlies the observed signals, which have many spatio-temporal scales and rough turbulent energy spectra near the resolved mesh scale. In this chapter, we develop theoretical criteria as guidelines to address issues for filtering turbulent signals in an idealized context. In particular, we consider the simplest turbulent model discussed in Chapter 5 with plentiful observations, that is, the observations are available at every model grid point.

In this idealized context, we will provide a useful insight into answering several practical issues, including:

- As the model resolution is increased, there is typically a large computational overhead in propagating the dynamical operator and this restricts the predictions to relatively small ensemble sizes. When is it possible to filter using standard explicit and implicit solvers for the original dynamic equations by using a large time step equal to the observation time (even violating the CFL stability condition with standard explicit schemes) to increase ensemble size, yet still retain statistical accuracy?
- If plentiful observations are available on refined meshes, what is gained by increasing the resolution of the operational model? How does this depend on the nature of the turbulent spectrum?

In particular, in this simplified context, we address the basic issues outlined in 1(a)–(d) of Chapter 1. Finally, we should point out that on top of addressing the above practical issues, our novel filtering strategy reduces filtering s-dimensional systems in $(2N+1)^d$ resolved grid points, which involves propagating a large covariance matrix of size $s(2N+1)^d \times s(2N+1)^d$ with classical Kalman filter or extended Kalman filter, to filtering N^d independently s-dimensional problems, each of which involves propagating a covariance matrix of size $s \times s$. We will show that the filtered solutions with this strategy are not only comparable to those using the standard ensemble Kalman filter, but they are also insensitive to the ensemble size, to the model resolution, and they are independent of any tunable parameters.

6.1 A mathematical theory for Fourier filter reduction

In Chapter 4, we reviewed the classical stability theory for a numerical solver which involves spectral analysis of a constant-coefficient linearized partial differential equation. Our reduced filtering strategy is motivated by this classical stability criterion but we now employ the spectral transform not only to the PDE but also to the noisy observations (see Fig. 6.1). We take advantage of the fact that the structure of our simplest turbulent model (the linear stochastic PDE, see Chapter 5) is completely decoupled into systems of Langevin equations. Together with the Fourier coefficients of the noisy observed signals, we have a reduced filtering strategy in Fourier space which we refer as the Fourier domain Kalman filter (FDKF). The distinct feature in this innovative strategy is that it ignores the correlations between different Fourier modes. In Chapters 7, 8 and 11, we shall see that this ignorance is in fact advantageous even for nonlinear systems with sparse regular observations. The mathematical theory developed here allows detailed Fourier analysis, akin to the difference scheme stability analysis of Chapter 4, to provide guidelines for filtering turbulent signals for stochastic PDEs provided the observations are plentiful. The parallel theory for sparse observations is described in Chapter 7.

To be more precise, consider the following canonical filtering problem in the real domain. The canonical stochastic PDE in (6.1) is an $s \times s$ dynamical system with a Gaussian spatially correlated $s \times s$ random field white noise forcing $\sigma(x)\vec{W}(t)$, constructed so that the observed climatological turbulent energy spectrum is reproduced as the stationary invariant measure of (6.1).

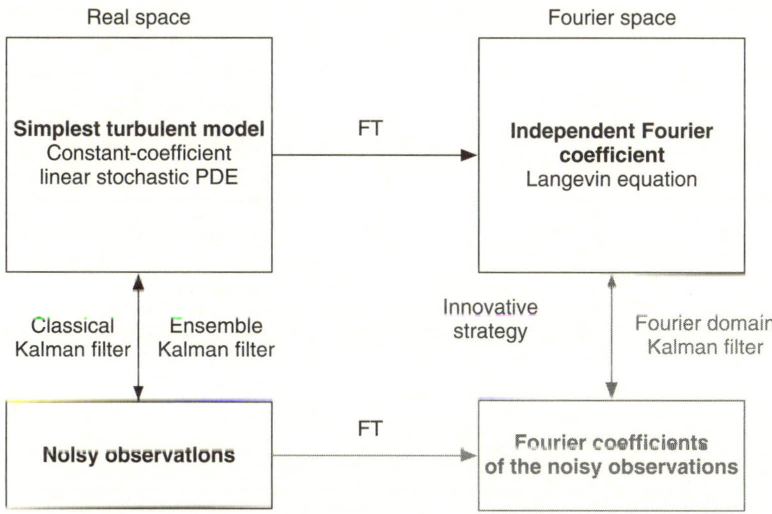

Figure 6.1 The Fourier domain Kalman filter (FDKF) blends the classical stability analysis of numerical PDEs and the classical spatial domain Kalman filter.

Canonical filtering problem: Plentiful observations

$$\frac{\partial}{\partial t}\vec{u}(x,t) = \mathcal{P}\left(\frac{\partial}{\partial x}\right)\vec{u}(x,t) - \gamma\left(\frac{\partial}{\partial x}\right)\vec{u}(x,t) + \sigma(x)\dot{\vec{W}}(t), \tag{6.1}$$

$$\vec{u}(x,0) = \vec{u}_0, \tag{6.2}$$

$$\vec{v}(x_j,t_m) = G\vec{u}(x_j,t_m) + \vec{\sigma}^o_{j,m}. \tag{6.3}$$

As in standard finite difference linear stability analysis, the problem in (6.1) is non-dimensionalized to a 2π-periodic domain so that continuous and discrete Fourier series can be used to analyze (6.1) and the related discrete approximations. In our canonical test problem, we realize the PDE (6.1) at $2N+1$ discrete points $\{x_j = jh, j = 0,\ldots,N\}$ such that $(2N+1)h = 2\pi$. This one-space dimension problem is chosen for simplicity in exposition but without loss of generality one can generalize it to d space dimensions.

We consider $q \le s$ observations $\vec{v}(x_j,t_m)$ in (6.3) which are attainable at every discrete time t_m and at every model grid point x_j with a fixed $q \times s$ observation matrix G. These plentiful observations are assumed to be imprecise, that is, they contain random measurement errors represented by zero-mean Gaussian random variables $\vec{\sigma}^o_m = \{\vec{\sigma}^o_{j,m}\}$ that are spatially and temporally independent at different grid points with covariance matrix

$$\langle \vec{\sigma}^o_m \otimes (\vec{\sigma}^o_m)^* \rangle = R^o, \tag{6.4}$$

where R^o is a block diagonal observation error covariance matrix with $q \times q$ block diagonal component $r^o\mathcal{I}$.

Recall the finite Fourier expansion as in Chapter 4,

$$\vec{u}(x_j,t_m) = \sum_{|k|\le N} \vec{\hat{u}}_k(t_m)e^{\mathrm{i}kx_j}, \quad \hat{u}_{-k} = \hat{u}^*_k \tag{6.5}$$

$$\vec{\hat{u}}_k(t_m) = \frac{h}{2\pi}\sum_{j=0}^{2N} \vec{u}(x_j,t_m)e^{-\mathrm{i}kx_j}. \tag{6.6}$$

Substituting (6.5) into (6.1)–(6.3) and using the identity (6.6), we obtain the following s-dimensional canonical filtering problem with q observations for each Fourier mode.

Fourier analogue of the canonical filtering problem: Plentiful observations

$$\vec{\hat{u}}_k(t_m+1) = F_k\vec{\hat{u}}_k(t_m) + \vec{\sigma}_{k,m+1}, \tag{6.7}$$

$$\vec{\hat{u}}_k(t_0) = \vec{\hat{u}}_{k,0}, \tag{6.8}$$

$$\vec{\hat{v}}_k(t_m) = G\vec{\hat{u}}_k(t_m) + \vec{\sigma}^o_{k,m}. \tag{6.9}$$

In (6.7), the operator F_k is an $s \times s$ diagonal matrix that solves or approximates the deterministic part of the s-dimensional Langevin equation in (5.8). In (6.7), the zero-mean complex Gaussian noises, $\vec{\sigma}_{k,m}$, are uncorrelated in time and their second moment is given by

$$\langle \vec{\sigma}_{k,m} \otimes (\vec{\sigma}_{k',m})^* \rangle = \delta_{k-k'}R_k, \quad |k|, |k'| \le N, \tag{6.10}$$

with R_k a strictly positive definite covariance matrix; for a vector Langevin equation, R_k is a diagonal $s \times s$ matrix.

In Fourier space, however, the observation noises have to be derived in a careful manner. Specifically, the observation error covariance is inversely proportional to the mesh size. In particular,

$$
\begin{aligned}
\langle \vec{\sigma}^o_{k,m} \otimes (\vec{\sigma}^o_{k',m})^* \rangle &= \left\langle \frac{h}{2\pi} \sum_{j=0}^{2N} \vec{\sigma}_{j,m} e^{-\mathrm{i}kx_j} \otimes \left(\frac{h}{2\pi} \sum_{j'=0}^{2N} \vec{\sigma}_{j',m} e^{-\mathrm{i}k'x_{j'}} \right)^* \right\rangle \\
&= \frac{h^2}{(2\pi)^2} \sum_{j=0}^{2N} \langle \vec{\sigma}_{j,m} \otimes \vec{\sigma}^*_{j,m} e^{-\mathrm{i}(k-k')x_j} \rangle \\
&= \frac{h^2}{(2\pi)^2} \sum_{j=0}^{2N} \langle \vec{\sigma}_{j,m} \otimes \vec{\sigma}^*_{j,m} \rangle \langle e^{-\mathrm{i}(k-k')x_j} \rangle \\
&= \frac{h^2}{(2\pi)^2} \sum_{j=0}^{2N} r^o \mathcal{I} \delta_{k-k'} \\
&= \frac{r^o \mathcal{I}}{2N+1} \delta_{k-k'}.
\end{aligned}
\tag{6.11}
$$

Here, we use the independence of observations at different locations as in (6.4) and the fact that $\{e^{\mathrm{i}kx_j}\}$ form a basis in k-space (see Chapter 4).

Consequently, we have the following (Majda and Grote, 2007):

Theorem 6.1 *If the observation points coincide with the discrete mesh points then for both the truth and any finite difference approximation:*

- *If the covariance matrix for the initial data, $\vec{\hat{u}}_{k,0}$, has the same block diagonal structure as in Eqns* (6.10) *and* (6.11) *for the system and observation noise, i.e. different Fourier modes are uncorrelated for $k \geq 0$, then the Kalman filtering test problem is equivalent to studying the independent $s \times s$ matrix Kalman filtering problems in Eqns* (6.7)–(6.11).
- *Provided that the $s \times s$ independent Kalman filtering problems in Eqns* (6.7)–(6.11) *are observable (see Chapters 2 and 3), then the unique steady-state limiting Kalman filter factors for the complete model into a block diagonal product of the limiting Kalman filters for each individual wavenumber, k.*

The practical significance of this result is that off-line tests for filter stability and model error for extremely complex PDEs can be developed for the simpler $s \times s$ matrix problems. For systems with $s > 1$ and observation matrix G with $q < s$ observations at each grid point, we are in the situation of filtering fewer observations than the actual dimension of the variables; this situation readily arises for the shallow water equation or geophysical primitive equations (Majda, 2003; Pedlosky, 1979) where only the pressure and temperature might be known at each observation point.

6.1.1 The number of observation points equals the number of discrete mesh points: Mathematical theory

In the above analysis, we studied the simplest situation where given $2N + 1$ discrete mesh points for the dynamical operator, $x_j = jh$, $j = 0, 1, \ldots, 2N$, the $2N + 1$ observation points, $\tilde{x}_j$, in (6.3) exactly coincide with these mesh points. Here we consider the situation where the $2N + 1$ observation points, $\tilde{x}_j$, do not coincide with the mesh point as often occurs in some applications, i.e. there are more observations over one area and fewer observations over another area. Here we present the mathematical theory (Majda and Grote, 2007) that establishes that suitable Fourier diagonal filters as in (6.7)–(6.9) provide upper and lower bounds on filter performance in the present setting. Lemmas 3.1 and 3.2 from Chapter 3 will be used in obtaining the rigorous upper and lower bound estimates.

For $\tilde{x}_j \neq \tilde{x}_k$, $0 \leq j, k \leq 2N$, consider the non-equispaced scalar real-valued trigonometric interpolation problem,

$$\begin{aligned} f(\tilde{x}_j) &= f_j, \quad j = 0, 1, \ldots, 2N, \\ f(x) &= \sum_{|k| \leq N} \hat{f}_k e^{ikx}. \end{aligned} \tag{6.12}$$

Write $\hat{f}_k = a_k + ib_k$, and the vector in $\mathbb{R}^{2N+1}$, $(a_0, a_1, b_1, \ldots, a_N, b_N)$, uniquely determines the finite Fourier series. For the $2N+1$ distinct points $\tilde{x}_j$, there is a unique invertible mapping $V : \mathbb{R}^{2N+1} \rightarrow \mathbb{R}^{2N+1}$ which solves the above trigonometric interpolation problem; with $\vec{f} = (f_0, \ldots, f_{2N})$,

$$V\vec{f} = (a_0, a_1, b_1, \ldots, a_N, b_N)^T. \tag{6.13}$$

There are even well-known explicit classical numerical analysis formulas for this unequally spaced trigonometric interpolation problem in a single space variable that define V (see chapter 5 of Isaacson and Keller (1966)). Obviously, the fixed matrix multiplication $G = \{G_{ij}\} \in \mathbb{R}^{s\times s}$ that defines the observations still commutes with the scalar transformation matrix, V, i.e.

$$\begin{aligned} V\sum_{j=1}^{s} G_{ij}\vec{u}_j(t_m) &= \sum_{j=1}^{s} G_{ij} V\vec{u}_j(t_m), \quad i = 1, \ldots s \\ &= \sum_{j=1}^{s} G_{ij}\big[a_{j,0}(t_m), a_{j,1}(t_m), b_{j,1}(t_m), \ldots, \\ &\qquad a_{j,N}(t_m), b_{j,N}(t_m)\big]^T \end{aligned} \tag{6.14}$$

where $\vec{u}_j(t_m) \in \mathbb{R}^{2N+1}$ is the jth variable of the true state at $2N + 1$ discrete points and $(a_{j,0}, a_{j,1}, b_{j,1}, \ldots, a_{j,N}, b_{j,N})$ is as defined in (6.13). Applying the interpolation formulas defined in (6.12) and (6.13) to the canonical observation (6.3), collecting the kth wavenumber from identity (6.14), and using the fact that $\hat{u}_{i,k} = a_{i,k} + ib_{i,k}$, we have an s-dimensional canonical observation for each wavenumber k:

$$\begin{aligned}\vec{\hat{v}}_k(t_m) &= (V\vec{v}_1(t_m), \ldots, V\vec{v}_s(t_m))_k \\ &= G\vec{\hat{u}}_k(t_m) + (V\vec{\sigma}^o_{m,1}, \ldots, V\vec{\sigma}^o_{m,s})_k. \end{aligned} \tag{6.15}$$

The formula in (6.15) can be continued trivially to k with $-N \leq k < 0$ by using the conjugate formula $\hat{u}_{-k} = \hat{u}^*_k$.

In the present case, the transformed observation noise for each $i = 1, \ldots, s$ in general has correlations across different Fourier modes with correlation matrix

$$\begin{aligned}\langle V\vec{\sigma}^o_{m,i} \otimes (V\vec{\sigma}^o_{m,i})^T \rangle &= V\langle \vec{\sigma}^o_{m,i} \otimes (\vec{\sigma}^o_{m,i})^T \rangle V^T \\ &= r^o V V^T. \end{aligned} \tag{6.16}$$

The full observation noise correlation matrix is an $s(2N+1) \times s(2N+1)$ matrix consisting of spatially correlated block diagonals (6.16) since the observation noise for each variable in the s-dimensional system is assumed to be uncorrelated. Therefore, the upper and lower bounds of the positive definite correlation matrix in (6.16)

$$r^o c^2_+ \mathcal{I}_{2N+1} \geq r^o V V^T \geq r^o c^2_- \mathcal{I}_{2N+1} \tag{6.17}$$

with $c^2_+ > c^2_- > 0$ depending on N, are also the upper and lower bounds of the full correlation matrix. Now consider the two Fourier diagonal problems with observations,

$$\vec{\hat{v}}_k(t_m) = G\vec{\hat{u}}_k(t_m) + \vec{\sigma}^{\pm}_{k,\,m}, \tag{6.18}$$

with the covariance given by

$$\langle \vec{\sigma}^{\pm}_{k,\,m} \otimes (\vec{\sigma}^{\pm}_{k,\,m})^* \rangle = r^o c^2_{\pm} \mathcal{I}_s. \tag{6.19}$$

With (6.17) and Lemmas 3.1 and 3.2 which guarantee the monotonicity of the Kalman filter covariance with respect to the observation noise, we immediately have:

Theorem 6.2 *If there are $2N + 1$ observation points $\tilde{x}_j$, which do not coincide with the grid points, then the upper and lower bounds on the discrete Kalman filtering process or the truth model as in Theorem 6.1 are achieved through the independent decoupled $s \times s$ filtering problems for each wavenumber k defined in* (6.7) *and* (6.18) *with the upper- and lower-bound diagonal noise covariances in* (6.19) *involving $c^2_{\pm}$.*

While Theorem 6.2 is interesting, it is often not practically useful since the condition numbers in (6.17) of the non equidistant discrete Fourier transform in (6.12) and (6.13) have $c^2_+/c^2_- \gg 1$ and this inversion problem is ill-conditioned. It is actually much better to use linear interpolation of the observations to a regular grid; surprisingly, it is a bad idea to use smoother interpolation algorithms for turbulent signals (Harlim, 2011).

6.2 Theoretical guidelines for filter performance under mesh refinement for turbulent signals

In many situations such as local regional numerical weather prediction over populated areas in developed countries, there are plentiful observations available for many successively refined discrete meshes. The nature of the turbulent spectra in the true dynamics might be white noise in space, reflecting physical processes such as moist convection over a few kilometers, or another steeper power law in the upper troposphere, reflecting gravity wave activity. Thus, a very interesting and relevant question for operational models is the following one: If plentiful observations are available, what is gained in filter performance by increasing the resolution of the operational model? How does this depend on the nature of the turbulent spectrum? Here, we provide theoretical guidelines for these important practical issues by answering the above question for filtering the general turbulent signals from the model in (6.1), (6.2) with the truth model itself.

Thus, a basic issue of practical interest in filtering a system like (6.1) is the following one: given the system in (6.1) with the energy spectrum E_k, what is the effect of increasing the number of grid points with plentiful observations on the Kalman gain? Under what circumstances should most of the weight be given to the observations alone for filtering at high spatial wavenumbers, and when should most of the weight be given to the dynamics alone at high spatial wavenumbers? As mentioned above, these are important issues to consider as guidelines for mesh refinement of the filtering problem in turbulent systems. Here we answer the above question completely for the scalar case with $s = 1$ by processing the explicit asymptotic Kalman gain matrix of the perfect filter in (6.1)–(6.3), which is simply a cross-product of the asymptotic Kalman gain for each individual wavenumber according to the second part of Theorem 6.1, in elementary analytic fashion using the formulas from Chapter 2.

As discussed earlier in Chapter 5, the dynamical operator for the stochastic PDE in (6.1) is given by the analytical solution of the Langevin equation in (5.8) for each wavenumber k with discrete form written in (6.7) with

$$F_k = e^{(\tilde{p}(ik)-\gamma(ik))\Delta t} \tag{6.20}$$

$$r_k = \frac{\tilde{\sigma}_k^2}{2\gamma(ik)}(1 - e^{-2\gamma(ik)\Delta t}). \tag{6.21}$$

Here F_k is a scalar complex number and r_k is a scalar real positive number and both are functions of the observation time Δt.

Consider the damping term $\gamma(ik)$ in the following form

$$\gamma(ik) = \gamma_o|k|^\alpha, \tag{6.22}$$

where γ_o is a positive real number. We refer to the case with $\alpha > 0$ as selective damping and with $\alpha = 0$ as uniform damping. In the remainder of this section, we consider both types of damping, $\alpha \geq 0$, satisying

$$0 < |F_k| = e^{-\gamma(ik)\Delta t} < 1, \tag{6.23}$$

and the typical turbulent energy spectrum in the form

$$E(k) = E_0|k|^{-\beta} \quad 0 \leq \beta < +\infty.$$

Consider the explicit formula for the asymptotic Kalman gain in (2.27) of Chapter 2 for the perfect model; for $2N + 1$ discrete modes we have

$$\tilde{z} = |F_k|^{-2} > 1 \tag{6.24}$$

while

$$\tilde{y} = A(k)\tilde{z}$$

with $A(k)$ the ratio of system noise to observational noise at wavenumber k; by using (5.9), i.e. $E_k = \tilde{\sigma}_k^2 / 2\gamma(ik)$, and (6.21), we have

$$A(k) = \frac{r_k(2N+1)}{r^o} = (2N+1)\frac{E_0}{r^o}|k|^{-\beta}\left[1 - e^{-2\gamma(ik)\Delta t}\right] \tag{6.25}$$

for $0 \leq |k| \leq N$. Note that in (6.25) we used the fact that the observational noise per mode decreases the observation noise by a factor of $2N + 1$. Our intuition for filtering the truth model suggests that if $A(k) \to 0$ as $|k|$ increases, there is more observational noise compared with decreasing system noise so that the filter should trust the dynamics alone; on the other hand, for $A(k) \to \infty$ the observation noise is relatively small and we should trust the additional observations alone in the filtering problem. Next we evaluate (6.25) asymptotically for $N \to \infty$ for the wavenumbers $N/2 \leq |k| \leq N$ (in the argument below, the lower bound with $1/2$ can be any factor smaller than one). From (6.25) we have the dichotomy

(A) For $1 - \beta < 0$

$$\max_{\frac{N}{2} \leq |k| \leq N} A(k) \leq \frac{E_0}{r^o} 2^\beta (2N+1) N^{-\beta} \sim N^{1-\beta} \to 0 \tag{6.26}$$

as $N \to \infty$.

(B) For $1 - \beta > 0$

$$\min_{\frac{N}{2} \leq |k| \leq N} A(k) \geq \frac{E_0}{r^o}(2N+1)N^{-\beta}(1 - |F^*|^2) \sim N^{1-\beta} \to \infty \tag{6.27}$$

as $N \to \infty$.

where $|F_k| < |F^*|$ for $N/2 \leq |k| \leq N$. Note that $|F^*| = |F_{N/2}| < 1$ for selective damping and $|F^*| - |F_k| < 1$ for any k for uniform damping. This is a quantitative estimate for the intuition mentioned earlier. This intuition is confirmed by the following:

Theorem 6.3 *For any damping $\gamma(ik)$ satisfying $0 < |F_k| < 1$ and energy spectrum (6.24), there are two different universal regimes of behavior for the filtering problem with plentiful observations for high wavenumbers $\frac{N}{2} \leq |k| \leq N$ as $N \to \infty$.*

(A) *For $\beta > 1$ the asymptotic Kalman gain matrix tends to zero uniformly for $\frac{N}{2} \leq |k| \leq N$ as $N \to \infty$. Thus in this regime even with plentiful observations, one can trust the dynamics alone on the large wavenumbers, with a refined mesh.*

(B) *For $\beta < 1$ the asymptotic Kalman gain matrix tends to one uniformly for $\frac{N}{2} \leq |k| \leq N$ as $N \to \infty$. Thus, in this regime with plentiful observations, one should primarily trust the observations on the large wavenumbers in the filtering problem with a refined mesh.*

Thus, for turbulent signals with $\beta < 1$, increasing mesh resolution only weights toward the additional observations while for $\beta > 1$, increased resolution improves the dynamics but the additional observations are not significant. This result depends on the spatial dimension; for d spatial dimensions, the proposition remains valid but the dichotomy $1 - \beta < 0$ or $1 - \beta > 0$ is replaced by $d - \beta < 0$ or $d - \beta > 0$. This arises simply because for plentiful observations in d dimensions we have the factor $(2N + 1)^d$ in (6.11), (6.25), (6.26), (6.27) replacing $2N + 1$. We state this as

Corollary 6.4 *In d space dimensions, Theorem 6.3 remains valid with $d - \beta$ replacing $1 - \beta$.*

With the above background from (6.24)–(6.27) the proof of the proposition is elementary through properties of the asymptotic Kalman gain formula in (2.23). In situation (A) with $\beta > 1$, we have from (6.26) the uniform asymptotic behavior for $\frac{N}{2} \leq |k| \leq N$

$$\tilde{y} \to 0$$

and (6.24) so by property (C) in (2.27) from Section 2.2 the asymptotic Kalman gain satisfies

$$K(F_k, \tilde{y}, \tilde{z}) \to 0$$

as $N \to \infty$ uniformly for $\frac{N}{2} \leq |k| \leq N$; this confirms the role of the dynamics on the high wavenumbers in the filtering.

In situation (B) with $\beta < 1$, we have from (6.27), the uniform asymptotic behavior for $\frac{N}{2} \leq |k| \leq N$

$$\tilde{y} \to \infty,$$

so by property (B) in Section 2.2 the asymptotic Kalman gain satisfies

$$K(F_k, \tilde{y}, \tilde{z}) \to 1$$

uniformly for $\frac{N}{2} \leq |k| \leq N$; this confirms that the high wavenumbers weight the filtered solution toward the observations.

In Fig. 6.2, we plot the asymptotic Kalman gain as a function of wavenumber k for different resolutions N for two spectra with the two sets of parameter values in the dissipative advection equation utilized in Sections 6.3 below. Theorem 6.3 is confirmed as one sees (in the first row of Fig. 6.2) that the asymptotic Kalman gain converges to one for large enough wavenumbers when $\beta = 0$ (see the first column of Fig. 6.2) and converges to zero for $\beta = 5/3$ (see the second column of Fig. 6.2). Similar trends are found for uniform damping case (see the second row) which confirm Corollary 6.4. In this last case, there are

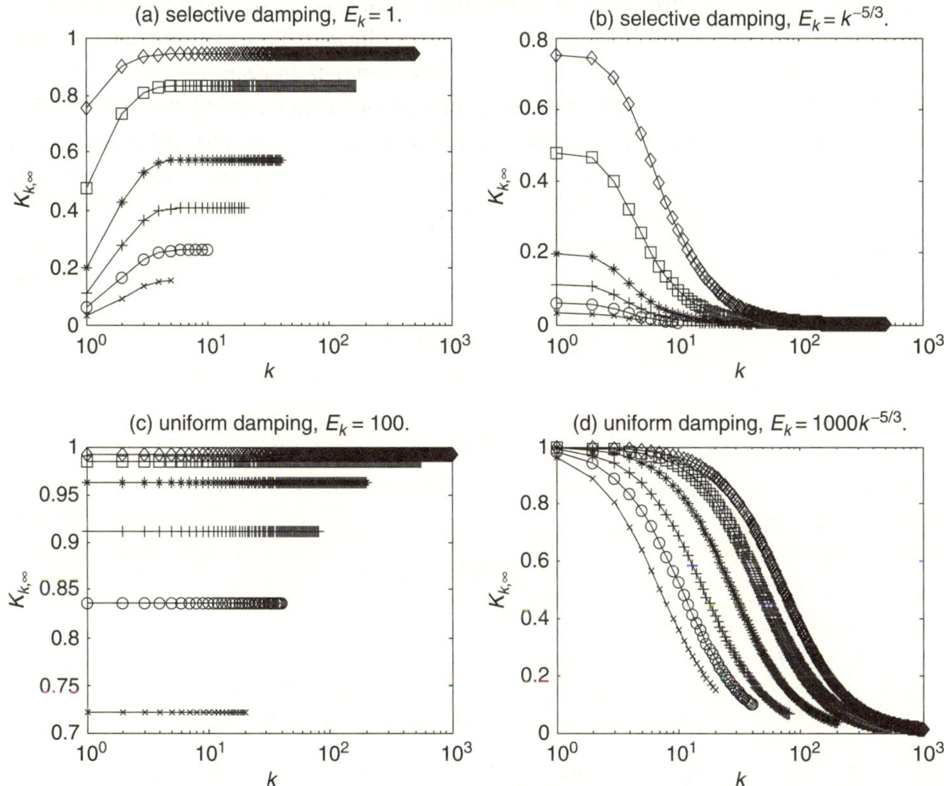

Figure 6.2 Kalman gains as a function of wavenumbers. The first row is for a selective damped signal with $r^o = 60$ and $\Delta t = 1$; in this regime, the model resolution is varied with wavenumbers $N = 5, 10, 20, 40, 150$ and 500. The second row is for a uniformly damped signal with $r^o = 1000$ and $\Delta t = 50$; in this regime, the model resolution is varied with wavenumbers $N = 20, 40, 80, 200, 500$ and 1000. The parameter values for selective decay and uniform damping are discussed in Section 6.3.

large prefactors in front of the power law so much larger values of N are needed to realize the theory.

6.3 Discrete filtering for the stochastically forced dissipative advection equation

Here we study filter performance for discretizations of the one-variable stochastically forced dissipative advection equation in (5.5) of Chapter 5 with zero mean forcing $\bar{F}$,

$$\frac{\partial u(x,t)}{\partial t} = -c\frac{\partial u(x,t)}{\partial x} - du(x,t) + \mu\frac{\partial^2 u(x,t)}{\partial x^2} + \sigma(x)\dot{W}(t). \tag{6.28}$$

One goal is to illustrate how rough turbulent signals generated by (6.28) can be filtered successfully by suitable discretization strategies with significant model error which respect the theoretical and computational guidelines established for each Fourier mode; in this example, these guidelines are nothing but those discussed earlier in Chapter 2 for the complex scalar problem. A second goal is to illustrate and compare these Fourier filters with an ensemble Kalman filter in physical space with the same approximation which generates errors which do not respect the Fourier diagonal covariance structure. Finally, the third goal is to point toward the potential practical use of Fourier diagonal filters with significant model errors (here these model errors are generated through spatio-temporal discretization), which are guided by the mathematical criteria in Chapter 2, as alternatives to the ensemble Kalman filter to overcome the "curse of ensemble size".

Here we consider discrete filter performance for the equation in (6.28) with **uniform damping** and without selective decay as a stringent test case. Thus, we generate truth signals for filtering (6.28) with $d > 0$ but $\mu = 0$. We utilize parameter values $c = 1$, $d = 0.01$, and $\Delta t = 50 = T_{\text{corr}}/2$ where $T_{\text{corr}} = 1/d = 100$ is the decorrelation time at each wavenumber. For this uniformly damped setting, the amplification factors at each wavenumber, F_k, satisfy $|F_k| = \mathrm{e}^{-d\Delta t} = \mathrm{e}^{-1/2} < 1$ so there is strong asymptotic stability in the perfect model filter in this regime. We consider truth signals generated from two extremely turbulent spectra, an equipartition spectrum with $E_k = 100$ and a $-5/3$ spectrum with $E_k = 1000k^{-5/3}$. These parameters are purposely chosen for evaluating the filter performance in a stiff regime with the temporal discretized schemes.

Next we consider a family of upwind discretizations as discrete filters. We consider a simple upwinding scheme

$$\frac{\partial}{\partial x}u(x_j, .) \sim \frac{u_j - u_{j-1}}{h}. \tag{6.29}$$

Using the discrete Fourier transform defined in (6.6) and as discussed in Chapter 4, we can rewrite the discrete spatial derivative as

$$\frac{\partial}{\partial x}u(x_j, .) \rightarrow \frac{1 - \mathrm{e}^{-ikh}}{h}\hat{u}_k.$$

As in Section 2.3.3 for the scalar test problem, we approximate the time derivative with the same three different time discretized schemes (forward Euler, backward Euler and trapezoidal) with

$$\lambda_k = -\gamma_k + \mathrm{i}\omega_k = -c\frac{1 - \mathrm{e}^{-ikh}}{h} - d.$$

In these three approximate filters, we replace the amplification factor F_k for the true dynamics as in (6.20) by a symbol $F_{h,k}$ for approximate dynamics. In a similar fashion, we replace the notation r_k for the true system noise variance as in (6.21) with approximate system noise variance $r_{h,k}$, which is computed directly from the corresponding time discretized scheme or from the information criteria discussed in Section 2.3.4. In total, we consider six approximate filters including three temporal discretized schemes and two different ways of choosing system noise variance.

For implicit Euler and trapezoidal schemes, the condition $|F_{h,k}| < 1$ is always satisfied for any resolution. In the trapezoidal schemes, however, almost every mode is marginally stable so that $|F_{h,k}| \cong 1$ for most modes. For the unstable forward Euler scheme, the amplitude satisfies $|F_{h,k}| > 1$ for the coarse mesh $N = 20$ and as we increase N, the magnitude of $F_{h,k}$ increases sharply.

6.3.1 Off-line test criteria

Here we use the theoretical off-line test criteria developed in Chapter 2 as a guideline for filter performance with these rough spectra and stiff regimes. Figures 6.3 and 6.4 show these off-line criteria for the backward Euler and trapezoidal methods, respectively.

In our off-line testing, we predict that the information criteria will improve filtering for both the backward Euler and trapezoidal schemes by giving practical controllability at all

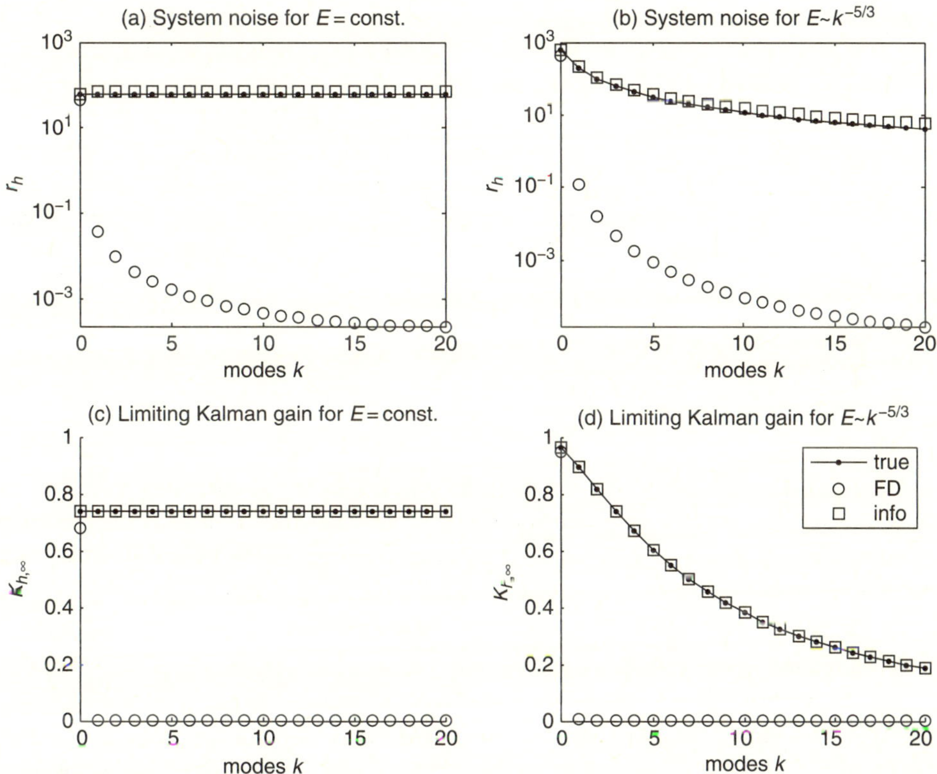

Figure 6.3 Off-line parameter set for the backward Euler method for equipartition spectrum $E_k = 100$ (left panels) and smooth spectrum $E_k = 1000k^{-5/3}$ (right panels). In panels (a) and (b), the system noise variance is plotted as a function of wavenumber. In panels (c) and (d), the asymptotic Kalman gain $K_{h,k,\infty}$ is plotted as a function of wavenumber.

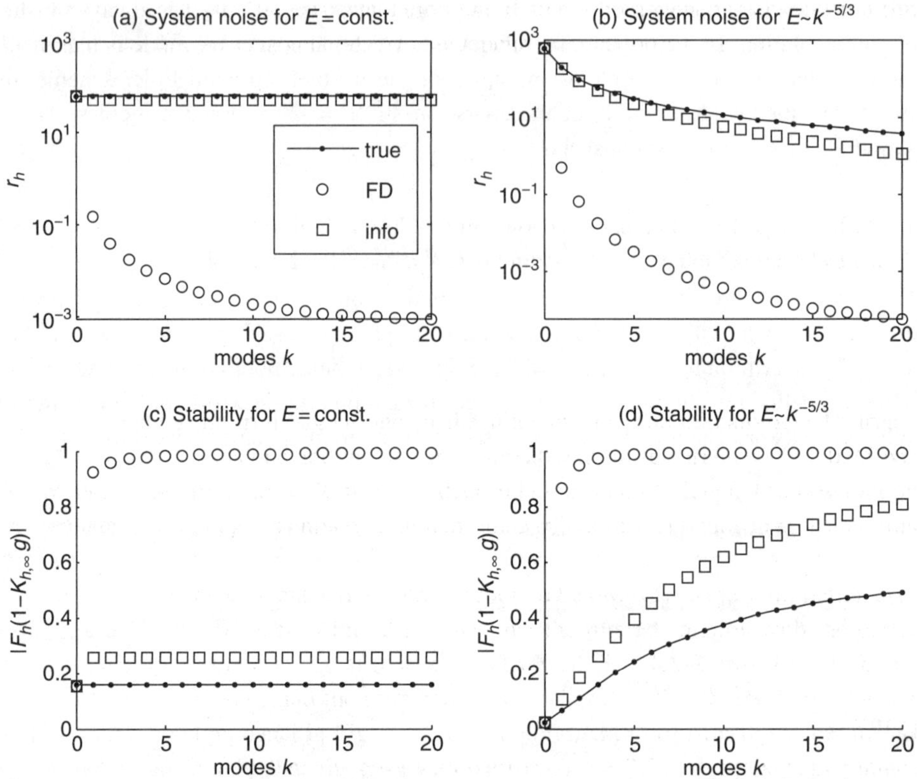

Figure 6.4 Off-line parameter set for the trapezoidal scheme for equipartition spectrum $E_k = 100$ (left panels) and smooth spectrum $E_k = 1000k^{-5/3}$ (right panels). In panels (a) and (b), the system noise variance is plotted as a function of wavenumber. In panels (c) and (d), the asymptotic stability factor $|F_{h,k}(1 - K_{h,k,\infty}g)|$ is plotted as a function of wavenumber.

wavenumbers through augmented system noise. In this regime, the Kalman gain $K_{k,\infty}$ is close to 1 for the constant-energy spectrum but it is much less than 1 for a smooth spectrum $E_k \sim k^{-\frac{5}{3}}$ (see panels c,d in Fig. 6.3). Thus, our theory predicts that the true filter is not much better than simply trusting the observations for the constant spectrum while it is better for the $-5/3$ spectrum.

When implicit Euler with finite discretized system noise is implemented, the violation of practical controllability significantly degrades the filter performance since the amplitude $|F_{h,k}| \approx 0$ for higher wavenumbers, but the Kalman gain is very small, $K_{h,k,\infty} = 0$ (see circles in Fig. 6.3) for almost every wavenumber. In other words, the filter overweighs the solutions to zero due to the over-damped dynamics with high certainty due to the tiny system noise variance.

For the trapezoidal method, the role of controllability is very similar to that in the backward Euler case. Here, the trapezoidal scheme is in a stiff regime with marginal stability

$|F_{h,k}| \approx 1$ for almost every wavenumber. When discretized system noise variance is used, the filter weights fully toward the dynamics but not to an over-damped dynamics as in the backward Euler scheme; instead it weights fully to an under-damped dynamics with high uncertainty and this yields an almost unity stability factor, $|F_{h,k}(1 - K_{h,k,\infty}g)| \approx |F_{h,k}| \approx 1$, which causes the filter solutions to only marginally converge.

To summarize, both the backward Euler and trapezoidal methods with standard finite difference noise discretization violate practical controllability criteria despite having filter stability (see circles in panels a,b in Figs 6.3 and 6.4); thus the off-line test criteria predict poor filter performance. The information criteria inflate the system noise appropriately and we already saw in Chapter 2 that it improves the filtered solutions significantly for a scalar filter, and as we shall see later, this improvement will translate to a better improvement at the PDE level. When we employ the information criteria, the better filter among the two implicit schemes is predicted by the asymptotic stability factor $|F_{h,k}(1 - K_{h,k,\infty}g)|$. In our simulations, we see that the backward Euler is a better scheme since $|F_{h,k}(1 - K_{h,k,\infty}g)| \approx 0$ for all k (not shown) while these factors for the trapezoidal method increase as functions of the wavenumber (see Fig. 6.4 panels c,d).

For unstable explicit Euler, the off-line criteria predict that the filtered solutions simply trust the observations with or without information criteria since every Fourier mode fully weights toward the observations for this numerically stiff regime, as discussed in Section 2.3.3 and Table 2.1.

6.3.2 Numerical simulations of filter performance

We check the actual filter performance with the prediction of the off-line testing shown earlier. In particular, we also compare results from filtering in the Fourier domain with ensemble filtering in real space. The real domain filter that we choose for comparison is the ensemble transform Kalman filter (ETKF) of Bishop *et al.* (2001) which will be discussed in detail in Chapter 9. The reason why we choose this ensemble filter is because it is easily implemented (Harlim and Hunt, 2007a) and for large ensemble size its accuracy is similar to a fully extended Kalman filter (Harlim, 2006). Note that for the numerical experiments with ETKF, the wave equation in (6.28) is integrated with the upwind scheme from (6.29) in the physical spatial domain.

For our numerical simulations, we generate the true trajectory by evolving an initial state that looks like a Gaussian hump, denoted as $\{\hat{u}_{k,0},\ k = 1, \ldots, 2N + 1\}$ in Fourier space, with the standard exact large time step discretization of (6.28) for $L = 100$ steps with time step $\Delta t = 50 = T_{\text{corr}}/2$. We simulate each observation by simply adding uncorrelated Gaussian random variables with mean 0 and variance $\hat{r}^o = r^o/(2N+1)$ to the true solution at each observation time Δt. In physical space, this reflects observations with variance r^o. As in the previous off-line testing, the observation noise is chosen to be $r^o = 1000$. We initiate each numerical simulation with randomly chosen initial states $u_{j,m|m}$ (or $\hat{u}_{k,m|m}$ in Fourier space).

In Figs 6.5 and 6.6, we plot (a) the RMS errors as functions of time for FDKF for an ensemble of size $K = 100$, (b) the RMS errors as functions of time for ETKF, also for $K = 100$, (c) the RMS errors as functions of ensemble size for FDKF, and (d) the ensemble error variances as functions of ensemble size for FDKF. In panels (a) and (b), the RMS error is averaged over space only. In panel (c), the RMS error is averaged over space and time (from $L_o = 50$ to $L = 100$). The ensemble error variance is defined as

$$\text{Error variance} = \text{Var}(u^k_{j,m|m} - u_{j,m}),$$

where the variance is taken over each ensemble member, space and time (also between $L_o = 50$ and $L = 100$).

From our simulations, we first notice that the forward Euler filter simply trusts the observations, as suggested by the off-line testing, regardless of the spectra. We clearly see that the backward Euler with time discretized noise $r_{h,k}$ over-damps the system and hence the prior forecast states deviate too far away from the noisy observations. For the trapezoidal scheme with time discretized noise, the filter is marginally unstable since $|F_{h,k}(1 - K_{h,k,\infty}g)| \approx 1$ (see Fig. 6.4). The information criteria improve the stability of both filters as predicted by the off-line testing with statistical accuracy nearly comparable to that of the truth filter with all the different spectra and resolutions. Thus, the off-line test criteria successfully predict all features of the filter performance.

Our numerical simulations confirm that both schemes (FDKF and ETKF) produce comparable filtered solutions in term of errors (e.g. compare also the filtered solutions of FDKF and ETKF in Figs 6.7 and 6.8, respectively). The ETKF is not robust in the sense that for an ensemble size of less than 50, the filter diverges (results not shown), while the FDKF with smaller ensemble size (even with only one realization) is performing as well as that with larger ensemble size in terms of RMS errors. In Figs 6.5(d) and 6.6(d), we notice that forward Euler has the smallest ensemble error variance since the Kalman gain is 1 and all ensemble members trust the observations. In contrast, the over-damped backward Euler with time discretized noise has Kalman gain 0 (see Fig. 6.3). In this situation, all ensemble members trust the dynamics and therefore it is obvious that the ensemble error variances are smaller than those of the truth model.

Now we check the filter performance for variations of model resolution $N = 20$, 40, and 80 (see Figs 6.9 and 6.10.). In a standard extended Kalman filter, the higher the model resolution, the larger the size of the error covariance matrix. In principle, the basic idea of an ensemble Kalman filter, including ETKF, is to approximate the error covariance matrix by the sample covariance from many realizations. Thus, the higher model resolution in ETKF requires larger ensemble size (see Harlim, 2006, chapter 3). In our ETKF simulations, we use $K = 200$ for $N = 80$ but $K = 100$ for $N = 20$ and 40 since the former simulation with 100 ensemble member suffers from a strong divergence. On the other hand, we see that the FDKF (including the approximate filters with time discretized schemes) are not sensitive at all to the variations of resolutions; in particular, the small ensemble size performs as well as the large ensemble size (see Figs 6.5 and 6.6.).

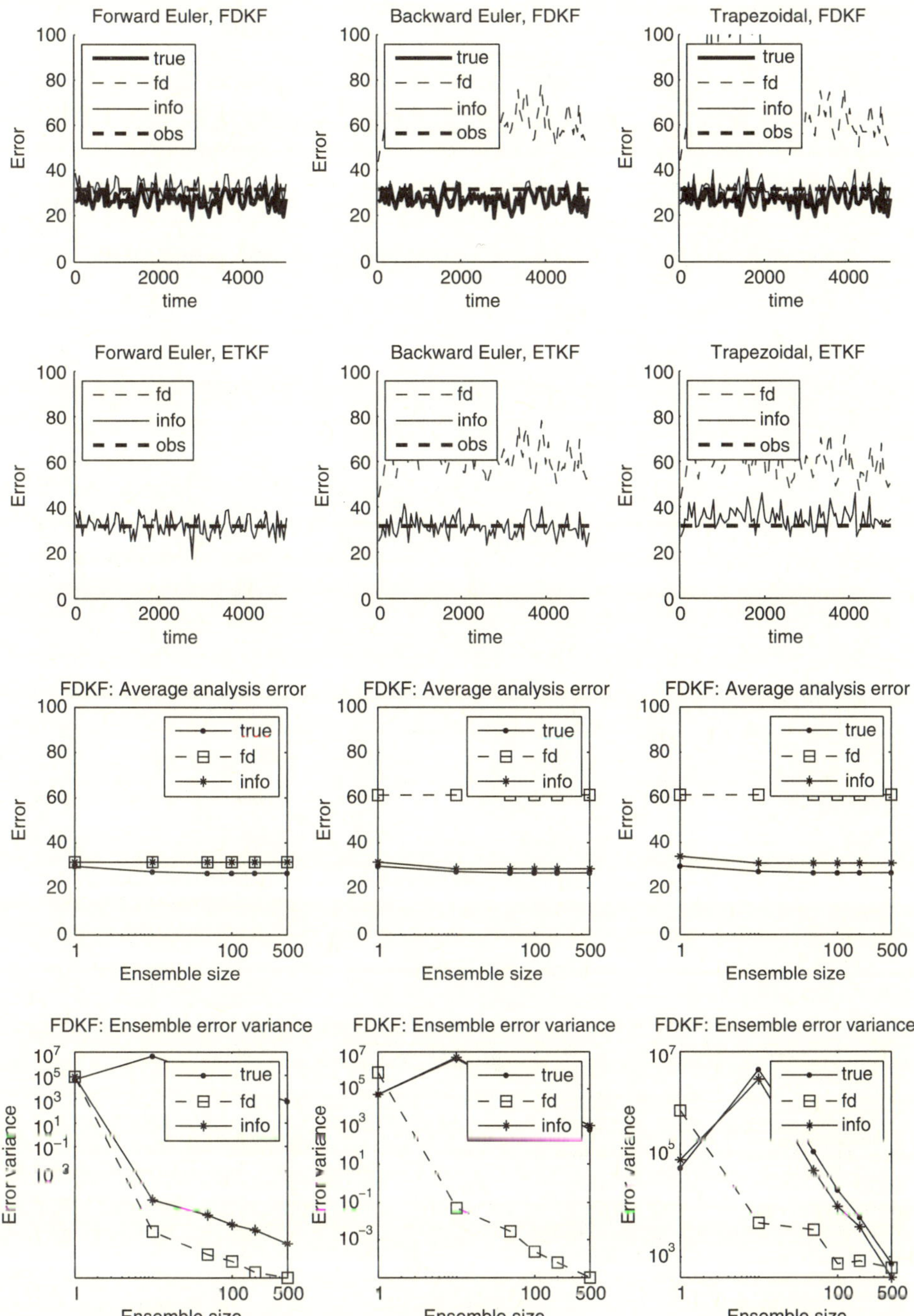

Figure 6.5 Uniform damping, $E_k = 100$ with $\Delta t = 50$ and $N = 20$: RMS errors as functions of time for FDKF with ensemble size $K = 100$ (first row), second row for ETKF also with $K = 100$, RMS errors as functions of ensemble size for FDKF (third row), and ensemble error variances as functions of ensemble size for FDKF (fourth row). The panels in the first column depict simulations with forward Euler, second column with backward Euler, and third column with trapezoidal. In each panel, "true" indicates the perfect filter with no model error, "fd" denotes the finite difference approximate filter, "info" denotes the approximate filter with information criterion noise variance, and "obs" denotes the observation error.

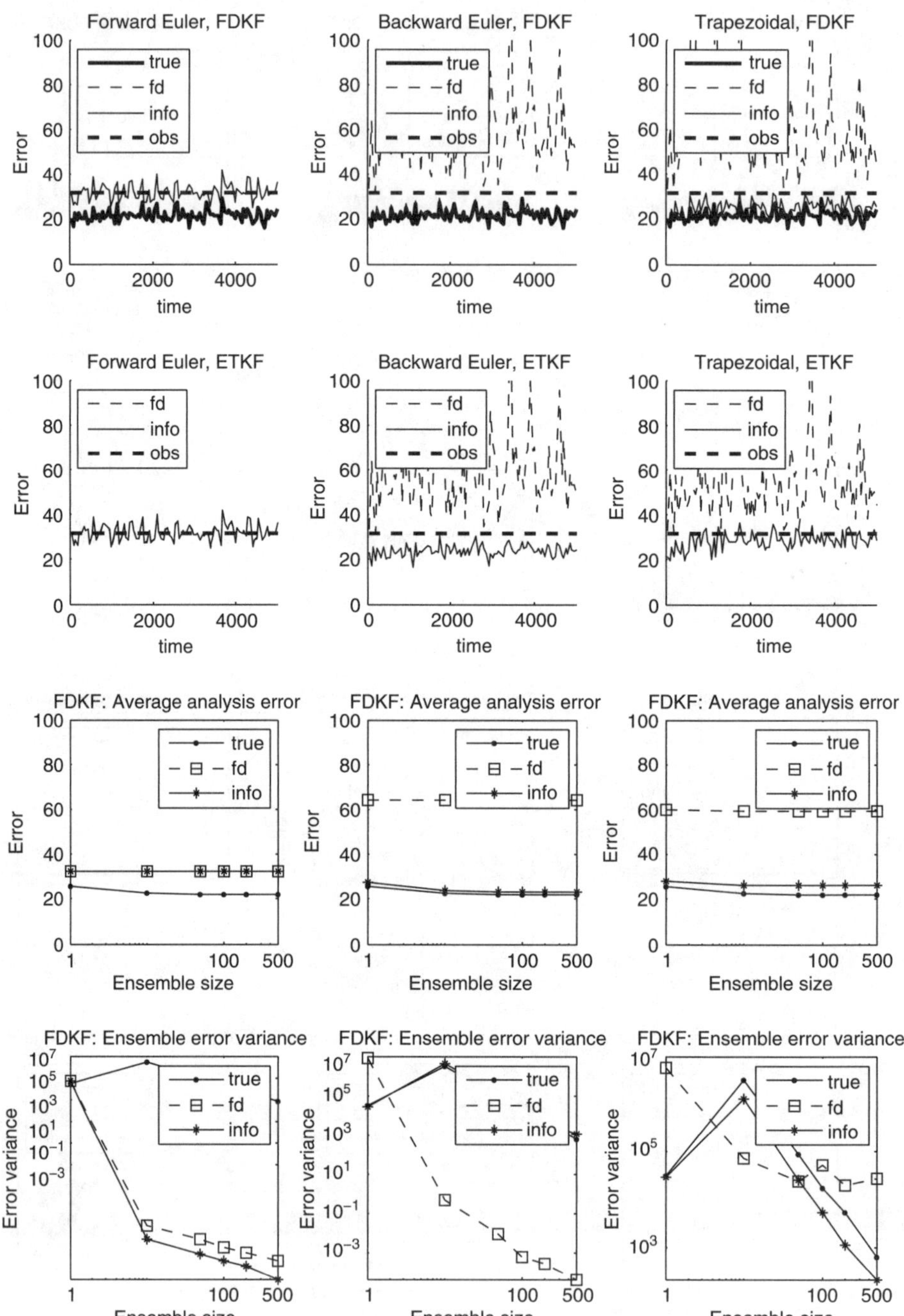

Figure 6.6 Uniform damping, $E_k \sim k^{-5/3}$ with $\Delta t = 50$ and $N = 20$: RMS errors as functions of time for FDKF with ensemble size $K = 100$ (first row), second row for ETKF also with $K = 100$, RMS errors as functions of ensemble size for FDKF (third row), and ensemble error variances as functions of ensemble size for FDKF (fourth row). The panels in the first column depict simulations with forward Euler, second column with backward Euler, and third column with trapezoidal. In each panel, "true" indicates the perfect filter, "fd" denotes the finite difference approximate filter, "info" denotes the approximate filter with information criterion noise variance, and "obs" denotes the observation error.

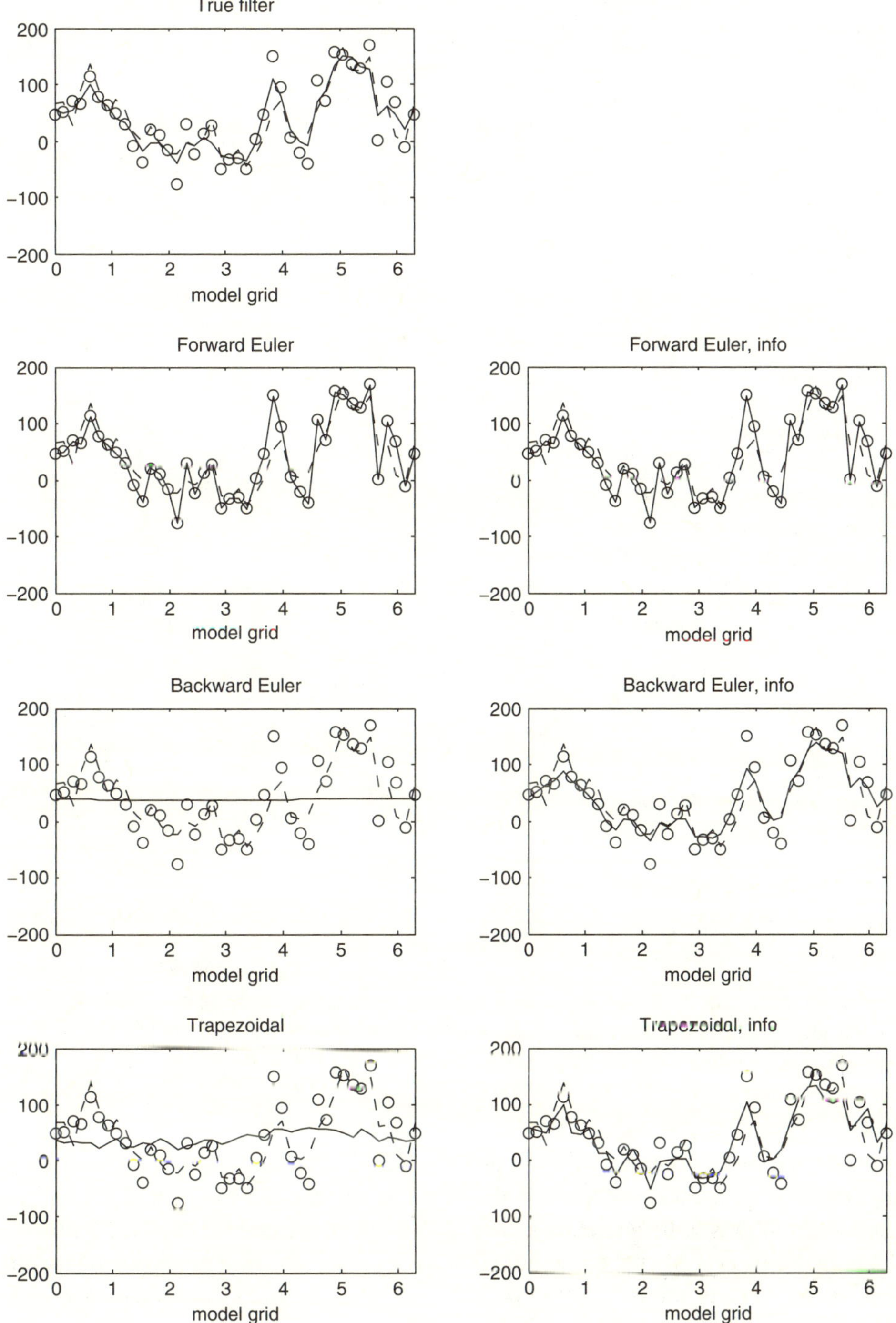

Figure 6.7 Snapshots of filtered solutions with FDKF as functions of model space after $L = 100$ assimilation cycles for uniform damping, $E_k \sim k^{-5/3}$ with $\Delta t = 50$ and $N = 20$. In each panel, we show the filtered solution (solid), the true signal (dashes) and the observations (circle).

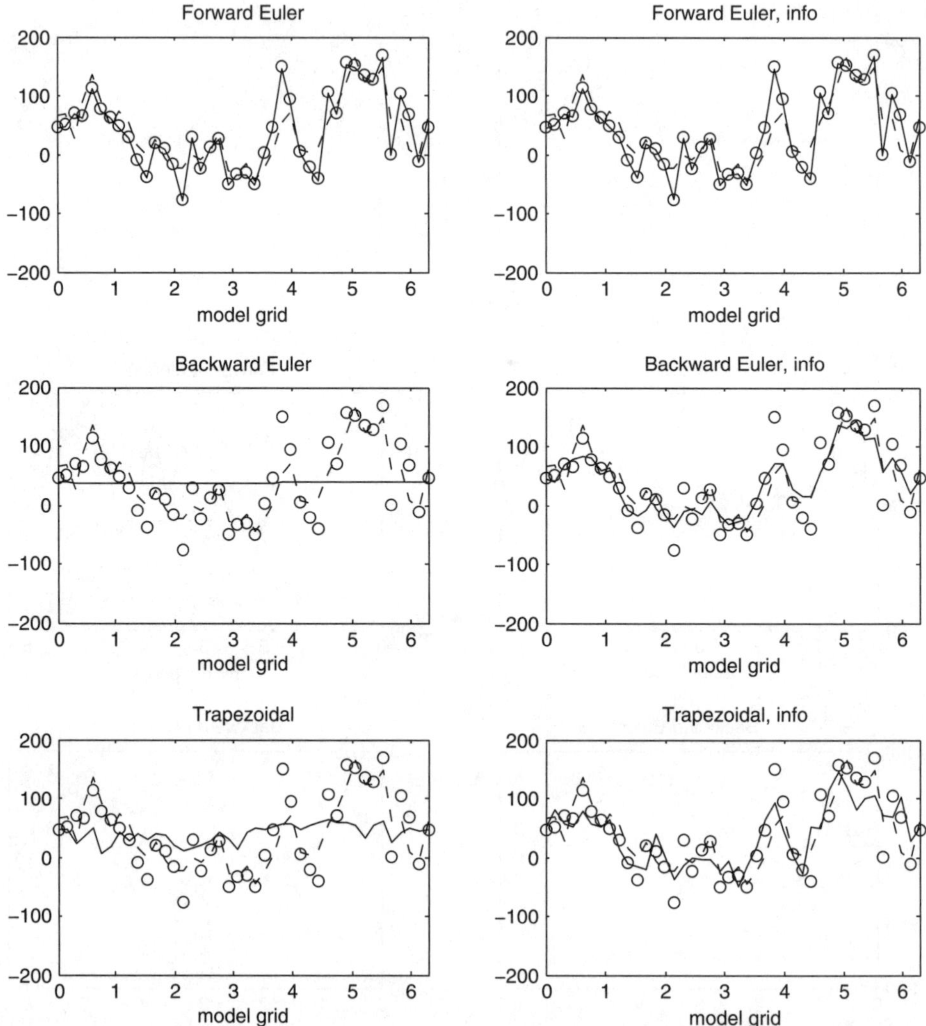

Figure 6.8 Snapshots of filtered solutions with ETKF as functions of model space after $L = 100$ assimilation cycles for uniform damping, $E_k \sim k^{-5/3}$ with $\Delta t = 50$ and $N = 20$. In each panel, we show the filtered solution (solid), the true signal (dashes) and the observations (circle).

From our numerical tests, we learn that the stable implicit schemes (backward Euler and trapezoidal) are the best filters in this stiff regime provided that their system noises are chosen to satisfy the information criteria. Between these two schemes, the backward Euler performs as well as the true filter. All these tendencies are fully predicted by off-line testing. In particular the superior performance of backward Euler over the trapezoidal method for the decaying spectrum can be traced to the decaying stability factor for large spatial wavenumbers in Fig. 6.4. The spectacular failure of the filtering performance of

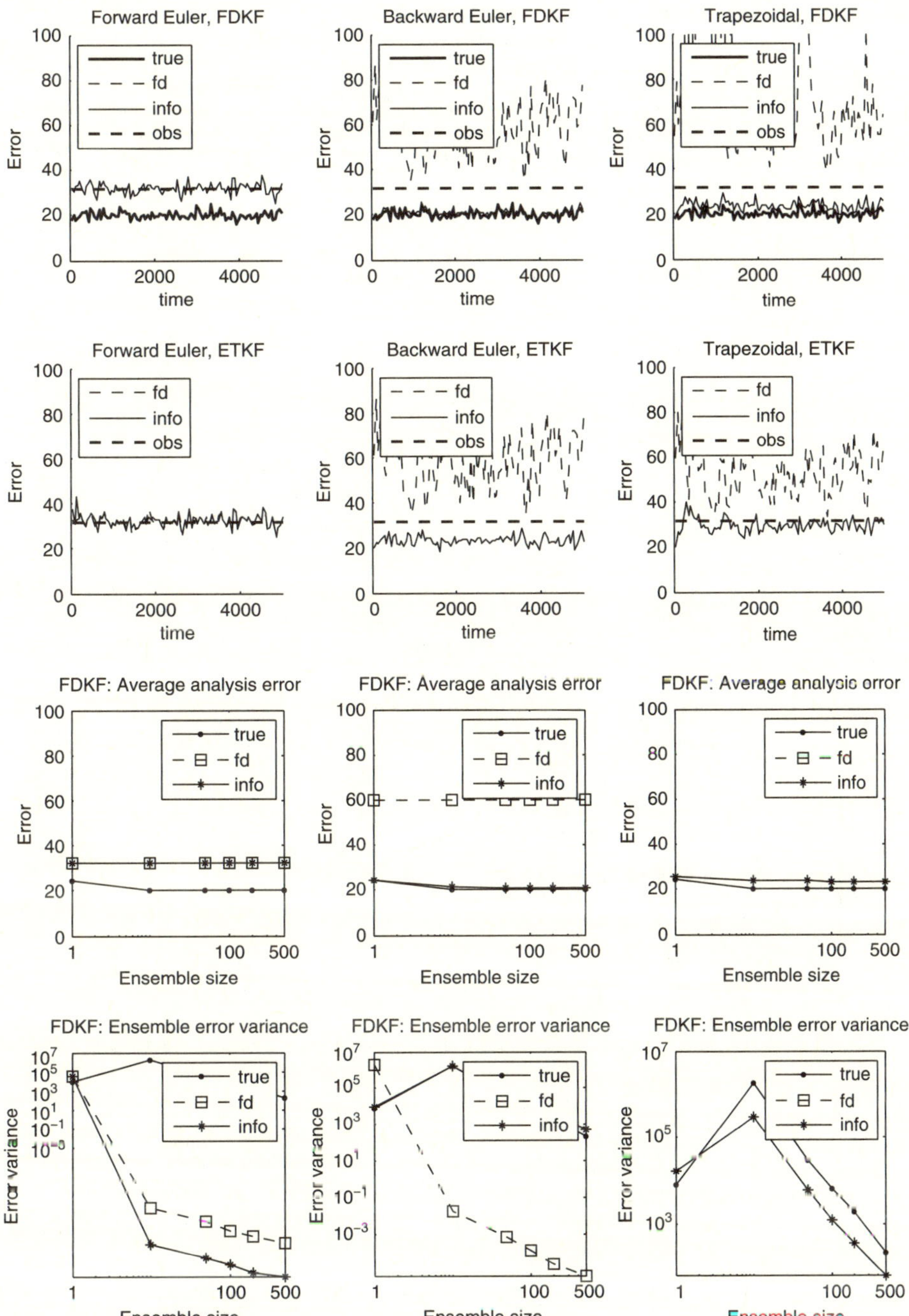

Figure 6.9 Uniform damping, $E_k \sim k^{-5/3}$ with $\Delta t = 50$ and $N = 40$: RMS errors as functions of time for FDKF with ensemble size $K = 100$ (first row), second row for ETKF also with $K = 100$, RMS errors as functions of ensemble size for FDKF (third row), and ensemble error variances as functions of ensemble size for FDKF (fourth row). The panels in the first column depict simulations with forward Euler, second column with backward Euler, and third column with trapezoidal. In each panel, "true" indicates the perfect filter, "fd" denotes the finite difference approximate filter, "info" denotes the approximate filter with information criterion noise variance, and "obs" denotes the observation error.

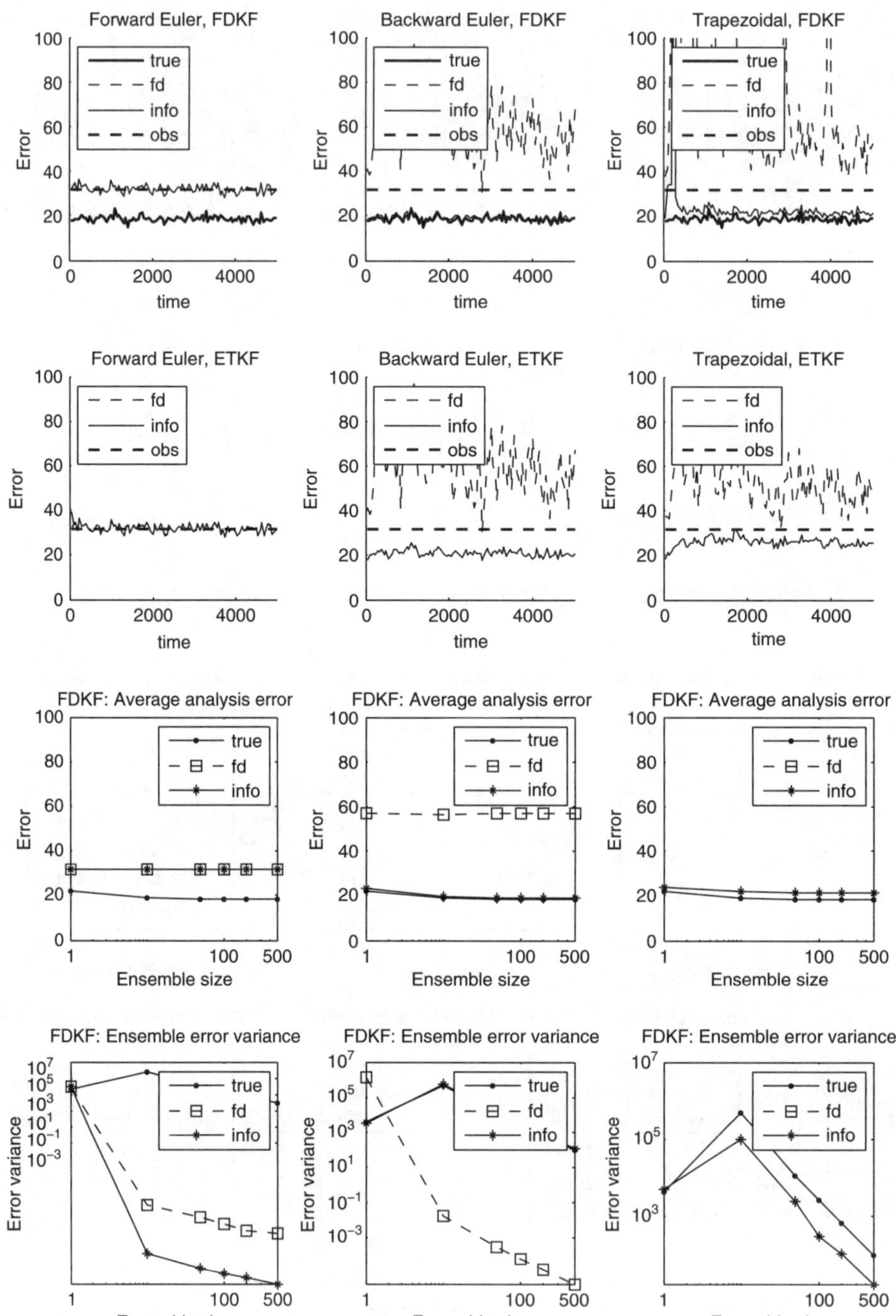

Figure 6.10 Uniform damping, $E_k \sim k^{-5/3}$ with $\Delta t = 50$ and $N = 80$: RMS errors as functions of time for FDKF with ensemble size $K = 200$ (first row), second row for ETKF also with $K = 200$, RMS errors as functions of ensemble size for FDKF (third row), and ensemble error variances as functions of ensemble size for FDKF (fourth row). The panels in the first column depict simulations with forward Euler, second column with backward Euler, and third column with trapezoidal. In each panel, "true" indicates the perfect filter, "fd" denotes the finite difference approximate filter, "info" denotes the approximate filter with information criterion noise variance, and "obs" denotes the observation error.

backward Euler and trapezoidal with standard time discretized noise is again predicted by the off-line Kalman gain which predicts full reliance on the highly inaccurate discrete dynamics without observation input.

Practically, both implicit filters in Fourier space are computationally inexpensive with such a giant time step and thus one can afford a large ensemble size. However, we also found that a large ensemble size is not necessary. As we have seen in Figs 6.5 and 6.6, even one realization is often acceptable. Moreover, the scalar Fourier domain filter is not sensitive to the variations of model resolution and is independent of tunable parameters (note that ETKF depends on variance inflation which is fixed to 10% for all the simulations above, see Chapter 9 for a detailed discussion).

In this chapter, we omit discussing the selective damping case since it produces similar conclusion as with uniform damping with much more accurate filtered solutions. The interested reader should consult Castronovo *et al.* (2008); there we also consider filtering the SPDE in (6.28) with $c = 1, d = 0, \mu = 0.1$ which is simply a stochastically forced scalar linear advection–diffusion equation with less energetic turbulent spectra $E_k = 1, k^{-5/3}$ and with $\Delta t = 1$ and $r^o = 60$. These parameters are used to generate panels a,b of Fig. 6.2. In this regime, the backward Euler and trapezoidal filters with and without information criteria have comparable performance.

7

Filtering turbulent signals: Regularly spaced sparse observations

The real-time filtering of noisy turbulent signals through sparse observations on a regularly spaced mesh is a notoriously difficult and important filtering problem; see Harlim and Majda (2008b) and Chapter 12. Here we study the real-time filtering of turbulent signals with sparse spatial observations in a canonical model where mathematical theory can be utilized to provide significant guidelines and approximations. The beginning mathematical developments parallel those in Chapter 6 for plentiful observations but a remarkable range of new phenomena occur which require significantly different and new ideas in filtering rough turbulent signals. In particular, here we will devise families of approximate filters which will be cheaper and more accurate than the Kalman filter for turbulent signals in suitable regimes. As discussed in Chapter 1, the skill of the Kalman filter and other approximate filters clearly depends on many features of the complex filtering problem for the turbulent signal such as

7(a) The specific underlying dynamics.

7(b) The energy spectrum at spatial mesh scales of the observed system and the system noise, i.e. decorrelation time, on these scales.

7(c) The number of observations and the strength of the observation noise.

7(d) The time-scale between observations relative to 7(a), (b).

7.1 Theory for filtering sparse regularly spaced observations

As in Chapter 6, we begin with the

Canonical filtering problem: Regularly spaced sparse observations

$$\frac{\partial}{\partial t}\vec{u}(x,t) = \mathcal{P}\left(\frac{\partial}{\partial x}\right)\vec{u}(x,t) - \gamma\left(\frac{\partial}{\partial x}\right)\vec{u}(x,t) + \sigma(x)\dot{\vec{W}}(t) + \vec{F}(x,t), \tag{7.1}$$

$$\vec{u}(x,0) = \vec{u}_0, \tag{7.2}$$

$$\vec{v}(\tilde{x}_j, t_m) = G\vec{u}(\tilde{x}_j, t_m) + \vec{\sigma}^o_{j,m}. \tag{7.3}$$

As in standard finite difference linear stability analysis, the problem in (7.1) is non-dimensionalized to a 2π-periodic domain so that continuous and discrete Fourier series can be used to analyze (7.1) and the related discrete approximations. In our canonical test problem, we realize the PDE (7.1) at $2N+1$ discrete points $\{x_j = jh, j = 0, 1, \ldots, 2N\}$ such that $(2N+1)h = 2\pi$. Thus, approximate solutions to the forward dynamical operator in (7.1) have the Fourier expansion,

$$\vec{u}(x_j, t_m) = \sum_{|k|\le N} \vec{\hat{u}}_k(t_m) e^{ikx_j}, \quad \hat{u}_{-k} = \hat{u}_k^*$$

$$\vec{\hat{u}}_k(t_m) = \frac{h}{2\pi} \sum_{j=0}^{2N} \vec{u}(x_j, t_m) e^{-ikx_j}.$$

For sparse regular observations, the observation points $\tilde{x}_j$ in (7.3) are less than the mesh points but are also regularly spaced. Thus, $\tilde{x}_j = j\tilde{h}, j = 0, 1, 2, \ldots, 2M$, with $(2M+1)\tilde{h} = 2\pi$ and $M < N$ for sparse observations. As an example, see Fig. 7.1 where there are only observations available at every third mesh point.

We consider $q \le s$ observations $\vec{v}(\tilde{x}_j, t_m)$ in (6.3) which are attainable at every discrete time t_m and sparsely at the grid points $\tilde{x}_j$ with a fixed $q \times s$ observation matrix G. These observations are assumed to be imprecise, that is, they contain random measurement errors

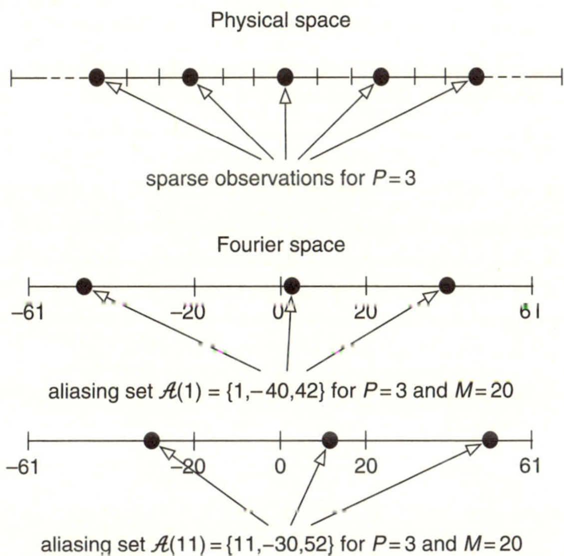

Figure 7.1 The top figure describes an example of physical space with regularly spaced sparse observations with $P = 3$. The bottom figures describe examples of two aliasing sets $\mathcal{A}(1)$, $\mathcal{A}(11)$, corresponding to the above physical space $P = 3$ for $M = 20$.

represented by zero-mean Gaussian random variables $\vec{\sigma}_m^o = \{\vec{\sigma}_{j,m}^o\}$ that are spatially and temporally independent with covariance matrix

$$\langle \vec{\sigma}_m^o \otimes (\vec{\sigma}_m^o)^* \rangle = R^o, \tag{7.4}$$

where R^o is a block diagonal observation error covariance matrix with $q \times q$ block diagonal component $r^o \mathcal{I}$.

Next we develop a canonical Fourier domain filtering problem for sparse regular observations following Majda and Grote (2007). Unlike the situation with plentiful observations, different Fourier modes will be coupled in the filtering problem across different aliasing sets arising from Fourier series on the regular computational mesh projected onto the sparse regular observation mesh. First, as in Chapters 4 and 6, approximations to the dynamical operator in (7.1), (7.2) on the finer computational mesh have the equivalent Fourier representation (stated for simplicity with the mean forcing $\bar{F}(\vec{x}, t) \equiv 0$):

Fourier dynamical operator

$$\begin{aligned} \vec{\hat{u}}_k(t_m + 1) &= F_k \vec{\hat{u}}_k(t_m) + \vec{\sigma}_{k,m+1}, \\ \vec{\hat{u}}_k(t_0) &= \vec{\hat{u}}_{k,0}, \quad \text{for } |k| \le N, \quad t_m = m\Delta t. \end{aligned} \tag{7.5}$$

In (7.5), the operator F_k is an $s \times s$ diagonal matrix that solves or approximates the deterministic part of the s-dimensional Langevin equation in (5.8). In (7.5), the zero-mean complex Gaussian noises, $\vec{\sigma}_{k,m}$, are uncorrelated in time and their second moments satisfy

$$\langle \vec{\sigma}_{k,m} \otimes (\vec{\sigma}_{k',m})^* \rangle = \delta_{k-k'} R_k, \quad |k|, |k'| \le N, \tag{7.6}$$

with R_k a strictly positive definite covariance matrix; for a vector Langevin equation, R_k is a diagonal $s \times s$ matrix.

Next, we treat the new feature which involves the sparse regularly spaced observations in (7.3) at the grid points $\tilde{x}_j = j\tilde{h}$, $j = 0, 1, \ldots, 2M$, $(2M + 1)\tilde{h} = 2\pi$. Apply the discrete Fourier transform on these $2M + 1$ points to the observation matrix in (7.3) to generate Fourier coefficients at wavenumbers ℓ with $|\ell| \le M$. Recall the aliasing formula from Proposition 4.3 of Chapter 4 which relates the Fourier coefficients for the sparse observation points with the Fourier coefficients from the finer regular computational mesh

$$\vec{\hat{f}}_{\tilde{h}}(\ell) = \sum_{q,\, |\ell + q(2M+1)| \le N} \vec{\hat{f}}_h(\ell + q(2M + 1)). \tag{7.7}$$

For a given ℓ with $|\ell| \le M$, define the aliasing set

$$\mathcal{A}(\ell) = \{k : |k| \le N, k = \ell + (2M + 1)q, \text{for any } q\}. \tag{7.8}$$

With the formulas in (7.7),(7.8), the sparse regularly spaced observations in (7.3) are equivalent to

Sparse regularly spaced observations in Fourier space

$$\hat{\vec{v}}_\ell(t_m) = G \sum_{k\in\mathcal{A}(\ell)} \hat{\vec{u}}_k(t_m) + \vec{\sigma}^o_{\ell,m}, \tag{7.9}$$

for $|\ell| \le M$. As in (6.11) of Chapter 6, the observational noise covariance in (7.9) is given by

$$\langle \vec{\sigma}^o_{\ell,m} \otimes (\vec{\sigma}^o_{\ell',m})^* \rangle = \frac{r_o \mathcal{I}}{2M+1}\delta_{\ell-\ell'},$$

for $|\ell|, |\ell'| \le M$. Clearly the aliasing sets, $\mathcal{A}(\ell)$, are disjoint for different ℓ. The Fourier dynamical operator in (7.5) and the Fourier representation of the observations in (7.6) define a Fourier space version of the original filtering problem at sparse regularly spaced observation points which is completely equivalent. The new feature here compared with Chapter 6 is that different Fourier modes in each disjoint aliasing set are coupled in the filtering process. With this structure we immediately have (Majda and Grote, 2007)

Theorem 7.1 (Sparse regularly spaced observations) *With the mesh size h for the difference scheme so that $(2N+1)h = 2\pi$ and sparse observations with regular spacing $\tilde{h}$ with $(2M+1)\tilde{h} = 2\pi$, assume $M < N$. Under these hypotheses, the basic filtering problem is equivalent to the filtering problem in Eqns* (7.5) *and* (7.9) *where the different modes in each disjoint observation aliasing set $\mathcal{A}(\ell)$ are coupled through the observation map in* (7.9). *In particular, if the initial covariance matrix has a block structure respecting the aliasing sets, the filter algorithm respects the same symmetries; if the filtering system is observable, the asymptotic limiting filter decomposes into a block structure along each of the disjoint observation aliasing sets $\mathcal{A}(\ell)$. Thus, under these hypotheses, the simpler filtering problems in Eqns* (7.5) *and* (7.9) *only need to be analyzed to develop off-line criteria.*

Compared with the situation with plentiful observations discussed in Chapter 6, there is a further significant compression of information in the present case of the actual turbulent signal which needs to be filtered because there are sparse regular observations. If we assume that the true signal to be filtered from the observations in (7.9) is generated, as in Chapter 5, from a truncation of the Langevin equation (7.1), i.e.

$$\vec{u}_{true} = \sum_{|k|\le N} \hat{\vec{u}}_{k,true},$$

then as in (7.7),

$$\hat{\vec{v}}_\ell - G \sum_{k\in\mathcal{A}(\ell)} \hat{\vec{u}}_{k,true}.$$

For signals with turbulent spectra such as $E_k \equiv 1$, there is substantial compression of the turbulent signal to be filtered due to the sparse observations and this can play a significant role in filter performance.

7.1.1 Mathematical theory for sparse irregularly spaced observations

Here we consider the general situation where there are sparse irregular spatial observations, at $2M + 1$ distinct points, $\tilde{x}_j$, $j = 0 \leq j \leq 2M$, where $M < N$ is smaller than the spatial discretization but these points $\tilde{x}_j$ do not coincide with the regularly spaced $2M + 1$ observation points, $j\tilde{h}, 0 \leq j \leq 2M$, with $(2M + 1)\tilde{h} = 2\pi$ considered in developing the canonical Fourier domain filtering problem in (7.5), (7.9). Akin to the developments in Theorem 6.2 from Chapter 6, we show that nevertheless, the canonical filtering problem in (7.5) and (7.9) provides upper and lower bounds. Let V_M from $\mathbb{R}^{2M+1} \to \mathbb{R}^{2M+1}$ be the same map defined in (6.13) which inverts the Fourier interpolation formula from the irregular mesh $\tilde{x}_j$, $j = 0 \leq j \leq 2M$, to the Fourier coefficients $\hat{u}_{\tilde{h},\ell}$, $|\ell| \leq M$, i.e. $M = N$ in (6.12) and (6.13). Repeating the same argument, the observations become

$$\begin{aligned}\vec{\hat{v}}_{\tilde{h},\ell}(t_m) &= (V_M \vec{v}_1(t_m), \ldots, V_M \vec{v}_s(t_m))_\ell \\ &= G\vec{\hat{u}}_{\tilde{h},\ell}(t_m) + (V_M \vec{\sigma}^o_{m,1}, \ldots, V_M \vec{\sigma}^o_{m,s})_\ell, \ |\ell| \leq M. \qquad (7.10)\end{aligned}$$

As in the proof of Theorem 6.2, exactly repeating that argument, the transformed observational noise is correlated but there are two families of observational noises which are decorrelated for different Fourier modes $0 \leq \ell_1, \ell_2 \leq M, \ell_1 \neq \ell_2$ yet provide upper and lower bounds on the observational noise covariance (see (6.16)–(6.19)). Now, use the aliasing formula

$$\vec{\hat{u}}_{\tilde{h},\ell}(t_m) = \sum_{k \in \mathcal{A}(\ell)} \vec{\hat{u}}_k(t_m)$$

in (7.10) in the same fashion as we did in Theorem 7.1. In this fashion we immediately have

Theorem 7.2 (Sparse irregularly spaced observations) *If there are $2M+1$ distinct sparse observation points, $\{\tilde{x}_j, j = 0, \ldots, 2M\}$, which do not coincide with the $2M + 1$ regular sparse observation points with $M < N$, then upper and lower bounds on the Kalman filtering matrices are achieved through the canonical $s \times s$ coupled Fourier domain filtering problem in* (7.5) *and* (7.9) *with upper and lower bound observational noise covariances involving $c^2_\pm$ exactly as in Theorem 6.2, that is,*

$$\vec{\hat{v}}_{\tilde{h},\ell}(t_m) = G \sum_{k \in \mathcal{A}(\ell)} \vec{\hat{u}}_k(t_m) + \vec{\sigma}^\pm_{\ell,m},$$

with observation noise covariance

$$\langle \vec{\sigma}^\pm_{\ell,\,m} \otimes (\vec{\sigma}^\pm_{\ell,\,m})^* \rangle = \frac{r^o}{2M+1} c^2_\pm \mathcal{I}_s.$$

As noted already in Chapter 6, the Fourier interpolation map V_M, for irregular mesh points, is often ill-conditioned so that $c^2_+/c^2_- \gg 1$ and Theorem 7.2 has limited practical skill. It is actually much better to use linear interpolation of the observations to a regular grid; surprisingly, it is a bad idea to use smoother interpolation algorithms for turbulent signals (Harlim, 2011).

7.2 Fourier domain filtering for sparse regular observations

In the above general discussion, the observations of the truth signal defined in (7.3) are taken at $2M+1$ grid points which are regularly spaced, i.e. $\tilde{x}_j = j\tilde{h}$, $j = 0, 1, \dots, 2M$, with $(2M+1)\tilde{h} = 2\pi$. When $M < N$ where $(2N+1)h = 2\pi$ and h denotes the mesh spacing for the finite difference approximation, we have sparse regular observations since there are fewer observations than discrete mesh points. In the rest of this chapter, for simplicity in exposition, we assume that M and N are related by

$$N = M + \tilde{P}(2M+1) \quad \text{for any fixed } \tilde{P} = 1, 2, 3, \dots$$

so that with $P = 2\tilde{P} + 1$ there are $P(2M+1)$ regular spaced mesh points and $2M+1$ regular spaced observation points. Thus, P defines the ratio of the total discrete mesh points available for the discretization to the number of sparse regular observation locations (see Figure 7.1 for a visual demonstration).

With the present simplification, given a discrete Fourier wavenumber, ℓ, $|\ell| \leq M$, associated with the sparse observation mesh $j\tilde{h}$, $j = 0, 1, 2, \dots, 2M$, the aliasing set $\mathcal{A}(\ell)$ on the fine discretization mesh is the set of P wavenumbers given by

$$\mathcal{A}(\ell) = \{k | k = \ell + (2M+1)q, |k| \leq M + \tilde{P}(2M+1)\}$$

where $q = 0, \pm 1, \pm 2, \pm 3, \dots$ and $P = 2\tilde{P} + 1$. For the remainder of this chapter, we consider a scalar field, u, i.e. $s = 1$, with observation coefficient $G \equiv 1$. In Section 7.1, we showed that the filtering problem defined by (7.1)–(7.3) is equivalent to the following family of complex P-dimensional Fourier domain filtering problems for the amplitudes at wavenumbers in each aliasing set, $\mathcal{A}(\ell)$:

$$\begin{array}{ll} \text{(A)} & \hat{u}^h_{k,m+1|m} = F_{h,k}\hat{u}^h_{k,m|m} + \bar{F}_{k,m} + \sigma_{h,k,m+1}, \quad \text{for } k \in \mathcal{A}(\ell) \\ \text{(B)} & \sum_{k\in\mathcal{A}(\ell)} \hat{u}_k((m+1)\Delta t) = \sum_{k\in\mathcal{A}(\ell)} \bar{\hat{u}}^h_{k,m+1|m+1} + \hat{\sigma}^o_{\ell,m+1} \end{array} \tag{7.11}$$

where, for each ℓ, $0 \leq \ell \leq M$, the observational noise is a zero-mean complex Gaussian random variable independent for different $\ell \geq 0$ with variance $\hat{r}^o = (2M+1)^{-1}r^o$, and r^o is the physical space observation noise variance. Since the original filter problem has only real coefficients, the reduced problem in (7.11) only needs to be studied or calculated for $0 \leq \ell \leq M$. Given a wavenumber ℓ, it is convenient to build the complex P-dimensional column vector

$$\vec{\hat{u}}(\ell) = \begin{pmatrix} \hat{u}^h_\ell \\ \hat{u}^h_{\ell+(2M+1)q_2} \\ \vdots \\ \hat{u}^h_{\ell+(2M+1)q_P} \end{pmatrix} = \begin{pmatrix} \hat{u}^h_{k_1} \\ \hat{u}^h_{k_2} \\ \vdots \\ \hat{u}^h_{k_P} \end{pmatrix} \tag{7.12}$$

where the various components in the aliasing set $\mathcal{A}(\ell)$ are indexed by the increasing magnitude of $k_j = \ell + (2M+1)q_j$, $j = 1, 2, \dots, P$; the primary wavenumber associated with

$\hat{u}^h_\ell$ is the first component of (7.12) and obviously has a special role. If we introduce the P-vector $\vec{g}_P = (1, \ldots, 1)^T$, with the notation in (7.12) we rewrite (7.11) (B) as given by

$$\vec{g}_P \cdot \vec{\hat{u}}((m+1)\Delta t) = \vec{g}_P \cdot \vec{\hat{u}}^h_{m+1|m+1} + \hat{\sigma}^o_{\ell,m+1}, \tag{7.13}$$

where we have made the aliasing of the truth signals to be filtered on the left-hand side of (7.13) explicit.

As discussed in Chapters 2 and 3, the standard simplest tests for stable filter performance in the canonical model are those for observability and controllability. Since the noises $\sigma_{h,k,m+1}$ in (7.11) (A) have nonzero variance, controllability is always satisfied but there are still subtle issues of practical controllability to be elucidated as already discussed in Chapters 2, 3 and 6; as noted already by Cohn and Dee (1988), the observability condition can be very subtle for discretizations of PDEs. Following Chapter 3, observability for the canonical model in (7.11) involves the invertibility of the $P \times P$ matrix,

$$\mathcal{O}_p = \begin{pmatrix} 1 & F_{h,k_1} & \ldots & F^{P-1}_{h,k_1} \\ 1 & F_{h,k_2} & \ldots & F^{P-1}_{h,k_2} \\ \vdots & \vdots & & \vdots \\ 1 & F_{h,k_P} & \ldots & F^{P-1}_{h,k_P} \end{pmatrix}. \tag{7.14}$$

The matrix in (7.14) is a Vandermonde matrix so that

$$\det \mathcal{O}_p = \prod_{i<j} \left(F_{h,k_i} - F_{h,k_j} \right). \tag{7.15}$$

It is important to test observability in the perfect model scenario for the scalar field where

$$F_{h,k} = \mathrm{e}^{\tilde{p}(\mathrm{i}k)\Delta t - \gamma(\mathrm{i}k)\Delta t}. \tag{7.16}$$

If there is selective damping then:

Remark 7.3 *If the dissipation $\gamma(ik)$ strictly increases with k for $k > 0$, then observability is always satisfied for the canonical model* (7.11) *since* $\det \mathcal{O}_p \neq 0$ *provided $\ell \neq 0$.*

It is left as an amusing exercise for the reader to check the observability conditions for selective damping and $\ell = 0$; in our example below, this is always satisfied because the successive aliased eigenvalues decrease in magnitude. On the other hand, if there is uniform damping, as often occurs with Ekmann friction or radiative cooling in geophysical models (Majda *et al.*, 2005; Majda and Wang, 2006), so that

$$\gamma(\mathrm{i}k) = d \tag{7.17}$$

then observability is readily violated for appropriate observation times Δt where $\det \mathcal{O}_p = 0$. With (7.15), (7.16) and (7.17) we calculate that for uniform damping,

$$\det \mathcal{O}_p = \mathrm{e}^{-\binom{P}{2}C)d\Delta t} \prod_{k_i \neq k_j \in \mathcal{A}(\ell)} \mathrm{e}^{\tilde{p}(\mathrm{i}k_i)\Delta t} - \mathrm{e}^{\tilde{p}(\mathrm{i}k_j)\Delta t}, \tag{7.18}$$

where $\binom{P}{2}C) \equiv P!((P-2)!2!)^{-1}$. Thus, from (7.18), observability fails in the perfect model with uniform damping at the observation time Δt, provided that, for some ℓ with $|\ell| < M$ and some q_j with $k_j = \ell + (2M+1)q_j$ belonging to the aliasing set, $\mathcal{A}(\ell)$, there is an integer Q so that

$$(\tilde{p}(\mathrm{i}k_i) - \tilde{p}(\mathrm{i}k_j))\Delta t = \mathrm{i}(\omega_{k_i} - \omega_{k_j})\Delta t = Q2\pi\mathrm{i}, \tag{7.19}$$

where $\tilde{p}(\mathrm{i}k_j) = \mathrm{i}\omega_{k_j}$. As we demonstrate next, the observability test often fails for wave-like systems with uniform damping. One important issue treated in detail by Harlim and Majda (2008b) and discussed briefly here is the significance of the failure of the classical observability condition and its role in actual filter performance for the turbulent filtering problems with sparse regular observations described at the beginning of this section.

The theoretical result in Remark 7.3 provides one possible computational filtering strategy to avoid the failure of observability for the truth model in the uniformly damped case, i.e.:

Remark 7.4 *Filter the uniformly damped turbulent signal generated from (7.17) with a modified model with selective damping $\tilde{\gamma}(ik)$ satisfying Remark 7.3 and use $F_{h,k} = e^{\tilde{p}(ik)\Delta t - \tilde{\gamma}(ik)\Delta t}$ in the computational filter.*

Next, we demonstrate all of these phenomena on the stochastically forced advection diffusion equation introduced in Chapter 5.

Here we illustrate these issues in filtering solutions of the general damped, stochastically forced advection equation

$$\frac{\partial u(x,t)}{\partial t} = -c\frac{\partial u(x,t)}{\partial x} - du(x,t) + \mu\frac{\partial^2 u(x,t)}{\partial x^2} + \bar{F}(x,t) + \sigma(x)\dot{W}(t). \tag{7.20}$$

In this example, $\tilde{p}(\mathrm{i}k) = \mathrm{i}\omega_k = -ick$ and the damping is given by

$$\gamma(\mathrm{i}k) = d + \mu k^2. \tag{7.21}$$

For a fixed $\mu > 0$ we have selective damping so that according to Remark 7.3 observability is always satisfied. On the other hand, for $d > 0$ and $\mu \equiv 0$, we have uniform damping and the observability condition fails if (7.19) is satisfied, that is,

$$\Delta t = \frac{Q2\pi}{cj(2M+1)} \tag{7.22}$$

for any positive integer Q and fixed j with $1 \le |j| \le P$. Note that for the uniformly damped advection equation observability fails simultaneously for all wavenumbers ℓ with $|\ell| \le M$ and simultaneously for many pairs of wavenumbers in a given aliasing set $\mathcal{A}(\ell)$. Thus, filtering turbulent solutions of (7.20) with small uniform damping at observation times satisfying (7.22) is a severe computational test problem discussed later in this chapter.

We demonstrate filter performance on this example in Section 7.4. We use the standard values $M = 20$, $P = 3$ so that there are observations of the turbulent signal from (7.20) at 41 sparse regular observation points with 123 discrete mesh points for the approximate filter. We study the filter performance for a case where the true signals are solutions of the advection–diffusion equation with small diffusion coefficient,

$$\mu = 10^{-2} \text{ and } d = 0. \tag{7.23}$$

We also study the filtering performance where the true signals are solutions of the uniformly damped advection equation with small damping

$$d = 10^{-2} \text{ and } \mu = 0. \tag{7.24}$$

In testing the strategy outlined in Remark 7.4 for filtering the true signal from the uniformly damped model with (7.24), we use $\mu = 10^{-2}$ while retaining $d = 10^{-2}$ so that according to (7.21), the filter model is selectively damped. The filter performance is also analyzed for power-law turbulent energy spectra in the truth model with $E_0 = 1$ and the exponent β varying for $\beta = 5/3$ and $\beta = 0$, so that we have either a $-5/3$ spectrum or equipartition of energy as in Chapter 5.

7.3 Approximate filters in the Fourier domain

With the notation in (7.12) and (7.13), the canonical Fourier domain filtering problem for regularly spaced sparse observations from (7.11) is written concisely for the complex P-vector $\vec{\hat{u}}^h$ as given by

$$\begin{aligned} &\text{(A)} \quad \vec{\hat{u}}^h_{m+1|m} = F_h \vec{\hat{u}}^h_{m|m} + \vec{\bar{F}}_{m+1} + \vec{\sigma}_{h,m+1} \\ &\text{(B)} \quad \bar{v}_{m+1} = \vec{g}_P \cdot \vec{\hat{u}}^h_{m+1|m+1} + \hat{\sigma}^o_{m+1} \end{aligned} \tag{7.25}$$

where

$$\begin{aligned} &\text{(A)} \quad \vec{g}_P = (1, 1, \ldots, 1)^T \in \mathbb{R}^P \\ &\text{B)} \quad \bar{v}_{m+1} = \vec{g}_P \cdot \vec{\hat{u}}((m+1)\Delta t) \end{aligned} \tag{7.26}$$

so that $\bar{v}_{m+1}$ is determined from the aliased components of the truth signal defined in Section 7.1. In (7.25) (A), F_h is the diagonal $P \times P$ complex matrix with entries determined by (7.11) (A).

The Fourier domain Kalman filter (FDKF) is simply the standard Kalman filter algorithm from Chapter 3 applied to all the disjoint aliasing sets $\mathcal{A}(\ell)$ for all $0 \le \ell \le M$. Thus, this filter neglects covariances across different aliasing sets compared with the standard Kalman filter in physical space.

As discussed in Chapter 3, in standard fashion, the $P \times P$ complex covariance matrix $R_{m|m}$ is updated in two steps: first, via the dynamics in (7.25) (A)

$$R_{m+1|m} = F_h R_{m|m} F_h^* + R_h \tag{7.27}$$

where R_h is the diagonal matrix with entries $r_{h,k}$; then $R_{m+1|m}$ is updated to $R_{m+1|m+1}$ via the single observation in (7.25) (B) so that

$$R_{m+1|m+1} = R_{m+1|m} - \Lambda(R_{m+1|m}, \hat{r}^o) R_{m+1|m} (\vec{g}_P \otimes \vec{g}_P^T) R_{m+1|m} \tag{7.28}$$

with the scalar factor $\Lambda(R_{m+1|m}, \hat{r}^o)$ given by

$$\Lambda(R_{m+1|m}, \hat{r}^o) = (\vec{g}_P \cdot R_{m+1|m} \vec{g}_P + \hat{r}^o)^{-1}. \tag{7.29}$$

In (7.28), $\vec{g}_P \otimes \vec{g}_P^T$ is the rank-one $P \times P$ matrix given by $(\vec{g}_P \otimes \vec{g}_P^T)\vec{w} = \vec{g}_P(\vec{g}_P \cdot \vec{w})$. The update of the mean, $\bar{\vec{u}}^h_{m+1|m+1}$, in the filtering process is given by

$$\vec{\bar{\hat{u}}}^h_{m+1|m+1} = \vec{K}_{m+1} \bar{v}_{m+1} + \vec{\bar{\hat{u}}}^h_{m+1|m} - \vec{K}_{m+1} \vec{g}_P \cdot \vec{\bar{\hat{u}}}^h_{m+1|m} \tag{7.30}$$

where the provisional update of the mean $\vec{\bar{\hat{u}}}^h_{m+1|m}$ is determined by the dynamics in (7.25) (A),

$$\vec{\bar{\hat{u}}}^h_{m+1|m} = F_h \vec{\bar{\hat{u}}}^h_{m|m} + \bar{F}_{m+1}. \tag{7.31}$$

The P-vector $\vec{K}_{m+1}$ is the Kalman gain vector (Chui and Chen, 1999)

$$\vec{K}_{m+1} = \Lambda(R_{m+1|m}, \hat{r}^o) R_{m+1|m} \vec{g}_P.$$

The FDKF algorithm is already a reduced filter and much less expensive and more stable than implementing the Kalman filter or the ensemble Kalman filter such as ETKF (Bishop *et al.*, 2001) directly on the spatial domain. Nevertheless, there are regimes of turbulent spectra and dynamics in (7.1) for the basic truth signal from where this filter can be ill-conditioned and less accurate; thus, it is desirable to develop simpler approximations to FDKF as both alternative computational filters and for theoretical purposes. This is the topic of the remainder of this section.

7.3.1 The strongly damped approximate filters (SDAF, VSDAF)

The SDAF filter is motivated by the following situation which readily arises in practice with selective damping. For a given P, assume that there are two separate groups in the dynamics (7.11) (A) with

$$\begin{aligned} &\text{(A) Moderate damping: } |F_{h,k_i}| = \mathcal{O}(1), \quad 1 \le i \le P_o \\ &\text{(B) Strong damping: } |F_{h,k_i}| = \mathcal{O}(\epsilon) \ll 1, \quad P_o + 1 \le i \le P. \end{aligned} \tag{7.32}$$

The assumption in (7.32) (B) means that there is strong damping in a subset of the aliased modes; in this situation, the dynamic covariance update in (7.27) can be inaccurate because the cross-covariances between the first P_o components and the last strongly damped $P - P_o$ components involve successive multiplications by large and small numbers to get order-one quantities. The SDAF algorithm eliminates this potential numerical difficulty. It approximates the covariance matrix, $R_{m+1|m}$, by the block diagonal covariance matrix

$$R_{m+1|m} = \left(\begin{array}{c|cccc} R_{m+1|m,P_o} & & 0 & & \\ \hline & r_{h,k_{P_o+1}} & & & \\ 0 & & r_{h,k_{P_o+2}} & & \\ & & & \ddots & \\ & & & & r_{h,k_P} \end{array}\right). \tag{7.33}$$

This is the covariance matrix that results from replacing the dynamics in (7.11) (A) on the strongly damped modes, $P_o + 1 \le i \le P$, by the "memoryless dynamics"

$$\hat{u}^h_{k_i,m+1|m} = \bar{F}_{k_i,m+1} + \sigma_{h,k_i,m+1}, \quad P_o + 1 \le i \le P. \tag{7.34}$$

When the situation in (7.32) is satisfied, it is intuitively clear that the approximation in (7.34) is a reasonable one. With this approximation in (7.33), the filter scaling factor Λ in (7.29) becomes

$$\begin{aligned}\Lambda(R_{m+1|m}, \hat{r}^o) &= \left(\vec{g}_{P_o} \cdot R_{m+1|m,P_o}\vec{g}_{P_o} + \sum_{i=P_o+1}^{P} r_{h,k_i} + \hat{r}^o\right)^{-1} \\ &\equiv \Lambda(R_{m+1|m,P_o}, \hat{r}_{P_o})\end{aligned}$$

with $\hat{r}_{P_o}$ the augmented noise

$$\hat{r}_{P_o} = \sum_{i=P_o+1}^{P} r_{h,k_i} + \hat{r}^o.$$

With the approximation in (7.32), the Kalman gain splits into contributions from moderately damped and memoryless (or strongly damped) modes and the gain vector is given by $\vec{K}_{m+1} = (\vec{K}_I, \vec{K}_{II})^T$ where

$$\begin{aligned}\vec{K}_I &= \Lambda(R_{m+1|m,P_o}, \hat{r}_{P_o}) R_{m+1|m,P_o}\vec{g}_{P_o} \\ \vec{K}_{II} &= \Lambda(R_{m+1|m,P_o}, \hat{r}_{P_o}) r_{h,k_i}, \quad P_o + 1 \le i \le P.\end{aligned} \tag{7.35}$$

Furthermore, the $P_o \times P_o$ block of the covariance matrix $R_{m+1|m+1,P_o}$ is updated by utilizing the same formula in (7.28) applied to the P_o-dimensional filtering problem. The important feature of the SDAF algorithm with the memoryless dynamics in (7.34) is that the entries of the covariance, $R_{m+1|m+1}$, corresponding to the memoryless modes never need to be computed. In the SDAF algorithm, the means of the memoryless modes are updated by (7.34) so

$$\bar{\hat{u}}^h_{k_i,m+1|m} = \bar{F}_{k_i,m+1}, \quad P_o + 1 \le i \le P.$$

This completes the construction of the SDAF algorithm since the mean state is updated through the general formula in (7.30) by utilizing (7.35) for the Kalman gain vector.

There is an important variant of the SDAF algorithm which needs to be mentioned here. Namely, one can use the memoryless approximation in (7.34) to update the covariance matrix in (7.33) but utilize the complete dynamics in (7.31) to update the mean state and calculate $\vec{\bar{u}}^h_{m+1|m}$; the mean state update $\vec{\bar{u}}^h_{m+1|m+1}$ is then achieved through the same formulas in (7.35) and (7.30). We call this filtering algorithm the variance strong damping approximate filter (VSDAF), since the strong damping is applied only for calculating the covariance matrix. Unlike SDAF, the VSDAF algorithm allows a weight for the dynamics of the memoryless modes in the filter.

Advantages of the strong damping algorithms

First, the strong damping algorithms (both SDAF and VSDAF) only require actual Kalman filtering on a reduced subspace of P_o dimensions yet automatically yield an update of the mean state of the system on all the variables including all of the $P - P_o$ memoryless variables. Furthermore, while we motivated these algorithms under the hypotheses in (7.32), the strong damping algorithms can be applied in any situation as a filtering approximation, even in circumstances where (7.32) is not satisfied (see Section 7.4). Another attractive feature of these algorithms is that the subtle features of violation of observability, discussed earlier in Sections 7.2, only need to be analyzed for the reduced P_o-dimensional filter; in particular, if $P_o = 1$, the strong damping algorithms always satisfy observability on the reduced subspace. Thus, the strong damping filters are an alternative computational strategy to the one proposed earlier in Remark 7.4 with artificial selective damping as a remedy for failure of observability. The filter performance of both strategies is compared in subsequent sections.

Another important attractive feature of the SDAF becomes apparent when we recall the nature of the turbulent truth signals described in Section 7.1 which we are attempting to filter. We claim that:

Remark 7.5 *The SDAF algorithm automatically provides a reasonably consistent estimate for the energy in the memoryless modes for any turbulent spectrum in the truth signal.*

To support the claim in Remark 7.5, we use the formula in (7.26) (B) for the mean signal from the truth model together with (7.35) and (7.30) to calculate that the strong damping algorithm produces the following estimate for the mean of the memoryless modes (assuming zero external forcing in (7.34)),

$$\text{(A)}\ \bar{\hat{u}}^h_{k_l,m+1|m+1} = \Gamma_i \sum_{k \in \mathcal{A}(\ell)} \hat{u}_k((m+1)\Delta t), \qquad P_o \leq i \leq P, \tag{7.36}$$

with

$$\text{(B)}\ \Gamma_i = r_{h,k_i} \Lambda(R_{m+1|m,P_o}, \hat{r}_{P_o}).$$

Note that in general the weights satisfy $\sum \Gamma_i < 1$. Consider, for illustration purposes, the difficult case of an equipartition spectrum where all the r_{h,k_i} are equal; then all the Γ_i are also identical and the SDAF approximation automatically distributes the compressed truth signal, $\sum_{k \in \mathcal{A}(\ell)} \hat{u}_k$, equally among the memoryless modes, which is intuitively consistent with equipartition of energy. Similarly, for a decaying spectrum, the SDAF approximation automatically provides less weight from the compressed truth signal in the memoryless modes with less turbulent energy in a weighted fashion roughly consistent with this spectrum.

7.3.2 The reduced Fourier domain Kalman filter

The RFDKF approximation is based on the intuitive idea that for a sufficiently rapid decay in the turbulent spectrum of the truth signal, the primary mode contains the most energy so only this mode should be actively filtered. Thus, RFDKF always trusts the dynamics in (7.25) for all the aliased modes $2 \le i \le P$ yielding a Kalman gain vector with the form

$$\vec{K}_{m+1} = (K_{m+1}, 0, 0, \ldots, 0)^T$$

while the primary mode, $\hat{u}^h_{k_1}$, is filtered by the one-dimensional algorithm,

$$\text{(A)}\ \hat{u}^h_{k_1,m+1|m} = F_{h,k_1}\hat{u}^h_{k_1,m|m} + \bar{F}_{k_1,m} + \sigma_{h,k_1,m+1}$$

$$\text{(B)}\ \bar{v}_{m+1} - \sum_{i=2}^{P} \bar{\hat{u}}^h_{k_i,m+1|m+1} = \bar{\hat{u}}^h_{k_1,m+1|m+1} + \hat{\sigma}^o_{m+1}.$$

This type of reduced filtering algorithm was introduced earlier for theoretical purposes in a different context by Grote and Majda (2006). Observability is always satisfied for RFDKF provide $0 < |F^h_1|$.

The computational advantage of RFDKF is apparent since only a one-dimensional Kalman filter is needed and the main input of the modes which are not observed by the sparse observations is to increase the resolution of the actual dynamics. In the rest of this chapter, we refer to the one-dimensional primary mode as the resolved mode since the single observation is available, and the trivially updated aliased modes as the unresolved modes since the observations are not available here. To simplify the jargon in the special case of the strong damping approximations with $P_o = 1$, we also refer to the moderately damped mode as the resolved mode and the memoryless modes as the unresolved modes.

7.3.3 Comparison of approximate filter algorithms

Compared with RFDKF which only trusts the dynamics for the unresolved (or aliased) modes, the SDAF approximation with $P_o = 1$ has an opposite design principle of "memoryless dynamics" but yields a nontrivial estimate for the mean on the unresolved modes based on the observations. For the special case with $P_o = 1$, the VSDAF uses the explicit Kalman gain vector in (7.36) to weight the unresolved dynamics and the effect of sparse observations in the filter dynamics as a natural blending between SDAF and RFDKF. Both the VSDAF with $P_o = 1$ and RFDKF algorithms require detailed Kalman filtering on the primary mode alone and satisfy observability. Furthermore, the VSDAF algorithm with $P_o = 1$ involves only modest additional processing through the explicit formulas in (7.36) beyond the RFDKF algorithm yet produces a reasonable state estimate for the unresolved Fourier modes. The filter performance of SDAF, VSDAF and RFDKF as well as FDKF is studied in Section 7.4 below.

7.4 New phenomena and filter performance for sparse regular observations

The filtering skill of the various filtering strategies depends on the observation time Δt, compared to the correlation time at each wavenumber in Fig. 7.2 as well as on the energy spectrum E_k. In this section, we highlight some of the most interesting results from our earlier work in Harlim and Majda (2008b). The setup has already been described in detail in the last two paragraphs of Section 7.2; there are 123 mesh points with observations at only 41 regularly spaced mesh points so $P = 3$ (see Fig. 7.1). We include some results from filtering the stochastically forced advection–diffusion equation for both filtering the damped stochastically forced advection–diffusion equation in (7.20) with fixed parameters

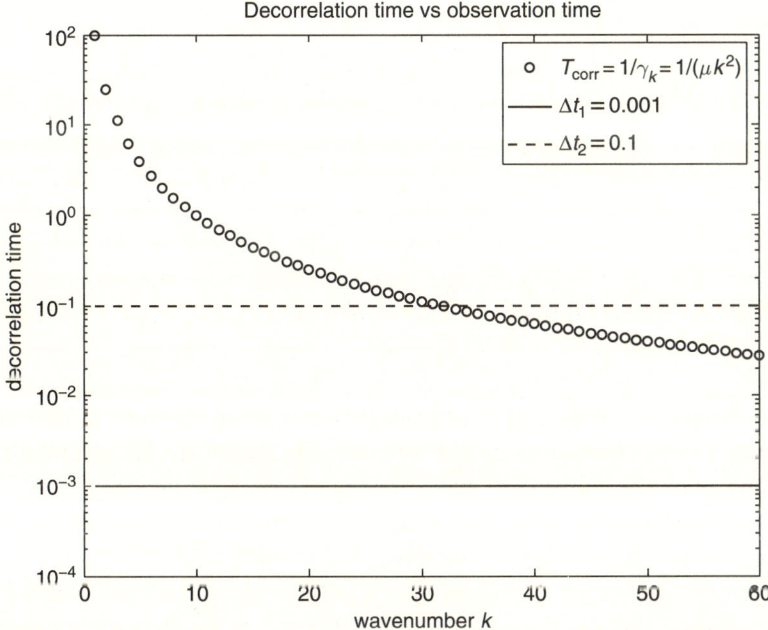

Figure 7.2 Decorrelation time of a selective damped signal from the advection–diffusion equation ($\gamma_k = \mu k^2$ for $\mu = 0.01$) compared with the observation times, both plotted as functions of model wavenumber k.

(7.23) and the stochastically forced weakly damped advection equation with fixed parameters (7.24) with the perfect model and with model error through an additional artificial diffusion as stated in Remark 7.4.

7.4.1 Filtering the stochastically forced advection–diffusion equation

Figure 7.2 provides an important diagnostic for filter skill which the reader should consult throughout this subsection as a guideline for the practical issues in 7(a)–(d). The general trend in filtering the advection–diffusion equation with fixed parameters $d = 0$, $\mu = 10^{-2}$ as in (7.23) is as follows:

- For the short observation time $\Delta t = 0.001$, where all Fourier modes in the true signal are substantially correlated, FDKF has significant skill and VSDAF has nearly comparable skill.
- For moderate observation times $\Delta t \geq 0.1$ so that some of the high-wavenumber modes in the truth signal are completely decorrelated, for the rough spectrum, $E_k = 1$, FDKF, VSDAF and SDAF all have comparable significant skill while RFDKF has significantly less skill.
- For $\Delta t \geq 0.1$ and the smoother spectrum $E_k = k^{-5/3}$, all the filters FDKF, VSDAF, SDAF and RFDKF have significant comparable skill.
- In almost every regime, the spatially based ensemble filter such as ETKF produces low skill (see Section 9.5 in Chapter 9).

Thus, if one is interested in computational efficiency, these results suggest the use of the alternative filters VSDAF, SDAF and RFDKF in various regimes of observation time and energy spectrum to address the practical issues in 7(a)–(d).

We find that, as shown in Fig. 7.3, ETKF with $K = 150, r = 40\%$ (see Chapter 8 for details of ETKF), $\Delta t = 0.1, r^o = 2.05$ (corresponds to 0.05 unit in Fourier space) diverges with resonant periodic forcing $\bar{F}_m$ as defined in the second term in the right-hand side of (5.16) with (5.17) even for the $k^{-5/3}$ spectrum and for the selectively damped advection–diffusion equation. What is intriguing is that in this large signal-to-noise ratio, the unfiltered solution is better than ETKF with RMS average 2.4 and average spatial correlation 0.93. In fact, the correlation skill converges to one as time grows. Hence, the Kalman filter seems to be redundant if one is interested only in the large-time asymptotic state. However, for practical real-time prediction problems the accurate transient filtering of the emerging location and amplitude of the large spike has obvious practical importance. As shown in Fig. 7.3, at transient times (after 100 and 500 assimilation cycles) as well as after 1000 assimilation cycles, SDAF has a tremendously high filtering skill.

In Fig. 7.4, we show numerical results for one and 10 realizations and compare them to the ideal situation with perfect realization for a regime with $E_k = 1$, decaying mean, and $\Delta t = 0.1$. This regime is considered to be a difficult test since the signal-to-noise ratio is rather small and the uniform spectrum will ensure rather high fluctuations. Filtering with a finite number of realizations corresponds to propagating each ensemble member

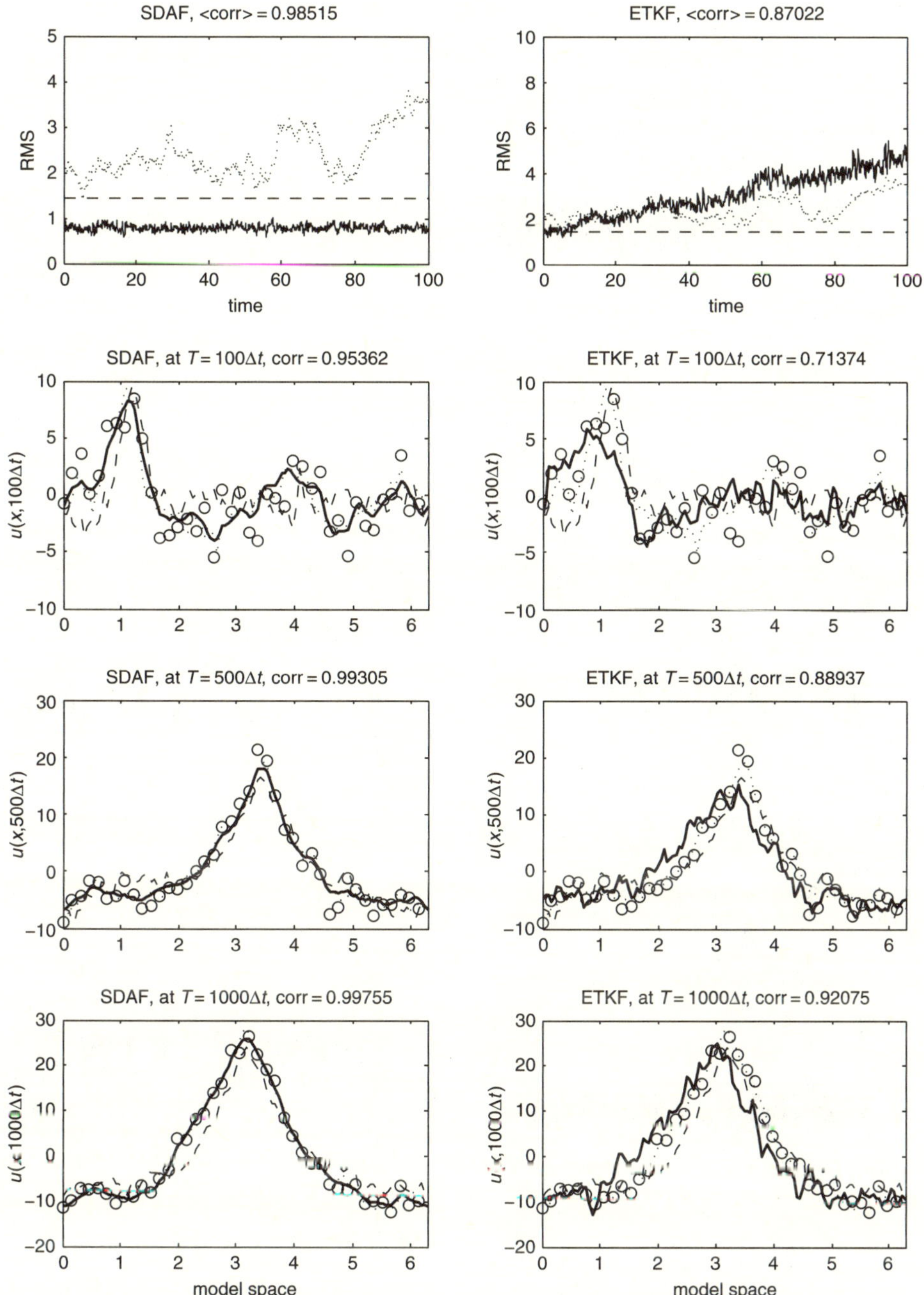

Figure 7.3 Advection–diffusion equation for $\Delta t = 0.1$, $E_k = k^{-5/3}$, and resonant periodic forcing. The first column shows the results with SDAF and the second column with ETKF. The first row shows the RMS errors as functions of time for no filter in solid line, the unfiltered solutions (dotted) and observation noise size (dashes). The last three rows show the snapshots of the filtered solution (thick solid) at time 10, 50, and 100, respectively, and compare it to the unfiltered solution (dashes), the true state (dotted) and the sparse observations (circle).

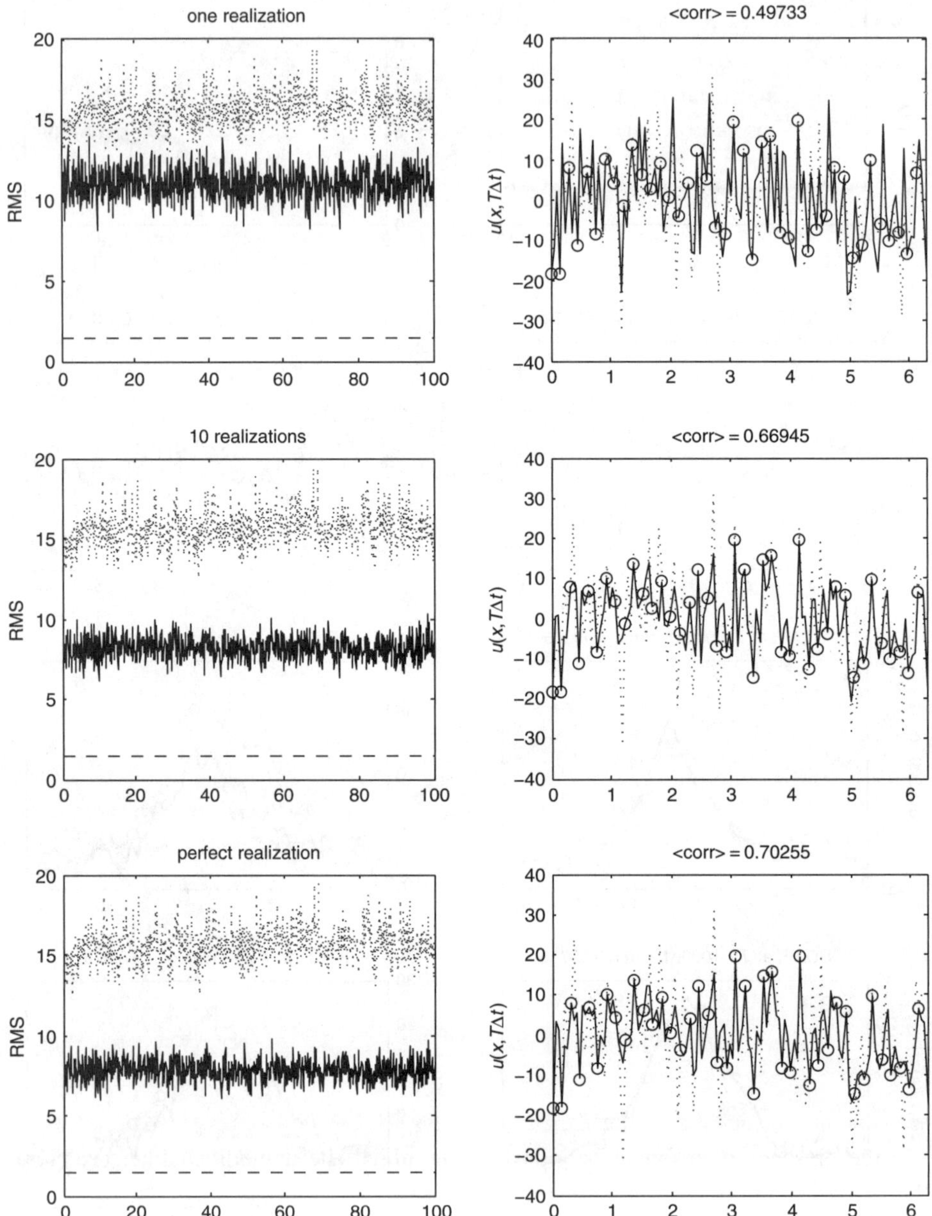

Figure 7.4 Advection–diffusion equation for $\Delta t = 0.1$, $E_k = 1$ and decaying mean. The first row shows the RMS errors as functions of time and a snapshot at time 100 for simulation with one realization, the second row for simulation with 10 realizations and the third row for simulation with perfect realization. The filtered solution in all panels is depicted in solid, the observation errors in the first column are denoted in dashes while the observations in the second column are denoted in circles. The dotted lines denote the unfiltered solutions in the first column and the true signals in the second column.

with (7.25) and to updating each ensemble member through the same observations with (7.30) with $\bar{\vec{\hat{u}}}$ replaced by the kth ensemble member $\vec{\hat{u}}^k$. The perfect realization corresponds to propagating the mean of (7.25), i.e. $\vec{\sigma} = 0$, and to updating the mean exactly with (7.30). In this sense, the mean in the perfect realization corresponds to the theoretical mean while the mean in the finite ensemble simulations corresponds to the ensemble average; when the ensemble size is small, the sampling errors prevent these two means from being identical. However, from Fig. 7.4, we see that the filter performance for the simulation with 10 realizations is almost indistinguishable from the simulation with perfect realization. The sampling effect is noticable when the numerical simulation is employed with one realization; however, the error difference is relatively small and acceptable considering that we only use a single realization. Thus, the sampling effect is negligible in filtering sparse observations with small ensemble size provided the ensemble has about 10 members. This result is robust for all of the alternative filters we test in Harlim and Majda (2008b) in all regimes.

7.4.2 Filtering the stochastically forced weakly damped advection equation: Observability and model errors

We now discuss the filter performance for sparse observations of a turbulent signal generated by the weakly damped advection equation in (7.20) with $d = 10^{-2}, \mu = 0$ as in (7.24). One additonal feature in this filtering problem is that for an appropriate Δt as in (7.19), the filtering problem with the perfect model is non-observable. Also with small uniform damping, unlike the situation in the selective damping case as displayed in Fig. 7.2, all observation times Δt here are below the correlation time at all spatial wavenumbers. These features combine to make this problem an extremely difficult test bed especially with an equipartition energy spectrum. Our simulations with non-resonant periodic forcing $\bar{F}_m$, as defined in the second term in the right-hand side of (5.16) with $\omega_o = \omega_k + 1/2$, suggest that filtering with an artificial diffusive term ($\mu = 0.01$) as pointed out in Remark 7.4 improves the filter skill compared to filtering with a perfect model ($\mu = 0$); this improved performance is illustrated in Fig. 7.5 for the observable time $\Delta t = 0.5$ and $E_k = k^{-5/3}$. Similar results hold for non-observable time for the decaying mean and non-resonant periodically forced signals (Harlim and Majda, 2008b).

In the presence of resonant periodic forcing, all of the unmodified filters including RFDKF, which fully satisfy the observability condition, have an average spatial correlation of 0.99; however, their RMS errors are much larger than those of the unfiltered solutions as illustrated for RFDKF in the upper panel of Fig. 7.6. This poor performance is caused by the lack of practical controllability. To justify this claim, in Fig. 7.6 we show the RMS errors as functions of time and also snapshots of filtered solutions after 1000 assimilation cycles in the regime with $\Delta t = 0.5$, $E_k = k^{-5/3}$, and resonant periodic forcing for RFDKF. The first row is for the unmodified filter with $\mu = 0$, where $r_{k_1} \sim 10^{-4}$ for aliasing sets $\mathcal{A}(\ell)$ with $\ell \geq 5$ when computed with (6.21); the RFDKF filter is unmodified

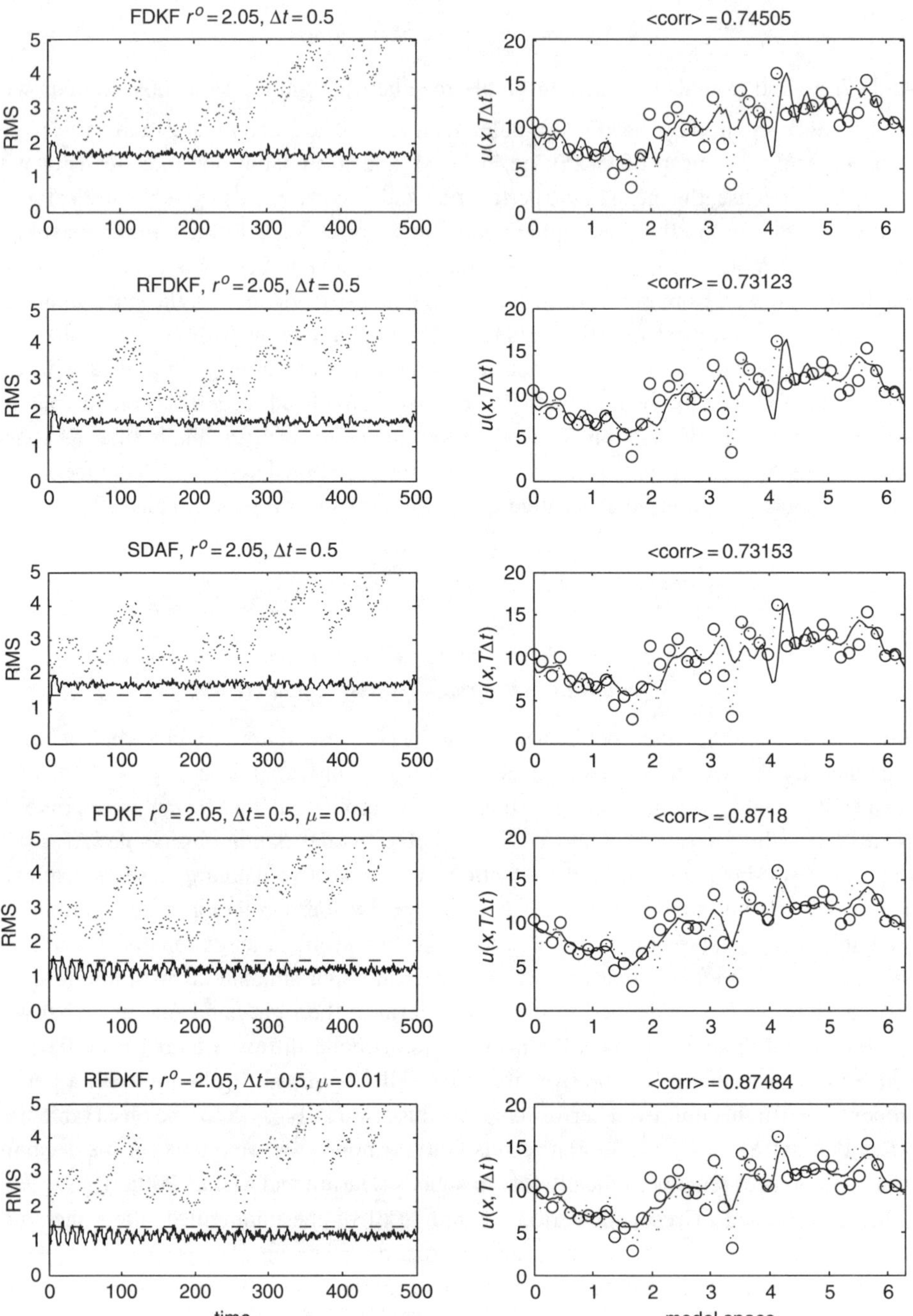

Figure 7.5 Weakly damped advection equation for $\Delta t = 0.5$, $E_k = k^{-5/3}$, and non-resonant periodic forcing. The first column shows RMS errors as functions of time for the unfiltered solution (dotted), the filtered solutions (solid) and observation noise size (dashes). The second column shows the snapshots of the filtered solution (corresponds to the filtering strategy stated in the subtitle of each panel in the first column for the same row) as functions of model spatial domain (solid), the true state (dotted) and the sparse observations (circle), all at time 500.

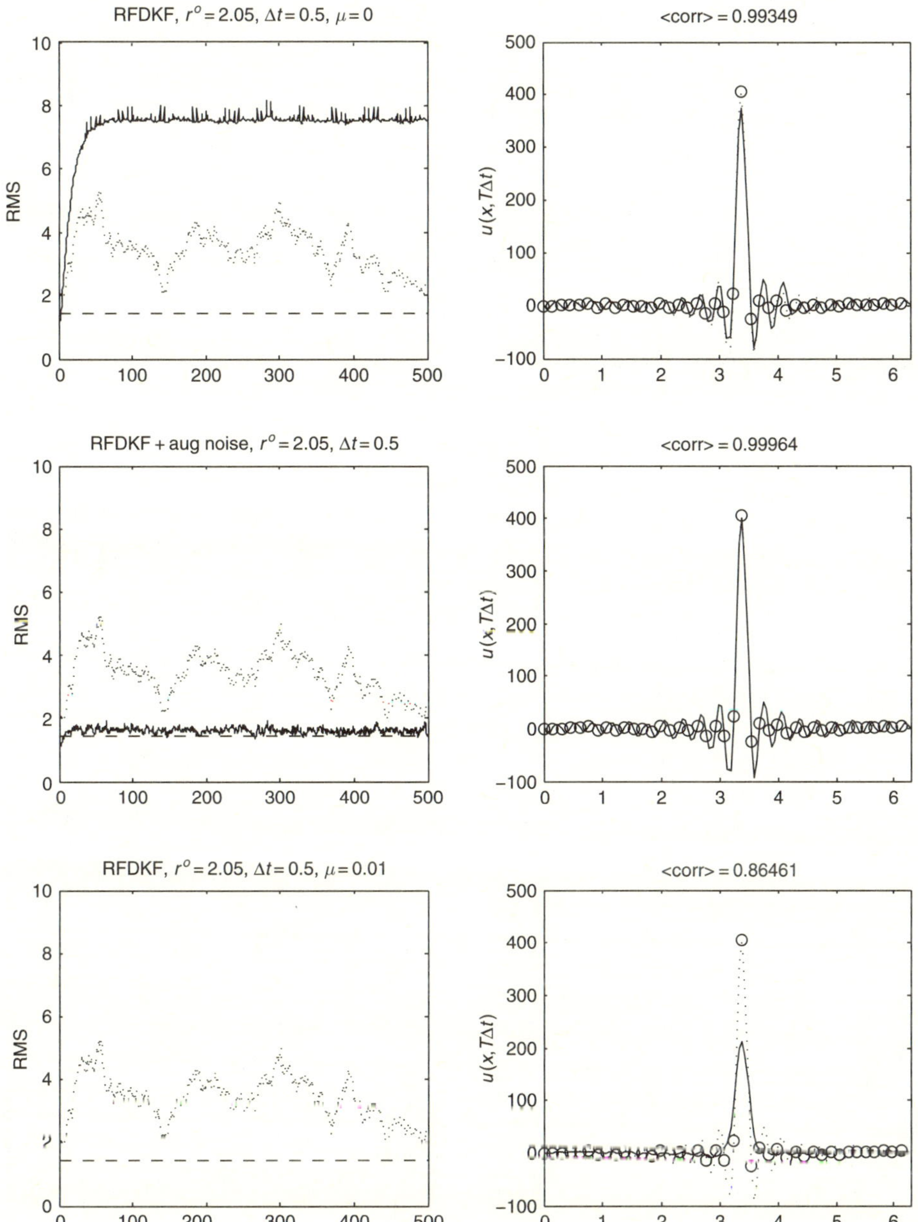

Figure 7.6 RFDKF (first row), RFDKF with increased noise (second row) and the modified RFDKF (third row) on the weakly damped advection equation for $\Delta t = 0.5$, $E_k = k^{-5/3}$, and resonant periodic forcing. The first column shows RMS errors as functions of time for the unfiltered solution (dotted), the filtered solutions (solid) and observation noise size (dashes). The second column shows the snapshots of the filtered solution (corresponds to the filtering strategy stated in the subtitle of each panel in the first column for the same row) as functions of model spatial domain (solid), the true state (dotted) and the sparse observations (circle), all at time 500. The RMS error for the modified RFDKF (third row) goes beyond the abscissas; it is on the order of 100.

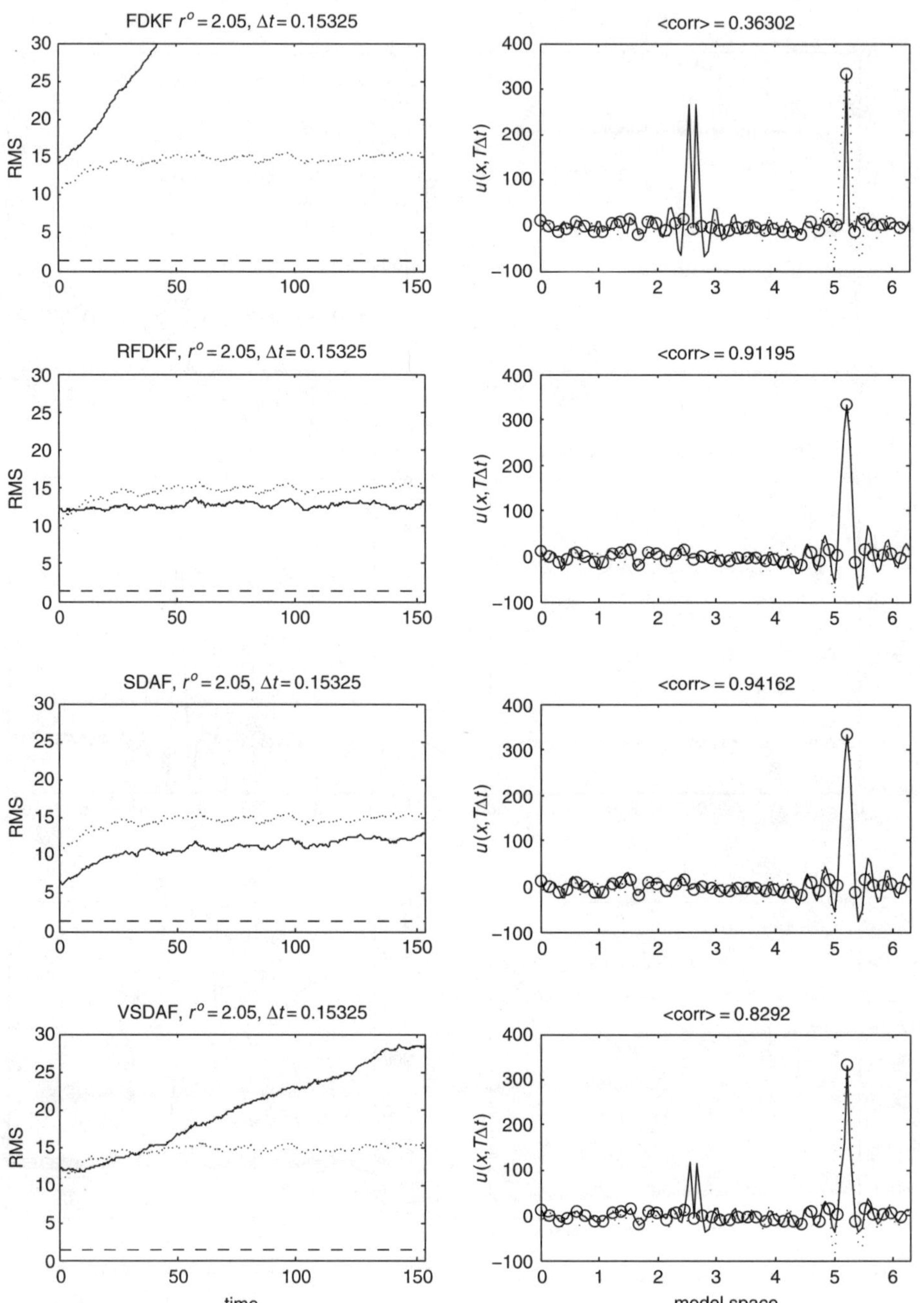

Figure 7.7 Weakly damped advection equation for $\Delta t = 2\pi/(2M + 1) = 0.1532$, $E_k = 1$, and resonant periodic forcing. The first column shows RMS errors as functions of time for the unfiltered solution (dotted), the filtered solutions (solid) and observation noise size (dashes). The second column shows the snapshots of the filtered solution (corresponds to the filtering strategy stated in the subtitle of each panel in the first column for the same row) as functions of model spatial domain (solid), the true state (dotted) and the sparse observations (circle), all at time 153.2.

as well in the second row; however, we now increase the system noise variance of mode $k = 5, \ldots, M$ (that is, the first component of aliasing sets $\mathcal{A}(\ell)$ with $\ell \geq 5$) by 0.01. Here, we see a significant improvement of the filter with larger system noise variances. This issue has already been pointed out in Chapters 2 and 6 where we utilized information criteria for inflating the system noise variances. The modified filter with $\mu = 0.01$ (see third row of Fig. 7.6) introduces large model errors in the presence of resonant periodic forcing. In this large signal-to-noise ratio regime for filtering, the modified models substantially damp the large localized spike in the mean solution as could be expected for any artificially diffusive modified model.

In Fig. 7.7, we consider a tough regime with non-observable time $\Delta t = 2\pi/(2M+1) = 0.1532$ and rough energy spectrum $E_k = 1$ with resonant periodic forcing so that the signal-to-noise ratio is large. Here, we see strong filter divergence for FDKF and VSDAF with unacceptably large ghost spikes in the filtered solution; the filter divergence in VSDAF was explained in detail in section 4.2 of Harlim and Majda (2008b) through scaling analysis and the interested reader can consult that reference. Notice that the filter divergence that occurs in FDKF, VSDAF and ETKF is avoided with the other alternative strategies, RFDKF and SDAF. In Fig. 7.7, we see that the better solutions are produced by the reduced filters, especially SDAF with average spatial correlation 0.94 and RMS error 10.96 whereas the unfiltered solutions have average spatial correlation of 0.89 and RMS error 14.14. In this regime, we do not show results with modified filters as in Fig. 7.6 since the presence of a large signal-to-noise ratio introduces significant model errors due to the artificial augmented diffusion as noted earlier (see Fig. 7.6).

It is useful to summarize the role of failure of observability in these complex PDE models. The results depend on the signal-to-noise ratio of the true mean signal being filtered (Harlim and Majda, 2008b); for a decaying mean signal with the perfect model, as in Chapter 3, the failure of observability does not affect filter performance. On the other hand, as shown in Fig. 7.7, FDKF suffers from strong filter divergence at unobservable times with resonant periodic forcing and a large signal-to-noise ratio. At the end of Chapter 3, we showed the lack of filtering skill with model error defined by forward Euler when practical observability is violated. In Chapter 10, we will show examples involving model error and the failure of observability for a prototype nonlinear system with multiple time-scales.

8

Filtering linear stochastic PDE models with instability and model error

A major difficulty in accurate filtering of noisy turbulent signals with many degrees of freedom is model error; the true signal from nature is processed for filtering and prediction through an imperfect model in which numerical approximations are imposed and important physical processes are parametrized due to inadequate resolution or incomplete physical understanding. In Chapters 2 and 6, we discussed the filtering performance on a canonical linear stochastic PDE model for turbulence in the presence of model error through discrete numerical solvers. Subsequently, in Chapter 7 we assessed the filtering performance with model error that arises from reduced filtering strategies that are used when the observations are sparse.

In this chapter, we discuss filtering with model error which arises from unstable turbulent processes that are often observed in nature such as the intermittent burst of baroclinic instability in the mid-latitude atmosphere and ocean turbulent dynamics and which might be hidden from the forecast model. In the study below, we will simulate an intermittent burst of instability with two-state Markov damping coefficients; we allow the damping coefficients of some Fourier modes in the turbulent model (SPDE in Chapter 5) to be negative (or unstable) on some random occasions with a positive long-time average damping (i.e. an overall stable dynamics). We will prescribe the switching times between the two (stable and unstable) states to be randomly drawn from exponential distributions.

Before we describe the turbulent model with instability, we review the two-state continuous-time Markov process. As a primary example we will add this two-state Markov instability/stability mechanism to the toy model of barotropic Rossby waves of Section 5.4 with stable stochastic dynamics. In this example, the model errors are introduced through our lack of information about the onset time and the duration of the instability regime. Subsequently, we will discuss the numerical results on filtering the intermittently unstable solutions of this newly described model with a perfectly specified damping model and with an imperfect model. The imperfect model, called the mean stochastic model (MSM), parametrizes the unknown damping coefficients with constant long-time average damping coefficients and thus is a stable stochastic model. We will check the filter performance for plentifully and sparsely observed signals with the Fourier domain Kalman filter (FDKF) of Chapters 6, and with the reduced Fourier domain Kalman filter (RFDKF) of Chapter 7. In the context of filtering turbulent dynamical systems, as we will see in Chapter 12, a suitable version of the MSM approach can have significant skill by fitting the parameters to some bulk climatological statistical feature of nature's signal; however in the context of real-time prediction, such a strategy performs poorly, as we will see below, when intermittent transitions to instability occur. As a remedy, we will consider a more sophisticated approach that corrects the model errors "on-the-fly" with an exactly solvable stochastic parametrization strategy in Chapter 13.

8.1 Two-state continuous-time Markov process

Here, we give a brief theoretical discussion about a two-state continuous-time Markov process. For a much more extensive discussion of finite state Markov chains, see Lawler (1995). Suppose a stochastic process X_t can take only one of the two values

$$S = \{s_{\mathrm{st}}, s_{\mathrm{un}}\},$$

which correspond to stable and unstable dynamics, for example. The process has the same value x until it changes to another value y, $x \neq y$, at some random time t. We consider a process with Markov property

$$P(X_t = y|X_r,\ 0 \leq r \leq s) = P(X_t = y|X_s),$$

and time-homogeneity

$$P(X_t = y|X_s = x) = P(X_{t-s} = y|X_0 = s),$$

for $x, y \in S$. Due to these two properties the process is fully determined by the transition probabilities $P(X_t = y|X_0 = x)$. Next, we define ν to be the rate of change from the stable state s_{st} to the unstable state s_{un}. Similarly, we define μ to be the rate of change from the unstable state s_{un} to the stable state s_{st}. These rates define the following local transition probabilities for small Δt

$$\begin{aligned}
P(X_{t+\Delta t} = s_{\mathrm{un}}|X_t = s_{\mathrm{st}}) &= \nu\Delta t + o(\Delta t),\\
P(X_{t+\Delta t} = s_{\mathrm{st}}|X_t = s_{\mathrm{un}}) &= \mu\Delta t + o(\Delta t),\\
P(X_{t+\Delta t} = s_{\mathrm{st}}|X_t = s_{\mathrm{st}}) &= 1 - \nu\Delta t + o(\Delta t),\\
P(X_{t+\Delta t} = s_{\mathrm{un}}|X_t = s_{\mathrm{un}}) &= 1 - \mu\Delta t + o(\Delta t),
\end{aligned}$$

where $o(\Delta t)$ denotes a function smaller than Δt, i.e. $o(\Delta t)/\Delta t \to 0$ as $\Delta t \to 0$. We write the differential equation for the transition probability from stable to unstable regimes $p_t(s_{\mathrm{st}}, s_{\mathrm{un}}) = P(X_t = s_{\mathrm{un}}|X_0 = s_{\mathrm{st}})$

$$\begin{aligned}
p_{t+\Delta t}(s_{\mathrm{st}}, s_{\mathrm{un}}) &= p_t(s_{\mathrm{st}}, s_{\mathrm{un}})p_{\Delta t}(s_{\mathrm{un}}, s_{\mathrm{un}}) + p_t(s_{\mathrm{st}}, s_{\mathrm{st}})p_{\Delta t}(s_{\mathrm{st}}, s_{\mathrm{un}})\\
&= p_t(s_{\mathrm{st}}, s_{\mathrm{un}})[1 - \mu\Delta t] + p_t(s_{\mathrm{st}}, s_{\mathrm{st}})\nu\Delta t + o(\Delta t).
\end{aligned}$$

After regrouping terms, we obtain a finite difference equation

$$\frac{p_{t+\Delta t}(s_{\mathrm{st}}, s_{\mathrm{un}}) - p_t(s_{\mathrm{st}}, s_{\mathrm{un}})}{\Delta t} = -\mu p_t(s_{\mathrm{st}}, s_{\mathrm{un}}) + \nu p_t(s_{\mathrm{st}}, s_{\mathrm{st}}) + o(1).$$

In the limit $\Delta t \to 0$, we obtain the following differential equation with initial condition

$$\partial_t p_t(s_{\rm st}, s_{\rm un}) = -\mu p_t(s_{\rm st}, s_{\rm un}) + \nu p_t(s_{\rm st}, s_{\rm st}),$$
$$p_t(s_{\rm st}, s_{\rm un})|_{t=t_0} = 0.$$

Similarly, we obtain the differential equations for all other transition probabilities, $p_t(s_{\rm un}, s_{\rm st})$, $p_t(s_{\rm st}, s_{\rm st})$ and $p_t(s_{\rm un}, s_{\rm un})$. We combine the transition probabilities into the transition probability matrix

$$P_t = \begin{pmatrix} p_t(s_{\rm st}, s_{\rm st}) & p_t(s_{\rm st}, s_{\rm un}) \\ p_t(s_{\rm un}, s_{\rm st}) & p_t(s_{\rm un}, s_{\rm un}) \end{pmatrix}.$$

Then, the differential equation for the matrix P_t and initial condition become

$$\frac{\partial P_t}{\partial t} = P_t A,$$
$$P_t|_{t=t_0} = I,$$

where the matrix A has the rates of change from one state to another

$$A = \begin{pmatrix} -\nu & \nu \\ \mu & -\mu \end{pmatrix}.$$

Matrix A is called an infinitesimal generator of the chain. Then the row vector of probabilities $\bar{p}(t)$ of each state satisfies the same equation

$$\partial_t \bar{p}(t) = \bar{p}(t)A$$

with solution

$$\bar{p}(t) = \bar{p}(0)e^{tA}.$$

Then, the probabilities for the system to be in stable and unstable regimes become

$$p_{\rm st}(t) = e^{-(\nu+\mu)t} p_{\rm st}(t_0) + \frac{\mu}{\nu+\mu}\left(1 - e^{-(\nu+\mu)t}\right),$$
$$p_{\rm un}(t) = e^{-(\nu+\mu)t} p_{\rm un}(t_0) + \frac{\nu}{\nu+\mu}\left(1 - e^{-(\nu+\mu)t}\right).$$

Now, it is easy to find the equilibrium distribution

$$p^{\rm eq} = \left(\frac{\mu}{\nu+\mu}, \frac{\nu}{\nu+\mu}\right). \tag{8.1}$$

On the other hand, we can give an interpretation of the Markov chain using an exponential distribution of switching times. Suppose the system is in the stable state $s_{\rm st}$ at the initial time t_0. We define

$$T_{\rm st} = \inf\{t : X_t = s_{\rm un}, t > t_0\}.$$

The Markov property can be used to see that $T_{\rm st}$ must have a loss of memory property and therefore it must have an exponential distribution (Lawler, 1995). Notice also that

$$P(T_{\mathrm{st}} \leq \Delta t) = P(X_{t+\Delta t} = s_{\mathrm{un}} | X_t = s_{\mathrm{st}}) = \nu \Delta t + o(\Delta t).$$

This implies that T_{st} is exponentially distributed with parameter ν, where ν is the rate of leaving the stable state. Note that an exponentially distributed random variable Z with rate b is characterized by the density function

$$p(z) = be^{-bz}, \quad z > 0.$$

The corresponding distribution is given by

$$P(Z < z) = \int_0^z p(s)ds = 1 - e^{-bz}.$$

Therefore, the probability of switching to the unstable state to occur before time t is

$$P(T_{\mathrm{st}} < t) = 1 - e^{-\nu t}.$$

Similarly, the probability of switching to the stable regime to occur before time t is

$$P(T_{\mathrm{un}} < t) = 1 - e^{-\mu t},$$

where μ is the switching rate to leave the unstable regime.

8.2 Idealized spatially extended turbulent systems with instability

Recall the prototype model for turbulent systems described in Chapter 5,

$$\frac{\partial u(x,t)}{\partial t} = \mathcal{P}\left(\frac{\partial}{\partial x}\right) u(x,t) - \gamma\left(\frac{\partial}{\partial x}\right) u(x,t) + f(x,t) + \sigma(x)\dot{W}(t). \tag{8.2}$$

In (8.2), the operators $\mathcal{P}(\partial_x)$ and $\gamma(\partial_x)$ are defined such that at a given wavenumber k, their eigenvalues are given by $\tilde{p}(ik)$ and $\gamma(ik)$, respectively, with eigenvector e^{ikx}. Here, $\sigma(x)\dot{W}(t)$ is a Gaussian statistically stationary spatially correlated scalar random field where $\dot{W}(t)$ denotes white noise in time, and $f(x,t)$ is a known deterministic forcing. In the numerical example below, we only consider $f = 0$. In Chapter 13, we will consider nonzero forcing and introduce a second source of model error by purposely not specifying the forcing.

Practically, we discretize (8.2) at $2N + 1$ equally spaced grid points $x_j = jh$, where $(2N + 1)h = 2\pi$. Thus, each Fourier mode, $\hat{u}_k$, solves the linear Langevin equation

$$d\hat{u}_k(t) = [\tilde{p}(ik) - \gamma(ik)]\,\hat{u}_k(t)\,dt + \sigma_k\,dW_k(t). \tag{8.3}$$

We assume that $\tilde{p}(ik) = i\omega_k$ is wavelike where $-\omega_k$ is the real-valued dispersion relation while $\gamma(ik)$ represents a stable or dissipative process when it is positive and an unstable process when it is negative. As in Section 5.4, we consider the barotropic Rossby wave dispersion relation $\omega_k = 8.91k^{-1}$; this choice of frequency suggests that the lowest wavenumber, $k = 1$, has an oscillation period of roughly $2\pi/\omega_1 = 17$ hours.

In Chapters 5–7, we considered stable systems, $\gamma(ik) > 0$, at all time with constant γ. Here, we allow $\gamma(ik)$ for modes 4 and 5 to alternate randomly in time between a positive and a negative value with a time average damping $\bar{d} > 0$. Simultaneously, we apply a stronger damping on modes 1–3 when modes 4–5 are unstable and a weaker damping on modes 1–4 when modes 5–6 are stable. We also choose the damping coefficient for modes 1–3 such that its average is exactly $\bar{d} > 0$ on each mode. Under these assumptions, the statistical equilibrium distribution for (8.3) exists provided $f(x,t) = 0$ and is Gaussian with mean zero and variance E_k, defining the climatological energy spectrum

$$E_k = \frac{\sigma_k^2}{2\bar{d}}. \tag{8.4}$$

Following Section 5.4, in our numerical experiments, we set the energy spectrum to $E_k = k^{-3}$ based on the observations (Majda, 2000b). Also, as in Section 5.4, we also choose an equilibrium damping strength $\bar{d} = 1.5$ such that the physical space correlation function

$$\mathcal{R}(\tau) = \frac{\sum_{k=-\infty}^{\infty} R_k(\tau)}{\sum_{k=-\infty}^{\infty} E_k} = e^{-\bar{d}\tau} \frac{\sum_{k=-\infty}^{\infty} E_k e^{-i\omega_k \tau}}{\sum_{k=-\infty}^{\infty} E_k} \tag{8.5}$$

decays after three days (Lorenz, 1996).

This two-state damping mechanism on modes 1–5 models the energy flux from the intermediate scales to the largest scales in the stable regime and vice versa in the unstable regime. For our numerical example below, we choose the damping for modes 1–3 to take values of either $d_w = 1.3$ (weak damping) or $d_s = 1.6$ (strong damping). For the modes 4–5, we consider damping coefficients that alternate between $d^+ = 2.27$ (stable phase) and $d^- = -0.04$ (unstable phase). In the stable regime, we apply the weaker damping, d_w, on modes 1–3 and the positive damping, d^+, on modes 4–5. In the unstable regime, we apply the stronger damping, d_s, on modes 1–3 and the negative damping, d^-, on modes 4–5. In both regimes, the damping coefficients for modes 6–N are constant, $\bar{d} = 1.5$. We also calibrate the two-state Markov process to spend on average of 10 days in the stable regime and five days in the unstable regime; these averaged switching times are parametrized by the switching rate $\nu = 0.1$ for switching from a stable to an unstable regime and vice versa with a switching rate $\mu = 0.2$. Then, we have the same average value of damping on every mode with the explicit equilibrium measure in (8.1). On modes 1–3, we have

$$\bar{d} = \frac{\nu d_w + \mu d_s}{\nu + \mu}.$$

In order to find the specific values of d_w and d_s, we use this equation together with the constraint $0 < d_w < d_s$ to find that

$$d_w = \frac{1}{\nu}((\nu + \mu)\bar{d} - \mu d_s),$$

where d_s can have any value from the interval

$$\bar{d} < d_s < \left(1 + \frac{\nu}{\mu}\right)\bar{d}.$$

Similarly, for modes 4 and 5, we find d^+ and d^- from

$$\bar{d} = \frac{\nu d^- + \mu d^+}{\nu + \mu}.$$

Therefore, we obtain

$$d^- = \frac{1}{\nu}((\nu + \mu)\bar{d} - \mu d^+).$$

In order to satisfy $d^- < 0 < d^+$, we set

$$d^+ > \left(1 + \frac{\nu}{\mu}\right)\bar{d}.$$

We tune these parameters such that the fluctuation of the total energy is more or less uniform since we do not expect the total energy to change its value drastically.

In Fig. 8.1, we demonstrate the evolution of modes $\hat{u}_1$ and $\hat{u}_5$ together with the corresponding damping parameters d_1 and d_5. Recall that d_1 takes values from $\{d_w, d_s\}$ and d_5 takes values from $\{d^+, d^-\}$. In Fig. 8.1, we see that the amplitude of mode 5 indeed grows

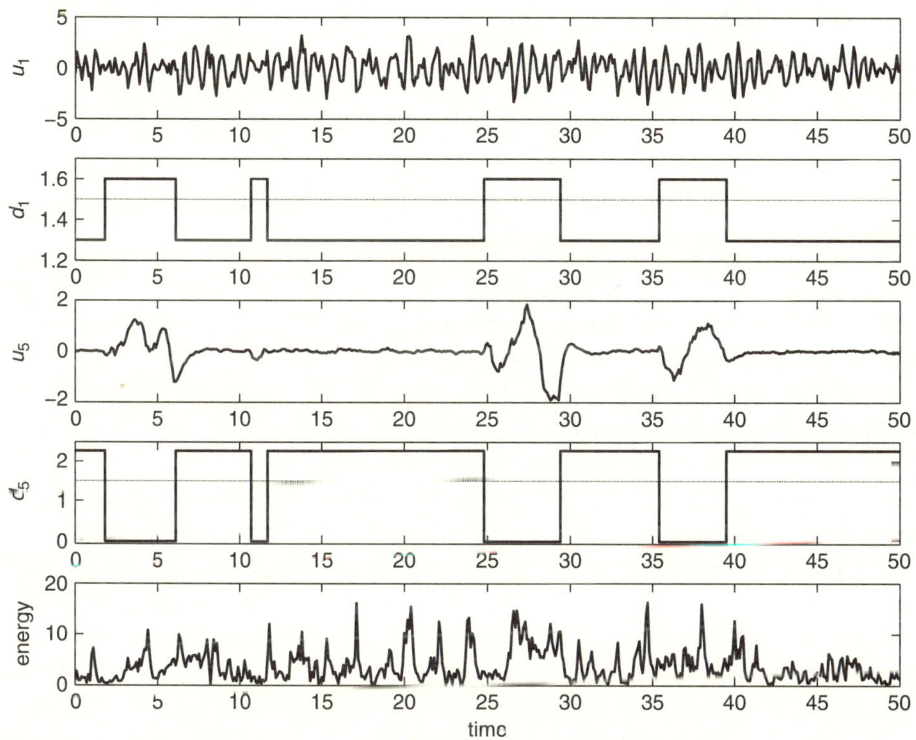

Figure 8.1 Panels 1–4: evolution of both stable $\hat{u}_1$ and unstable $\hat{u}_5$ modes together with the corresponding damping parameters. Panel 5: evolution of the total energy.

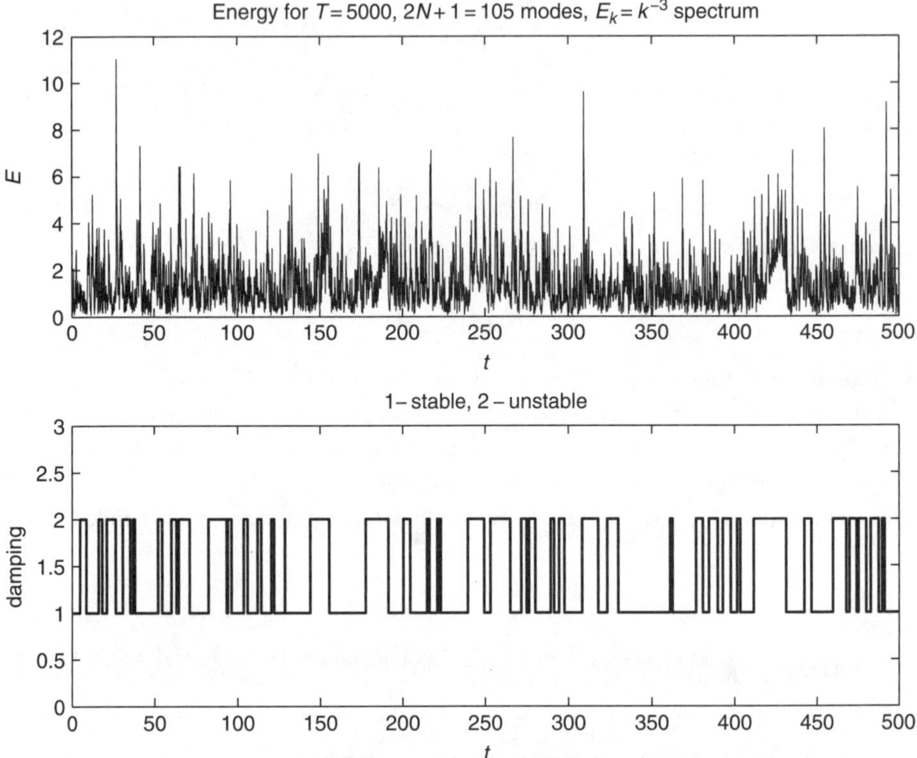

Figure 8.2 Upper panel: total energy $E(t)$; lower panel: damping regime.

in the unstable regime with negative damping $d^- = -0.04$. In Fig. 8.2, we observe the evolution of the total system energy, numerically simulated with $N = 52$. We again note that the system gains a significant amount of energy during the unstable regime and loses excessive energy while being strongly damped in the stable damping regime. In Fig. 8.3, we plot the physical space picture of the coherent Rossby wave train that emerges from a random background during an unstable phase of the dynamics.

In Fig. 8.4, we show the time-average energy spectrum of the model together with the spectra averaged only over stable and unstable regimes. We note that the peak that corresponds to the switching modes 4 and 5 is higher in the unstable regime when these modes are effectively driven by negative damping. On the other hand, there are no peaks at these modes in the stable regime since all modes are damped with positive damping. It is also interesting to point out the ratio of energy budgets that correspond to the stable and unstable modes. On average (over both stable and unstable regimes), 89% of the total energy is contained in modes 1, 2 and 3, which are always damped with the positive damping. The switching modes 4 and 5 contain around 10% of the total energy and the remaining 1% of total energy corresponds to modes 6–52. We will see that we can obtain accurate filtered solutions by weighting the posterior solutions fully to the dynamics (or prior states)

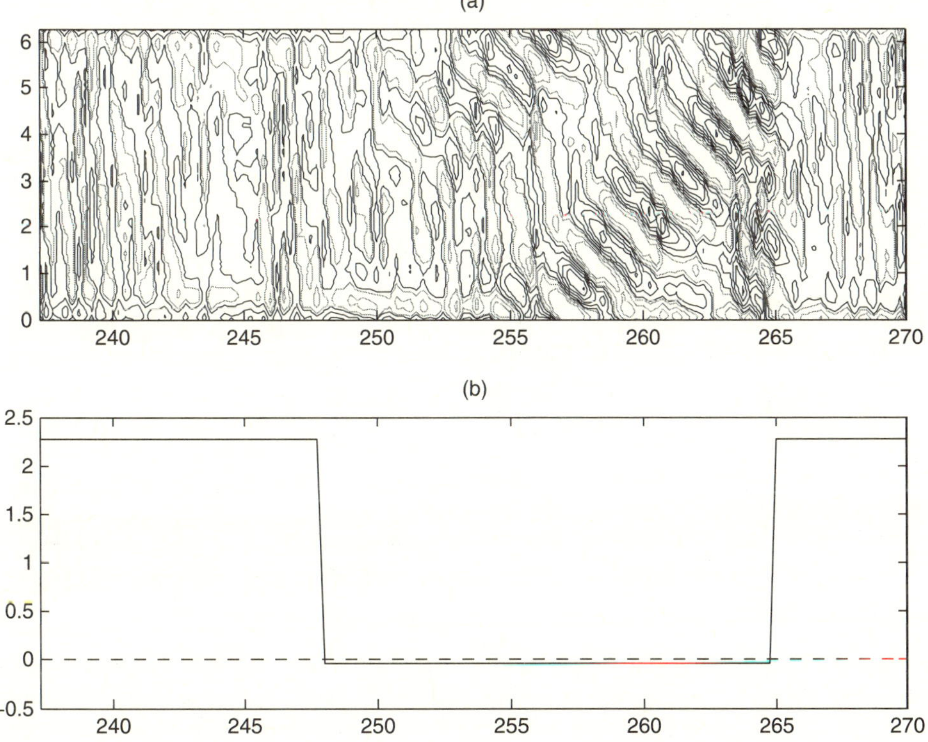

Figure 8.3 Spatial pattern for a turbulent system of the externally forced barotropic Rossby wave equation with instability through intermittent negative damping. Note the coherent wave train that emerges during the unstable regime.

on the least energetic modes with the cheap RFDKF from Chapter 7. It is also worthwhile mentioning that when the energy is averaged over the unstable regime (which takes roughly one-third of the total time), the unstable modes contain a significant fraction, roughly 24%, of the total energy, whereas modes 1–3 take 75% of the total energy. As a consequence of this, we will see later that it will be crucial to include the switching modes in the set of modes that are filtered, or otherwise the overall filter performance degrades significantly even though these modes are far from being the most energetic in an averaged sense; this is a manifestation of intermittency.

8.3 The mean stochastic model for filtering

As described in Chapter 5 in detail, a simple approach for filtering signals with intermittent instability is to use the mean stochastic model (MSM) (Harlim and Majda, 2008a, 2010a), which is a model that is based on the two equilibrium statistical quantities, the energy spectrum and the damping time; MSM is exactly the climatological stochastic model (CSM) in Harlim and Majda (2008a, 2010a). As in Chapter 5, the mean stochastic model (MSM) solves

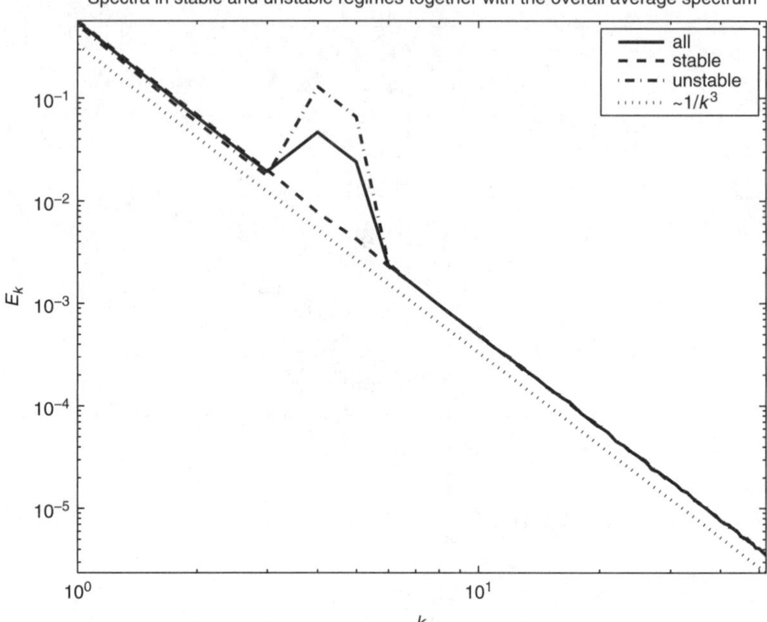

Figure 8.4 Energy spectrum for the model with switching damping. The solid line corresponds to the overall spectrum. The dash-dotted line corresponds to the spectrum averaged over all the unstable regimes (hence the high peak at modes 4 and 5). The dashed line corresponds to the spectrum averaged over all the stable regimes (no peaks). The dotted line represents the slope of $\sim 1/k^3$ and is shown for reference.

$$\frac{d\hat{u}_k(t)}{dt} = (-\bar{d} + \mathrm{i}\omega_k)\hat{u}_k(t) + \sigma_k \dot{W}_k(t) \tag{8.6}$$

for prediction and its first- and second-order statistics are used for filtering. Here $\bar{d} = 1.5$ is the average damping constant for all modes and σ_k is the noise strength specified by the equilibrium spectrum in (8.4); model errors are introduced naturally through the unknown two-state continuous-time Markov damping coefficient on the switching modes (1–5 in our example above). As a benchmark, we will also simulate the perfectly specified filter in which we prescribe the underlying true damping coefficients used in generating the true signals.

Recall the filtering of turbulent signals from Chapter 7 where sparse regularly spaced observations are available at every Pth grid point (Harlim and Majda, 2008b, 2010a). Note that when $P = 1$, we have observations at every grid point and this is the case of plentiful observation discussed in Chapter 6. Suppose simulated observations of the true signal that solves (8.2) are taken at $2M + 1$ regularly spaced grid points, i.e. at $\tilde{x}_j = j\tilde{h}$, $j = 0, 1, \ldots, 2M$, with $(2M+1)\tilde{h} = 2\pi$. When $M \le N$, where $(2N+1)h = 2\pi$ and h denotes the mesh spacing for the finite difference approximation, we have sparse regularly spaced observations since there are fewer observations than discrete mesh points. Here, as in Chapter 7, for simplicity in exposition, we assume that $\tilde{x}_j$ coincides with x_{q+jP} for some $q \in \mathbb{Z}$, and

$$P = \frac{2N+1}{2M+1} \tag{8.7}$$

defines the ratio of the total number of mesh points available to the number of sparse regular observation locations in (8.2). For example, consider $N = 52$ and $P = 5$. Then, we have $2N + 1 = 105$ model grid points. Accordingly, we have observations only at $(2N + 1)/P = 21$ spatial locations. This corresponds to having only $M = 10$ primary (most energetic) modes, where $2M + 1 = 21$. The sparse observation model in Fourier space, see Chapter 7, is given by

$$\hat{v}_{\ell,m} = \sum_{k \in \mathcal{A}(\ell)} \hat{u}_{k,m} + \hat{\sigma}^o_{\ell,m}, \quad |\ell| \le M, \tag{8.8}$$

where $\hat{\sigma}^o_{\ell,m}$ is a Gaussian noise with mean zero and variance $\hat{r}^0 = r^o/(2M + 1)$ and

$$\mathcal{A}(\ell) = \{k | k = \ell + (2M + 1)q, |k| \le N, q \in \mathbb{Z}\} \tag{8.9}$$

is the aliasing set of wavenumber ℓ, which has P components (see chapter 7 or Majda and Grote (2007) and Harlim and Majda (2008b, 2010a) for details). In the example above, we have $\mathcal{A}(1) = \{1, 22, 43, -20, -41\}$ for $P = 5$ and $M = 10$ and the primary (or the most energetic) mode in $\mathcal{A}(1)$ is mode 1 since $E_1 > k^{-3}$ for all $k > 1$.

The important point emphasized in Chapter 7 is that modes belonging to different aliasing sets are completely independent. However, modes within one aliasing set are coupled through the observation model in (8.8). Therefore, the filtering problem is reduced into $2M + 1$ independent filtering problems; each involves P components in each aliasing set. In Table 8.1, we show the number of aliasing sets and the number of observations for a given P.

In our numerical experiments, we apply the standard Kalman filter algorithm to observation (8.8) with a perfectly specified model and with model errors through MSM in (8.6). We denote these two algorithms as KF and KFME (KF with model error), respectively. In our numerical examples below, we also consider the reduced Fourier domain Kalman filter (RFDKF) in which we set the posterior mean state of modes 6–52 to equal the prior mean state (see Chapter 7 for details). We implement RFDKF both with the perfect model and MSM. This is motivated by the fact that the energy of the high modes in our barotropic Rossby wave toy model constitutes only 1% of the total energy. The RFDKF simplifies the filtering problem to only M scalar filters; each corrects the primary mode of each aliasing set with the following adjusted observation model

$$v'_{\ell,m} = \hat{v}_{\ell,m} - \sum_{k \in A(\ell), k \ne \ell} \bar{\hat{u}}_{k,m|m} + \hat{\sigma}^o_{\ell,m}, \tag{8.10}$$

where ℓ with $|\ell| \le M$ denotes the primary mode. Notice that when $P \ge 15$ (see Table 8.1), the switching modes 4–5 are not the primary modes and RFDKF sets these modes to always

Table 8.1 The number of aliasing sets M and number of observations for a given P. The last two lines correspond to the case when the switching modes 4 and 5 are not the primary (most energetic) modes of their aliasing sets. In the reduced Fourier domain KF, including these modes plays a crucial role in the filter performance as we will see below.

P	# of obs	# of aliasing sets M
1	105	52
3	35	17
5	21	10
7	15	7
15	7	3
21	5	2

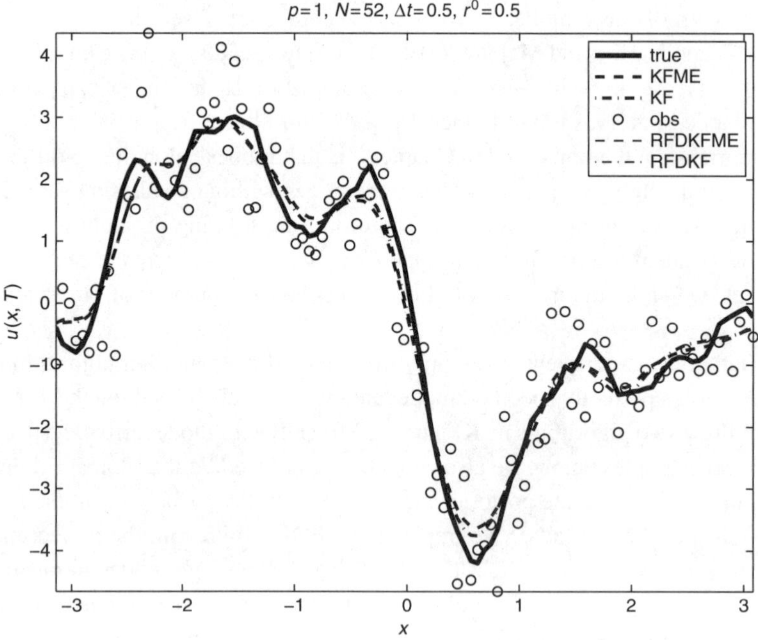

Figure 8.5 Filter performance for $P = 1$ at $T = 5000$. The thick solid line is the truth signal, the thick dashed-dotted line is the filtered signal with the KF with exact damping, the thick dashed line is the filtered signal with the KF with the constant damping for all modes, the thin dashed-dotted line is the filtered signal with the RFDKF with the exact damping, the thin dashed line is the filtered signal with the RFDKF with constant damping for all modes, the circles show the observations.

be the prior mean state or solution of (8.6) with constant mean damping $\bar{d} = 1.5$. Therefore, we expect the RFDKF to be significantly worse than the KF for such extremely sparse observation networks when the true dynamics exhibits many instability transitions even when the perfect model is used for RFDKF.

8.4 Numerical performance of the filters with and without model error

First, we study the path-wise performance for fixed observation time $\Delta t = 0.5$, and observation variance $r^0 = 0.5$, for both plentiful $P = 1$ and sparse $P > 1$ observations. We show the filtered solutions for both cases with the perfectly specified damping model and with model error through the mean damping model, MSM, using KF and RFDKF. Therefore, we have a total of four filtering strategies to compare with the true signal. In Fig. 8.5, we compare the true with the four filtered solutions for the plentiful observations, $P = 1$. Note that for $P = 1$, there is no difference between KF and RFDKF. We also note that the model error due to using the time-averaged value of damping is not large; the filtered solutions with and without model error are quite close to each other. In Fig. 8.6, we demonstrate

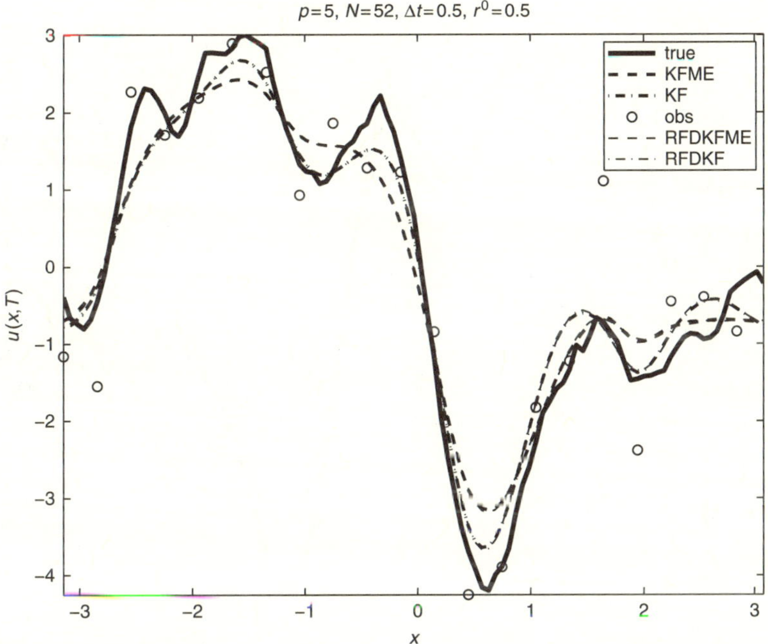

Figure 8.6 Filter performance for $P = 5$ at $T = 5000$. The thick solid line is the truth signal, the thick dashed-dotted line is the filtered signal with the KF with exact damping, the thick dashed line is the filtered signal with the KF with the constant damping for all modes, the thin dashed-dotted line is the filtered signal with the RFDKF with the exact damping, the thin dashed line is the filtered signal with the RFDKF with constant damping for all modes, the circles show the observations.

the filter performance for regularly spaced sparse observations with $P = 5$. Here, we note that the filtered solution from the exact Kalman filter (KF) is closer to the true signal than the filtered solution with model error (KFME). We also find that RFDKF gives practically the same solutions as the standard KF in this case. Therefore, we conclude that there is no advantage of using the full Fourier domain KF over the RFDKF when the observations are relatively dense. In Fig. 8.7, we demonstrate the filter performance for a very sparse observation network with $P = 15$ so that some of the instabilities for $k = 4, 5$ are hidden from the primary modes. Here, the exact KF performs better than KFME which treats $k = 4, 5$ as stable modes. Moreover, RFDKF produces a solution that is different from that of the full Fourier domain KF. Below, we check the bulk skill through RMS errors of all the discussed filtering techniques as a function of observation time and observation variance which confirms the trends illustrated above.

In Fig. 8.8, we demonstrate the dependence of the filter skill on the observation time Δt for fixed $r^0 = 0.1$ and various values of P. For relatively dense observations ($P = 1, 3, 5, 7$), the full Fourier domain KF performs almost as well as the RFDKF. We also

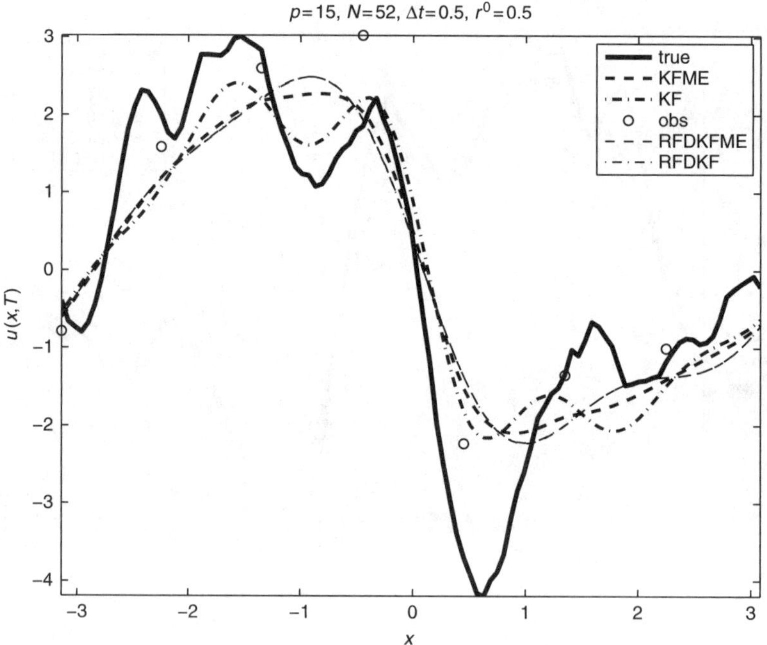

Figure 8.7 Filter performance for $P = 15$ at $T = 5000$. The thick solid line is the truth signal, the thick dashed-dotted line is the filtered signal with the KF with exact damping, the thick dashed line is the filtered signal with the KF with the constant damping for all modes, the thin dashed-dotted line is the filtered signal with the RFDKF with the exact damping, the thin dashed line is the filtered signal with the RFDKF with constant damping for all modes, the circles show the observations.

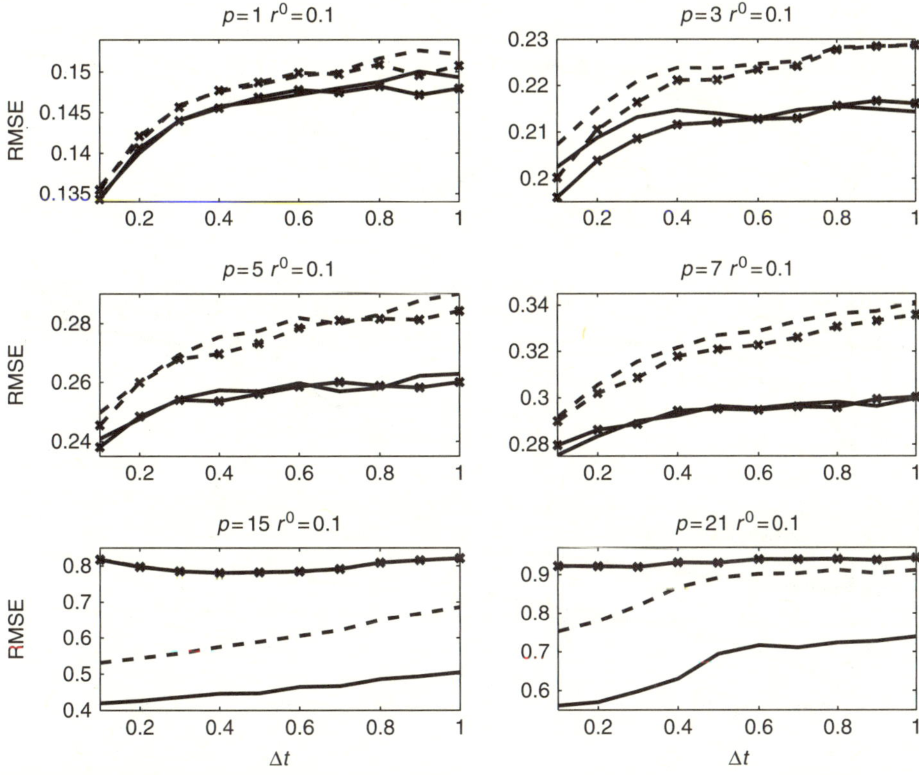

Figure 8.8 RMS error as a function of observation time Δt for $r^0 = 0.1$ and various values of P. The solid lines correspond to the KF with exact damping and the dashed lines correspond to the KF with model error. The solid lines with pluses correspond to RFDKF and the dashes with pluses correspond to RFDKF with model error.

observe examples ($P = 3$ and small Δt) when the RFDKF with model errors performs better than the full Fourier domain KF. Moreover, we also note that the RMS error grows as the number of observations decreases. On the other, for very sparse observations ($P = 15, 21$), the full Fourier domain KF even with (large) model error performs much better than the RFDKF in the perfect model. For all the values of P and all the considered filtering techniques, we note that the RMS error grows as a function of Δt which is intuitively clear.

In Fig. 8.9, we demonstrate the dependence of the filter skill on the observation variance r^0 for fixed $\Delta t = 0.1$ and various values of P. Here, we note that the filter skill is better when the observation variance is smaller. Here, again the RFDKF works as well as the full Fourier domain KF for relatively dense observations ($P \leq 7$), while for very sparse observations, the full Fourier domain KF has much more skill than the RFDKF.

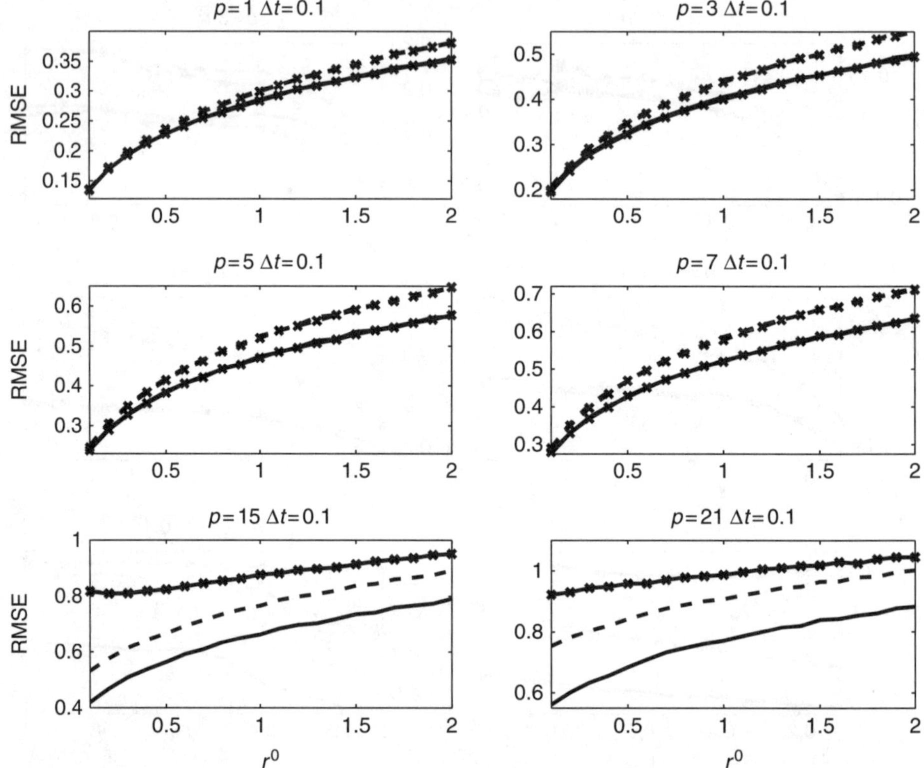

Figure 8.9 RMS error as a function of observation variance r^0 for $\Delta t = 0.1$ and various values of P. The solid lines correspond to the KF with exact damping and the dashed lines correspond to the KF with model error. The solid lines with pluses correspond to RFDKF and the dashes with pluses correspond to RFDKF with model error.

The explanation of the unsatisfying performance of the RFDKF on the sparsely observed network, $P \geq 15$, is because the switching modes 4 and 5 are not the primary modes and hence the filtered solutions for these modes are simply the prior dynamics, i.e. solutions of (8.6) with constant damping $\bar{d} = 1.5$ (again, see Table 8.1 to determine the primary modes). In the situation with the denser observations ($P \leq 7$), the RFDKF filters modes 4 and 5 since they are the primary modes. Hence their filtered solutions are not too different from the full KF. Note that on average modes 4 and 5 contain only 10% of the total system energy (see Fig. 8.4 for the energy spectrum). Naively, one might think that an accurate filtered solution can be obtained by only filtering the most energetic modes and omitting the least energetic modes. However, as our simulations show, the switching modes 4 and 5 have to be included in the set of filtered modes even though these modes are not the most energetic, especially when observations are very sparse.

Part III

Filtering turbulent nonlinear dynamical systems

9

Strategies for filtering nonlinear systems

In this chapter, we discuss several approaches for filtering nonlinear systems. In particular, we consider filtering the following discrete

Canonical nonlinear filtering problem

$$\vec{u}_{m+1} = \vec{f}(\vec{u}_m) + \vec{\sigma}_{m+1}, \tag{9.1}$$

$$\vec{v}_{m+1} = \vec{g}(\vec{u}_{m+1}) + \vec{\sigma}^o_{m+1}, \tag{9.2}$$

where $\vec{f}$ is an arbitrary deterministic nonlinear operator that propagates the signal $\vec{u} \in \mathbb{R}^N$ forward in time and $\vec{\sigma}_m \in \mathbb{R}^N$ is a vector of Gaussian white noise with zero mean and covariance

$$\langle \vec{\sigma}_m \otimes (\vec{\sigma}_m)^T \rangle = R. \tag{9.3}$$

In (9.2), observation $\vec{v} \in \mathbb{R}^M$ is associated with the true signal through a nonlinear observation operator $\vec{g}$ and independent Gaussian random measurement errors $\vec{\sigma}^o_m = \{\sigma^o_{j,m}\}$ with zero mean and diagonal covariance matrix

$$\langle \vec{\sigma}^o_m \otimes (\vec{\sigma}^o_m)^T \rangle = R^o = r^o \mathcal{I}. \tag{9.4}$$

We start this chapter by discussing a standard approach which approximates the nonlinear filtering problem in (9.1), (9.2) with a suboptimal filter known as *the extended Kalman filter* (EKF). Here we study EKF in its standard linearized form; in Chapters 10 and 14, we will introduce a fully nonlinear extended Kalman filter (NEKF) algorithm without linearization on prototype test models with multiple time-scales and utilize such a NEKF algorithm for stochastic parametrization "on-the-fly" in Chapter 13. We point out the drawback of the EKF approach for large-dimensional systems (typically $N \approx 10^7$–10^9 in the current global circulation models due to increasing computing power in the last four decades). We then briefly mention a simple remedy, called *optimal interpolation* (Gandin, 1965) or *3D-Var* (Parrish and Derber, 1992; Courtier *et al.*, 1998); the latter technique has been operationally used by the US National Weather Service (National Centers for Environmental Prediction or NCEP) since 1996 (Kalnay, 2003). Subsequently, we discuss a Monte Carlo approach for approximating the Kalman filter formula (Evensen,

1994; Burgers *et al.*, 1998), *the ensemble Kalman filter* (EnKF). In contrast, we also present two deterministic ensemble-based approaches for approximating the Kalman filter formula without Monte Carlo simulation. These deterministic approaches are *the ensemble transform Kalman filter* (ETKF) (Bishop *et al.*, 2001) and *the ensemble adjustment Kalman filter* (EAKF) (Anderson, 2001); they are also known as *the ensemble square-root filters* (EnSRF). In Chapter 11, we discuss their performance for high-dimensional turbulent nonlinear dynamical systems. An alternative Monte Carlo approach that accounts for nonlinearity and non-Gaussianity is *particle filtering* (Bain and Crisan, 2009); this approach is theoretically a well-established strategy but is practically only accurate for low-dimensional systems or systems with low-dimensional attractors. Recent mathematical theory strongly supports this curse of dimensionality for particle filtering (Bengtsson *et al.*, 2008; Bickel *et al.*, 2008). How to utilize the attractive aspects of particle filtering for high-dimensional dynamical systems is an important contemporary research topic. This is the topic in Chapter 15.

Subsequently, we present numerical results with various nonlinear filters discussed above on the three-dimensional Lorenz-63 model (Lorenz, 1963). In particular, we are interested in the filtering skill with low ensemble size (three members) to mimic some of the subtle issues that arise in filtering much higher-dimensional systems. From our numerical study, we will find that both EAKF and ETKF produce the most accurate filtered solutions despite the inherent linearization in these algorithms. At the end of this chapter, we will also show the results of using EAKF and ETKF to filter turbulent signals from the linear stochastic models from Chapters 5–7. Numerical results with higher-dimensional turbulent nonlinear dynamical systems will be shown in Chapter 11. We end this chapter with a discussion of the potential advantages and disadvantages of the ensemble approaches.

9.1 The extended Kalman filter

The basic idea of the extended Kalman filter (EKF) is to approximate the nonlinear operators $\vec{f}, \vec{g}$ by Taylor series up to order one, assuming that both operators are sufficiently smooth $\vec{f} \in C^1(\mathbb{R}^N)$, $\vec{g} \in C^1(\mathbb{R}^M)$, about the posterior and prior mean states consecutively

$$\vec{f}(\vec{u}) \approx \vec{f}(\bar{\vec{u}}_{m|m}) + F_m(\vec{u} - \bar{\vec{u}}_{m|m}), \quad F_m = \nabla \vec{f}(\vec{u})|_{\vec{u}=\bar{\vec{u}}_{m|m}}, \tag{9.5}$$

$$\vec{g}(\vec{u}) \approx \vec{g}(\bar{\vec{u}}_{m+1|m}) + G_m(\vec{u} - \bar{\vec{u}}_{m+1|m}), \quad G_m = \nabla \vec{g}(\vec{u})|_{\vec{u}=\bar{\vec{u}}_{m+1|m}}. \tag{9.6}$$

In (9.5), (9.6), one should notice that the linearized operators F_m, G_m depend on the posterior and prior estimates $\bar{\vec{u}}_{m|m}, \bar{\vec{u}}_{m+1|m}$ and any formulas that depend on these linearized terms are no longer off-line.

The linearized operator F_m maps the perturbation vector $\delta\vec{u} = \vec{u} - \bar{\vec{u}}_{m|m}$ at time t_m to time t_{m+1}. We illustrate this procedure of linearization of EKF on the L-63 model in (9.38) and (9.39) in Section 9.4 below; there is a vast literature on how to approximate this linear operator including *the tangent linear model* (Lorenz, 1965) which does not involve derivatives or the Jacobian as in (9.5). Note that the transpose (or adjoint in complex variables) of F_m is called *the adjoint linear model* which maps the perturbation vector $\delta\vec{u}$ backward in time from time t_{m+1} to time t_m.

With these approximations, the extended Kalman filter essentially estimates the filtering problem (9.1), (9.2) with

$$\vec{u}_{m+1} = \vec{f}(\bar{\vec{u}}_{m|m}) + F_m(\vec{u}_m - \bar{\vec{u}}_{m|m}) + \vec{\sigma}_{m+1}, \tag{9.7}$$

$$\vec{v}_{m+1} = \vec{g}(\bar{\vec{u}}_{m+1|m}) + G_m(\vec{u}_{m+1} - \bar{\vec{u}}_{m+1|m}) + \vec{\sigma}^o_{m+1}. \tag{9.8}$$

This approximation yields the following statistics in the forecast step

$$\bar{\vec{u}}_{m+1|m} = \vec{f}(\bar{\vec{u}}_{m|m}) \tag{9.9}$$

$$\begin{aligned} R_{m+1|m} &\equiv \langle(\vec{u}_{m+1} - \bar{\vec{u}}_{m+1|m}) \otimes (\vec{u}_{m+1} - \bar{\vec{u}}_{m+1|m})^T\rangle \\ &= \langle(F_m(\vec{u}_m - \bar{\vec{u}}_{m|m}) + \vec{\sigma}_{m+1}) \otimes (F_m(\vec{u}_m - \bar{\vec{u}}_{m|m}) + \vec{\sigma}_{m+1})^T\rangle \\ &= F_m\langle(\vec{u}_m - \bar{\vec{u}}_{m|m}) \otimes (\vec{u}_m - \bar{\vec{u}}_{m|m})^T\rangle F_m^T + \langle\vec{\sigma}_{m+1} \otimes \vec{\sigma}^T_{m+1}\rangle \\ &= F_m R_{m|m} F_m^T + R, \end{aligned} \tag{9.10}$$

where the first term of (9.10) follows from the definition of the posterior error covariance matrix $R_{m|m}$ as in (3.4) in Chapter 3 and the second term through (9.3).

As in Chapter 3, the Kalman filter formula is obtained by minimizing the following cost function

$$\begin{aligned} J(\vec{u}) &= (\vec{u} - \bar{\vec{u}}_{m+1|m})^T R^{-1}_{m+1|m}(\vec{u} - \bar{\vec{u}}_{m+1|m}) \\ &\quad +(\vec{v}_{m+1} - \vec{g}(\vec{u}))^T (R^o)^{-1}(\vec{v}_{m+1} - \vec{g}(\vec{u})), \\ &= \|\vec{u} - \bar{\vec{u}}_{m+1|m}\|^2_{R^{-1}_{m+1|m}} + \|\vec{v}_{m+1} - \vec{g}(\vec{u})\|^2_{(R^o)^{-1}} \end{aligned} \tag{9.11}$$

$$\begin{aligned} &= \|\vec{u} - \bar{\vec{u}}_{m+1|m}\|^2_{R^{-1}_{m+1|m}} \\ &\quad +\|\vec{v}_{m+1} - \vec{g}(\bar{\vec{u}}_{m+1|m}) - G_m(\vec{u} - \bar{\vec{u}}_{m+1|m})\|^2_{(R^o)^{-1}}, \end{aligned} \tag{9.12}$$

where we use the notation $\|\vec{u}\|^2_A = \vec{u}^T A\vec{u}$. The minimum of (9.12) is given as follows

$$\bar{\vec{u}}_{m+1|m+1} = \bar{\vec{u}}_{m+1|m} + K_{m+1}(\vec{v}_{m+1} - \vec{g}(\bar{\vec{u}}_{m+1|m})), \tag{9.13}$$

where

$$K_{m+1} = (R^{-1}_{m+1|m} + G_m^T(R^o)^{-1}G_m)^{-1}G_m^T(R^o)^{-1} \tag{9.14}$$

$$R_{m+1|m+1} = (\mathcal{I} - K_{m+1}G_m)R_{m+1|m}. \tag{9.15}$$

The extended Kalman filter is closed under Eqns (9.9), (9.10), (9.13)–(9.15).

Notice that this is a suboptimal filter since the evolution of both the mean and covariance contains errors introduced through the first-order truncation in the Taylor series approximation for the nonlinear operators $\vec{f}$ and $\vec{g}$. In Chapter 10, we will discuss a nonlinear filtering algorithm (NEKF) on a three-dimensional toy model with statistics which are analytically computable without Taylor expansion. Secondly, it is worth while to mention that each equation in the dynamics of the second-order statistics involves either F_m or G_m and therefore it is clear that the covariance matrix cannot be estimated off-line as in the linear Kalman filter problem, discussed in Chapters 2 and 3. Aside from these drawbacks, the extended Kalman filter has been used with success in many disciplines involving small dimensional state variables.

In numerical weather prediction (NWP), however, the extended Kalman filter is practically useless since the current state of art GCM has $N \approx 10^7$–10^9 state variables. As an illustration of why EKF is practically useless, consider implementing EKF for an NCEP-GCM with resolution 300 km, which has roughly three million state variables, with six-hour observation time frequency. The six-hour forecast of this model takes roughly six minutes of computing time in a CPU with a Pentium 1 Ghz processor and 2 GB RAM. Thus a six-hour forecast of a prior covariance matrix of size 3 million $\times$ 3 million, computed using Eqn (9.10), requires about 3 million $\times$ 6 minutes $\approx$ 300,000 hours of computing time.

A natural remedy is to let the prior error covariance matrix $R_{m+1|m}$ be a constant B in time and hence the expensive computational overhead in (9.10) is ignored. This strategy is known as optimal interpolation (OI) (Gandin, 1965) which simply uses Eqns (9.9), (9.13)–(9.15) with a constant matrix B replacing $R_{m+1|m}$. On the other hand, Lorenc (1986) showed that OI is equivalent to the three-dimensional variational approach (3D-Var) (Parrish and Derber, 1992; Courtier *et al.*, 1998), which numerically minimizes the cost function $J(u)$ in (9.11) with iterative methods such as the conjugate gradient or the quasi-Newton methods. Unfortunately, the filtering skill of these static covariance approaches is suboptimal and very sensitive to the choice of B. Interested readers should consult Parrish and Derber (1992); Courtier *et al.* (1998); Kalnay (2003). We close this section by mentioning that typically B is chosen to represent a certain asymptotic statistical quantity, which is sometimes referred to as the climatological statistics in the atmospheric and ocean science literature. However, as we are all aware of, the daily or weekly fluctuation in the weather pattern is not ignorable and this motivates the discussion in the remainder of this chapter. The family of finite ensemble filters that we discuss next are useful primarily in Sections 9.2–9.4 when there is no system noise in the nonlinear dynamics in (9.1).

9.2 The ensemble Kalman filter

The basic idea of the ensemble Kalman filter (EnKF) is to approximate both the prior and the posterior error covariance matrices $R_{m+1|m}$ and $R_{m+1|m+1}$ with ensemble covariance matrices around the corresponding ensemble mean, $\bar{\vec{u}} = K^{-1} \sum_{k=1}^{K} \vec{u}^k$,

$$R_{m+1|m} \approx \tilde{R}_{m+1|m} = \frac{1}{K-1} U_{m+1|m} U_{m+1|m}^T \tag{9.16}$$

$$R_{m+1|m+1} \approx \tilde{R}_{m+1|m+1} = \frac{1}{K-1} U_{m+1|m+1} U_{m+1|m+1}^T \tag{9.17}$$

where

$$
\begin{aligned}
U_{m+1|m} &= [\vec{u}^1_{m+1|m} - \bar{\vec{u}}_{m+1|m}; \vec{u}^2_{m+1|m} \\
&\quad -\bar{\vec{u}}_{m+1|m}, \ldots; \vec{u}^K_{m+1|m} - \bar{\vec{u}}_{m+1|m}] \\
U_{m+1|m+1} &= [\vec{u}^1_{m+1|m+1} - \bar{\vec{u}}_{m+1|m+1}; \vec{u}^2_{m+1|m+1} \\
&\quad -\bar{\vec{u}}_{m+1|m+1}, \ldots; \vec{u}^K_{m+1|m+1} - \bar{\vec{u}}_{m+1|m+1}].
\end{aligned}
\tag{9.18}
$$

From now on, we omit the "$\sim$" sign in R and it is understood that the covariance matrix in the rest of the discussion is the ensemble covariance matrix. With these ensemble approximations, we avoid the expensive computation in (9.10) and replace it by integrating each ensemble member in time through the original model in (9.1). In this fashion, the ensemble Kalman filter does not require any tangent approximation F_m of the nonlinear operator $\vec{f}$ as in the extended Kalman filter.

Substituting (9.16) into the Kalman gain update formula as in (3.6) or (9.14), we obtain

$$
\begin{aligned}
K_{m+1} &= (R^{-1}_{m+1|m} + G^T (R^o)^{-1} G) G^T (R^o)^{-1} \\
&= R_{m+1|m} G^T (G R_{m+1|m} G^T + R^o)^{-1} \\
&= (K-1)^{-1} U (GU)^T ((K-1)^{-1} (GU)(GU)^T + R^o)^{-1},
\end{aligned}
\tag{9.19}
$$

where we omit subscript $m+1|m$ on every matrix U to avoid cumbersome notation; in the remainder of this chapter it is understood that U is the $U_{m+1|m}$ defined in (9.18).

Notice that every time the linearized matrix G appears, it is next to the perturbation matrix U. For the nonlinear observation operator $\vec{g}$, we can replace GU by the following approximation

$$
V = [\vec{g}(\vec{u}^1) - \bar{\vec{v}}; \vec{g}(\vec{u}^2) - \bar{\vec{v}}, \ldots; \vec{g}(\vec{u}^K) - \bar{\vec{v}}],
\tag{9.20}
$$

where in practice one can choose

$$
\bar{\vec{v}} = \vec{g}(\bar{\vec{u}}), \text{ or } \quad \bar{\vec{v}} = \overline{\vec{g}(\vec{u})}.
$$

The first choice is nothing but the Taylor expansion about the mean state $\bar{\vec{u}}$ as in (9.6); the second choice of $\bar{\vec{v}}$ yields a zero column sum of matrix V which is consistent with the definition of U. Notice that with this ensemble approximation, we avoid having to linearize $\vec{g}$ on the entire model space and hence there is no need to compute $\nabla \vec{g}$ as in the EKF.

To complete one iteration of an update, we need to generate a posterior ensemble with mean equal to the mean posterior state, obtained by substituting (9.19) into (9.13), and covariance satisfying (9.15) with approximate sample covariances as in (9.16), (9.17). The EnKF (Evensen, 1994; Burgers *et al.*, 1998; Evensen, 2003) updates each ensemble member by the following

$$
\vec{u}^k_{m+1|m+1} = \vec{u}^k_{m+1|m} + K_{m+1}(\vec{v}^k_{m+1} - \vec{g}(\vec{u}^k_{m+1|m})),
\tag{9.21}
$$

where

$$K_{m+1} = (K-1)^{-1}UV^T((K-1)^{-1}VV^T + R^o)^{-1}, \tag{9.22}$$

and the observation $\vec{v}_m$ is perturbed

$$\vec{v}_m^k = \vec{v}_m + \vec{\eta}_m^k \tag{9.23}$$

by a Gaussian random noise $\vec{\eta}_m^k$ with zero mean and covariance R^o. This random perturbation is in some sense a Monte Carlo method applied to the Kalman filter formula and it yields an asymptotically correct analysis error covariance estimate for large ensembles. Interested readers should consult Burgers *et al.* (1998) to convince themselves that perturbing the observation as in (9.23) is necessary, otherwise this ensemble will never satisfy (9.15) and hence the covariance matrix is underestimated.

However, as pointed out by Whitaker and Hamill (2002), when the ensemble size is small, this stochastic perturbation (9.23) is an additional source of sampling errors and is responsible for suboptimal filtering. A simple remedy to avoid this suboptimal filtering is to inflate the covariance matrix (Anderson and Anderson, 1999; Whitaker and Hamill, 2002) a priori,

$$U \leftarrow (\sqrt{1+r})U, \tag{9.24}$$

by a positive coefficient $r > 0$, where the optimal r is obtained empirically. Note that this commonly used strategy is called multiplicative variance inflation (Anderson and Anderson, 1999; Whitaker and Hamill, 2002); other types of inflation strategies, such as additive variance inflation (Ott *et al.*, 2004) and adaptive inflation (Anderson, 2007), show comparable performance and a slight advantage over the multiplicative strategy.

To conclude this section, we provide step-by-step guidance for generating the analysis ensemble $\{\vec{u}^k_{m+1|m+1}, k = 1, \ldots, K\}$ through EnKF, given the prior ensemble $\{\vec{u}^k_{m+1|m}, k = 1, \ldots, K\}$, observation $\vec{v}_{m+1} \in \mathbb{R}^M$, observation operator $\vec{g}$, and observation error covariance matrix $R^o \in \mathbb{R}^{M\times M}$.

1. Compute the prior ensemble average, $\bar{\vec{u}}_{m+1|m}$, form the matrices $U \in \mathbb{R}^{N\times K}$ and $V \in \mathbb{R}^{M\times K}$, as in (9.18) and (9.20) respectively.
2. Inflate U and V by a factor of $\sqrt{1+r}$.
3. Compute the Kalman gain matrix as in (9.22).
4. Generate K M-dimensional random vectors $\vec{\eta}^k_{m+1}$ from a Gaussian distribution with zero mean and covariance R^o. In our numerical simulation, we generate these random vectors by first randomly drawing a $M \times K$ size matrix A and taking the eigenvalue decomposition of $(K-1)^{-1}AA^T = F\Sigma F^T \in \mathbb{R}^{M\times M}$. Thus, we have the unbiased random vectors $\vec{\eta}^k_{m+1}$ which is nothing but the column vectors of the matrix $T = (R^o)^{1/2}F\Sigma^{-1/2}F^TA$. When $K < M$, Σ and F in T correspond to the nonzero eigenvalues and eigenvectors.
5. Update each ensemble member through (9.21) with perturbed observations (9.23).

In the next section, we discuss an alternative strategy for generating the ensemble $\{\vec{u}^k_{m|m}\}$ without any stochastic perturbations.

9.3 The ensemble square-root filters

The main idea of the deterministic finite ensemble filters is to generate a posterior ensemble $\{\vec{u}^k_{m|m}\}$ in the following form

$$\begin{aligned}\vec{u}^k_{m|m} &= \bar{\vec{u}}_{m|m} + U^k_{m|m}. \\ &= \bar{\vec{u}}_{m|m} + \left(\sqrt{(K-1)R_{m|m}}\right)^k,\end{aligned} \tag{9.25}$$

where the superscript k in the right-hand side denotes the kth column of its corresponding matrix. In this approach, the posterior mean, $\bar{\vec{u}}_{m|m}$, and the posterior error covariance, $R_{m|m}$, are updated with the basic Kalman filter formula (9.13), (9.15) as in the extended Kalman filter with Kalman gain matrix through (9.22). This approach is sometimes called the ensemble square-root filter (EnSRF) since it involves taking the square root of covariance matrices. In the last decade, there are several different approaches that have been introduced for taking the square root; numerically these different approaches basically transform the covariance matrix into a space where the covariance matrix becomes well-conditioned and hence it is easily diagonalized or inverted; important approaches include Whitaker and Hamill's EnSRF for a scalar observation (Whitaker and Hamill, 2002), Bishop's ensemble transform Kalman filter (ETKF) (Bishop *et al.*, 2001), Anderson's ensemble adjustment Kalman filter (EAKF) (Anderson, 2001), Ott *et al.*'s local ensemble Kalman filter (LEKF) (Ott *et al.*, 2004), etc. Recently, Thomas *et al.* (2007) proposed a generalized least-square–square-root filter (GLS-SRF) that does not require square-rooting any matrix nor singular value/eigenvalue decompositions for generating the posterior ensemble perturbation matrix $U_{m|m}$; instead it relies on the orthogonal rotations. In the remainder of this section, we discuss only ETKF and EAKF in detail. In the formulation for ETKF, we only discuss the efficient algorithm introduced by Hunt *et al.* (2007) instead of the original ETKF of Bishop *et al.* (2001).

9.3.1 The ensemble transform Kalman filter

For ETKF, the basic idea is to find a "transformation" matrix T (this jargon was coined by Bishop *et al.* (2001)) such that

$$U_{m+1|m+1} = U_{m+1|m}T = UT \quad \text{and} \quad UT(UT)^T = (K-1)R_{m+1|m+1},$$

where $R_{m+1|m+1}$ is the sample posterior covariance matrix solved by enforcing the Kalman filter formula in (9.15). Each ensemble member is then generated by adding each column of $U_{m+1|m+1}$ to the posterior mean state $\bar{\vec{u}}_{m+1|m+1}$ as in (9.25).

To find T, notice that with the following identity

$$A^T(AA^T + R^o)^{-1} = (\mathcal{I} + A^T(R^o)^{-1}A)^{-1}A^T(R^o)^{-1}$$

for $A = V$, the Kalman gain matrix in (9.22) can be rewritten as

$$K_{m+1} = \frac{1}{K-1}U\left(\mathcal{I} + \frac{1}{K-1}V^T(R^o)^{-1}V\right)^{-1}V^T(R^o)^{-1}$$

so that by (9.15),

$$\begin{aligned}
R_{m+1|m+1} &= \left(\mathcal{I} - \frac{U}{K-1}\left(\mathcal{I} + \frac{V^T(R^o)^{-1}V}{K-1}\right)^{-1}V^T(R^o)^{-1}G\right)\frac{UU^T}{K-1} \\
&= \frac{1}{K-1}U\left(\mathcal{I} - \left(\mathcal{I} + \frac{V^T(R^o)^{-1}V}{K-1}\right)^{-1}\frac{V^T(R^o)^{-1}V}{K-1}\right)U^T \\
&= \frac{1}{K-1}U\left(\mathcal{I} - (\mathcal{I} + B)^{-1}B\right)U^T, \qquad (9.26)
\end{aligned}$$

where

$$B = \frac{V^T(R^o)^{-1}V}{K-1}.$$

Notice that in (9.26), the term $\mathcal{I} - (\mathcal{I} + B)^{-1}B = (\mathcal{I} + B)^{-1}$, hence we obtain

$$\begin{aligned}
R_{m+1|m+1} &= U\left((K-1)\mathcal{I} + V^T(R^o)^{-1}V\right)^{-1}U^T \\
&= U\Sigma U^T = U\frac{TT^T}{K-1}U^T. \qquad (9.27)
\end{aligned}$$

Numerically, T can be obtained by applying eigenvalue decomposition to $\Sigma^{-1} = X\Gamma X^T$ where X is a unitary matrix and Γ is a diagonal matrix with diagonal components which are eigenvalues of Σ^{-1} and setting

$$T = \sqrt{K-1}X\Gamma^{-1/2},$$

as in the original ETKF (Bishop *et al.*, 2001). Another choice of T, which is also referred as the spherical simplex ETKF (Wang *et al.*, 2004), is to let

$$T = \sqrt{K-1}X\Gamma^{-1/2}X^T \qquad (9.28)$$

so that it remains symmetric. In all of the numerical simulations involving ETKF in this book, we adopt the symmetric square root as in (9.28).

As in the previous section, we provide step-by-step guidance for generating the analysis ensemble $\{\vec{u}^k_{m+1|m+1}, k = 1, \ldots, K\}$ through ETKF, given the prior ensemble $\{\vec{u}^k_{m+1|m}, k = 1, \ldots, K\}$, observation $\vec{v}_{m+1} \in \mathbb{R}^M$, observation operator $\vec{g}$ and observation error covariance matrix $R^o \in \mathbb{R}^{M\times M}$. Note that this step-by-step algorithm follows the formulation suggested by Hunt *et al.* (2007) and it is numerically more efficient than the original formulation which is noted by Whitaker and Hamill (2002). In particular, we do not take the eigenvalue decomposition of an $N \times N$ matrix with this efficient algorithm.

1. Compute the prior ensemble average, $\bar{\vec{u}}_{m+1|m}$, form the matrices $U \in \mathbb{R}^{N \times K}$ and $V \in \mathbb{R}^{M \times K}$, as in (9.18) and (9.20) respectively.
2. Compute the singular value decomposition of the $K \times K$ matrix
$$J = \frac{(K-1)}{1+r}\mathcal{I} + V^T (R^o)^{-1} V = X \Gamma X^T.$$
3. The Kalman gain matrix is computed as follows
$$K_{m+1} = U(X\Gamma^{-1}X^T)V^T(R^o)^{-1},$$
where $X\Gamma^{-1}X^T$ is the pseudo-inverse of J. Alternatively, one can simply solve the following $K \times K$ linear algebra problem,
$$Jx = V^T(R^o)^{-1}(\vec{v}_{m+1} - \vec{g}(\bar{\vec{u}}_{m+1|m})),$$
for x.
4. Update the posterior mean with
$$\begin{aligned}\bar{\vec{u}}_{m+1|m+1} &= \bar{\vec{u}}_{m+1|m} + K_{m+1}(\vec{v}_{m+1} - \vec{g}(\bar{\vec{u}}_{m+1|m})),\\ &= \bar{\vec{u}}_{m+1|m} + Ux.\end{aligned} \tag{9.29}$$
5. Compute the transformation matrix $T = \sqrt{K-1}X\Gamma^{-1/2}X^T$, and the posterior perturbation matrix $U_{m+1|m+1} = UT$.
6. The posterior ensemble $\{\vec{u}^k_{m+1|m+1}, k = 1, \ldots, K\}$ is obtained by adding the posterior mean, $\bar{\vec{u}}_{m+1|m+1}$, to each column vector of $U_{m+1|m+1}$.

9.3.2 The ensemble adjustment Kalman filter

The main idea here is to find an "adjustment" matrix A (Anderson, 2001) such that
$$U_{m+1|m+1} = AU_{m+1|m} = AU \text{ and } AUU^T A^T = (K-1)R_{m+1|m+1}.$$

To obtain A, we first take the eigenvalue decomposition of the prior ensemble error covariance
$$R_{m+1|m} = F\Sigma^2 F^T, \tag{9.30}$$
where each column vector of F is an eigenvector corresponding to the eigenvalue in the diagonal matrix Σ^2; F is also a unitary matrix, i.e. $FF^T = F^T F = \mathcal{I}$. Here, we notice that
$$\Sigma^{-1}F^T R_{m+1|m} F \Sigma^{-1} = \mathcal{I}. \tag{9.31}$$

Let us take a second eigenvalue decomposition of the matrix
$$\Sigma F^T G^T (R^o)^{-1} G F \Sigma,$$

that is,

$$X^T \Sigma F^T G^T (R^o)^{-1} G F \Sigma X = D. \tag{9.32}$$

where X is unitary and D is diagonal. Thus, we can also rewrite (9.31) as

$$X^T \Sigma^{-1} F^T R_{m+1|m} F \Sigma^{-1} X = \mathcal{I}. \tag{9.33}$$

Now let us start from the first line in Eqn (9.27) and using (9.31), (9.32) and (9.33), we obtain

$$\begin{aligned}
R_{m+1|m+1} &= U \left((K-1)\mathcal{I} + V^T (R^o)^{-1} V \right)^{-1} U^T \\
&= \left(R_{m+1|m}^{-1} + G^T (R^o)^{-1} G \right)^{-1} \\
&= F \Sigma X X^T \Sigma^{-1} F^T \left(R_{m+1|m}^{-1} + G^T (R^o)^{-1} G \right)^{-1} F \Sigma^{-1} X X^T \Sigma F^T \\
&= F \Sigma X \left(X^T \Sigma F^T R_{m+1|m}^{-1} F \Sigma X + X^T \Sigma F^T G^T (R^o)^{-1} G F \Sigma X \right)^{-1} X^T \Sigma F^T \\
&= F \Sigma X \left((X^T \Sigma^{-1} F^T R_{m+1|m} F \Sigma^{-1} X)^{-1} + X^T \Sigma F^T G^T (R^o)^{-1} G F \Sigma X \right)^{-1} X^T \Sigma F^T \\
&= F \Sigma X \left(\mathcal{I} + D \right)^{-1} X^T \Sigma F^T \\
&= F \Sigma X \left(\mathcal{I} + D \right)^{-1/2} \left(\mathcal{I} + D \right)^{-1/2} X^T \Sigma F^T \\
&= F \Sigma X \left(\mathcal{I} + D \right)^{-1/2} \Sigma^{-1} F^T R_{m+1|m} F \Sigma^{-1} \left(\mathcal{I} + D \right)^{-1/2} X^T \Sigma F^T.
\end{aligned} \tag{9.34}$$

Equation (9.34) is basically

$$R_{m+1|m+1} = A R_{m+1|m} A^T$$

with adjustment matrix

$$A = F \Sigma X \left(\mathcal{I} + D \right)^{-1/2} \Sigma^{-1} F^T. \tag{9.35}$$

The step-by-step guidance for generating the analysis ensemble

$$\{\vec{u}^k_{m+1|m+1}, k = 1, \ldots, K\}$$

through EAKF, given the prior ensemble

$$\{\vec{u}^k_{m+1|m}, k = 1, \ldots, K\},$$

observation $\vec{v}_{m+1} \in \mathbb{R}^M$, linear observation operator G, and observation error covariance matrix $R^o \in \mathbb{R}^{M \times M}$, is given as follows:

1. Compute the prior ensemble average, $\bar{\vec{u}}_{m+1|m}$, and form the matrix $U \in \mathbb{R}^{N \times K}$ as in (9.18).
2. Compute the eigenvalue decomposition of the $N \times N$ matrix

$$\frac{1+r}{K-1} U U^T \approx (1+r) R_{m+1|m+1} = F \Sigma^2 F^T.$$

3. Compute the eigenvalue decomposition of the matrix

$$\Sigma F^T G^T (R^o)^{-1} G F \Sigma = X D X^T.$$

4. Form the adjustment matrix $A = F\Sigma X(\mathcal{I} + D)^{-1/2}\Sigma^{-1}F^T$ and perturbation matrix $U_{m+1|m+1} = AU$.
5. Solve the $N \times N$ linear algebra problem $Lx = y$ for $x = \bar{\vec{u}}_{m+1|m+1}$ with

$$L = F\Sigma^{-2}F^T + G^T(R^o)^{-1}G$$
$$y = F\Sigma^{-2}F^T\bar{\vec{u}}_{m+1|m} + G^T(R^o)^{-1}\vec{v}.$$

6. The posterior ensemble $\{\vec{u}^k_{m+1|m+1}, k = 1, \ldots, K\}$ is obtained by adding the posterior mean, $\bar{\vec{u}}_{m+1|m+1}$, to each column vector of $U_{m+1|m+1}$.

To summarize, EAKF requires two eigenvalue decompositions in Steps 2 and 3 to compute the adjustment matrix A as in (9.35); the former involves an $N \times N$ matrix (which is very expensive when N is large). In many situations, the ensemble size is less than the total state variables, $K \leq N$, so that Rank$(R_{m+1|m}) = K - 1$. In this situation, the adjustment matrix A is computed similarly except that Steps 3 and 4 are carried out with the nonzero truncated $\Sigma \in \mathbb{R}^{(K-1)\times(K-1)}$ and its corresponding eigenvectors so that D and X are also $(K-1) \times (K-1)$.

9.4 Ensemble filters on the Lorenz-63 model

The three-dimensional L-63 model (Lorenz, 1963) is a radically truncated approximation to the Navier–Stokes equations describing convective movement in the atmosphere. It is a toy model which does not describe actual atmospheric dynamical features. On the other hand, this model is found to be useful to describe laser physics (Haken, 1975) and it is well-known as the first example of a simple deterministic dynamical system with solutions that are sensitive to initial conditions; this behavior is coined as deterministic chaos or simply chaos. The governing equations of the L-63 model are given as follows,

$$\begin{aligned}\frac{dx}{dt} &= \sigma(y - x)\\ \frac{dy}{dt} &= \rho x - y - xz\\ \frac{dz}{dt} &= xy - bz.\end{aligned} \tag{9.36}$$

with the parameter set (σ, b, ρ), where in its original derivation (see e.g. Lorenz, 1963; Solari *et al.*, 1996), σ is called the Prandl number and ρ is the Rayleigh number. Note that the L-63 model is symmetric about the z-axis, i.e. it is invariant under the transformation $(x, y, z) = (-x, -y, z)$.

In our numerical simulations, we generate the true state by running the model for 40 nondimensional time units (with Runge-Kutta (RK4) with step size $\Delta t = 0.01$) with $\sigma = 10$,

$b = 8/3$, $\rho = 28$, whose trajectory looks like a "butterfly" wing (Lorenz, 1963); the solutions are fluctuating about the z-axis. Linear stability analysis (Sparrow, 1982) suggested that there are three saddles; all of them are stable in the z-direction and unstable in the (x, y)-direction; one of them is exactly the origin and the other two saddles are symmetric about the z-axis with outward-spiraling flows. To characterize the chaos in a dynamical system, one typically verifies whether the system has positive Lyapunov exponents (the rate of separation of infinitesimally close trajectories). With these parameters, the attractor has one positive Lyapunov exponent 0.906 that corresponds to a doubling time of 0.78 time units. In the L-63 model, the shortest correlation time is 0.16 time units; in our simulations, we generate "observations" every 0.08 and 0.25 time units by adding Gaussian noise with mean 0 and variance 2 to each coordinate of the true state. For the fully observed system, the linear observation operator G is an identity matrix and the observation error covariance matrix R^o is a diagonal matrix with all diagonal components equal to 2. This choice of error variance implies that a typical observation error is $\sqrt{2}$; by comparison, the natural variabilities (standard deviations over time) of x, y, and z are 7.91, 8.98, and 8.60, respectively. A necessary condition for a filter to have any skill whatsoever on the L-63 model is that the square roots of the RMS errors are smaller than the standard deviations from this attractor size. In the numerical experiments below, we also consider partially observed systems, $\{x, y\}, \{x, z\}, \{y, z\}, \{x\}, \{y\}, \{z\}$. Note from our discussion at the beginning of this paragraph that observing x, y projects nicely on the chaotic dynamics while observing z alone projects poorly on the chaotic dynamics.

For our simulation with the extended Kalman filter, we consider the Euler temporal discrete approximation of (9.36),

$$\vec{u}_{m+1} = \vec{f}(\vec{u}_m), \tag{9.37}$$

where $\vec{u} = (x, y, z)$ and

$$\vec{f}(x, y, z) = \begin{pmatrix} x + \Delta t\sigma(y - x) \\ y + \Delta t(\rho x - y - xz) \\ z + \Delta t(xy - bz) \end{pmatrix}. \tag{9.38}$$

Let $L_m \equiv \nabla \cdot \vec{f}(\vec{u})|_{\vec{u}=\vec{u}_m}$ and let $\vec{u}_m$ be the solution of the RK4 discrete time approximation at time t_m, solved with time step $\Delta t = 0.01$. For assimilation with observation time 0.08, we estimate the linear tangent approximation as follows

$$F_m = L_{m+7} \dots L_{m+1} L_m, \tag{9.39}$$

where the right-hand side is a composite of the Jacobian L evaluated at time steps $\{t_{m+j} \equiv t_m + j\Delta t : j = 0, 1, \dots 7\}$. For observation time 0.25 units, a similar approximation is used with $j = 0, 1, \dots, 24$.

In all our numerical simulations, we generate the true signal from $\vec{u} = (1.5089, -1.5313, 25.4609)$ and each filtered solution is initiated from a (or an ensemble of) perturbation around this initial true state with a Gaussian noise with mean zero and variance $r^o = 2$. In Fig. 9.1, we show the evolution of the posterior solution variable x

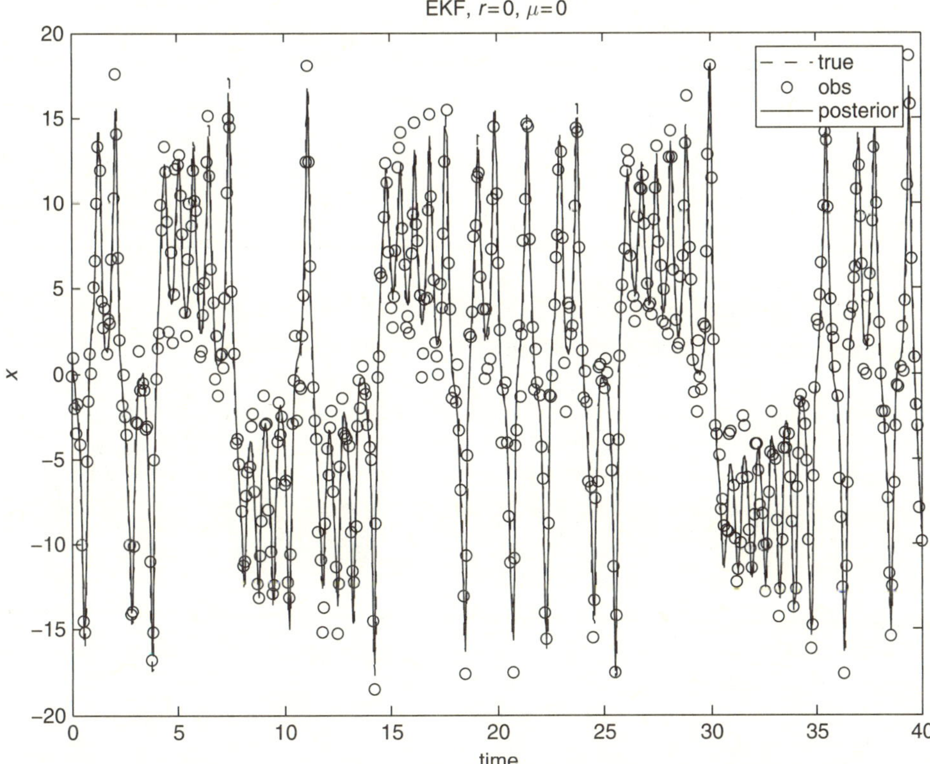

Figure 9.1 Extended Kalman Filter for observation time 0.08 units and full observations: filtered solution x variable as a function of time.

(solid line), assimilated every 0.08 time units, with full observations as a function of time and compare it with the observations (circle), and the true signal (dashes). In this time interval, the true signal oscillates about the z-axis for about 10 times and the filtered solution is very skillful with RMS error of 0.28, well below the observation error $\sqrt{r^o} = 1.41$.

For EnKF, ETKF and EAKF, we consider numerical experiments with a small ensemble size to mimic practical filtering issues with higher-dimensional turbulent dynamical systems as discussed in Chapter 11. In the simulation with the ensemble Kalman filter (EnKF) with the same observation time of 0.08 units and full observation, we find that assimilation with ensemble size $K = 3$ is not enough to guarantee a convergent solution. We see that when the ensemble size is increased to $K = 10$, substantial skill is gained with RMS error 0.33 (see also Table 9.1). A simple way to deal with filter divergence due to the small ensemble size is through variance inflation (Anderson, 2001), that is, we inflate the prior ensemble perturbation $U_{m+1|m}$ in (9.18) with a factor of $\sqrt{1+r}$, $r > 0$, before each assimilation cycle as discussed in (9.24); this technique is called multiplicative variance inflation. In our numerical simulation, we find that when $r = 4\%$ is added with ensemble

Table 9.1 Average RMS errors for numerical simulations with observation time of 0.08 units. EnKF is performed with $K = 10$, $r = 4\%$; ETKF and EAKF are performed with $K = 3$, $r = 4\%$.

	$\{x, y, z\}$	$\{x, y\}$	$\{x, z\}$	$\{y, z\}$	$\{x\}$	$\{y\}$	$\{z\}$
EKF	0.28	0.37	1.34	0.31	1.14	0.47	9.51
EnKF	0.26	0.30	0.48	0.29	0.99	0.32	4.09
ETKF	0.29	0.37	0.46	0.29	0.73	0.45	5.11
EAKF	0.26	0.39	0.45	0.29	0.88	0.48	5.38

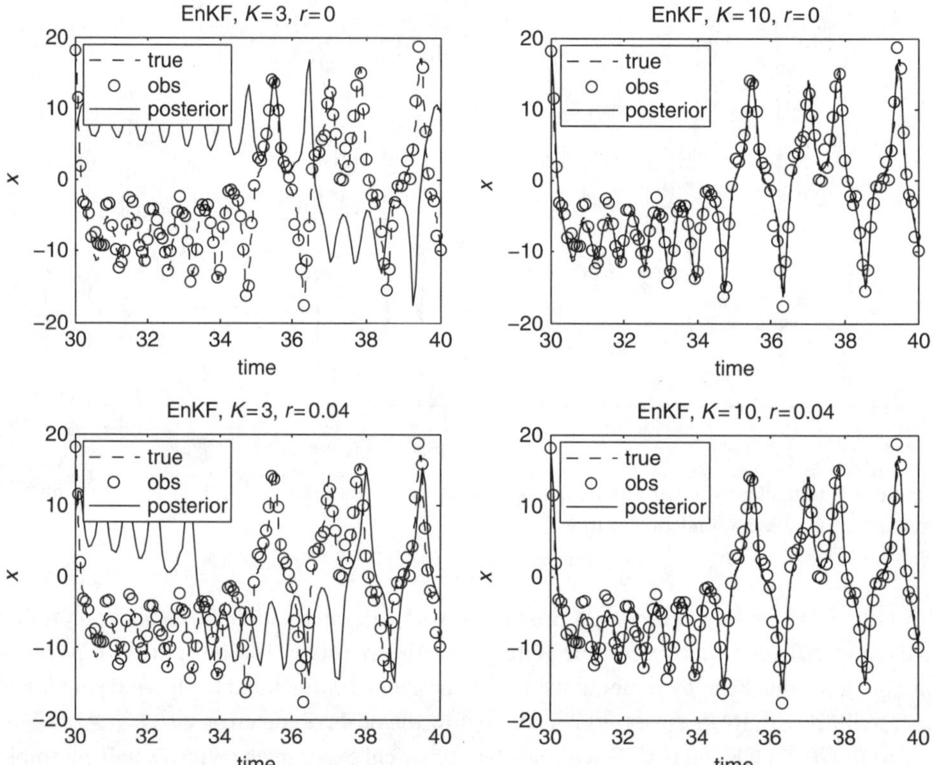

Figure 9.2 Ensemble Kalman filter for observation time of 0.08 units for various ensemble sizes K and variance inflation coefficient $r = 0.04$. Note the divergence filter with $K = 3$.

size 3, the solution still does not converge (see Fig. 9.2 for the last 10 time units); the additional sampling errors due to the perturbations of the observations are significant. However, when variance inflation of $r = 0.04$ is implemented with an ensemble size of 10, the RMS error decreases from 0.33 to 0.26 and this is slightly lower than that of EKF, 0.28 (see Table 9.1). For EAKF and ETKF with the small ensemble size 3, $r = 4\%$ variance inflation is sufficient for comparable skillful filtered solutions (see Fig. 9.3). In Table 9.1, we

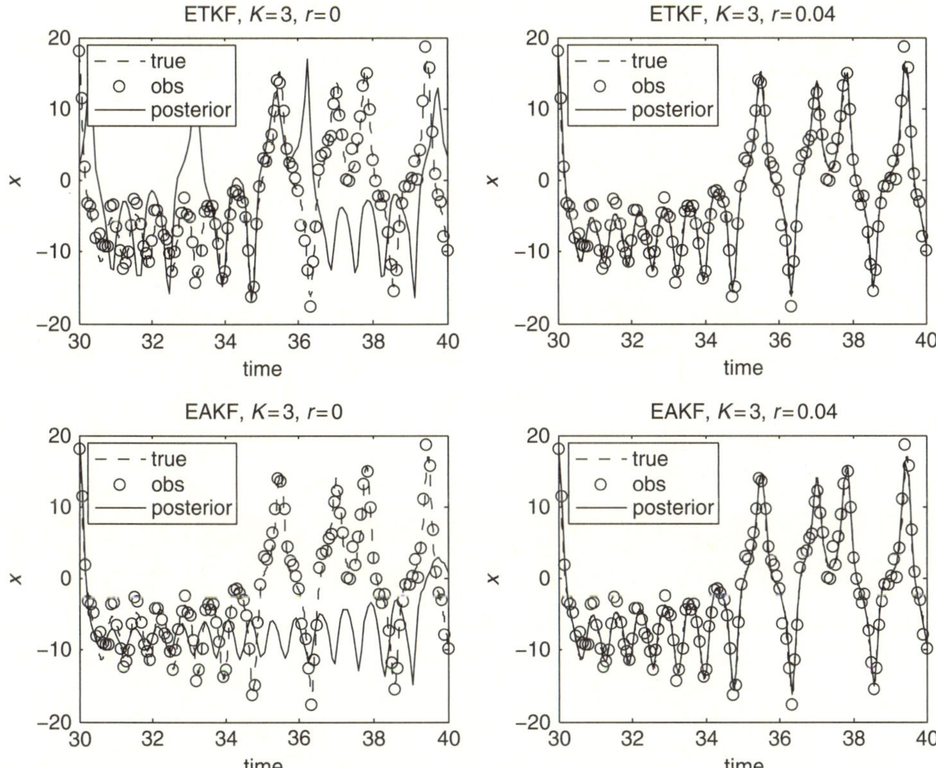

Figure 9.3 ETKF and EAKF for observation time of 0.08 units for ensemble size $K = 3$. Left panels with no variance inflation and right panels with $r = 4\%$ of inflation. Note the filter divergence without variance inflation and the high skill with small variance inflation.

also include RMS errors for simulations with partial observations. From these simulations, we find that observing variables x, y is essential for obtaining reasonably accurate filtered solutions since these variables identify the solution location that is symmetric about the z-axis. We find that ETKF and EAKF with small ensemble size are the most robust strategies among these.

Next, we discuss results from simulations with the longer observation time of 0.25 units which is beyond the shortest correlation time in L-63. As pointed out by Yang *et al.* (2006) and Kalnay *et al.* (2007), we also find that EKF with $r = 5\%$ does not produce a convergent filtered solution (see Fig. 9.4) between 30 and 34 time units. A second type of inflation is to add the diagonal components of the posterior covariance matrix $R_{m|m}$ with uniformly distributed random numbers drawn from the interval $[0, \mu]$ (Yang *et al.*, 2006; Kalnay *et al.*, 2007). Significant improvement (as shown in Fig. 9.4 and Table 9.2) is obtained through this second type of variance inflation with or without the multiplicative variance inflation; the RMS error drops to 0.68. In our next simulations with EnKF, ETKF and

Table 9.2 Average RMS errors for numerical simulations with observation time of 0.25 units. EKF is performed with additive variance inflation, $\mu = 0.1$. EnKF is performed with $K = 10$, $r = 5\%$, ETKF and EAKF are performed with $K = 3$, $r = 5\%$.

	$\{x, y, z\}$	$\{x, y\}$	$\{x, z\}$	$\{y, z\}$	$\{x\}$	$\{y\}$	$\{z\}$
EKF	0.68	1.17	2.88	1.42	7.39	4.11	21.21
EnKF	0.75	0.92	4.50	0.84	4.55	1.05	4.60
ETKF	0.59	0.81	2.35	0.60	2.41	1.08	7.23
EAKF	0.62	0.74	3.68	0.78	2.84	1.30	4.22

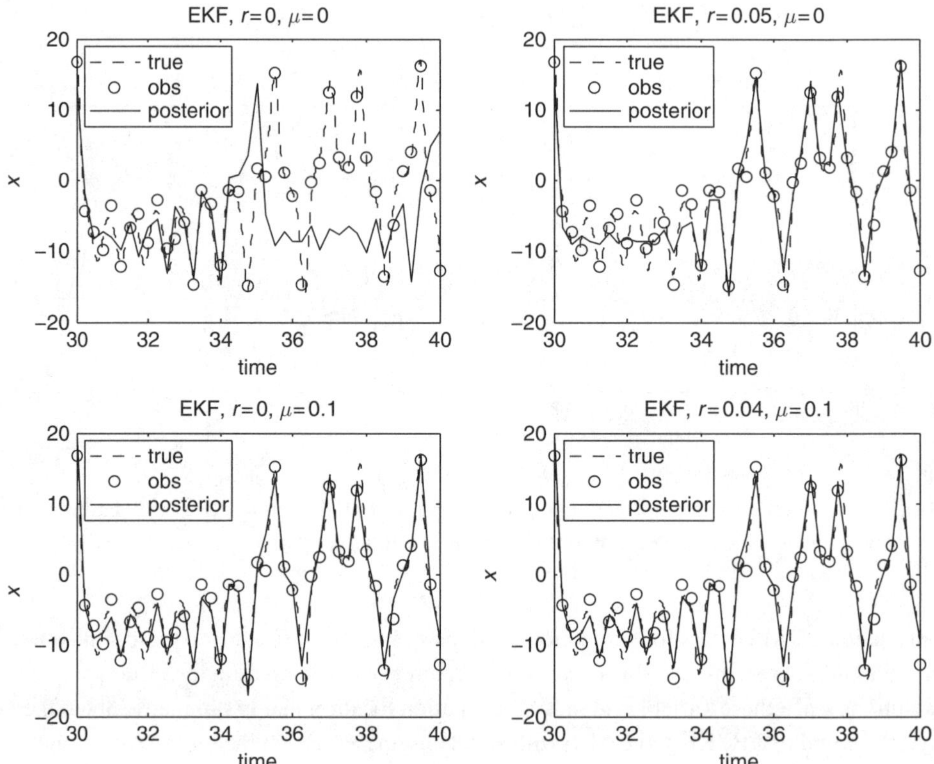

Figure 9.4 Extended Kalman filter for observation time of 0.25 units, implemented with various kinds of variance inflations.

EAKF (see Fig. 9.5), we only employ the multiplicative variance inflation $r = 5\%$ and ensembles of sizes $K = 10$, 3, and 3, consecutively. We find that both ETKF and EAKF provide lower average RMSs compared to EKF and EnKF with large ensemble size for various parameters (see Table 9.2). When observations are sparse, the ETKF filter skill is slightly better than EAKF except for the subtle situation where z alone is observed where EAKF has much better skill.

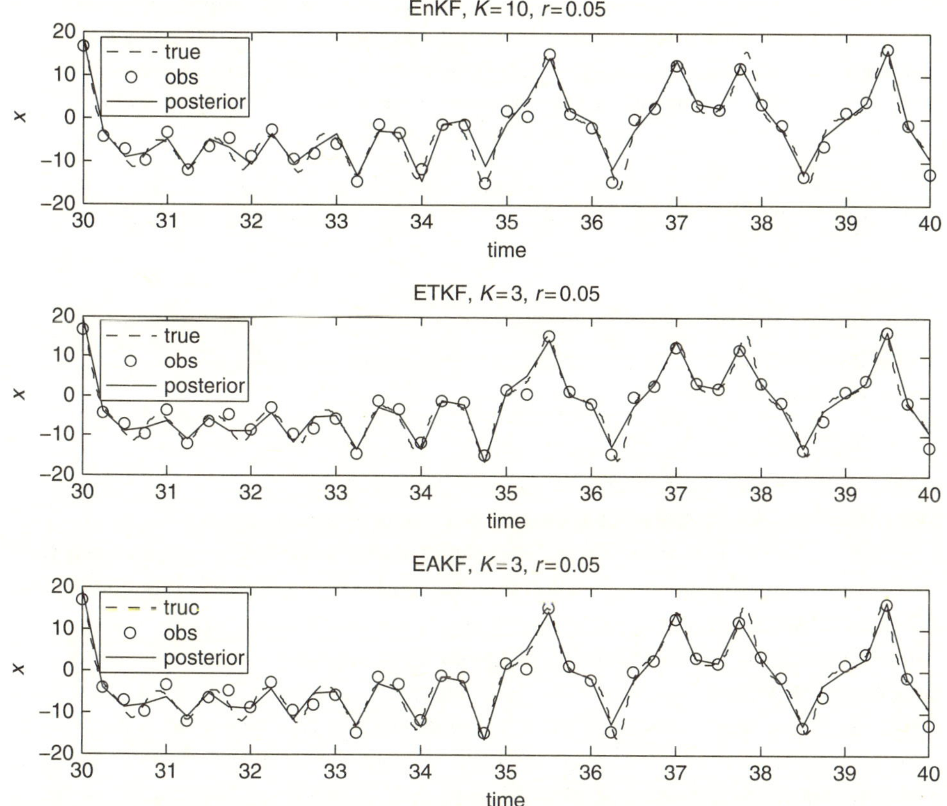

Figure 9.5 EnKF, ETKF and EAKF for observation time of 0.25 units, implemented with ensemble size $K = 10$ for EnKF and $K = 3$ for ETKF, EAKF, and $r = 5\%$ for all three methods.

9.5 Ensemble square-root filters on stochastically forced linear systems

In this section, we check the performance of ETKF and EAKF on the stochastically forced linear advection–diffusion equation from (7.20),

$$\frac{\partial u(x,t)}{\partial t} = -\frac{\partial u(x,t)}{\partial x} - du(x,t) + \mu \frac{\partial^2 u(x,t)}{\partial x^2} + \bar{F}(x,t) + \sigma(x)\dot{W}(t), \tag{9.40}$$

discretized at $2N + 1 = 123$ grid points with obervations available at every $P = 3$ grid points so in total we have $2M + 1 = 41$ sparse regularly spaced observations. These parameters are fixed exactly as in Section 7.2 of Chapter 7. In our experiments, we test both filters with various ensemble sizes $K = 100, 150, 300$ and multiplicative variance inflations $r = 0, 10\%, \ldots, 40\%$.

In Table 9.3, we show the average RMS errors of extensive runs for a regime where the energy spectrum is smooth, $E_k = k^{-5/3}$, the observation frequency is rather short,

Table 9.3 RMS errors of filtering the stochastically driven advection–diffusion equation with ETKF and EAKF in a regime with smooth spectrum $E_k = k^{-5/3}$, decaying mean and observable time $\Delta t = 0.1$ for various ensemble size K and covariance inflation r. The maximum temporal average spatial correlation is 0.74 and is achieved by the experiment with EAKF with $K = 150$ and $r = 40\%$.

RMS error	$r = 0\%$	$r = 10\%$	$r = 20\%$	$r = 30\%$	$r = 40\%$
ETKF $K = 100$	1.41	1.34	1.31	1.26	1.22
ETKF $K = 150$	1.37	1.30	1.29	1.26	1.24
ETKF $K = 300$	1.35	1.46	1.54	1.59	–
EAKF $K = 100$	1.98	1.85	1.71	1.60	1.50
EAKF $K = 150$	1.92	1.77	1.64	1.51	1.44
EAKF $K = 300$	1.59	1.48	1.38	1.30	–

$T_{\text{obs}} = 0.1$ (see Fig. 7.2 for comparison with the decorrelation time) and with decaying mean $\bar{F}(x, t) = 0$, for the stochastically forced selectively damped advection–diffusion equation, i.e. $d = 0$, $\mu = 0.01$. We find that the RMS errors of ETKF are slightly below the observation noise $\sqrt{r^o} = 1.43$ (recall that $r^o = 2.05$ is chosen to reflect observation error variance of 5% in each Fourier mode, see (7.4) in Chapter 7) for experiments with ensemble sizes 100 and 150. The highest average spatial correlation is 0.74 and it is achieved by the experiment with EAKF with $r = 40\%$ and $K = 150$ for which the RMS, 1.44, is not the lowest. In fact, experiments with EAKF produce larger RMS errors in this regime. Repeating the result shown in Fig. 7.3 in Chapter 7, there can be filter divergence for ETKF with resonant periodic forcing ($\bar{F}$ defined as in (5.15) in Chapter 5 with $\omega_o(k) = \omega_k$ for $|k| \leq M$) even for the $k^{-5/3}$ spectrum and for the selectively damped advection–diffusion equation. For the rough spectrum $E_k = 1$, the situation is worse and both filters have very little skill even for a decaying mean state and for an ensemble size of $K = 500$. The lowest average RMS errors is roughly 11.28 and the highest average spatial correlation is only 0.38. These rather poor results of filter performance for ETKF and EAKF should be contrasted with the high skill of those already reported in Chapter 7 by the reduced filters FDKF, SDAF, VSDAF and RFDKF.

Next we consider the stochastically forced weakly damped advection equation in (9.40) with $d = 0.01$, $\mu = 0$. As discussed in Chapter 7, this model is considered to be the hardest test bed when it is simultaneously non-observable, has rough equipartition kinetic energy spectrum and is dominated by a resonant periodic forcing signal. Here, we will show that even when these three conditions aren't met, both ETKF and EAKF fail to produce a reasonably skillful solution. We did extensive simulations in the favorable regime with an observable time $\Delta t = 0.5$, decaying mean $\bar{F}(x, t) = 0$, and a smooth spectrum $E_k = k^{-5/3}$. In Table 9.4, the best average RMS error from various ensemble sizes and covariance inflation is still worse than the observation error $\sqrt{r^o} = \sqrt{(2M+1)\hat{r}^o} = 1.43$; it is achieved by EAKF with $K = 300$ with no variance inflation.

When $E_k = 1$, the results are even worse; the lowest average RMS error is about 12.65 and the highest average spatial correlation is 0.22. Again, we emphasize that these results

Table 9.4 RMS errors of filtering the stochastically driven weakly damped advection equation with ETKF and EAKF in a regime with smooth spectrum $E_k = k^{-5/3}$, decaying mean and observable time $\Delta t = 0.5$ for various ensemble size K and covariance inflation r.

RMS error	$r = 0\%$	$r = 10\%$	$r = 20\%$	$r = 30\%$	$r = 40\%$
ETKF $K = 100$	1.90	2.09	2.17	2.23	2.33
ETKF $K = 150$	2.02	2.01	2.08	2.16	2.31
ETKF $K = 300$	2.19	2.47	2.58	2.69	–
EAKF $K = 100$	1.84	1.89	2.01	2.11	2.22
EAKF $K = 150$	1.83	1.88	2.02	2.14	2.26
EAKF $K = 300$	1.49	1.56	1.68	1.79	–

are not yet in the severe regime, since the observability is not violated and at the same time non-periodic forcing is not considered. In Chapter 7, we showed that much better filtered solutions are obtainable with the Fourier domain filtering strategies.

9.6 Advantages and disadvantages with finite ensemble strategies

From the formulation in Sections 9.2–9.4, it is clear that the ensemble approach does not require any linear tangent or adjoint of the nonlinear operator $\vec{f}$; it simply evolves each ensemble member through the full nonlinear dynamics (9.1). When the observation time frequency is short enough, the prior ensemble distribution is not too different from a Gaussian distribution since the nonlinearity is weak; on the other hand, when the nonlinearity is strong, the update becomes suboptimal since the prior ensemble is no longer Gaussian. Now suppose that the prior ensemble samples exactly a Gaussian distribution; a nonlinear observation operator $\vec{g}$ can also lead to a completely non-Gaussian posterior distribution. In the Kalman-based ensemble approach, this non-Gaussianity is not fully described since it implicitly linearizes the observation operator through (9.20) and it forces a Gaussian posterior solution.

An interesting question with the ensemble approach is how many ensemble members are needed for accurate filtering? From the computational point of view, we hope that the ensemble size is much less than the model size (e.g. an ensemble of less than 100 members for a model of 10^7 variables is typically used/proposed in numerical weather prediction). On such a high-dimensional problem, we can rule out particle filtering since the ensemble size required for convergent solutions is practically impossible at this point. On the other hand, when the ensemble size is small, the sampling errors in the covariance estimate underestimate the covariance even in the perfect model situation. This is largely caused by the strong nonlinearity of $\vec{f}$, which produces an ensemble whose members point in very similar directions in phase space and eventually collapse into a single direction. A simple remedy to this ensemble collapse is to inflate the covariance matrix (Anderson, 2001) as in our example in Section 9.4 or perhaps to split the ensemble into two, such that the update of the first half is performed with the covariance estimated from the second half and vice versa (Houtekamer and Mitchell, 2001).

Furthermore, when the ensemble size is small, there are significant sampling errors due to spurious correlations from distant grid points (Lorenc, 2003). In the current generation of ensemble filters, various physical space localized updates are implemented (Keppenne, 2000; Mitchell *et al.*, 2002; Ott *et al.*, 2004; Hunt *et al.*, 2007) to prevent such spurious correlations. When the filter is updated sequentially (one observation at a time assuming that each observation is spatially uncorrelated with others) as adopted by Anderson (2001) and Whitaker and Hamill (2002), the covariance is localized (centered at the corresponding observation) by a Schur product with a correlation function with compact support as in Gaspari and Cohn (1999) so that the local update only impacts the model variables locally. This localized approach is computationally expensive when the observations are dense and it also introduces dynamical imbalances (Houtekamer and Mitchell, 2001) when two independent adjacently local filters assign different Kalman gain. In Chapter 11, we will utilize this sequential filtering strategy on EAKF for filtering sparse observations in large-dimensional turbulent dynamical systems. Alternatively, we can update each model grid point by processing all of the observations within the neighborhood of the associated grid point (Keppenne, 2000; Ott *et al.*, 2004; Hunt *et al.*, 2007). This approach has the advantage that it allows the local filters to be computed independently by parallel processors and it produces accurate filter solutions with ensemble sizes that are much less than the number of local state variables. At the same time, the physical imbalance is not a problem since adjacent grid points have overlapped observations and therefore the Kalman weights are not completely distinct. However, when the ensemble size K is smaller than the size of the local state space, there is a limitation to fit more than K data, e.g. the moisture field is typically available in a horizontally denser observation network relative to the wind or temperature field.

As pointed out in Chapter 1, the high-dimensional turbulent dynamical systems which define the true signal from nature have a fundamentally different statistical character than the Lorenz-63 model discussed in Section 9.4. The Lorenz-63 model is weakly mixing with one unstable direction on an attractor with high symmetry. This characteristic is signified by the bi-modality of the distribution (or "butterfly"-like attractors). In contrast, realistic turbulent dynamical systems have a large phase space dimension, a large dimensional unstable manifold on the attractor and are strongly mixing with exponential decay of correlations. In Chapter 11, we will discuss the filtering performance of the finite ensemble filter approach on the two simple prototype examples of turbulent dynamical systems introduced in Chapter 1: the Lorenz-96 model (Lorenz, 1996) and the two-layer quasi-geostrophic (QG) model (Salmon, 1998; Vallis, 2006). In particular, we will see "catastrophic filter divergence" with ETKF formulated as in Section 9.3 above for filtering the sparsely observed L-96 model. We will also encounter filter divergence with EAKF when the filter model is numerically stiff as in the ocean with a small Rossby radius for filtering a sparsely observed QG model.

10

Filtering prototype nonlinear slow–fast systems

Many contemporary problems in science and engineering involve large-dimensional turbulent nonlinear systems with multiple time-scales, i.e. slow-fast systems. Here we introduce a prototype nonlinear slow-fast system with exactly solvable first- and second-order statistics which despite its simplicity, mimics crucial features of realistic vastly more complex physical systems. The exactly solvable features facilitate new nonlinear filtering algorithms (see Chapter 13 for another use in stochastic parametrizations) and allow for unambiguous comparison of the true signal with an approximate filtering strategy to understand model error. First we present an overview of the models, the main issues in filtering slow–fast systems, and filter performance. In the second part of this chapter we give more pedagogical details.

10.1 The nonlinear test model for filtering slow–fast systems with strong fast forcing: An overview

The dynamic models for the coupled atmosphere–ocean system are prototype examples of slow–fast systems where the slow modes are advective vortical modes and the fast modes are gravity waves (Salmon, 1998; Embid and Majda, 1998; Majda, 2003). Depending on the spatio-temporal scale, one might need only a statistical estimate through filtering of the slow modes, as in synoptic scales in the atmosphere (Daley, 1991) or both slow and fast modes such as for squall lines on mesoscales due to the impact of moist convection (Khouider and Majda, 2006). In either situation, the noisy partial observations of quantities such as temperature, pressure and velocity necessarily mix both the slow and fast modes (Daley, 1991; Cohn and Parrish, 1991; Majda, 2003). Moreover, tropical moist convection involves strong forcing of the fast gravity waves and this forcing involves rapid spontaneous bursts of activity; the emerging fast gravity waves have a major impact on tropical weather and intraseasonal climate forecasting (Khouider and Majda, 2007) so successful filtering of such signals is a central contemporary science issue. Furthermore, the dynamical models often suffer from significant model errors due to lack of resolution or inadequate parametrization of physical processes such as clouds, moisture, boundary layers or topography. Here, we briefly review the development (Gershgorin and Majda, 2008, 2010) of a three-dimensional stochastic test model which has transparent analogues for all of the scientific issues with slow–fast systems described above. The advantage of using this low-dimensional test model for filtering is that on the one hand it is simple enough to have exactly solvable first- and second-order statistics needed for the dynamic forecast in a nonlinear extended Kalman filter (NEKF). On the other hand, the test model carries important properties of realistic systems such as multiple time-scales, nonlinear non-Gaussian dynamics and strong fast forcing.

The test model is given by a three-dimensional system of stochastic differential equations for the slow real mode u_1 and complex fast mode u_2

$$du_1 = (-\gamma_1 u_1 + f_1(t))dt + \sigma_1 dW_1, \tag{10.1}$$

$$du_2 = \Big((-\gamma_2 + \mathrm{i}\omega_0/\varepsilon + \mathrm{i}a_0 u_1)u_2 + f_2(t)\Big)dt + \sigma_2 dW_2. \tag{10.2}$$

where γ_1, γ_2 and σ_1, σ_2 are damping and white noise coefficients, respectively, that represent the interaction of the model with the unresolved modes; the small parameter ϵ measures the ratio between the deterministic time-scales of the fast and slow modes; a_0 is the nonlinearity coefficient; and f_1 and f_2 represent the forcing of the slow and fast modes, respectively. The structure of the model is motivated by studies of geophysical systems that demonstrate that the central feature of the slow–fast interactions are a slow vortical mode, represented by u_1 here, and fast gravity waves, represented by u_2 here (Gershgorin and Majda, 2008). Moreover, tropical moist convection as another major dynamical property in the system is represented by the strong fast forcing f_2. The fast forcing has a direct impact on the modulation of the fast wave amplitude $|u_2|$. Moreover, filtering mixes the fast modes with the slow mode through observations and, therefore, indirect impact of the fast forcing on the slow mode can also occur. Exact statistical formulas for the nonlinear non-Gaussian statistical solutions of the test model in (10.1), (10.2) are developed (Gershgorin and Majda, 2008, 2010) by generalizing and extending the exact solutions for the Kubo oscillator from statistical physics for non-stationary and cross-correlated dynamics. These details of the exact solution methodology and examples of the non-Gaussian features of the models are discussed in detail below in this chapter and more can be found in Gershgorin and Majda (2008, 2010). Figure 10.1 depicts a typical realization of the turbulent dynamical system in (10.1), (10.2). The lowest panel depicts the sequence of randomly located bursts of spontaneous fast forcing $f_2(t)$ generated by a Markov jump process as in Chapter 8. The top panel shows the slow mode to be filtered which sees only the slow time periodic forcing from $f_1(t)$ in (10.1). The second panel shows $Re[u_2]$ for the nonlinear test model with $a_0 = 1$ and the strong nonlinear response while the third panel shows $Re[u_2]$ for the weaker response to fast forcing in the linear system for (10.2) with $a_0 \equiv 0$ with a superficial resemblance to the nonlinear case. The wave envelopes, $|u_2|$, associated with the nonlinear and linear cases are compared in the fourth panel of Fig. 10.1 to quantify the differences in behavior. Such a complex signal with multiple scales defines the typical turbulent signal to be filtered in the slow–fast system in (10.1), (10.2). For example, a typical observation of the true system which mixes the slow

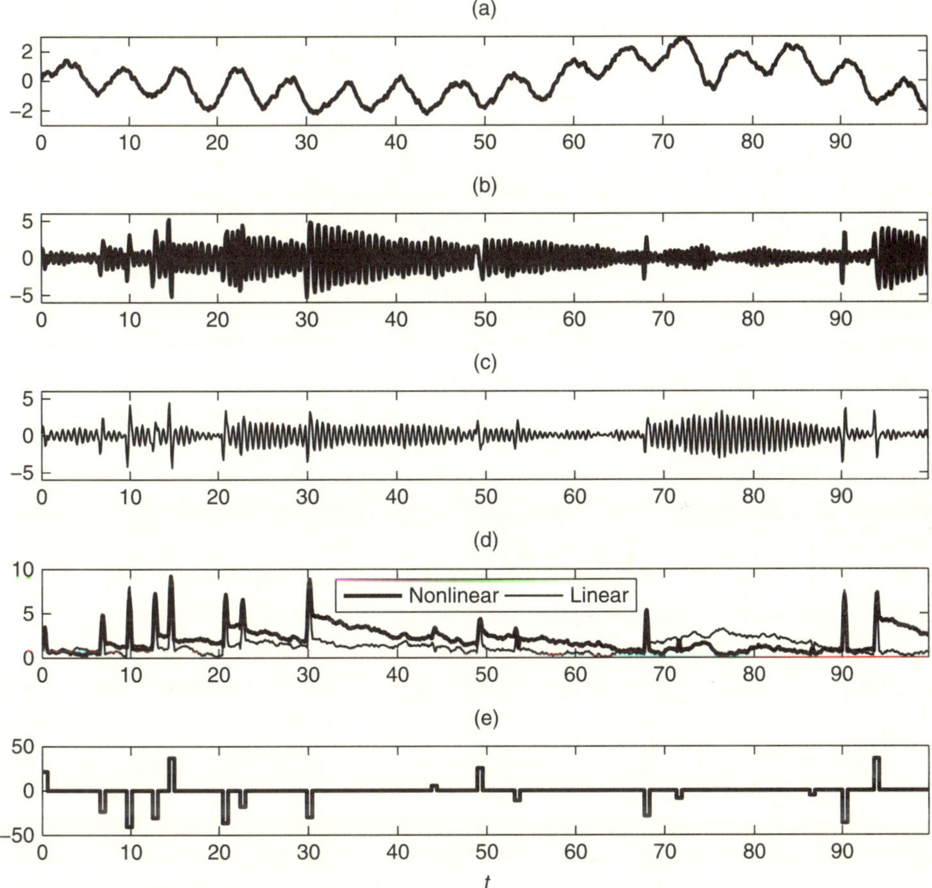

Figure 10.1 (a) Slow mode u_1, (b) fast mode u_2 (its real part) for the nonlinear case with $a_0 = 1$, (c) fast mode u_2 (its real part) for the linear case with $a_0 = 0$, (d) amplitude $|u_2|$ for the nonlinear case (thick line) and linear case (thin line) and (e) fast forcing f_2 with randomly located spontaneous bursts

and fast modes involves a weighted sum of the signals in (a) and (b) from Fig. 10.1 with random noise added. Central practical issues for the filtering skill for such signals involve

- recovery of the slow mode amplitude, u_1, from sparse noisy observations which mix the slow and fast modes;
- recovery of the fast wave u_2 and the large-scale envelope $|u_2|$ for the fast wave in response to the strong fast forcing.

It is important to note that for a reasonably long observation time Δt, it is unrealistic to expect high skill in full recovery of the phase of u_2; on the other hand, the wave envelope,

$|u_2|$, is the most important crude feature of the signal to capture. Clearly, the observation time, Δt, and observation noise variance are central parameters to vary to test the skill of filtering algorithms for the nonlinear slow–fast system. Next, we briefly discuss this topic.

10.1.1 NEKF and linear filtering algorithms with model error for the slow–fast test model

In the test model, we have exact nonlinear statistics for the first and second moments. These attractive features allow us to define a nonlinear extended Kalman filter (NEKF) in the following fashion. We use the notation developed in Section 1.1 of Chapter 1 to explain the NEKF algorithm. First assume that $p_{m,+}(u) = \mathcal{N}(u_{m,+}, R_{m,+})$, then take these means and covariances and propagate them by the exact statistical formulas for the mean and covariance for the nonlinear dynamics in (10.1), (10.2) for the observation time Δt to generate $u_{m+1,-}$, $R_{m+1,-}$; use these values to build an approximate Gaussian $p_{m+1,-}(u)$ given by $\mathcal{N}(u_{m+1,-}, R_{m+1,-})$, and then use the Kalman filter formulas in (1.16) to update the effect of the observations to generate $\mathcal{N}(u_{m+1,+}, R_{m+1,+})$. This completes the definition of the NEKF algorithm. This nonlinear filtering algorithm has a very attractive feature compared with the conventional extended Kalman filters (Anderson and Moore, 1979) which are usually proposed for nonlinear problems as discussed in Chapter 9. The extended Kalman filter (EKF) requires a linear tangent model and local linearization which can be completely inaccurate for a slow–fast system with multiple time-scales such as (10.1), (10.2); the NEKF algorithm circumvents this issue completely by using exact first- and second-order statistics for the nonlinear problem. Of course, for highly non-Gaussian nonlinear models, the mean and covariance are not necessarily an accurate reflection of the true statistical dynamics; however, the models here in the slow–fast regime are mildly non-Gaussian (Gershgorin and Majda, 2008, 2010).

There is another linear filter algorithm for the system (10.1), (10.2) which is akin to the one in Chapter 12: the judicious linear stochastic models for filtering nonlinear turbulent dynamical systems. Namely, set the coefficient of nonlinearity, a_0, in (10.2) to zero, $a_0 = 0$, and use the Kalman filter for the linear system with model error to filter nonlinear signals such as those in Fig. 10.1. Below, we call this the linear stochastic model filter (LSMF) algorithm.

The mathematical properties and filtering skill for NEKF and LSMF are discussed below in this chapter and more extensively by Gershgorin and Majda (2008, 2010). In this overview section, we briefly highlight some of the interesting results for filtering skill with strong fast forcing (Gershgorin and Majda, 2010). We illustrate filtering skill in the slow–fast system for the difficult sparse case of a single observation which mixes the slow and fast modes with equal weights at various observation times, Δt, with varying observational noise variance, r^o (Gershgorin and Majda, 2010).

Recall that the true signal being filtered has all the complexity depicted in Fig. 10.1. In Fig. 10.2 and Table 10.1 we show the skill in recovering the slow mode for both the NEKF

Table 10.1 RMSE of the slow mode u_1 of NEKF and the linear Kalman filter with model error (in parentheses). Segments of sample trajectories are shown in Fig. 10.1

$r^o \backslash \Delta t$	0.1	1.0
0.1	0.14(0.16)	0.33(0.33)
1.0	0.22(0.22)	0.35(0.36)
10.0	0.34(0.34)	0.49(0.5)

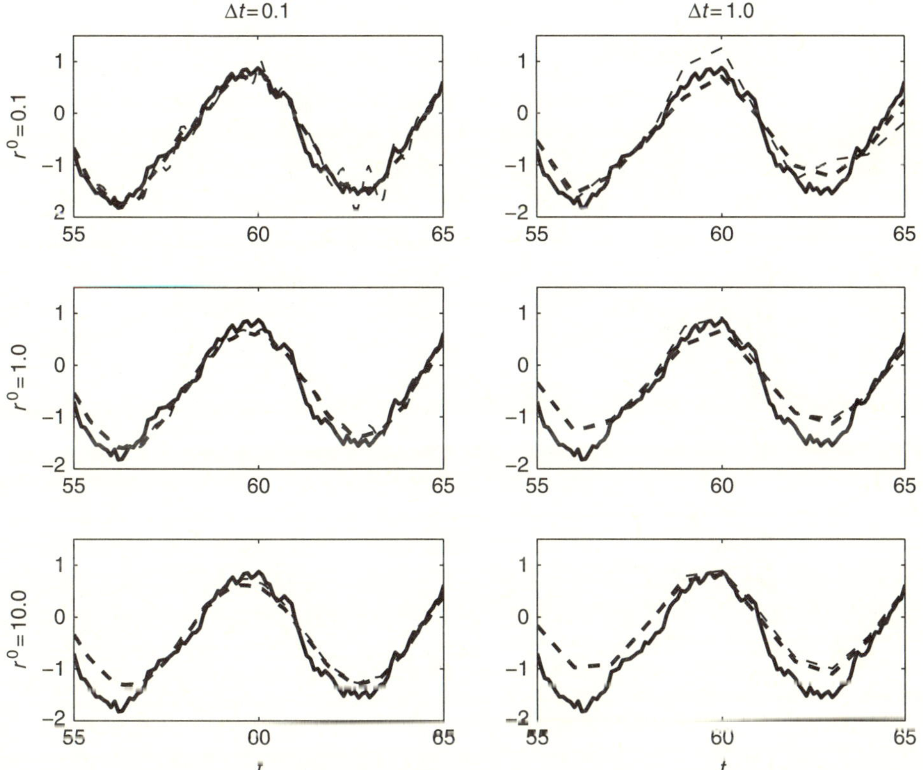

Figure 10.2 Slow mode u_1: truth signal (solid line), NEKF (thick dashed line) and linear Kalman filter (thin dashed line). The corresponding RMSE for a long trajectory are shown in Table 10.1

and LSMF algorithms; it is clear from both Fig. 10.2 and Table 10.1 that the two algorithms have comparable and excellent skill in recovering the slow mode except for long observation times and large noise where the filtering skill deteriorates a bit. Table 10.2 illustrates through RMS errors the much higher skill of NEKF compared with LSMF for

Table 10.2 RMSE of the fast mode u_2 of NEKF and the linear Kalman filter with model error (in parentheses).

$r^o \backslash \Delta t$	0.1	1.0
0.1	0.24(0.61)	0.62(1.10)
1.0	0.44(0.97)	0.98(1.30)
10.0	1.05(1.28)	1.25(1.53)

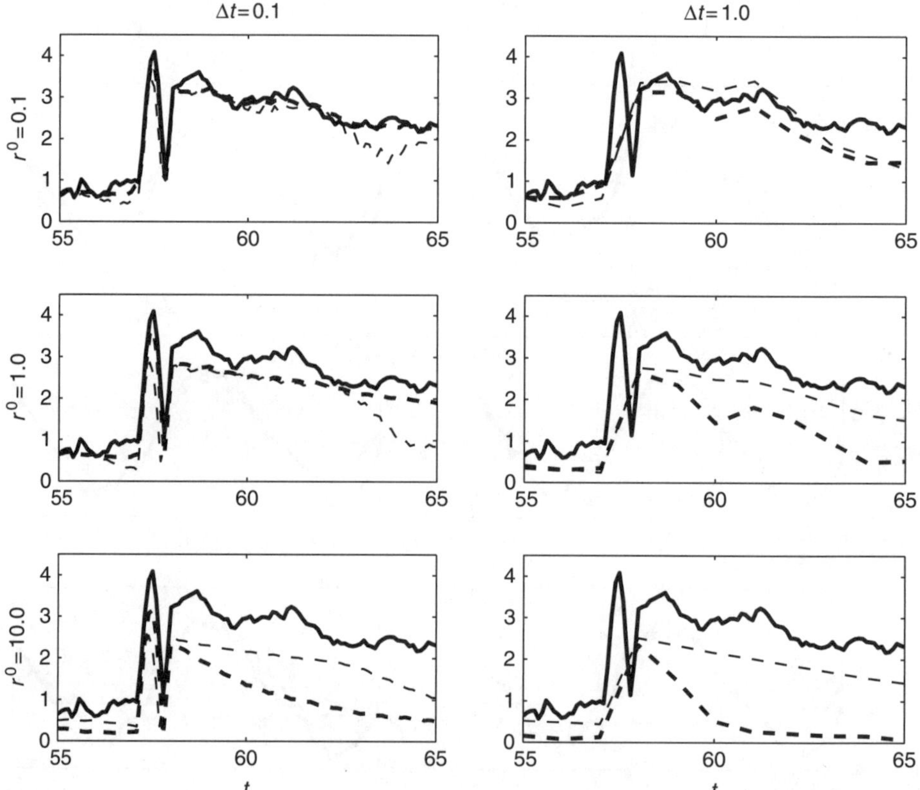

Figure 10.3 Amplitude $|u_2|$ of the fast mode: truth signal (solid line), NEKF (thick dashed line) and linear Kalman filter (thin dashed line). The corresponding RMSEs for a long trajectory are shown in Table 10.3

filtering the entire signal u_2 for all observation times and noise levels; the effect of model error in LSMF is significant for less skillful capturing of the correct phase in u_2. On the other hand, as depicted in Fig. 10.3 and Table 10.3, if one is only interested in the filtering skill for the fast wave envelope, $|u_2|$, the LSMF algorithm is nearly as skillful as NEKF

Table 10.3 RMSE for the amplitude $|u_2|$ of the fast mode of NEKF and the linear Kalman filter with model error (in parentheses). Numbers in bold indicate the situations when the linear Kalman filter performs better than NEKF. Segments of sample trajectories are shown in Fig. 10.3

$r^o \backslash \Delta t$	0.1	1.0
0.1	0.17(0.28)	0.44(0.52)
1.0	0.36(0.58)	0.91(**0.77**)
10.0	1.03(**0.82**)	1.20(**0.93**)

for short observation times and signal noise and, remarkably, the LSMF algorithm has significant high skill superseding NEKF for more noisy, sparsely observed regimes! Thus, we have another instance of the theme developed already in Chapter 7 and which continues in Chapters 12 and 13 for turbulent dynamical systems; namely, for noisy sparsely observed turbulent signals, cheap filters with judicious model error can have high skill, often exceeding that with the perfect model.

10.2 Exact solutions and exactly solvable statistics in the nonlinear test model

As presented in the overview in Section 10.1, consider a system of one slow wave represented by a real-valued function u_1 and two fast waves represented by a complex-valued function u_2. For simplicity in exposition, here and in the remainder of the chapter we drop the interesting effect of fast forcing so that $f_2(t) \equiv 0$. We model this system via a system of coupled stochastic differential equations of the form

$$du_1 = (-\gamma_1 u_1 + f_1(t))dt + \sigma_1 dW_1,$$
$$du_2 = (-\gamma_2 + \mathrm{i}\omega_0/\varepsilon + \mathrm{i}a_0 u_1)u_2 dt + \sigma_2 dW_2,$$

where γ_1 and γ_2 are the damping coefficients of the slow and fast waves, respectively, $f_1(t)$ is forcing of the slow wave, σ_1 and σ_2 represent the strengths of the noise of the slow and fast waves, respectively, ε is a small parameter that characterizes the time-scale difference between the slow and the fast modes, ω_0 is the typical frequency of the fast wave in units of ε, and a_0 is a nonlinearity parameter. We choose the oscillatory forcing $f_1(t) = A\sin(\omega t)$. System (10.1) and (10.2) is to be solved together with the initial conditions

$$u_1(t_0) = u_{10},$$
$$u_2(t_0) = u_{20},$$

where u_{10} and u_{20} are correlated Gaussian random variables with known parameters: $\langle u_{10}\rangle$, $\langle u_{20}\rangle$, $\mathrm{Var}(u_{10})$, $\mathrm{Var}(u_{20})$, $\mathrm{Cov}(u_{20}, u_{20}^*)$ and $\mathrm{Cov}(u_{20}, u_{10})$. As usual, $\langle\cdot\rangle$ is expectation, $\mathrm{Var}(\cdot)$ is variance and $\mathrm{Cov}(\cdot,\cdot)$ is covariance.

10.2.1 Path-wise solution of the model

It is not difficult to develop path-wise solutions of the model equations. Such path-wise solutions provide the signals, which we attempt to filter. First, we solve Eqn (10.1) since it is independent of the fast wave u_2. As in Chapter 2, the slow wave u_1 is easily found using an integrating factor

$$u_1(t) = u_{10}e^{-\gamma_1(t-t_0)} + F_1(t) + \sigma_1 \int_{t_0}^{t} e^{\gamma_1(s-t)} dW_1(s), \tag{10.3}$$

where

$$\begin{aligned} F_1(t) &= \int_{t_0}^{t} f_1(s) e^{-\gamma_1(t-s)} ds \\ &= \frac{A}{\gamma_1^2 + \omega^2}\Big(e^{-\gamma_1(t-t_0)}\big(\omega\cos(\omega t_0) - \gamma_1 \sin(\omega t_0)\big) - \omega\cos(\omega t) + \gamma_1 \sin(\omega t)\Big). \end{aligned}$$

Note that $u_1(t)$ is Gaussian random process and thus is fully defined by its mean and variance, which will be computed below.

Now, to solve Eqn (10.2) we treat u_1 as a known function. Using the integrating factor for the equation with time-dependent frequency, we obtain u_2

$$u_2(t) = e^{-\gamma_2(t-t_0)}\psi(t_0, t)u_{20} + \sigma_2 \int_{t_0}^{t} e^{-\gamma_2(t-s)}\psi(s, t) dW_2(s), \tag{10.4}$$

where we defined new functions

$$\begin{aligned} \psi(s, t) &= e^{\mathrm{i}J(s,t)}, \\ J(s, t) &= \int_{s}^{t} \Big(\omega_0/\varepsilon + a_0 u_1(s')\Big) ds' = (t - s)\omega_0/\varepsilon + a_0 \int_{s}^{t} u_1(s') ds' \\ &= J_D(s, t) + J_W(s, t) + b(s, t)u_{10}, \end{aligned}$$

where the deterministic part of $J(s, t)$ is

$$\begin{aligned} J_D(s, t) &= (t - s)\omega_0/\varepsilon \\ &\quad + \frac{Aa_0}{\gamma_1^2 + \omega^2}\left(\frac{\gamma_1}{\omega}\Big(\cos(\omega s) - \cos(\omega t)\Big) + \Big(\sin(\omega s) - \sin(\omega t)\Big)\right. \\ &\quad \left. + \Big(e^{-\gamma_1(s-t_0)} - e^{-\gamma_1(t-t_0)}\Big)\left(\frac{\omega}{\gamma_1}\cos(\omega t_0) - \sin(\omega t_0)\right)\right), \end{aligned}$$

the noisy part of $J(s, t)$ is

$$J_W(s, t) = \sigma_1 a_0 \int_{s}^{t} ds' \int_{t_0}^{s'} e^{\gamma_1(s''-s')} dW_1(s''),$$

and the prefactor of u_{10} is

$$b(s,t) = \frac{a_0}{\gamma_1}\left(e^{-\gamma_1(s-t_0)} - e^{-\gamma_1(t-t_0)}\right).$$

Note that $J(s,t)$ depends linearly on u_1 and, therefore, $J(s,t)$ is also a sum of independent Gaussian random fields. We also introduce the following notation for later use

$$\psi_D(s,t) = e^{\mathrm{i}J_D(s,t)},$$
$$\psi_W(s,t) = e^{\mathrm{i}J_W(s,t)}.$$

On the other hand, u_2 is not a Gaussian random variable in the case $a_0 \neq 0$ because the solution formula involves exponentials of Gaussian random variables. Below, we will confirm non-Gaussianity of u_2 numerically. Nevertheless, we will be able to compute the statistics of u_2 analytically due to the special structure of the solution.

10.2.2 Invariant measure and choice of parameters

Here, we obtain the invariant measure of system (10.1) and (10.2) without forcing. We separate system (10.1) and (10.2) into two parts: the deterministic part

$$du_1 = 0, \tag{10.5}$$
$$du_2 = \mathrm{i}(\omega_0/\varepsilon + a_0 u_1)u_2 dt, \tag{10.6}$$

and the randomly fluctuating and damped part

$$du_1 = -\gamma_1 u_1 dt + \sigma_1 dW_1, \tag{10.7}$$
$$du_2 = -\gamma_2 u_2 dt + \sigma_2 dW_2. \tag{10.8}$$

Now, we easily find the invariant measure for both systems. System (10.5) and (10.6) has the invariant (Liouville) measure

$$p_{\mathrm{det}}(u_1, u_2) = p_1(u_1)p_2(|u_2|), \tag{10.9}$$

for any probability measures p_1 and p_2. As in Chapters 2 and 3, it is well known (Gardiner, 1997) that the unique invariant measure for the Langevin equation is Gaussian. Therefore, the unique invariant measure for system (10.7) and (10.8) is a product of Gaussian measures

$$p_{\mathrm{rand}}(u_1, u_2) = \frac{\sqrt{2\gamma_1\gamma_2}}{\pi\sigma_1\sigma_2}\exp\left(-\frac{\gamma_1 {u_1}^2}{\sigma_1^2} - \frac{2\gamma_2|u_2|^2}{\sigma_2^2}\right). \tag{10.10}$$

Note that $p_{\mathrm{rand}}(u_1, u_2)$ satisfies Eqn (10.9) and, therefore, the invariant measure for system (10.1) and (10.2) without forcing is given by Eqn (10.10).

From Eqn (10.10), we conclude that the average energies of the three modes (which are equal to the corresponding variances) are

$$E_{u_1} = \frac{\sigma_1^2}{2\gamma_1},$$
$$E_{\mathrm{Re}[u_2]} = E_{\mathrm{Im}[u_2]} = \frac{\sigma_2^2}{4\gamma_2}. \tag{10.11}$$

Now, we can choose parameters of the model in order to control the average energies. For example, for the energy equipartition case, we choose parameters to satisfy

$$E_{u_1} = E_{\mathrm{Re}[u_2]} = E_{\mathrm{Im}[u_2]}.$$

which is the same as

$$\frac{\sigma_1^2}{\gamma_1} = \frac{\sigma_2^2}{2\gamma_2}.$$

Recall from Chapters 2 and 5 that the decorrelation time of a mode is proportional to the inverse of the damping coefficient. We will consider two different regimes with strong and weak damping.

For the case of strong damping, we choose all the parameters to be of order 1, i.e. $\gamma_1 = 1$, $\sigma_1 = 1$, $\gamma_2 = 1.5$ and $\sigma_2 = 1$. Then, we have $E_{u_1} = 0.5$ and $E_{Re[u_2]} = E_{Im[u_2]} = 0.5625$.

For the case of weak damping, which is the relevant physical regime for slow–fast systems, we choose the decorrelation time of the slow mode to be much longer than the oscillation period of the fast mode. The oscillation period of the slow mode is of order 1 while the oscillation period of the fast mode is of order $1/\varepsilon$. We also choose the decorrelation time of the fast mode to be of the same order as the decorrelation time of the slow mode. We take $\varepsilon = 0.1$. Then, the oscillation time of the fast mode is of order $T_0 = 2\pi/\varepsilon \approx 0.6$. Now we can choose $\gamma_1 = 0.09$ and $\gamma_2 = 0.08$ such that $T_0 \ll 1/\gamma_1$ and $T_0 \ll 1/\gamma_2$. Suppose the average energy of the first mode is $E_{u_1} = 1$. Then, for the energy equipartition case, we have $\sigma_1 = \sqrt{\gamma_1} = 0.3$ and $\sigma_2 = \sqrt{2\gamma_2} = 0.4$. The remaining parameters are chosen to be the same for both strong and weak damping: $\omega_0 = 1$, $a_0 = 1$, $A = 1$ and $\omega = 1$.

Numerical test of the analytical path-wise solutions

Here, we confirm that the numerical solution of Eqns (10.1) and (10.2) converges to the analytical solution given by Eqns (10.3) and (10.4). We use the standard Euler–Maruyama method for solving Eqns (10.1) and (10.2) numerically (Gardiner, 1997). We denote by u_j^n a numerical approximation of $u_j(t_n)$ for $j = \{1, 2\}$ at time grid point $t_n = hn$, where h is a time step. Then, the numerical scheme becomes

$$u_1^{n+1} = u_1^n + h(-\gamma_1 u_1^n + f_1(t_n)) + \sigma_1 \Delta W_1^n, \tag{10.12}$$
$$u_2^{n+1} = u_2^n + h\Big(-\gamma_2 + \mathrm{i}\omega_0/\varepsilon + i a_0 u_1^n\Big)u_2^n + \sigma_2 \Delta W_2^n, \tag{10.13}$$

where ΔW_1^n are independent real Gaussian random variables with mean 0 and variance h, and ΔW_2^n are independent complex Gaussian random variables with mean 0 and variance $h/\sqrt{2}$ of both real and imaginary parts. Note that even though we have an exact analytical formula for u_2, we still need to evaluate the stochastic integral in Eqn (10.4). We perform this evaluation numerically and, therefore, the solution u_2 strictly speaking becomes *semi*-analytical. However, we use a very fine time step $h = 10^{-3}$ (the same as for the numerical solution) and, therefore, evaluate the integral in Eqn (10.4) with a high precision. The convergence study of the numerical solution that we present below confirms this approximation.

In Fig. 10.4, we demonstrate both the analytical solution (Eqns (10.3), (10.4)) and the numerical approximation (Eqns (10.12), (10.13)). We note that there is an excellent correspondence between them. Moreover, we observe that for the case of strong damping, u_2 makes only a few oscillations within its typical decorrelation time ($\sim 1/\gamma_2$), whereas

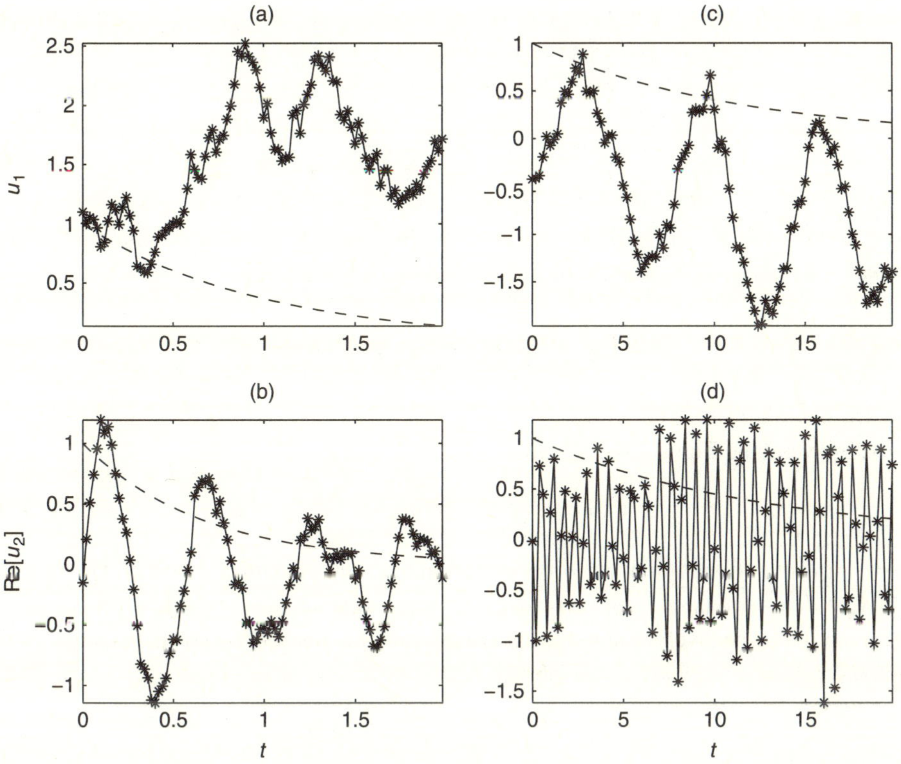

Figure 10.4 Analytical solutions (solid line) computed via Eqns (10.3), (10.4) and numerical solutions (asterisks) computed via Eqns (10.12), (10.13). Panels (a) and (b) show u_1 and $Re[u_2]$ for the case of strong damping. Panels (c) and (d) show u_1 and $Re[u_2]$ for the case of weak damping. The dashed line represents the exponential decay $e^{-\gamma_j t}$ with the corresponding γ_j with $j = \{1, 2\}$. Note the different scales of the x-axis due to different rates of damping.

for the case of weak damping, u_2 makes many oscillations until it decorrelates. Next, we compute closed formulas for the mean and covariance of u_1 and u_2, which are needed in an extended Kalman filter for the nonlinear system.

10.2.3 Exact statistical solutions: Mean and covariance

We start with u_1 where ideas from Chapters 2 and 5 easily apply. By taking the average of Eqn (10.3), we obtain

$$\langle u_1 \rangle = \langle u_{10} \rangle e^{-\gamma_1 (t-t_0)} + \frac{A}{\gamma_1^2 + \omega^2} \times \Big(e^{-\gamma_1 (t-t_0)} \big(\omega \cos(\omega t_0) - \gamma_1 \sin(\omega t_0) \big) - \omega \cos(\omega t) + \gamma_1 \sin(\omega t) \Big), \tag{10.14}$$

where we used the fact that the mean of the stochastic integral is always zero (Gardiner, 1997). Furthermore, the variance of u_1 becomes

$$\mathrm{Var}(u_1) = \langle (u_1 - \langle u_1 \rangle)^2 \rangle = \mathrm{Var}(u_{10}) e^{-2\gamma_1 (t-t_0)} + \frac{\sigma_1^2}{2\gamma_1} (1 - e^{-2\gamma_1 (t-t_0)}). \tag{10.15}$$

In order to compute the variance we used the Itô isometry formula (Gardiner, 1997) as already illustrated in detail in Section 2.1.1 from Chapter 2:

$$\left\langle \Big(\int g(t) dW(t) \Big)^2 \right\rangle = \int g^2(t) dt,$$

for any deterministic $g(t)$.

After averaging Eqn (10.4), we have

$$\langle u_2 \rangle = e^{-\gamma_2 (t-t_0)} \langle \psi(t_0, t)\, u_{20} \rangle. \tag{10.16}$$

For simplicity of notation we drop the parameters in the functions $J(s,t)$, $J_D(s,t)$, $J_W(s,t)$, $b(s,t)$, $\psi(s,t)$, $\psi_D(s,t)$, $\psi_W(s,t)$ for $s = t_0$, and denote them as J, J_D, J_W, b, ψ, ψ_D, ψ_W, respectively. Using the assumption that the noise $W_1(t)$ is independent of the initial conditions u_{10} and u_{20}, we obtain

$$\langle u_2 \rangle = e^{-\gamma_2 (t-t_0)} \psi_D \langle \psi_W \rangle \langle u_{20} \exp(i b u_{10}) \rangle. \tag{10.17}$$

The averages on the right-hand side of Eqn (10.17) can be computed via the characteristic function of a Gaussian random variable. For any probability distribution, we define a characteristic function as

$$\phi_{\mathbf{v}}(\mathbf{s}) = \langle \exp(i\mathbf{s}^T \mathbf{v}) \rangle, \tag{10.18}$$

where $\mathbf{s} \in \mathbb{C}^d$ and d is the number of dimensions. For a Gaussian distribution, the characteristic function is known (Gardiner, 1997) to have the following form

$$\phi_{\mathbf{v}}(\mathbf{s}) = \exp\left(\mathrm{i}\mathbf{s}^T\langle\mathbf{v}\rangle - \frac{1}{2}\mathbf{s}^T\Sigma\mathbf{s}\right), \tag{10.19}$$

where Σ is the covariance matrix. In particular for $\langle\psi_W\rangle$, we obtain

$$\begin{aligned}\langle\psi_W\rangle = \langle e^{\mathrm{i}J_W}\rangle &= \exp\left(\mathrm{i}\langle J_W\rangle - \frac{1}{2}\mathrm{Var}(J_W)\right)\\ &= \exp\left(-\frac{1}{2}\mathrm{Var}(J_W)\right).\end{aligned}$$

Computation of $\mathrm{Var}(J_W)$ is straightforward since J_W is Gaussian and the result is

$$\mathrm{Var}(J_W) = -\frac{\sigma_1^2 a_0^2}{2\gamma_1^3}\left(3 - 4e^{-\gamma_1(t-t_0)} + e^{-2\gamma_1(t-t_0)} - 2\gamma_1(t-t_0)\right).$$

Next, we compute $\langle u_{20}\exp(ibu_{10})\rangle$. Here, it is convenient to use the triad real-valued representation of $(u_1(t), u_2(t))$

$$\begin{aligned}x &= u_1,\\ y &= \mathrm{Re}[u_2],\\ z &= \mathrm{Im}[u_2].\end{aligned}$$

Then, we just need to find $\langle y_0\exp(ibx_0)\rangle$ and $\langle z_0\exp(ibx_0)\rangle$ and afterwards combine them using $u_{20} = y_0 + \mathrm{i}z_0$ (where the zero subscript refers to the initial values at $t = t_0$) to obtain the second average on the right-hand side of Eqn (10.16). Consider a three-dimensional vector $\mathbf{v} = (x_0, y_0, z_0)$ and the corresponding characteristic function given by Eqn (10.19). By its definition, the characteristic function is a Fourier transform of the corresponding probability density function (pdf)

$$\phi_{\mathbf{v}}(\mathbf{s}) = \frac{1}{(2\pi)^3}\int \exp(\mathrm{i}\mathbf{s}^T\mathbf{v})g(\mathbf{v})d\mathbf{v}. \tag{10.20}$$

Next, we use the basic property of the Fourier transform, i.e. multiplication by a variable in physical space (e.g. y_0) corresponds to differentiation over the dual variable (e.g. s_2). We have

$$\frac{\partial\phi_{\mathbf{v}}(\mathbf{s})}{\partial s_2} = \frac{1}{(2\pi)^3}\int \mathrm{i}y_0\exp(\mathrm{i}\mathbf{s}^T\mathbf{v})g(\mathbf{v})d\mathbf{v} = \mathrm{i}\langle y_0\exp(\mathrm{i}\mathbf{s}^T\mathbf{v})\rangle. \tag{10.21}$$

Therefore, we obtain

$$\langle y_0\exp(\mathrm{i}bx_0)\rangle = -\mathrm{i}\left.\frac{\partial\phi_{\mathbf{v}}(\mathbf{s})}{\partial s_2}\right|_{\mathbf{s}=(b,0,0)^T}. \tag{10.22}$$

Similarly, we find

$$\langle z_0 \exp(\mathrm{i}bx_0)\rangle = -\mathrm{i}\frac{\partial \phi_{\mathbf{v}}(\mathbf{s})}{\partial s_3}\bigg|_{\mathbf{s}=(b,0,0)^T}.$$

Using the particular form of a Gaussian pdf of $\mathbf{v}$, we find that

$$\frac{\partial \phi_{\mathbf{v}}(\mathbf{s})}{\partial s_2} = \Big(\mathrm{i}\langle y_0\rangle - \mathrm{Var}(y_0)s_2 - \mathrm{Cov}(x_0, y_0)s_1 - \mathrm{Cov}(y_0, z_0)s_3\Big)\phi_{\mathbf{v}}(\mathbf{s}),$$
$$\frac{\partial \phi_{\mathbf{v}}(\mathbf{s})}{\partial s_3} = \Big(\mathrm{i}\langle z_0\rangle - \mathrm{Var}(z_0)s_3 - \mathrm{Cov}(x_0, z_0)s_1 - \mathrm{Cov}(y_0, z_0)s_2\Big)\phi_{\mathbf{v}}(\mathbf{s}).$$

After evaluating the partial derivatives at $\mathbf{s} = (b, 0, 0)^T$, we obtain

$$\langle y_0 \exp(\mathrm{i}bx_0)\rangle = \Big(\langle y_0\rangle + \mathrm{i}\mathrm{Cov}(x_0, y_0)b\Big)\exp\left(\mathrm{i}b\langle x_0\rangle - \frac{1}{2}\mathrm{Var}(x_0)b^2\right),$$
$$\langle z_0 \exp(\mathrm{i}bx_0)\rangle = \Big(\langle z_0\rangle + \mathrm{i}\mathrm{Cov}(x_0, z_0)b\Big)\exp\left(\mathrm{i}b\langle x_0\rangle - \frac{1}{2}\mathrm{Var}(x_0)b^2\right).$$

Combining these two equations yields

$$\langle u_{20} \exp(\mathrm{i}bu_{10})\rangle = \Big(\langle u_{20}\rangle + \mathrm{i}\mathrm{Cov}(u_{20}, u_{10})b\Big) \times \exp\Big(\mathrm{i}b\langle u_{10}\rangle - \frac{1}{2}b^2\mathrm{Var}(u_{10})\Big).$$

Therefore, we have found all the components of the right-hand side of Eqn (10.17) using the initial data. In a similar manner (see Gershgorin and Majda, 2008, for details), we find the variance of u_2, and the cross-covariance of u_2 with u_2^* and of u_2 with u_1. We have

$$\mathrm{Var}(u_2) = e^{-2\gamma_2(t-t_0)}\Big(\mathrm{Var}(u_{20}) + |\langle u_{20}\rangle|^2 - |\langle u_{20}\psi\rangle|^2\Big) + \frac{\sigma_2^2}{2\gamma_2}\Big(1 - e^{-2\gamma_2(t-t_0)}\Big), \tag{10.23}$$

where

$$|\langle u_{20}\psi\rangle|^2 = e^{-\mathrm{Var}(J_W)-b^2\mathrm{Var}(u_{10})}|\langle u_{20}\rangle + \mathrm{i}b\mathrm{Cov}(u_{20}, u_{10})|^2.$$

We also obtain formulas for the cross-covariances

$$\mathrm{Cov}(u_2, u_1) = \psi_D\langle\psi_W\rangle e^{-(\gamma_1+\gamma_2)(t-t_0)} \times\Big(\langle u_{10}u_{20}\exp(\mathrm{i}bu_{10})\rangle - \langle u_{10}\rangle\langle u_{20}\exp(\mathrm{i}bu_{10})\rangle\Big), \tag{10.24}$$

$$\mathrm{Cov}(u_2, u_2^*) = \exp(-2\gamma_2(t-t_0) + 2\mathrm{i}J_D - 2\sigma_{J_W}^2)\langle u_{20}^2\exp(\mathrm{i}2bu_{10})\rangle - \langle u_2\rangle^2, \tag{10.25}$$

where $\langle u_{10}u_{20}\exp(\mathrm{i}bu_{10})\rangle$ and $\langle u_{20}^2\exp(\mathrm{i}2bu_{10})\rangle$ are given by Eqns (A.8) and (A.9) from Gershgorin and Majda (2008).

Testing the analytical formulas via Monte Carlo simulations

It is very instructive to provide a visual comparison of the analytical formulas for the various statistics of u_1 and u_2 that we derived above with the numerically averaged values using Monte Carlo simulations. We used an ensemble of $M = 10^4$ members in Monte Carlo averaging. As an object of study, let us choose the mean of u_2.

In Fig. 10.5, the solid line represents $\mathrm{Re}[\langle u_2 \rangle]$ which we computed via Eqn (10.17). The circles, on the other hand, were obtained using Monte Carlo averaging of an ensemble of trajectories that were computed via Eqn (10.4). The upper and lower panels correspond to the case of strong and weak damping, correspondingly. The plots for $\mathrm{Im}[\langle u_2 \rangle]$ are not displayed here because they are similar to the plots of $\mathrm{Re}[\langle u_2 \rangle]$. We observe excellent agreement between the analytically obtained mean of u_2 and the result of the averaging using Monte Carlo simulation. Next in Fig. 10.6, we demonstrate the time evolution of the cross-covariance between u_2 and u_2^*. Here, we also have very good agreement between the analytical prediction and Monte Carlo averaging.

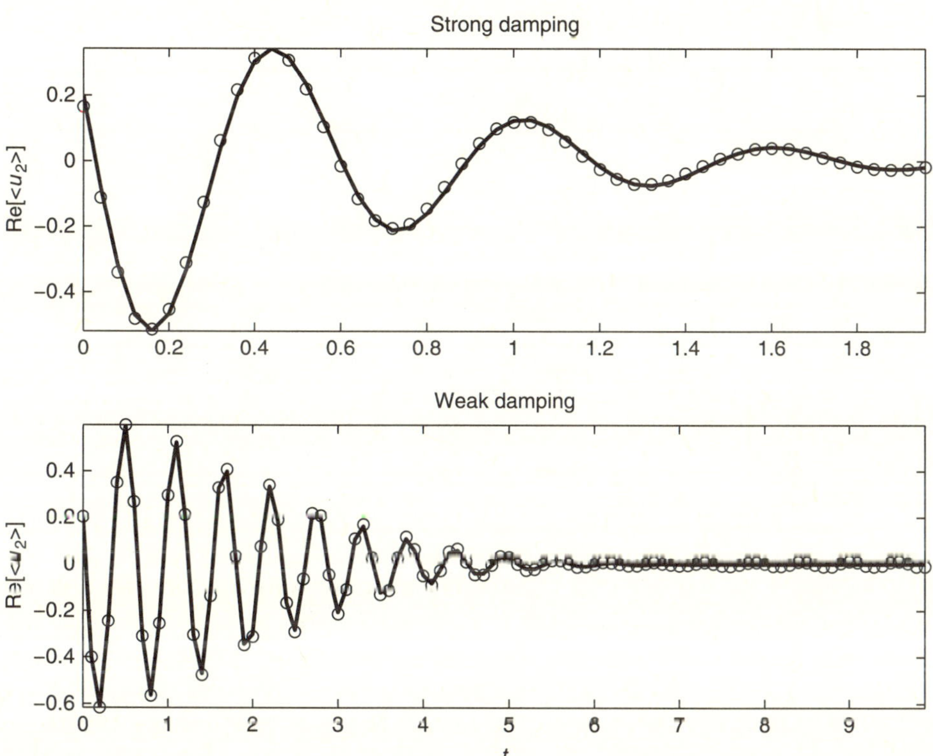

Figure 10.5 The solid line corresponds to $\langle u_2 \rangle$ computed via Eqn (10.17), and circles correspond to Monte Carlo averaging of an ensemble of solutions u_2 each computed via Eqn (10.4). Note the different scales of the x-axis due to different rates of damping

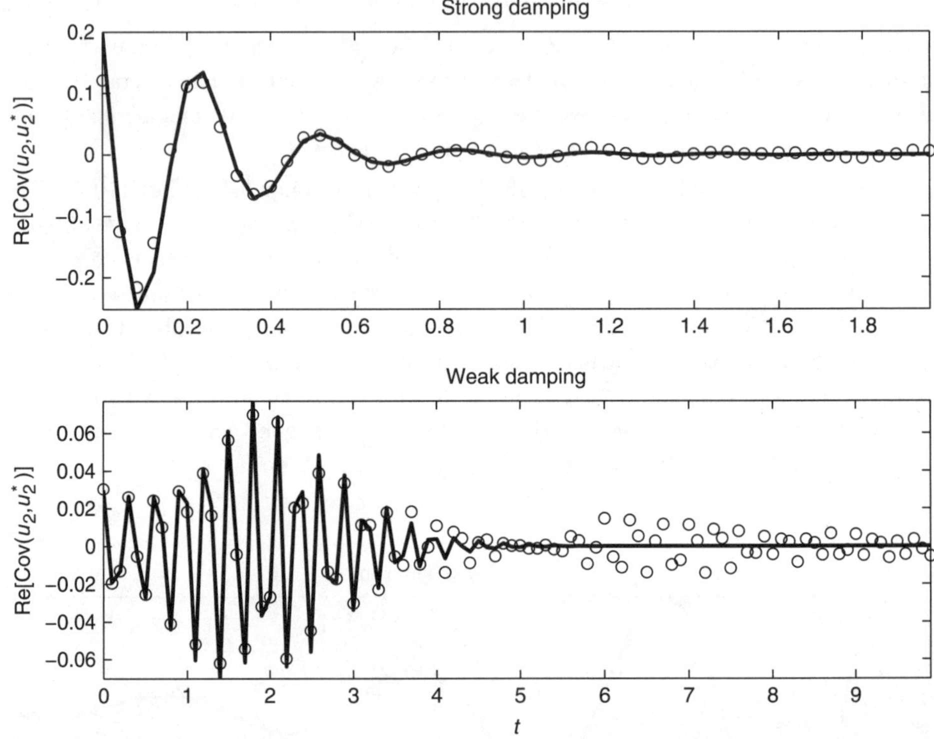

Figure 10.6 The solid line corresponds to $\mathrm{Cov}(u_2, u_2^*)$ computed via Eqn (10.25), and circles correspond to Monte Carlo averaging of an ensemble of solutions u_2 each computed via Eqn (10.4). Note the different scales of the x-axis due to different rates of damping.

Next, we present the comparison of the analytical formula with the results of Monte Carlo averaging for the correlator $\langle u_2 u_1 \rangle$ shown in Fig. 10.7. Note that here again we have excellent agreement.

Non-Gaussianity of u_2

Here, we present the evidence of non-Gaussianity of the fast wave u_2. Due to the nonlinear structure of the governing equation (10.2), it is natural to expect that u_2 is not Gaussian. However, as we will see below depending on the regime of the system we can observe both strongly non-Gaussian and almost Gaussian statistics of u_2. Note that, as we discussed earlier, the invariant measure of our system is always Gaussian. Therefore, non-Gaussianity can only appear in the transient. In order to detect the deviation from Gaussian statistics, we measure the skewness and kurtosis excess. The skewness and kurtosis are defined as the third and the fourth normalized central moments, correspondingly. For a random variable ξ, the skewness is

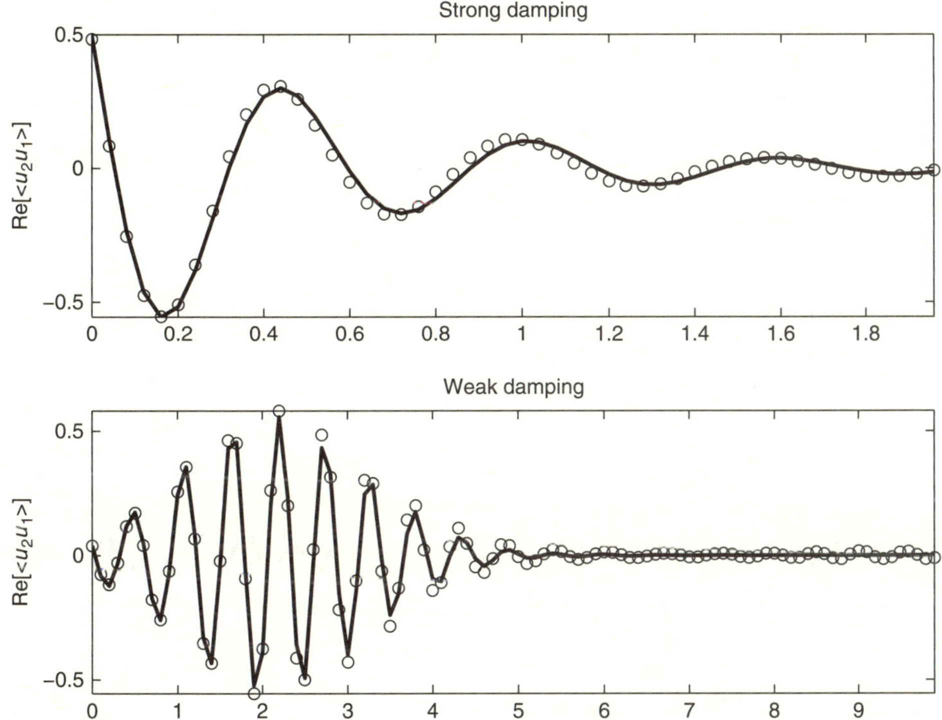

Figure 10.7 The solid line corresponds to $\langle u_2 u_1 \rangle$ computed via Eqns (10.24), (10.14) and (10.17), and circles correspond to Monte Carlo averaging of an ensemble of solutions u_1 each computed via Eqn (10.3) and u_2 each computed via Eqn (10.4). Note the different scales of the x-axis due to different rates of damping.

$$\text{Skewness} = \frac{\left\langle \left(\xi - \langle \xi \rangle\right)^3 \right\rangle}{\text{Var}(\xi)^{3/2}},$$

and the kurtosis is

$$\text{Kurtosis} = \frac{\left\langle \left(\xi - \langle \xi \rangle\right)^4 \right\rangle}{\text{Var}(\xi)^2}.$$

For Gaussian ξ, we obtain skewness $= 0$ and kurtosis $= 3$. In general, the skewness and the kurtosis excess, which is equal to

$$\text{kurtosis excess} = \text{kurtosis} - 3,$$

measure the non-Gaussian properties of a given random variable.

In Figs 10.8 and 10.9, we demonstrate the time evolution of the skewness and kurtosis excess, respectively, for both the weakly damped and the strongly damped cases. Note

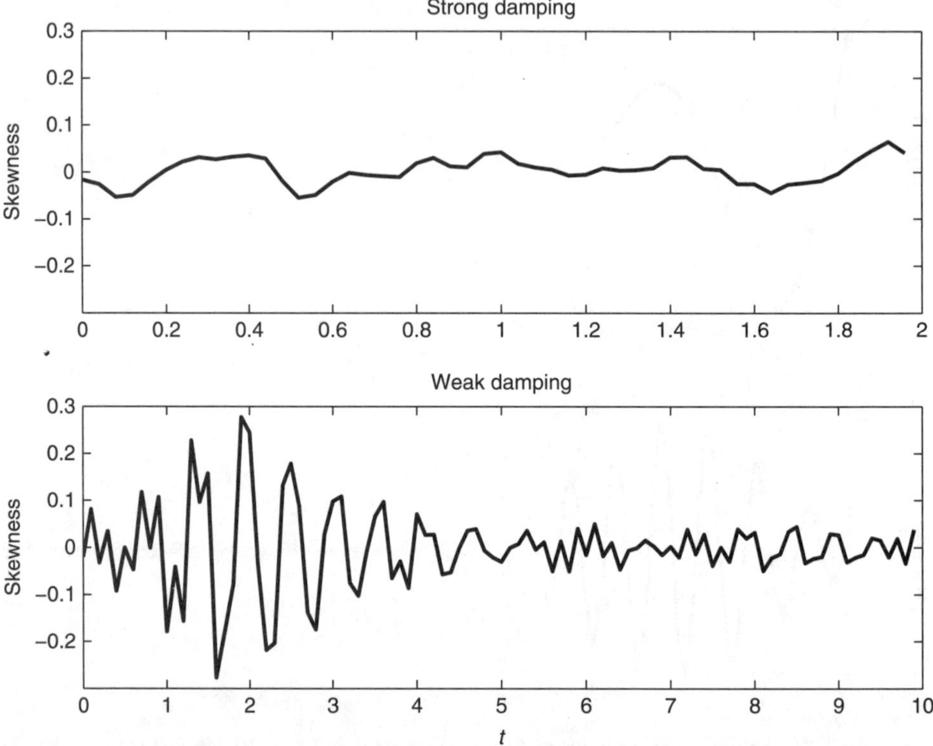

Figure 10.8 Skewness evolution for the strongly damped (upper panel) and weakly damped (lower panel) cases. The weakly damped system is more non-Gaussian than the strongly damped system. Note the different scales of the x-axis due to different rates of damping.

that in the strongly damped case (Figs 10.8 and 10.9 upper panels), both the skewness and kurtosis excess have values close the Gaussian ones. In the strongly damped case, the transient time is short and the statistics of the system do not deviate much from their Gaussian values. On the other hand, in the weakly damped regime (Figs 10.8 and 10.9 lower panels) we observe strong non-Gaussianity in both the skewness and kurtosis excess. However, as predicted by the invariant measure the statistics converge to their Gaussian values after the decorrelation time. We note that a sufficiently large ensemble should be used in Monte Carlo simulations for higher moments in order to obtain a precise result. In our simulation, we used the ensemble of $M = 10^4$ members, which is large enough to get an accurate qualitative picture. The effect of strong fast forcing produces much larger values of skewness than in Fig. 10.8 and more significant departures from Gaussianity (Gershgorin and Majda, 2010).

Finally, we end this section with some comments. In principle, analytic expressions for the higher-order moments can be computed in a similar fashion as for the second-order

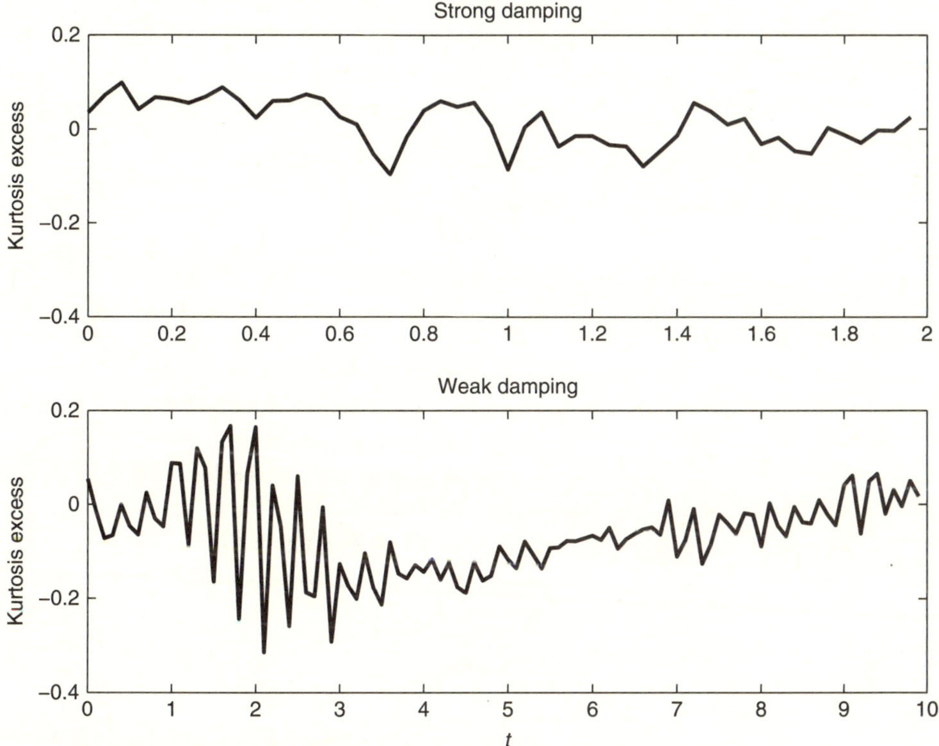

Figure 10.9 Kurtosis excess evolution for the strongly damped (upper panel) and weakly damped (lower panel) cases. The weakly damped system is more non-Gaussian than the strongly damped system. Note the different scales of the x-axis due to different rates of damping.

statistics; however, the explicit formulas become extremely lengthy. Also, the perceptive reader will note that we could use non-Gaussian initial data in the exact solution for the mean and covariance provided that we know the characteristic function of this random variable explicitly.

10.3 Nonlinear extended Kalman filter (NEKF)

Here, we briefly introduce the extended Kalman filter algorithm for the test model (10.1) and (10.2). Suppose that at time $t_m = m\Delta t$, where $m \geq 0$ is a time step index and Δt is the observation time step, the truth signal is denoted as $\mathbf{u}_m$, which is a realization of (u_1,u_2) computed via Eqns (10.3) and (10.4). However, we assume that $\mathbf{u}_m$ is unknown and instead we are given some linear transformation of $\mathbf{u}_m$ distorted by some Gaussian noise

$$\mathbf{v}_m = G\mathbf{u}_m + \sigma_m^0,$$

where $\mathbf{v}_m$ is called the observation, G is a rectangular matrix of size $q \times 3$ with the number of observations $q = \{1, 2, 3\}$, $\mathbf{u} = (x, y, z)^T \equiv (u_1, \text{Re}[u_2], \text{Im}[u_2])^T$ and σ_m^0 is the observation noise. The goal of filtering is to find the filtered signal $\mathbf{u}^f$, which is as close as possible to the original truth signal $\mathbf{u}$. The information that can be used in filtering is limited to:

- the model for the dynamical evolution of $\mathbf{u}_m$;
- the matrix G;
- the mean and covariance of the Gaussian noise σ_m^0.

As discussed in Chapters 2 and 3, when the evolution of $\mathbf{u}_m$ is described by a linear equation, the best approximation to the truth signal in the least-square sense is given via the Kalman filter algorithm (Anderson and Moore, 1979; Chui and Chen, 1999). The Kalman filter consists of two steps: (i) forecast using the dynamics and (ii) correction using observations. If we assume Gaussian initial conditions then, due to the linear dynamics, the solution stays linear for all times and can be fully described by the mean and covariance matrix. Denote the mean and covariance of the filtered signal at time t_m as $\langle \mathbf{u} \rangle_{m|m}$ and $\Gamma_{m|m}$, respectively. Then the forecast step gives us the following, so called prior, values of the mean and covariance at the next time step t_{m+1}

$$
\begin{aligned}
\langle \mathbf{u} \rangle_{m|m} &\to \langle \mathbf{u} \rangle_{m+1|m}, \\
\Gamma_{m|m} &\to \Gamma_{m+1|m}.
\end{aligned}
\tag{10.26}
$$

Note that $\langle \mathbf{u} \rangle_{m+1|m}$ and $\Gamma_{m+1|m}$ depend solely on the prior information up to time t_m. In order to utilize the observations $\mathbf{v}_{m+1}$ at time t_{m+1}, the least-squares correction method is used, which yields the posterior values of the mean and covariance

$$
\begin{aligned}
\langle \mathbf{u} \rangle_{m+1|m+1} &= \langle \mathbf{u} \rangle_{m+1|m} + K_{m+1}(\langle \mathbf{v} \rangle_{m+1} - G \langle \mathbf{u} \rangle_{m+1|m}), \\
\Gamma_{m+1|m+1} &= (I_3 - K_{m+1} G)\Gamma_{m+1|m}, \\
K_{m+1} &= \Gamma_{m+1|m} G^T (G \Gamma_{m+1|m} G^T + R^0)^{-1},
\end{aligned}
\tag{10.27}
$$

where K_{m+1} is a Kalman gain matrix of size $3 \times q$ and I_3 is the identity 3×3 matrix. The posterior distribution is the Gaussian distribution with the mean and covariance given in Eqn (10.27).

In the more general case, when the dynamics is given by nonlinear equations, the procedure described in Eqns (10.26) and (10.27) is called the nonlinear extended Kalman filter (NEKF). Due to nonlinearity, the Gaussianity of the signal can be lost and this filter may not be optimal anymore. Note that we are propagating the mean and covariance of the exact nonlinear dynamics between observation times; unlike the EKF algorithm described in Section 9.1 from Chapter 9, no linearization approximation is utilized here. One purpose here is to investigate the skill of the NEKF using our test model. The advantage of studying the test model (10.1) and (10.2) is in the fact that exact analytical formulas can be used for the mean and covariance (see Section 10.2) in order to make the prior forecast (10.26).

In the NEKF, we use exact evolution equations for the mean and, therefore, in the simulation, we use the mean value of the observation $\langle \mathbf{v} \rangle_m$ in the first equation in (10.27). The effect of the observation noise size is accounted for in computing the Kalman gain matrix K_{m+1}.

Structure of observations

In a typical slow–fast system such as the shallow water equations, observation of pressure, temperature and velocity automatically mixes the fast and slow components and can corrupt the filtering of the slow component. Below, we introduce prototype observations with these features. We will consider three different types of observations with the corresponding observation matrices G and covariances R^0

One observation

$$G = \begin{pmatrix} 1 & \frac{1}{\sqrt{2}} & \frac{1}{\sqrt{2}} \end{pmatrix},$$
$$v_1 = x + \frac{1}{\sqrt{2}}(y+z) + \sigma_1^0,$$
$$R^0 = 2r^0,$$

Two observations

$$G = \begin{pmatrix} 1 & \frac{1}{\sqrt{2}} & \frac{1}{\sqrt{2}} \\ 1 & \frac{1}{\sqrt{2}} & -\frac{1}{\sqrt{2}} \end{pmatrix},$$

$$\begin{cases} v_1 = x + \frac{1}{\sqrt{2}}(y+z) + \sigma_1^0, \\ v_2 = x + \frac{1}{\sqrt{2}}(y-z) + \sigma_2^0, \end{cases}$$

$$R^0 = \begin{pmatrix} 2r^0 & 0 \\ 0 & 2r^0 \end{pmatrix},$$

Three observations

$$G = \begin{pmatrix} 1 & \frac{1}{\sqrt{2}} & \frac{1}{\sqrt{2}} \\ 1 & \frac{1}{\sqrt{2}} & -\frac{1}{\sqrt{2}} \\ 1 & 0 & 0 \end{pmatrix},$$

$$\begin{cases} v_1 = x + \frac{1}{\sqrt{2}}(y+z) + \sigma_1^0, \\ v_2 = x + \frac{1}{\sqrt{2}}(y-z) + \sigma_2^0, \\ v_3 = x + \sigma_3^0, \end{cases}$$

$$R^0 = \begin{pmatrix} 2r^0 & 0 & 0 \\ 0 & 2r^0 & 0 \\ 0 & 0 & r^0 \end{pmatrix}.$$

Here, we assumed that the observation noise components σ_1^0, σ_2^0 and σ_3^0 are independent mean-zero Gaussian with variances $2r^0$, $2r^0$ and r^0, respectively.

10.4 Experimental designs

We generate the truth signal using the following procedure. We consider any random initial data as discussed in Section 10.2 and obtain a realization of the trajectory (u_1, u_2) using the exact solutions given by Eqns (10.3) and (10.4). Note that u_1 can be easily computed since its random part is Gaussian with known statistics. However, u_2 is not Gaussian and its random part depends on the evolution of u_1. Therefore, even if we need the true trajectory only at discrete times with a time step Δt, we still need to compute u_1 with a much finer resolution with time step h. This fine trajectory of u_1 is then used to compute u_2.

10.4.1 Linear filter with model error

Now, suppose that the prior forecast $\mathbf{u}_{m+1|m}$ is made using the linearized version of the analytical equations that we obtained in Section 10.2. We set $a_0 = 0$ in Eqn (10.2) and thus we introduce model error. In this case, the linearization is made at the climatological mean state. As we have seen from the invariant measure, in the long run the correlation between the slow wave u_1 and fast wave u_2 vanishes and so does the effect of nonlinear coupling through a_0. On the other hand, for shorter observation time steps Δt, the non-Gaussianity as an effect of nonlinearity can be sufficiently strong (see Section 10.2.3) and the model error can be rather large. The advantage of using a linear model as an approximation of the true dynamics is for practical application. In real physical problems, the true dynamics of the model is often unknown and ensemble approximations to the Kalman filter are very expensive for a large-dimensional system. Thus, the performance of the linear filter in the nonlinear test model for the slow–fast system is interesting for several reasons (Harlim and Majda, 2008a, 2010a). We will discuss such algorithms for high-dimensional nonlinear dynamical systems in Chapter 12. Note that the truth signal is always produced via the nonlinear version of Eqns (10.1) and (10.2) with $a_0 \neq 0$. Therefore, if we use the linear approximation to the original system, we may not obtain the optimal filtered signal due to model error. Below, we will compare the error in filtering the test problem using the perfect model assumption (forecast via the first two moments for the nonlinear equation) and the linear dynamics equations with model error. In our test model, linearization only affects the fast mode u_2. Substituting $a_0 = 0$ into Eqn (10.17) yields the following linear equation

$$\langle u_2 \rangle = \exp\Big((-\gamma_2 + \mathrm{i}\omega_0/\varepsilon)\Delta t\Big)\langle u_{20} \rangle.$$

Therefore, the forecast is made according to

$$\langle \mathbf{u} \rangle_{m+1|m} = B\langle \mathbf{u} \rangle_{m|m} + C, \tag{10.28}$$

where

$$B = \begin{pmatrix} e^{-\gamma_1 \Delta t} & 0 & 0 \\ 0 & e^{-\gamma_2 \Delta t}\cos(\alpha) & -e^{-\gamma_2 \Delta t}\sin(\alpha) \\ 0 & e^{-\gamma_2 \Delta t}\sin(\alpha) & e^{-\gamma_2 \Delta t}\cos(\alpha) \end{pmatrix},$$

with

$$\alpha = \Delta t \omega_0 / \varepsilon,$$

and

$$C = \begin{pmatrix} F_1(t) \\ 0 \\ 0 \end{pmatrix}.$$

Equation (10.28) is used as a prior forecast for the mean. Similarly by substituting $a_0 = 0$ into Eqns (10.23), (10.24) and (10.25) we obtain the prior covariance of the linearized model.

Observability

Another advantage of using linearized equations is that it then becomes possible to strictly address the issue of observability as discussed in Chapters 2, 3 and 7. We study the observability of the system for the case of observation type 1. In this case, the observability matrix is defined by

$$\begin{aligned} \mathbf{O} &= \begin{bmatrix} G \\ GB \\ GB^2 \end{bmatrix} \\ &= \begin{pmatrix} 1 & 1/\sqrt{2} & 1/\sqrt{2} \\ e^{-\gamma_1 \Delta t} & \frac{e^{-\gamma_2 \Delta t}}{\sqrt{2}}\Big(\cos(\alpha) + \sin(\alpha)\Big) & \frac{e^{-\gamma_2 \Delta t}}{\sqrt{2}}\Big(\cos(\alpha) - \sin(\alpha)\Big) \\ e^{-2\gamma_1 \Delta t} & \frac{e^{-2\gamma_2 \Delta t}}{\sqrt{2}}\Big(\cos(2\alpha) + \sin(2\alpha)\Big) & \frac{e^{-2\gamma_2 \Delta t}}{\sqrt{2}}\Big(\cos(2\alpha) - \sin(2\alpha)\Big) \end{pmatrix}. \end{aligned} \tag{10.29}$$

This is a 3×3 matrix and in order for the system (10.28) to be fully observable, matrix $\mathbf{O}$ should have rank 3. It is easy to conclude from Eqn (10.29) that whenever we have $\sin(\alpha) = 0$ or equivalently

$$\Delta t = 2\pi l \varepsilon / \omega_0, \tag{10.30}$$

for any integer l, the two last columns become equal and, therefore, matrix $\mathbf{O}$ becomes singular. This results in losing observability. We find the determinant of $\mathbf{O}$

$$\begin{aligned} \det(\mathbf{O}) = -e^{-\gamma_2 \Delta t}\sin(\alpha)\Big(\big[e^{-\gamma_2 \Delta t} - e^{-\gamma_1 \Delta t}\cos(\alpha)\big]^2 \\ + e^{-2\gamma_1 \Delta t}\big[1 - \cos^2(\alpha)\big]\Big). \end{aligned}$$

Therefore, in Eqn (10.30) we have all the values Δt for which observability is lost. However, practically we can still lack observability even if the matrix $\mathbf{O}$ has rank 3, but it is close to being singular, which means that $\det(\mathbf{O})$ is close to zero.

The analysis of observability of the linearized model will also be useful for the nonlinear regime because linear dynamics still plays an important role in the system evolution in the nonlinear regime. In the nonlinear case, we might also expect deteriorating filter performance around the values of Δt described by Eqn (10.30).

Details of numerical simulation of filtering

We turn to the discussion of the results of the numerical simulation of NEKF on our test model (10.1) and (10.2). Here, we describe the parameters that were used in the numerical simulations. The procedure of generating the truth signal was given in Section 10.2. In order to test the filter performance, we compare the truth signal $\mathbf{u}_m$ with the filtered mean $\langle\mathbf{u}\rangle_{m|m}$. We also estimate the role of the prior forecast and observations by studying the prior mean $\langle\mathbf{u}\rangle_{m|m-1}$. We use the root mean square error (RMSE) and cross-correlation (XC) to measure filter skill. RMSE is defined via

$$\mathrm{RMSE}(\mathbf{z}-\mathbf{w}) = \sqrt{\frac{1}{N}\sum_{j=1}^{N}|z_j - w_j|^2},$$

where $\mathbf{z}$ and $\mathbf{w}$ are the complex vectors to be compared and N is the length of each vector. Looking ahead at Figs 10.10–10.13, where we show individual trajectories that are being filtered, the typical amplitude of $\mathbf{u}$ is below 2 in magnitude so RMSE is roughly twice as big as the normalized percentage error in the study below. For real vectors $\mathbf{x}$ and $\mathbf{y}$, XC is defined by

$$\mathrm{XC}(\mathbf{x},\mathbf{y}) = \frac{\sum_{j=1}^{N} x_j y_j}{\sqrt{\sum_{j=1}^{N} x_j^2 \sum_{j=1}^{N} y_j^2}}.$$

For the complex-valued vectors $\mathbf{z}$ and $\mathbf{w}$, the cross-correlation is computed via

$$\mathrm{XC}(\mathbf{z},\mathbf{w}) = \frac{1}{2}\Big(\mathrm{XC}(\mathrm{Re}[\mathbf{z}],\mathrm{Re}[\mathbf{w}]) + \mathrm{XC}(\mathrm{Im}[\mathbf{z}],\mathrm{Im}[\mathbf{w}])\Big).$$

Note that, if two signals are close to each other, the RMSE of their difference is then close to zero and their XC is close to one. On the other hand, for two different signals, the RMSE diverges from zero and in principle is unbounded and XC approaches zero.

In all the simulations we use $N = 10^4$ data filtering cycles with the given observation time step Δt. The type of observation, observation variance r^0 and observation time step Δt will be varied and specified in each particular situation. Here, we will only discuss the weak damping case since it has more physical importance for filtering slow–fast systems.

10.5 Filter performance

Here, we demonstrate how the NEKF works on individual trajectories. We chose two values of the observation time step: $\Delta t = 0.13$ and $\Delta t = 1.43$. The first one is considerably smaller than the typical oscillation period $T_2 = 2\pi\omega_0/\varepsilon \approx 0.63$ of the fast mode u_2 and the second one is larger than T_2. We also chose two values of the observation variance $r^0 = 0.1$ and $r^0 = 2.0$. The first value of r^0 is chosen to be smaller than the average energy of each mode $E = 1$ and the second value is chosen to be larger than $E = 1$. Therefore, we have four pairs $(\Delta t, r^0)$ for which we have studied the performance of the NEKF. We have used observations of type 1 in this numerical experiment so that slow and fast modes are mixed in the single observation.

The four panels in Fig. 10.10 correspond to the four pairs of parameters $(\Delta t, r^0)$. In each panel in Fig. 10.10, we demonstrate a segment of the truth trajectory u_1 together with the prior forecast and the posterior signal. In Table 10.4, we also show the cross-correlation

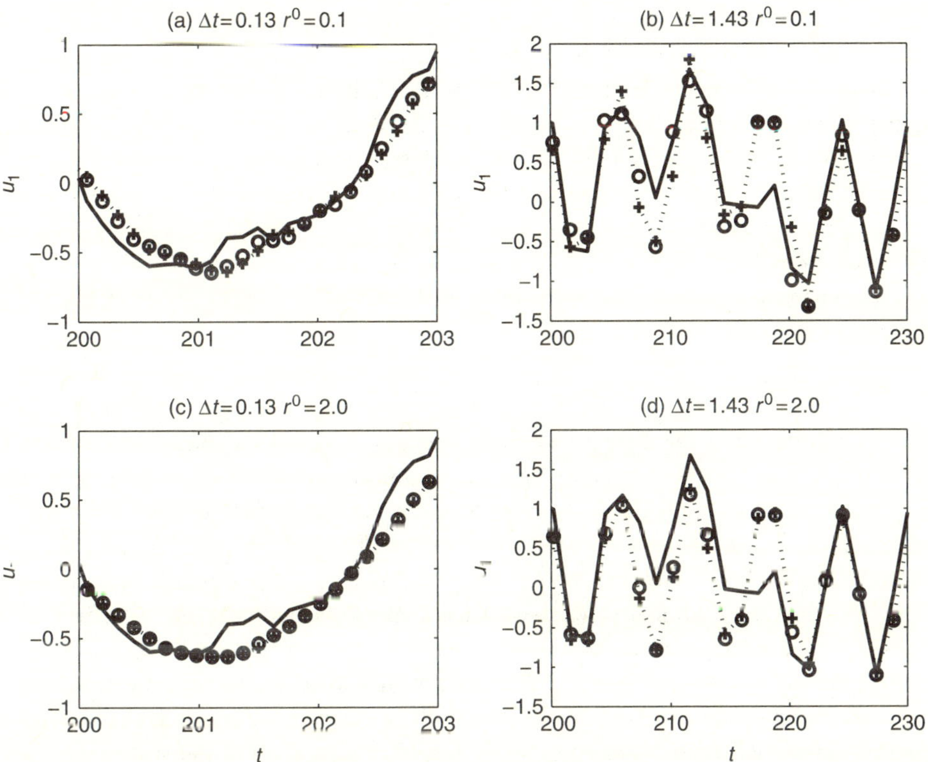

Figure 10.10 The truth signal u_1 is shown with a solid line, the prior forecast is shown with pluses connected with a dotted line, and the posterior signal is shown with circles. The values of Δt and r^0 are shown on the top of each panel. The corresponding cross-correlations XC_f and XC_p are given in Table 10.4. Note the different time-scales for different Δt.

Table 10.4 NEKF performance on u_1: XC_p and XC_f (in parentheses). The segments of the corresponding trajectories are shown in Fig. 10.10.

$r^0 \backslash \Delta t$	0.13	1.43
0.1	0.988 (0.982)	0.936 (0.888)
2.0	0.951 (0.945)	0.855 (0.823)

XC_f between the truth signal and prior forecast and cross-correlation XC_p between the truth signal and posterior signal. Table 10.4 is made for the same set of parameters as Fig. 10.10, and the cross-correlations were measured along the trajectories of $n = 10^4$ observation time steps Δt. Note that for all four cases we have $XC_f < XC_p$, which means that the correction step of the NEKF improves the forecast using observations. Comparing the columns of Table 10.4, we observe that the skill of the NEKF decreases significantly when we increase Δt. On the other hand, comparing the rows of Table 10.4, we also observe that the skill of the NEKF decreases significantly when we increase r^0.

Figure 10.11 shows the evolution of u_2 together with the prior forecast and posterior signal. Table 10.5, shows the cross-correlations XC_p and XC_f that correspond to the trajectories of u_2 (shown partially in Fig. 10.11). Here, we also note the decrease in the skill of the NEKF if larger Δt or r^0 are used. However, we should point out that if the time step Δt becomes significantly larger than the oscillation time T_2 (in our case $\Delta t = 1.43$ and $T_2 \approx 0.63$) we observe very poor filter skill in u_2 but not in u_1, which is still filtered quite well (compare $XC_p = 0.855$ for u_1 in Table 10.4, with $XC_p = 0.622$ for u_2 in Table 10.5 for $\Delta t = 1.43$ and $r^0 = 2.0$).

Filtering with model error

Here, we discuss how the filter performance changes if we use the linearized model with $a_0 = 0$ as a prior forecast instead of the exact nonlinear model with $a_0 \neq 0$. In Figs 10.12 and 10.13, we show segments of the trajectories of the truth signal together with two filtered signals, where one was obtained using the exact model, and another one using the linearized model. Here, we had the same set of parameters that we used for generating Figs 10.10 and 10.11. Similarly, we constructed Tables 10.6 and 10.7 with the values of the cross-correlation XC_p already discussed above and cross-correlation XC_{me} between the truth signal and the posterior signal obtained via the linearized model, i.e. with model error.

By comparing the filtered signals with the truth trajectories, we conclude that the model error is quite small in the filtering of u_1 (Table 10.6) regardless of the size of r^0 or Δt. On the other hand, using the linearized model for the prior forecast drastically affects the performance of filtering u_2 – from Table 10.7, we conclude that $XC_p > XC_{me}$ for u_2. Below, we will study how the model error depends on the type of observations, observation variance and observation time step.

Table 10.5 NEKF performance on u_2: XC_p and XC_f (in parentheses). The segments of the corresponding trajectories are shown in Fig. 10.11.

$r^0 \backslash \Delta t$	0.13	1.43
0.1	0.967 (0.955)	0.740 (0.544)
2.0	0.911 (0.897)	0.622 (0.357)

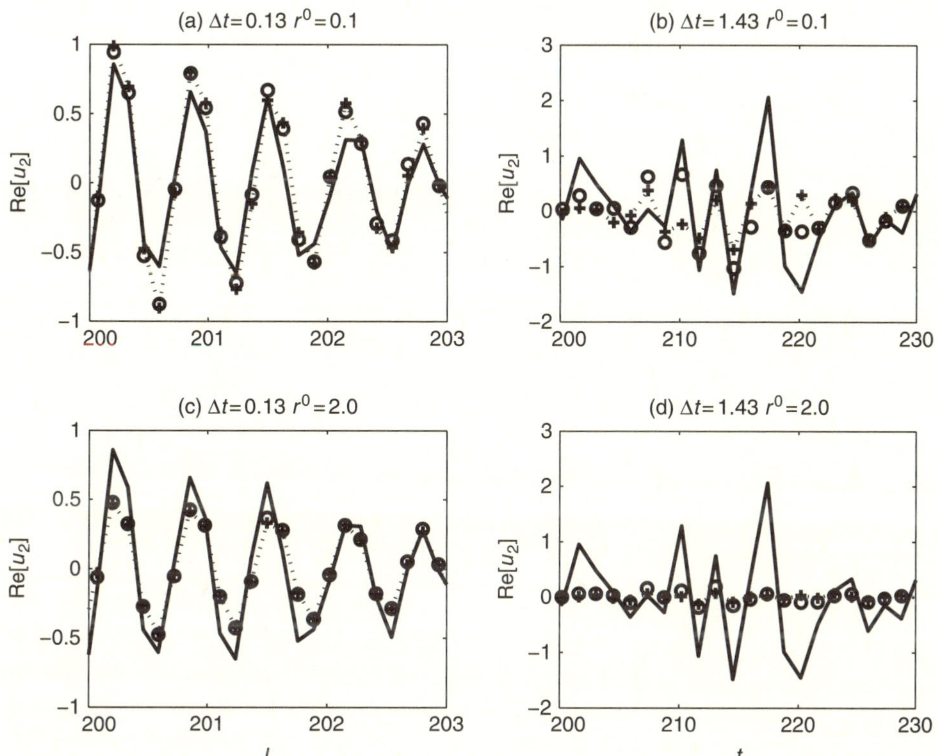

Figure 10.11 The truth signal u_2 is shown with a solid line, the prior forecast is shown with pluses connected with a dotted line, and the posterior signal is shown with circles. The values of Δt and r^0 are shown on the top of each panel. The corresponding cross-correlations XC_f and XC_p are given in Table 10.5. Note the different time-scales for different Δt.

Filtering the linear model and the test of observability

In Fig. 10.14, we present the results of numerical filtering of the linearized problem with $a_0 = 0$. In this simulation, the truth was also computed via the linearized version of Eqns (10.3) and (10.4), i.e. with $a_0 = 0$. Therefore, the Kalman filter produces the optimal filtered signal and there is no model error in this simulation. In

Table 10.6 NEKF performance on u_1: cross-correlations XC_p and XC_{me} (in parentheses). The segments of the corresponding trajectories are shown in Fig. 10.12.

$r^0 \backslash \Delta t$	0.13	1.43
0.1	0.988(0.985)	0.936(0.921)
2.0	0.951(0.950)	0.855(0.856)

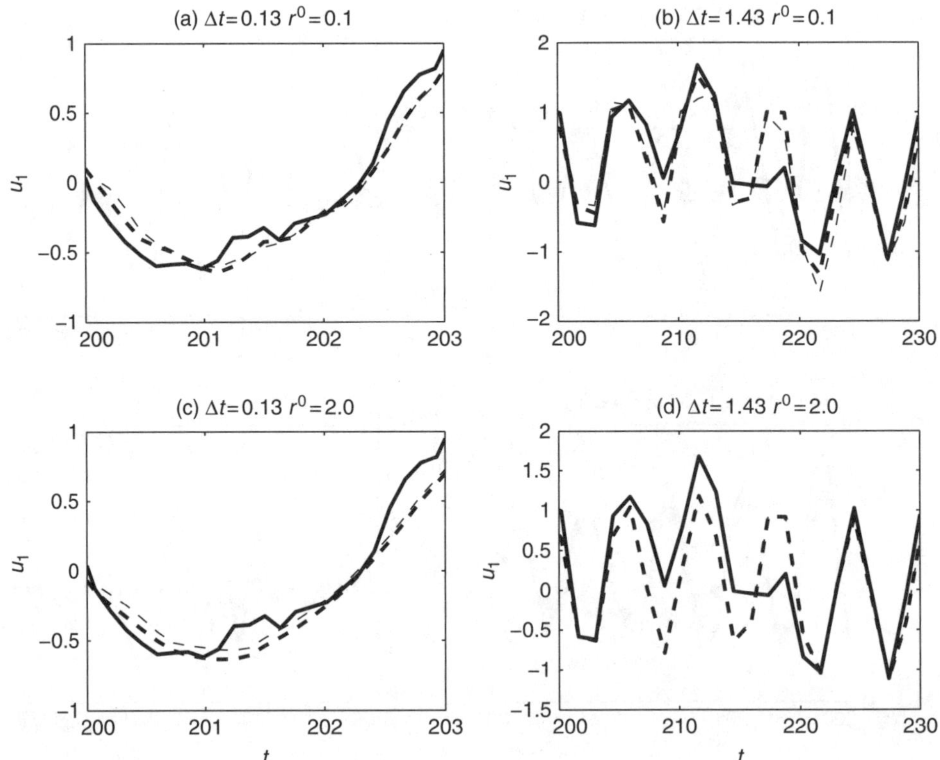

Figure 10.12 The truth signal u_1 is shown with a solid line, the posterior signal computed via a nonlinear forecast is shown with a bold dashed line, and the posterior signal computed via a linear forecast is shown with a thin dashed line (note that in panel (d) the two dashed lines almost coincide). The corresponding values of Δt and r^0 are shown on the top of each panel. The corresponding cross-correlations XC_p and XC_{me} are given in Table 10.6. Note the different time-scales for different Δt.

Fig. 10.14(a), we show the dependency of the RMSE on the observation time step Δt. In Fig. 10.14(b), we demonstrate the corresponding dependency of $\det(\mathbf{O})^{-1}$ on Δt. We note the strong correlation between the two plots. The peaks that were predicted by observability analysis (Fig. 10.14(b)) are observed at the same locations on the plot of RMSE

Table 10.7 NEKF performance on u_2: cross-correlations XC_p and XC_{me} (in parentheses). The segments of the corresponding trajectories are shown in Fig. 10.13.

$r^0 \backslash \Delta t$	0.13	1.43
0.1	0.967(0.864)	0.740(0.543)
2.0	0.911(0.618)	0.622(0.436)

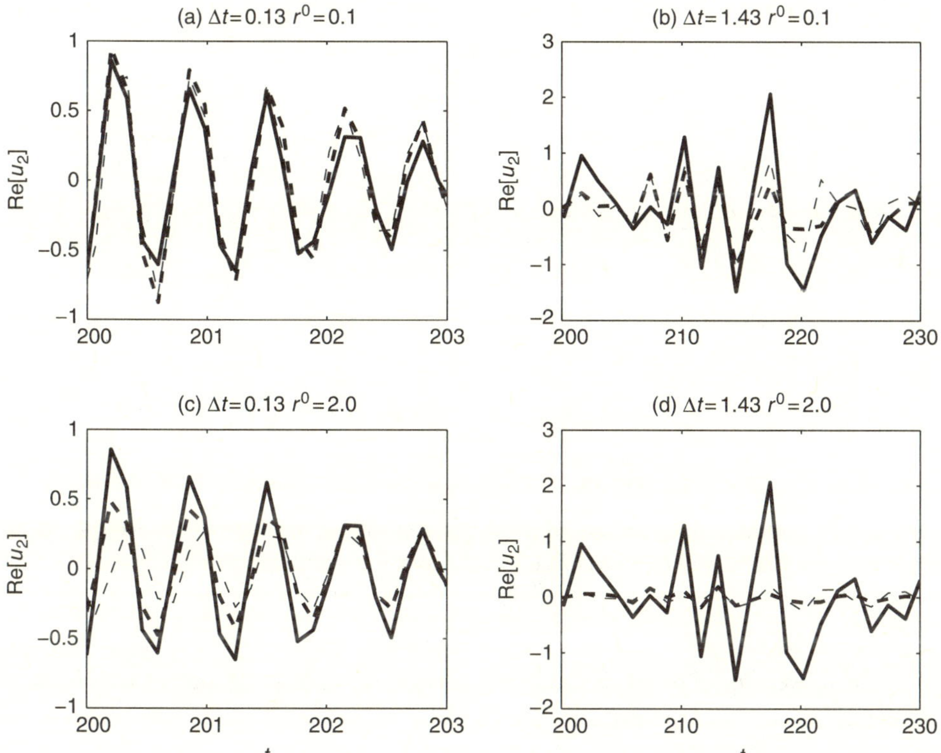

Figure 10.13 The truth signal u_2 is shown with a solid line, the posterior signal computed via a nonlinear forecast is shown with a bold dashed line, and the posterior signal computed via a linear forecast is shown with a thin dashed line (note that in panel (d) the two dashed lines almost coincide). The corresponding values of Δt and r^0 are shown on the top of each panel. The corresponding cross-correlations XC_p and XC_{me} are given in Table 10.7. Note the different time-scales for different Δt.

of u_1 (Fig. 10.14(a)). Here, we have taken a very small observation variance $r^0 = 0.002$. Thus we ensure that the filter tends to trust the observations and, therefore, observability becomes crucial. We note that for the fully observable case, i.e. with observations of type 3, there are no peaks at all. In that case, the observation matrix G always has rank 3, regardless

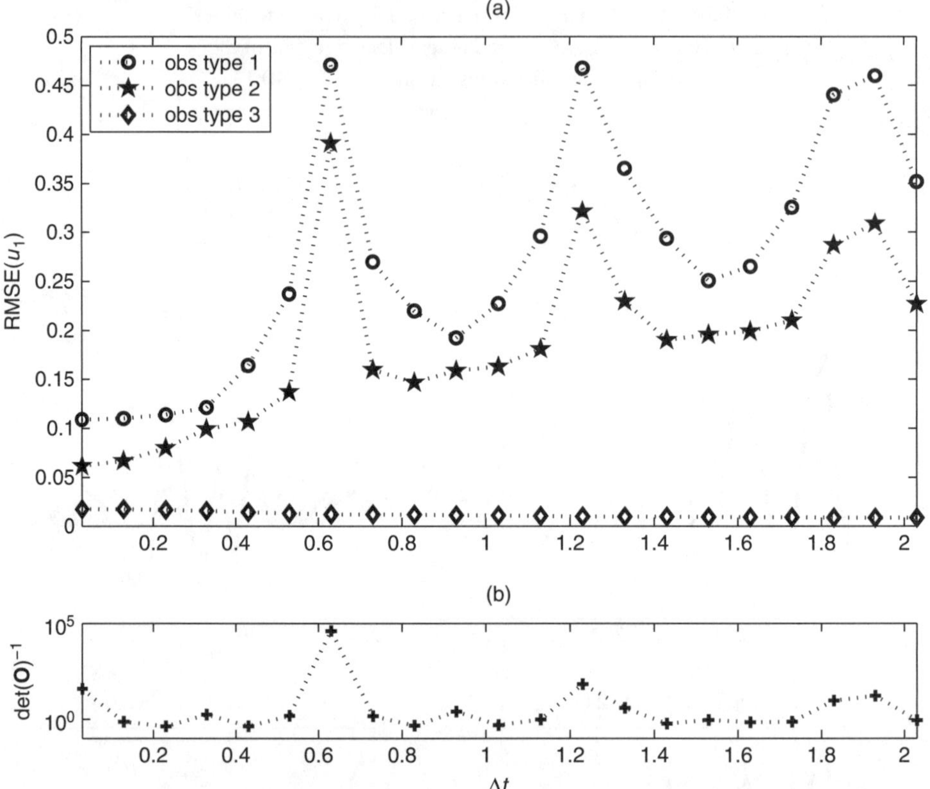

Figure 10.14 (a) RMSE of u_1 for three types of observations and observation variance $r^0 = 0.002$ as a function of Δt, (b) det$(\mathbf{O})^{-1}$ as a function of Δt (note the logarithmic scale of the y-axis). Weak damping was used.

of Δt. However, for observations of types 1 and 2 we clearly see lack of observability around the values of Δt predicted by Eqn (10.30).

Filter performance as a function of observation time step

Next, we study how performance of the NEKF depends on the observation time step Δt under various conditions such as type of observations, observation variance and the model used in the prior forecast. As we studied above, for the linear model ($a_0 = 0$) the dependence of the NEKF performance on Δt is strongly correlated with the observability properties of the NEKF. Here, we will see how that result is reflected in filtering the nonlinear model.

In Fig. 10.15, we show the RMSE of the filtered solution of u_1 compared with the truth signal of u_1. We have chosen three fixed values of observation variance: $r^0 = 0.002$, $r^0 = 0.256$, $r^0 = 2.048$ (panels (a), (b) and (c) in Fig. 10.15). For each r^0, we consider all

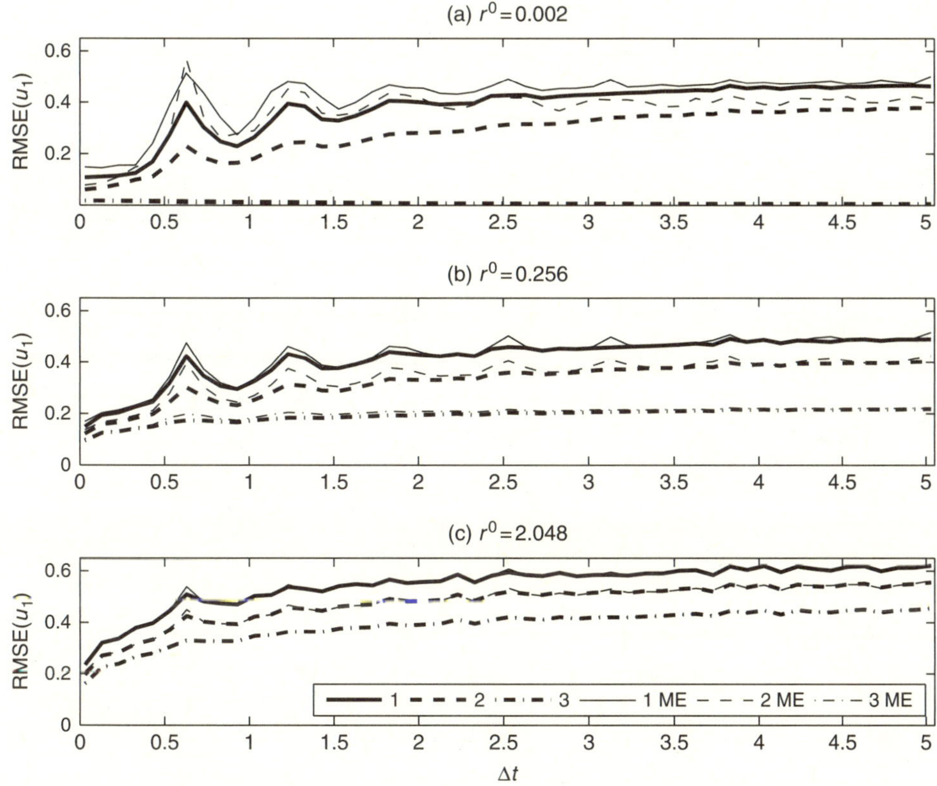

Figure 10.15 RMSE of u_1 as a function of Δt. Observations of type 1 (solid line), type 2 (dashed line), and type 3 (dashed-dotted line) with the nonlinear model (bold line) and linearized model with model error (ME, thin line) were used in the simulation. Fixed r^0 (shown on top of the corresponding panel) was used in each simulation.

three types of observations. Moreover, for each type of observations, we used two models for the prior forecast: a perfect model with $a_0 = 1$ (the same a_0 that was used for creating the truth signal), and a linearized model with $a_0 = 0$ (with model error). As a result of this simulation, we draw the following conclusions:

1. the RMSE of u_1 approaches some finite value as Δt increases and r^0 is fixed;
2. for observations of types 1 and 2, the NEKF performance becomes very poor around the values of Δt, where the linearized system has lack of observability, whereas observations of type 3 make the NEKF always fully observable;
3. a larger number of observations leads to better NEKF performance regardless of r^0;
4. for observations of type 3, the RMSE is almost negligible for small r^0 and monotonically increases for larger r^0;

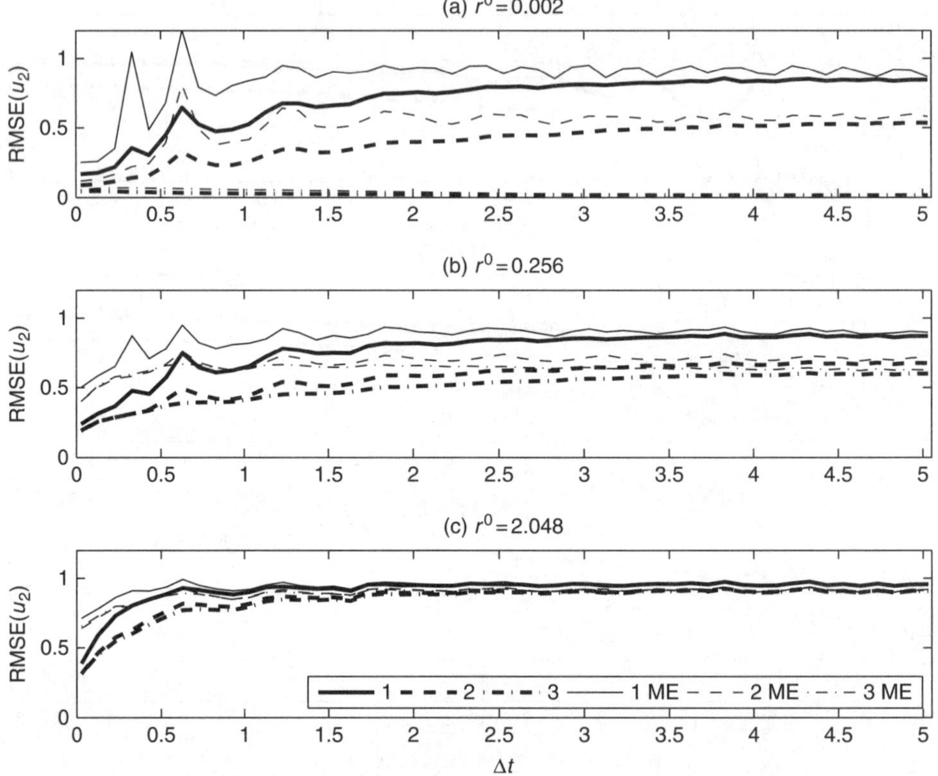

Figure 10.16 RMSE of u_2 as a function of Δt. Observations of type 1 (solid line), type 2 (dashed line), and type 3 (dashed-dotted line) with the nonlinear model (bold line) and linearized model with model error (ME, thin line) were used in the simulation. Fixed r^0 (shown on top of the corresponding panel) was used in each simulation.

5. the RMSE with model error is much larger with smaller r^0 and it becomes almost negligible for larger r^0 as Δt increases;
6. for larger r^0, the peaks that correspond to poor filter performance are less sharp than for smaller r^0, which is explained by the fact that observability is less important with a larger observation error than with a smaller one since greater weight is given to the dynamics.

We now study the dependence of the NEKF performance on mode u_2, which is shown in Fig. 10.16. We make the following observations after examining Fig. 10.16:

1. the RMSE of u_2 approaches some finite value as Δt increases and r^0 is fixed;
2. the RMSE of u_2 is smaller when the number of observations is larger;

3. observations of type 1 and 2 show lack of observability at the values of Δt predicted by observability analysis (Eqn (10.30));
4. observations of type 1 and 2 have an extra peak at $\Delta t \approx T_2/2$, which also corresponds to a lack of observability at this time step (see the small peak at $\Delta t \approx T_2/2$ in Fig. 10.14(b)); this peak was absent in the corresponding plot for u_1;
5. for all Δt the difference among the types of observations becomes smaller as r^0 increases;
6. for all Δt the model error becomes smaller as r^0 increases.

Statistics of mean model error and individual realizations

As described in detail in Chapter 3, mean model error statistics are a standard part of offline testing for Kalman filter performance in linear models (Anderson and Moore, 1979). How useful are mean model error statistics as a qualitative guideline for filter performance in nonlinear slow–fast systems? The exact solution formulas for statistics in the test model allow us to address these issues here. We now study the mean model error due to linearization. In Section 10.4.1, we estimated the model error by following individual trajectories for a given observation type and the set of parameters Δt and r^0, and then by measuring averaged model error along these trajectories.

Now, we consider averaging over all possible ensembles of the random signal trajectories as well as the random noise in observations. This way, we will obtain a fully deterministic system for estimating the mean model error. The solution of this system requires much less computational work than resolution of a long individual trajectory and then filtering it. As a result of the mean model error analysis, we expect to obtain some indications of the model error along the individual trajectories.

Next, we describe the algorithm of measuring the mean model error. Note that, since we have the exact analytical formulas for the time evolution of the mean $\langle \mathbf{u} \rangle$ of the model in a nonlinear regime ($a_0 \neq 0$), we can compute it and, therefore, obtain the truth signal for the mean. Moreover, we use the same formulas for $\langle \mathbf{u} \rangle$ together with the formulas for the evolution of the covariance but in the linearized form ($a_0 = 0$) to produce a prior forecast.

$$\bar{\mathbf{u}}_{m|m} \xrightarrow{a_0=0} \bar{\mathbf{u}}_{m+1|m},$$

where we used the notation $\bar{\mathbf{u}}$ instead of $\langle \mathbf{u} \rangle$ to distinguish between filtering the mean signal and filtering of the individual random trajectories. The posterior forecast in the mean model error estimation is made via the following reasoning. The observation errors have mean zero, therefore, after averaging over all possible trajectories, we have $\bar{\mathbf{v}}_m = G\bar{\mathbf{u}}_m$. The correction step of the filter becomes

$$\begin{aligned}
\bar{\mathbf{u}}_{m+1|m+1} &= \bar{\mathbf{u}}_{m+1|m} + K_{m+1}G(\mathbf{u}_{m+1} - \mathbf{u}_{m+1|m}),\\
\Gamma_{m+1|m+1} &= (I_3 - K_{m+1}G)\Gamma_{m+1|m},\\
K_{m+1} &= \Gamma_{m+1|m}G^T(G\Gamma_{m+1|m}G^T + R^0)^{-1}.
\end{aligned}$$

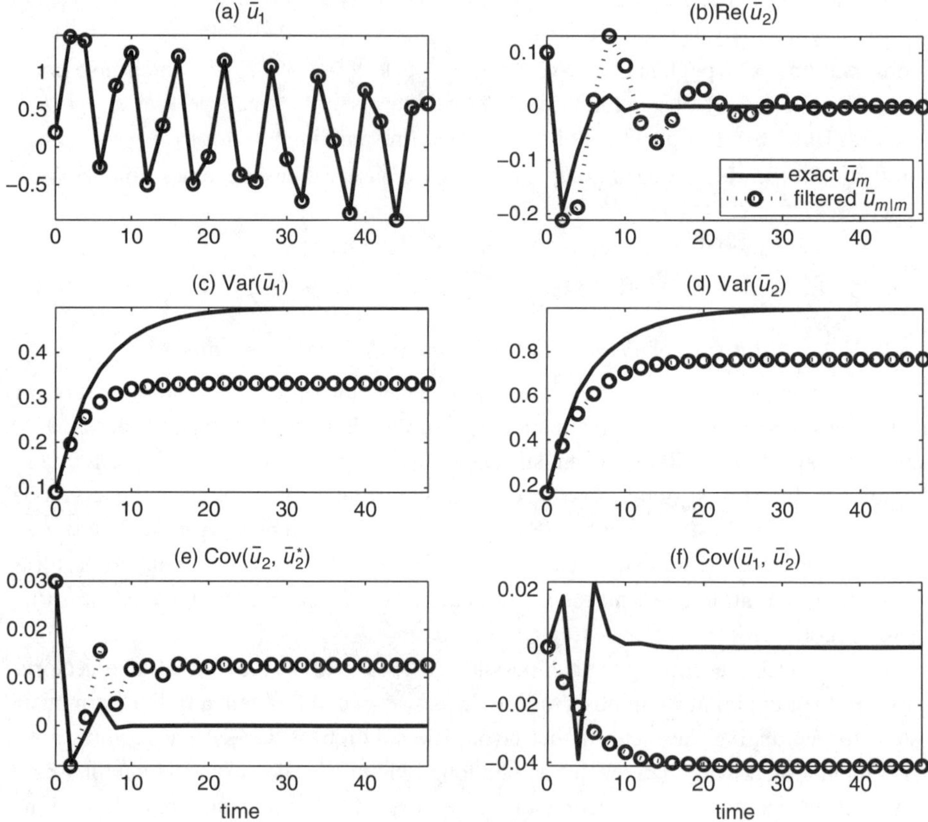

Figure 10.17 Mean model error. The Solid line shows the truth values and circles show the filtered values. $r^0 = 2.048$, $\Delta t = 2$, and observations of type 1 were used.

Note that the so-called off-line statistics Γ and K are computed by the same formulas as for the individual trajectories, and only the update of the mean is different. It is very important to stress here that even though the Kalman filter algorithm for computing the mean model error seems to be the same as for computing the filtered signal of individual trajectories. In the mean model error analysis we filter an averaged signal while previously, we estimated individual trajectories with averaged posterior filtered solutions.

In Fig. 10.17, we demonstrate the time evolution of the mean (panels (a) and (b)) and covariance (panels (c)–(f)) of $(\bar{u}_1, \bar{u}_2)$ and their corresponding filtered values. Observations of type 1 were used. From Fig. 10.17(a), we see that the error in filtering $\bar{u}_1$ is very small. On the other hand, the error in filtering $\bar{u}_2$ is rather large within the decorrelation time $(\sim 1/\gamma_1 \approx 1/\gamma_2)$ until both the truth signal $\bar{u}_{2,m}$ and the filtered signal $\bar{u}_{2,m|m}$ converge to zero. The mean model error is characterized by the measure of the difference between $\bar{u}_m$

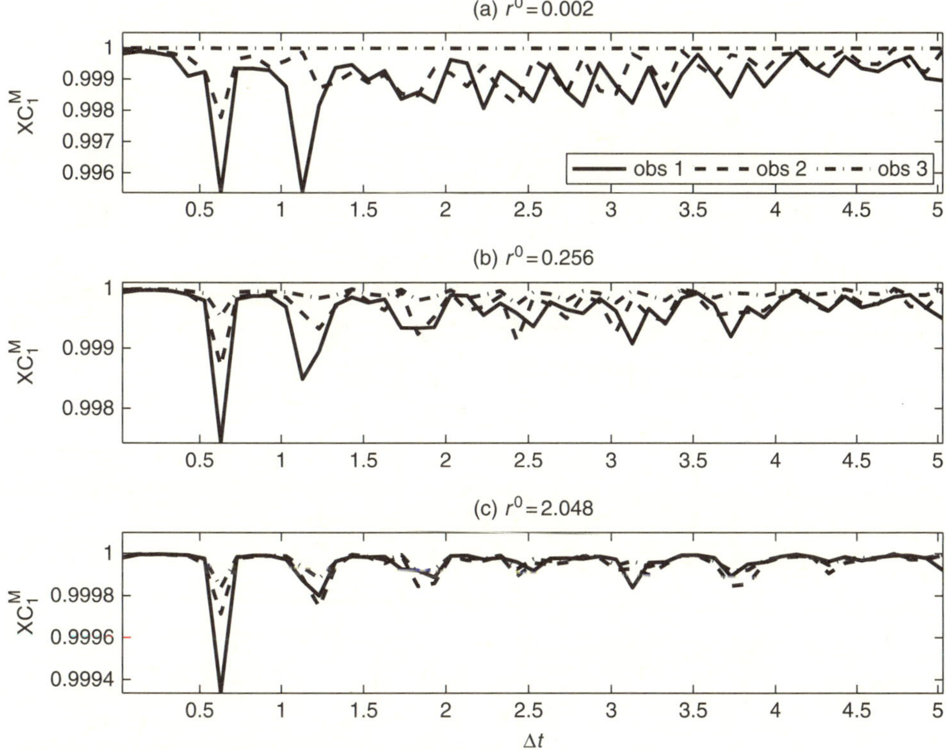

Figure 10.18 Mean model error of u_1. Observations of type 1 (solid line), type 2 (dashed line), and type 3 (dashed-dotted line) are shown.

and $\bar{u}_{m|m}$, e.g. the cross-covariance between the two of them. The various components of the truth covariance and posterior covariance reach constant values within the decorrelation time. The difference between the corresponding components of the truth and posterior covariance can be another characteristic of the mean model error.

Next, we study what information about the mean model error is relevant to estimating the model error in filtering individual trajectories, and what information can be misleading. In Fig. 10.18, we show the dependence of the mean model error of the slow mode u_1 on the observation time step Δt for three fixed values of the observation variance r^0. As a measure of the mean model error, we have chosen the cross-correlation XC_1^M between $\bar{u}_{1,m}$ and $\bar{u}_{1,m|m}$ along the typical decorrelation time (chosen to be $\tau = 30$ from studying Fig. 10.17 as it appears to be the typical relaxation time). On the other hand, in Fig. 10.19, we demonstrate the cross-correlation XC_1 between the two signals $\langle u_{1,m|m} \rangle$ filtered from the same individual trajectory: one computed via the nonlinear dynamics ($a_0 \neq 0$) and the other one computed via linearized dynamics ($a_0 = 0$). Moreover, in Figs 10.20 and 10.21, we show similar dependencies of XC_2^M and XC_2 on Δt for the

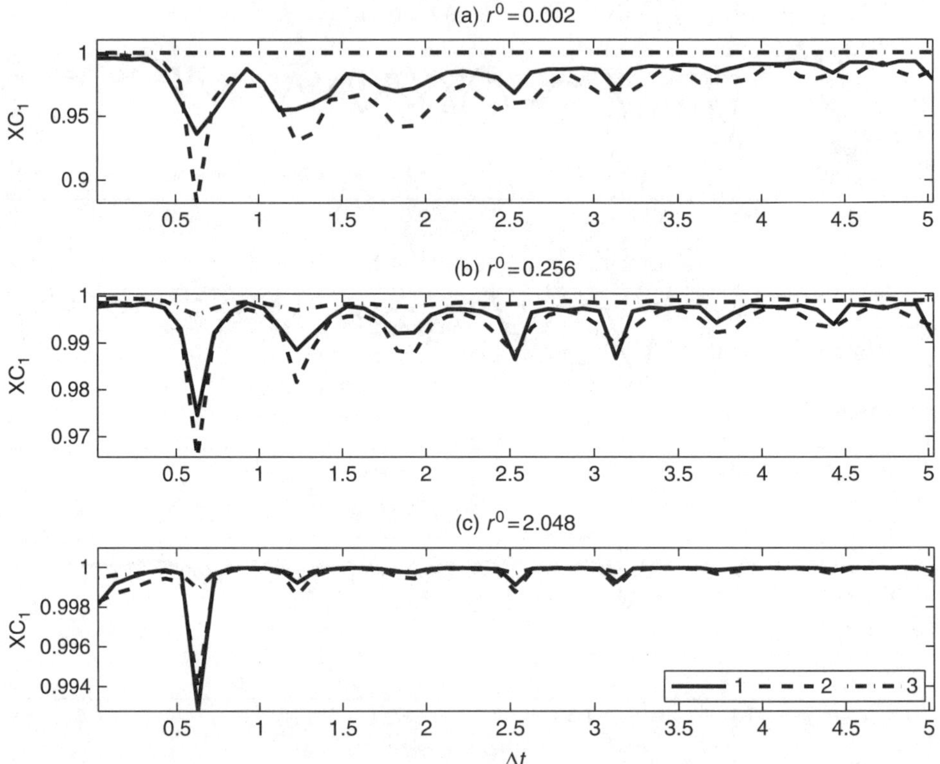

Figure 10.19 Model error of u_1. Observations of type 1 (solid line), type 2 (dashed line), and type 3 (dashed-dotted line) are shown.

fast mode u_2. It is very instructive to compare Fig. 10.18 with Fig. 10.19 and Fig. 10.20 with Fig 10.21.

Below, we present the ideas about the model error that we can predict from the mean model error analysis, i.e. the similarities between XC_1^M and XC_1 and between XC_2^M and XC_2:

1. larger model error at the values of Δt that were predicted by observability analysis;
2. the model error of filtering with observations of type 3 is much less than with observations of type 1 and 2;
3. for larger r^0, the model error is less dependent on the type of observations;
4. disregarding local oscillations, the model error for u_2 first increases as Δt increases and then decreases. This does not mean that the filter skill improves as Δt increases, it just means that the model error due to linearization becomes less important;
5. for larger r^0, observations of types 2 and 3 give almost the same model error, while observations of type 1 lead to larger model error.

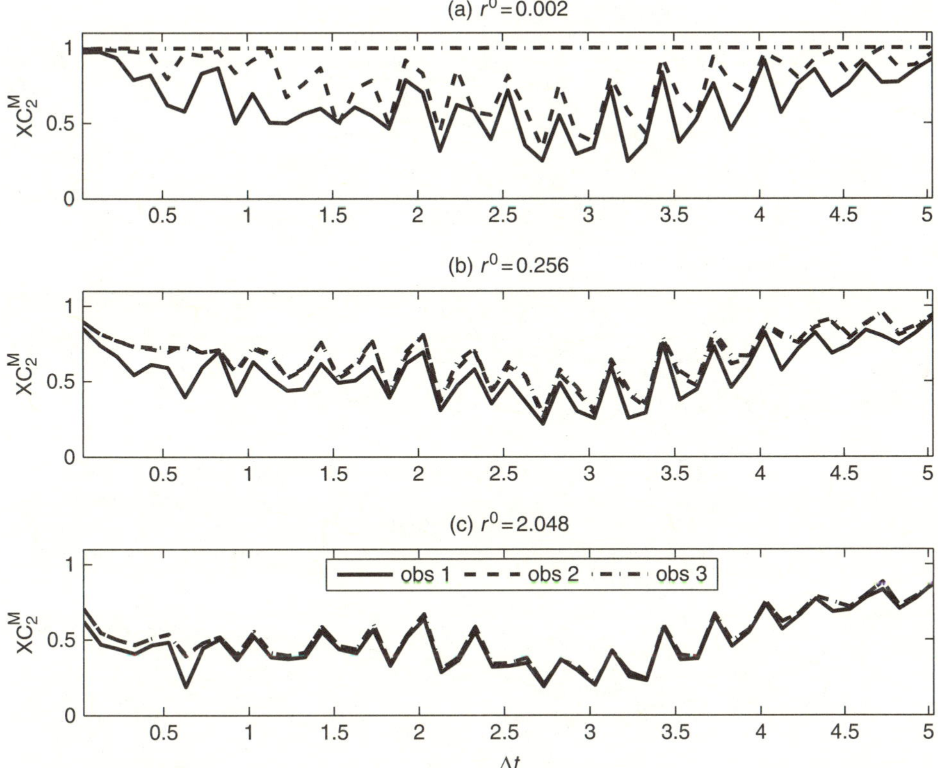

Figure 10.20 Mean model error of u_2. Observations of type 1 (solid line), type 2 (dashed line), and type 3 (dashed-dotted line) are shown.

Now, we point out the properties of the model error along individual trajectories that are not predicted by the mean model error analysis:

1. for smaller r^0, observations of type 2 have larger model error than observations of type 1;
2. the measure of the model error is not predicted well quantitatively although the qualitative behavior is similar;
3. the dependence on Δt appears to be quite smooth (especially for observations of types 2 and 3) while the mean model error analysis predicts it to be more oscillatory.

What is particularly striking here is the strong correlation of the mean model error for the fast mode, u_2, from Fig. 10.20, at the two larger observational noises with the non-Gaussianity in the true signal display in Fig. 10.5. With larger observational noise, the Kalman filter trusts the Gaussian dynamics of the imperfect model significantly so there are larger mean model errors on the fast modes, even with three observations here!

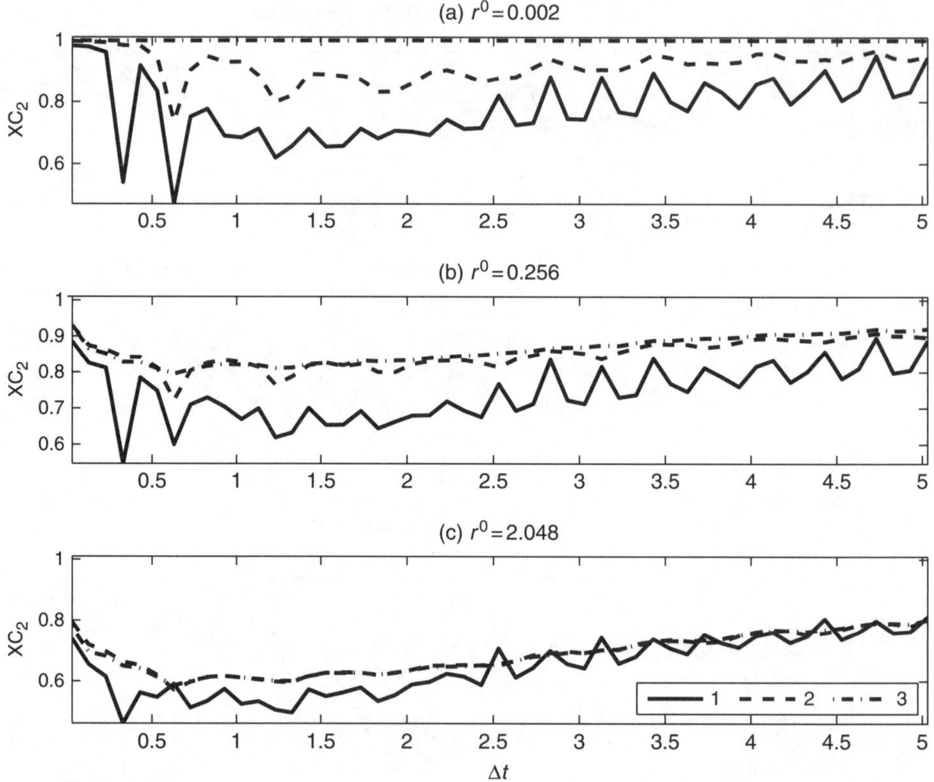

Figure 10.21 Model error of u_2. Observations of type 1 (solid line), type 2 (dashed line), and type 3 (dashed-dotted line) are shown.

10.6 Summary

In the overview of this chapter, we motivated the need to develop a simple nonlinear test model for filtering slow–fast systems and introduced a nonlinear three-dimensional real stochastic triad model as the simplest prototype model for slow–fast systems. We also presented an overview of results for the case with strong fast forcing. In Section 10.2, we established analytic non-Gaussian nonlinear statistical properties of the model including exact formulas for the propagating mean and covariance. An exact nonlinear extended Kalman filter (NEKF) for the slow–fast system and a linear Kalman filter with a non-trivial model error on the fast dynamics through linearization at the climitological mean state were introduced in Section 10.3. Various aspects of filter performance for these two filters were discussed in Section 10.5. While there were detailed summaries in Section 10.5, it is useful to summarize several features of the filter performance.

First, the partial observations of type 1 and 2 were designed to mix both slow and fast modes as occurs in many practical applications. The theoretical analysis of observability in Section 10.4.1 leads us to predict deteriorating filter skill in the vicinity of the non-observable times for observations of type 1 and 2 predicted by Eqn (10.30); this prediction is confirmed for both the linear and nonlinear filters in Figs 10.14–10.16 for small and moderate observational noise. On the other hand, for observation time steps Δt, which are away from the non-observable ones, either shorter or longer than the fast oscillation period, as reported in Section 10.5, the NEKF always has significant skill on the slow mode u_1 for observations of type 1, which mix the slow and fast modes, and retains non-trivial skill for the fast modes (Tables 10.4 and 10.5); on the other hand, the linear filter with model error retains the significant filter skill of the nonlinear model for the slow mode (Table 10.6) but the filter skill for the fast mode deteriorates significantly (Table 10.7). The slow mode in the nonlinear test model has an exact linear slow manifold and the filter skill with model error on the slow mode reflects this fact; nevertheless, these results here suggest interesting possibilities for filtering slow–fast systems by linear systems with model error when only estimates of the slow dynamics are required in an application. Finally, we compared the exact nonlinear analysis for filtering skill through off-line (super)ensemble mean model error in the test model with the actual filter performance for individual realizations discussed throughout the chapter, since such off-line tests can provide important guidelines for filter performance (see chapter 2 and Anderson and Moore (1979); Majda and Grote (2007); Castronovo *et al.* (2008); Harlim and Majda (2008b)). It is established above that the off-line mean model error analysis for the nonlinear test model provides very good qualitative guidelines for the actual filter performance on individual signals with some discrepancies discussed in detail there.

11

Filtering turbulent nonlinear dynamical systems by finite ensemble methods

In this chapter, we review the L-96 model (Lorenz, 1996) and its dynamical properties. This model is a 40-dimensional nonlinear chaotic dynamical system mimicking the large-scale behavior of the mid-latitude atmosphere. This model has a wide variety of different chaotic regimes as the external forcing varies, ranging from weakly chaotic to strongly chaotic to fully turbulent. We check the performance of the ensemble square-root filters described in Chapter 9 on filtering this model for various chaotic regimes, observation times, observation errors for plentiful and regularly spaced sparse observations, and we discuss the phenomenon of "catastrophic filter divergence" for the sparsely observed L-96 model. As a more realistic example of geophysical turbulence, we also review the two-layer quasi-geostrophic (QG) model (Salmon, 1998; Vallis, 2006) that mimics the waves and turbulence of both the mid-latitude atmosphere and the ocean in suitable parameter regimes. We present numerical results of filtering true signals from the QG model with the ensemble adjustment Kalman filter (EAKF), implemented sequentially with the local least-squares framework (Anderson, 2003). In Chapters 12 and 13, we will revisit these models and study much cheaper filters with high skill based on suitable linear stochastic models (Chapter 12) combined with the stochastic parametrized extended Kalman filter (SPEKF) in Chapter 13.

11.1 The L-96 model

The L-96 model was introduced by Lorenz (1996) to represent an "atmospheric variable" u at $2N$ equally spaced points around a circle of constant latitude; thus, it is natural for u to be solved with a periodic boundary condition. The jth component is propagated in time following the differential equation

$$\frac{du_j}{dt} = (u_{j+1} - u_{j-2})u_{j-1} - u_j + F \tag{11.1}$$

where $j = 0, ..., 2N - 1$ represent the spatial coordinates ("longitude"). Note that this model is not a simplification of any atmospheric system, however, it is designed to satisfy three basic properties: it has linear dissipation (the $-u_j$ term) that decreases the total

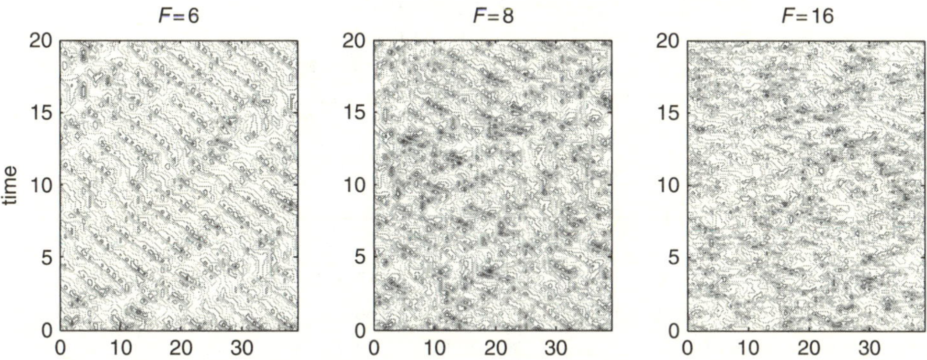

Figure 11.1 Filtering: Time series of the L-96 model in various regimes ranging from weakly chaotic ($F = 6$), to strongly chaotic ($F = 8$), to fully turbulent ($F = 16$).

energy defined as $E = \frac{1}{2}\sum_{j=0}^{2N-1} u_j^2$, an external forcing term $F > 0$ that can increase or decrease the total energy, and a quadratic discrete advection-like term that conserves the total energy (i.e. it does not contribute to dE/dt) just like many atmospheric models (see Majda and Wang, 2006). Combining all these terms together, u_j is always bounded for every j and any time $t \in \mathbb{R}$, and hence one can show that the L-96 model (11.1) is a dissipative dynamical system with the "absorbing ball property" (Constantin *et al.*, 1988).

Following Lorenz (1996), Majda *et al.* (2005), and Majda and Wang (2006), p. 239, we set $N = 20$ so that the distance between two adjacent grid points roughly represents the mid-latitude Rossby radius (≈ 800 km), assuming the circumference of the mid-latitude belt is about 30,000 km. In Majda *et al.* (2005), it was shown in detail that this system, (11.1), has analogues to the most important mid-latude weather waves, the Rossby waves, which have westward phase velocity but eastward group velocity. In Fig. 11.1, we can qualitatively observe this weather pattern for different regimes ranging from weakly chaotic $F = 6$ to highly chaotic $F = 8$, to fully turbulent $F = 16$. When $F = 6$ and 8, we see a clear pattern of westward waves with peaks (the darker contour) propagating eastward. When $F = 16$, the system is fully turbulent and this weather-like pattern becomes less obvious.

As in Abramov and Majda (2007), we show four quantities (in Table 11.1) in the non-dimensional units described in (11.4) to characterize the dynamical behavior of system (11.1): the largest Lyapunov exponent λ_1 which reflects the rate of which two nearby trajectories diverge from each other, the number of positive Lyapunov exponents N^+ which reflects the dimension of the expanding subspace of the attractor, the Kolmogorov–Sinai KS entropy (the sum of all the positive Lyapunov exponents) which measures the information loss on the chaotic attractor (Young, 2002), and the autocorrelation function decay time for the rescaled u_j, computed as follows

$$T_{\mathrm{corr}} = \int_0^\infty |\langle (u_j(t) - \bar{u})(u_j(t+\tau) - \bar{u})\rangle_t| d\tau, \tag{11.2}$$

Table 11.1 Dynamical properties of L-96 model for regimes with $F = 6, 8, 16$. λ_1 denotes the largest Lyapunov exponent, N^+ denotes the dimension of the expanding subspace of the attractor, KS denotes the Kolmogorov–Sinai entropy and T_{corr} denotes the correlation time of the energy-rescaled time correlation function.

F	λ_1	N^+	KS	T_{corr}
6	1.02	12	5.54	8.23
8	1.74	13	10.94	6.70
16	3.94	16	27.94	5.59

where $|\langle\cdot\rangle_t|$ denotes the absolute value of the temporal average. Here, T_{corr} is a rescaled quantity since it is the autocorrelation function of a rescaled quantity $\tilde{u}_j$, which has zero mean and unit energy of perturbations E_p, defined as follows (Majda *et al.*, 2005)

$$u_j - \bar{u} = E_p^{1/2}\tilde{u}_j \text{ and } \tilde{t} = E_p^{1/2}t, \tag{11.3}$$

where

$$E_p = \lim_{T\to\infty} \frac{1}{2T} \sum_{j=0}^{2N-1} \int_{T_o}^{T_o+T} (u_j(t) - \bar{u})^2 dt. \tag{11.4}$$

Practically, we compute the integrals in (11.2) for sufficiently large times T and τ. As shown below and in Chapter 12 as well as in Majda *et al.* (2005) and Majda and Wang (2006), the L-96 system has exponentially decaying correlations and is a mixing ergodic dynamical system for the values of F considered here (Young, 2002) as determined by numerical experiments. Thus, the finite time empirical averages as in (11.2) and (11.4) are reasonable estimates for the correlation time and variance, respectively, of the attractor. From Table 11.1, we see that as we increase the external forcing F, the three measures λ_1, N^+ and KS increase while the correlation time T_{corr} decreases. These quantitative facts reflect the visual evidence in Fig. 11.1 that the forcing values, $F = 6, 8, 16$, define weakly chaotic, strongly chaotic, and fully turbulent dynamical systems respectively.

It was shown by Abramov and Majda (2004) that the structure of the correlation function suggests stronger mixed chaotic waves for larger F's. They also showed the near Gaussian density for each individual Fourier mode for regimes $F = 6, 8$. In Fig. 11.2, we plot the marginal probability distribution of a single node u_7 from a long-time trajectory of a solution up to $11,000$ time units (dashes) omitting the first 1000 time units and compare it to the Gaussian distribution with the same mean and variance (solid). Here, we see that for all three values of F the probability distributions are weakly skewed from a Gaussian distribution. Many comprehensive models for the mid-latitude atmosphere have such weakly

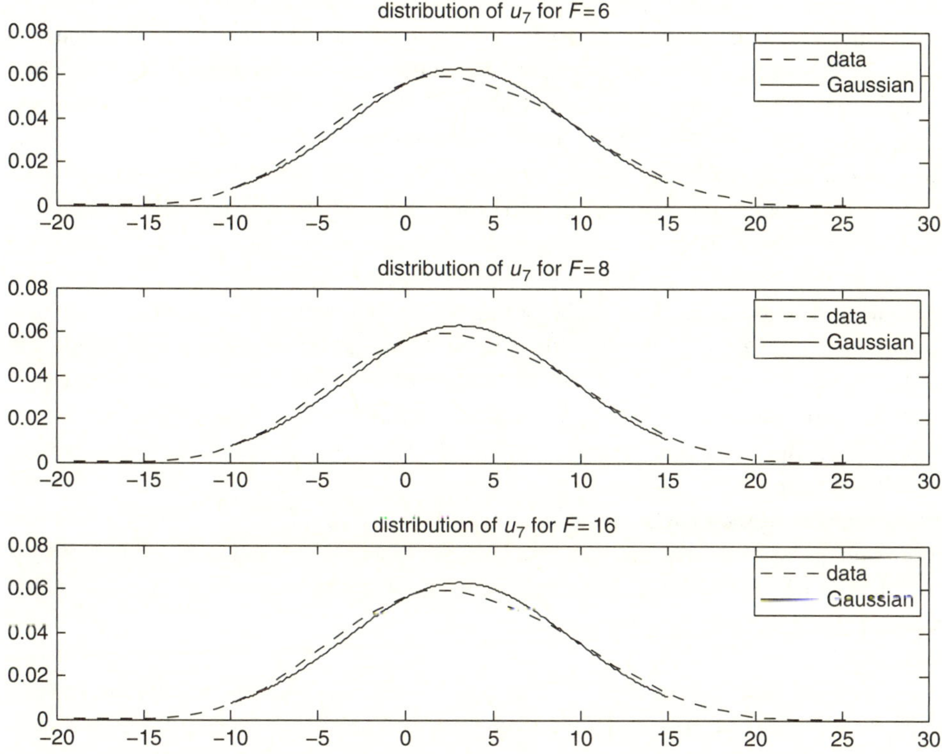

Figure 11.2 Long-time distribution of the L-96 model for the weakly chaotic ($F = 6$), to strongly chaotic ($F = 8$), to fully turbulent ($F = 16$) regime.

skewed probability distributions for the most energetic large-scale components which are nearly Gaussian (Majda *et al.*, 2008).

11.2 Ensemble square-root filters on the L-96 model

In this section, we discuss the numerical results of implementing both ensemble square root filters (ETKF and EAKF), described in Section 9.3 of Chapter 9, on the L-96 model. In our numerical experiments, we always start the data assimilation with an ensemble of size $K = 80$; each ensemble member is generated by perturbing the initial true state with a Gaussian noise with climatological variance E. To avoid ensemble collapse (see Chapter 9), we fix the variance inflation coefficient to be $r = 0.05$. Each simulation is run for 5000 assimilation cycles with time-independent diagonal observation error covariance matrix R^o with diagonal components with variance $r^o\mathrm{I}$.

We define $P = N/M$ to be the ratio between the number of positive Fourier modes $N = 20$ of the model and the number of positive Fourier modes M of the observation; we assume that the number of observations is always less than or equal to the total model grid points, i.e. $M \leq N$. In physical space, as in Chapter 7, this setting reflects a total of $2M$ observations regularly spaced at every P model grid points; the case $P = 1$ corresponds to plentiful observations discussed in Chapter 6 and the case $P > 1$ corresponds to regularly spaced sparse observations discussed in Chapter 7. In our simulations, we vary P from 1 to 5 and notice that when $P = 3$ there are two adjacent observations since N is not divisible by 3. In our numerical simulations, we integrate the model with the fourth-order Runge–Kutta (RK4) method for every $\Delta t = 1/64$ time units and show results from simulations with observations at every $n = 5$ steps ($T_{\text{obs}} = n\Delta t = 0.078$) and $n = 15$ steps (or $T_{\text{obs}} = 0.234$). When $F = 8$, Lorenz suggests that 0.05 non-dimensionalized units are equivalent to 6 hours based on doubling time in a global weather model (Lorenz, 1996). Thus, our choices of n's correspond to roughly 9 and 28 hours, respectively. Compared to the correlation time T_{corr} in Table 11.1, these observation times with $n = 5, 15$ are relatively short since $\tilde{T}_{\text{obs}} = n\Delta t E_p^{1/2}$ is always smaller than T_{corr} except when $F = 16$ and $n = 15$. To measure the performance, we show the RMS average and the temporal average spatial correlation as in Chapter 7. In each chaotic regime, we also show the average RMS error of the unfiltered solution in the last row in Tables 11.2–11.4; this quantity reflects the approximate size of the chaotic attractor. Thus, for example, when the observation error is $\sqrt{r^o} = \sqrt{3} = 1.73$, this is relatively high (60%) for the weakly chaotic regime $F = 6$ and relatively low (10%) for fully turbulent regime $F = 16$, compared to the size of the chaotic attractor.

For plentiful observation, $P = 1$, ETKF and EAKF are comparable except for the fully turbulent regime $F = 16$ with observation error variance $r^o = 3$ where ETKF is significantly better than EAKF. As P increases, i.e. the number observations decreases, EAKF is very skillful for the weakly chaotic regime $F = 6$. For the strongly chaotic regime, $F = 8$, EAKF is skillful except when $P = 5$ with large r^o and/or T_{obs}. In Fig. 11.3, we show snapshots for a skillful regime $F = 8$, $P = 2$, $T_{\text{obs}} = 0.078$, $r^o = 3$ after 2500 and 5000 assimilation cycles. In this regime, the average spatial correlation of the posterior mean state with the true signal increases from 0.96 at assimilation cycle 2500 to 0.98 at assimilation cycle 5000. In the fully turbulent regime $F = 16$, EAKF is not very skillful for $r^o = 3$; this filtering failure is often called filter divergence in the data assimilation literature. Earlier in Chapters 2, 3 and 7, we have given other examples of filter divergence; the filter is stable but has very little skill and remains far from the true signal. In Fig. 11.4, we show snapshots for the unskillful regime for EAKF with $F = 16$, $P = 2$, $T_{\text{obs}} = 0.234$ after 2500 and 5000 assimilation cycles. In this regime, the average spatial correlation of the posterior mean state with the true signal drops from 0.80 at assimilation cycle 2500 to 0.38 at assimilation cycle 5000. In the same figure, we also show the prior mean state and see that there is reasonable improvement of skill in each assimilation step; when the observations are taken into account in each assimilation step, the average spatial correlation increases by about 0.2 units. We also find that the filter divergence is unavoidable in this regime for various variance inflation coefficients; we tried 10%, 15%, 20%, ...,40% and obtained roughly similar RMS average and average spatial correlation. Overall, EAKF is more accurate compared to ETKF for sparse observations ($P > 1$); furthermore, EAKF

Table 11.2 $F = 6$, $T_{obs} = 0.078$ and 0.234, $r^o = 0.01, 3$. The quantity in brackets near ∞ denotes the number of cycles the filter goes through before it blows up.

p	$T_{obs} = 0.078$, EAKF RMS	$r^o = 0.01$ corr	ETKF RMS	corr	$T_{obs} = 0.234$, EAKF RMS	$r^o = 0.01$ corr	ETKF RMS	corr
1	0.01	1	0.01	1	0.02	1	0.23	1
2	0.02	1	0.06	0.99	0.03	1	0.04	1
3	0.03	0.99	0.12	0.99	0.04	0.99	0.08	0.99
4	0.03	0.99	∞(16)	–	0.05	0.99	∞(7)	–
5	0.04	0.99	∞(9)	–	0.06	0.99	∞(7)	–
	$T_{obs} = 0.078$,	$r^o = 3$			$T_{obs} = 0.234$,	$r^o = 3$		
1	0.34	0.99	0.35	0.99	0.65	1	0.62	1
2	0.54	0.98	∞(10)	–	1.08	0.92	∞(8)	–
3	0.77	0.96	∞(9)	–	1.39	0.87	∞(7)	–
4	1.14	0.91	∞(6)	–	1.66	0.81	∞(4)	–
5	1.54	0.84	∞(4)	–	1.98	0.71	∞(4)	–
Number of observations	2.84	–0.01						

Table 11.3 $F = 8$, $\Delta t = 0.078$ and 0.234, $r^o = 0.01, 3$. The quantity in brackets near ∞ denotes the number of cycles the filter goes through before it blows up.

p	$T_{\text{obs}} = 0.078$, EAKF RMS	$r^o = 0.01$ corr	ETKF RMS	corr	$T_{\text{obs}} = 0.234$, EAKF RMS	$r^o = 0.01$ corr	ETKF RMS	corr
1	0.02	1	0.02	1	0.03	1	0.03	1
2	0.03	1	0.06	0.99	0.04	0.99	0.06	0.99
3	0.03	0.99	0.10	0.99	0.06	0.99	∞(11)	–
4	0.04	0.99	∞(23)	–	0.07	0.99	∞(9)	–
5	0.05	0.99	∞(9)	–	∞(2932)	–	∞(7)	–
	$T_{\text{obs}} = 0.078$,	$r^o = 3$			$T_{\text{obs}} = 0.234$,	$r^o = 3$		
1	0.49	0.99	0.46	0.99	1.01	1	0.87	1
2	1.05	0.95	∞(17)	–	1.62	0.89	∞(7)	–
3	1.52	0.91	∞(7)	–	2.12	0.81	∞(5)	–
4	2.58	0.74	∞(5)	–	2.68	0.68	∞(5)	–
5	2.88	0.66	∞(4)	–	2.95	0.59	∞(4)	–
Number of observations	3.66	0.01						

Table 11.4 $F = 16$, $\Delta t = 0.078$ and 0.234, $r^o = 0.01, 3$. The quantity in brackets near ∞ denotes the number of cycles the filter goes through before it blows up.

p	$T_{obs} = 0.078$, EAKF RMS	$r^o = 0.01$ corr	ETKF RMS	corr	$T_{obs} = 0.234$, EAKF RMS	$r^o = 0.01$ corr	ETKF RMS	corr
1	0.02	1	0.03	1	0.04	1	0.04	1
2	0.04	1	0.06	1	0.07	1	0.09	1
3	0.05	1	0.09	0.99	0.10	0.99	$\infty(8)$	–
4	0.07	0.99	$\infty(16)$	–	5.42	0.55	$\infty(7)$	–
5	0.09	0.99	$\infty(11)$	–	5.63	0.48	$\infty(6)$	–
	$T_{obs} = 0.078$,	$r^o = 3$			$T_{obs} = 0.234$,	$r^o = 3$		
1	3.28	0.86	0.66	0.99	2.43	0.92	1.85	0.95
2	6.68	0.56	$\infty(10)$	–	4.14	0.77	$\infty(4)$	–
3	6.43	0.57	$\infty(6)$	–	4.94	0.64	$\infty(4)$	–
4	7.07	0.43	$\infty(4)$	–	5.49	0.51	$\infty(4)$	–
5	6.20	0.45	$\infty(5)$	–	5.69	0.45	$\infty(4)$	–
Number of observations	6.35	0						

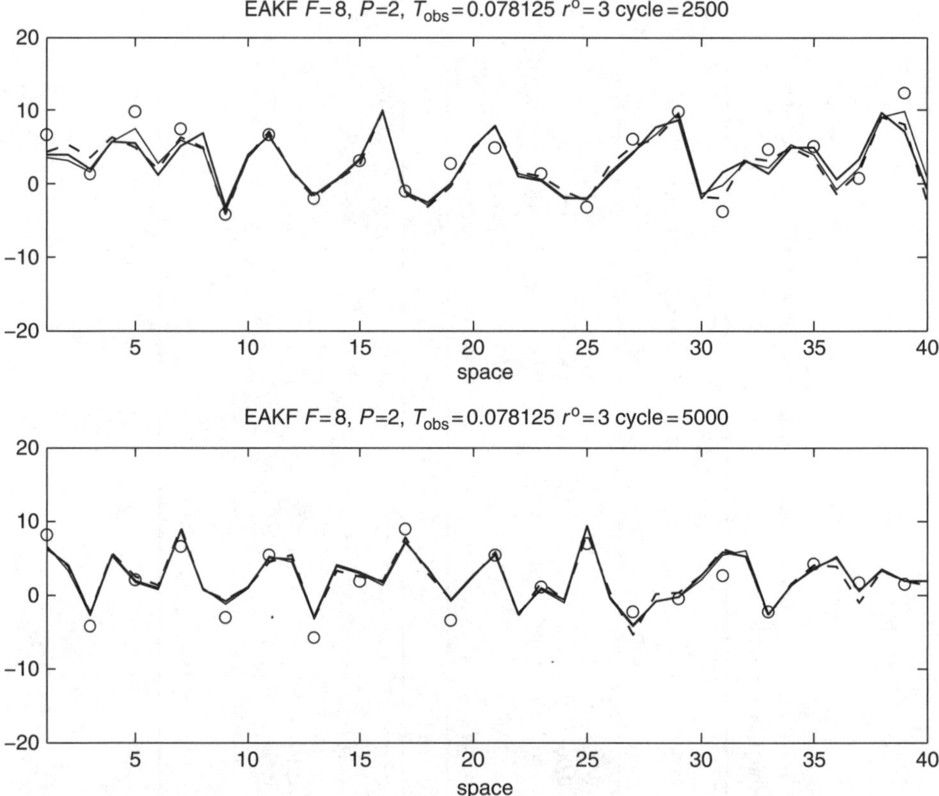

Figure 11.3 Snapshots of the prior (thick solid line) and posterior (thin solid line) mean state after 2500 and 5000 assimilation cycles for EAKF with parameter set $F = 8$, $P = 2$, $T_{\rm obs} = 0.078$, $r^o = 3$. The true signal is represented by dashes and the observations are indicated by circles.

rarely suffers from catastrophic filter divergence (see Tables 11.2–11.4) while ETKF is very prone to this problem with sparse observations in any regime. Note that the ETKF used here is reformulated as in Hunt *et al.* (2007) and Section 9.3 and its computational cost is less than the original ETKF of Bishop *et al.* (2001).

11.3 Catastrophic filter divergence

Catastrophic filter divergence is a new phenomenon in filtering dissipative turbulent dynamical systems where the filtering process strongly diverges to machine infinity in finite time (Harlim and Majda, 2010a). In Tables 11.2–11.4, we find that ETKF suffers from catastrophic filter divergence when the observation error variance is $r^o = 3$ for any sparse observed network $P > 1$ regardless of whether the dynamics is weakly chaotic $F = 6$, strongly chaotic $F = 8$, or fully turbulent $F = 16$. In every regime, we see this filtering

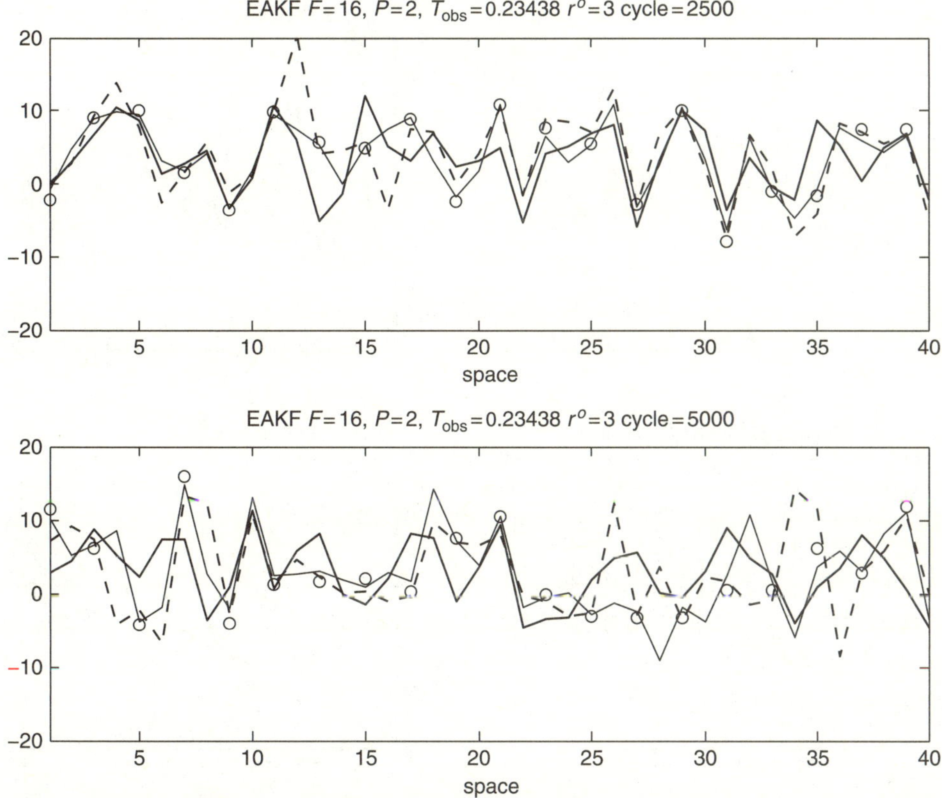

Figure 11.4 Snapshots of the prior (thick solid line) and posterior (thin solid line) mean state after 2500 and 5000 assimilation cycles for EAKF with parameter set $F = 16$, $P = 2$, $T_{\text{obs}} = 0.234$, $r^o = 3$. At assimilation cycle 2500, the spatial correlation between the prior mean state and the true state is 0.6127 and increases to 0.8063 after the observation is assimilated. At cycle 5000, the spatial correlation between the prior mean state and the true state is 0.1129 and increases to 0.3875 after the observation is assimilated. The true signal is represented by dashes and the observations are indicates by circles.

failure after less than 20 steps, and the solution diverges to machine infinity. In Fig. 11.5, we show an example of catastrophic filter divergence with ETKF, exhibited by one of its ensemble members among a total of 80 ensemble members; this occurs after only three steps of the data assimilation cycle when filtering the L-96 model with $F = 16$, $P = 2$, observation time $T_{\text{obs}} = 0.234$, and observation noise size $r^o = 3$; this is exactly the same regime as in Fig. 11.4 and we start the assimilation with exactly the same ensemble that is used in EAKF in Fig. 11.4.

When observations are very sparse $P = 5$, infrequent $T_{\text{obs}} = 0.234$, and relatively accurate with noise variance $r^o = 0.01$, EAKF suffers from a catastrophic filter divergence.

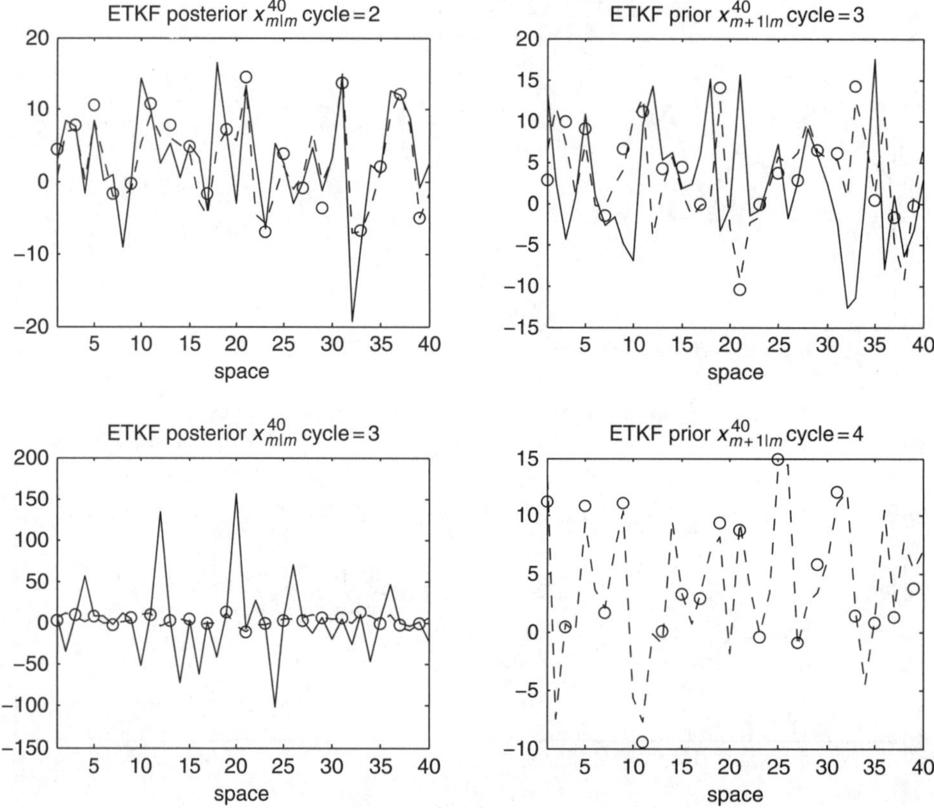

Figure 11.5 Snapshots of a single ensemble member exhibiting catastrophic filter divergence from the 80-member ensemble in ETKF at the fourth assimilation cycle. The parameters are $F = 16$, $P = 2$, $T_{\text{obs}} = 0.234$, $r^o = 3$. In each panel, we show the posterior or prior state (solid), the true signal (dashes) and observations (in circle), as functions of model space. In the last panel, the prior state is not plotted since it diverges to ∞.

In Fig. 11.6, we show snapshots of 27 ensemble members exhibiting catastrophic filter divergence from the 80-member ensemble in EAKF at assimilation cycle 2931. All the posterior ensemble members on the left panels are collapsing on the same exploding large-amplitude state.

We have shown that catastrophic filter divergence can occur in filtering the dissipative nonlinear L-96 model (Lorenz, 1996) with the absorbing ball property and unimodal quasi-Gaussian distributions (see Fig. 11.3) when the number of observations is at most one-half of the model spatial dimension and when these observations are located at regularly spaced grid points even in the perfect model with various turbulent regimes. Moreover, perfect model simulations with EAKF, which is a more stable and very skillful filtering scheme compared to ETKF, is not immune from catastrophic filter divergence. In Chapter 12, we

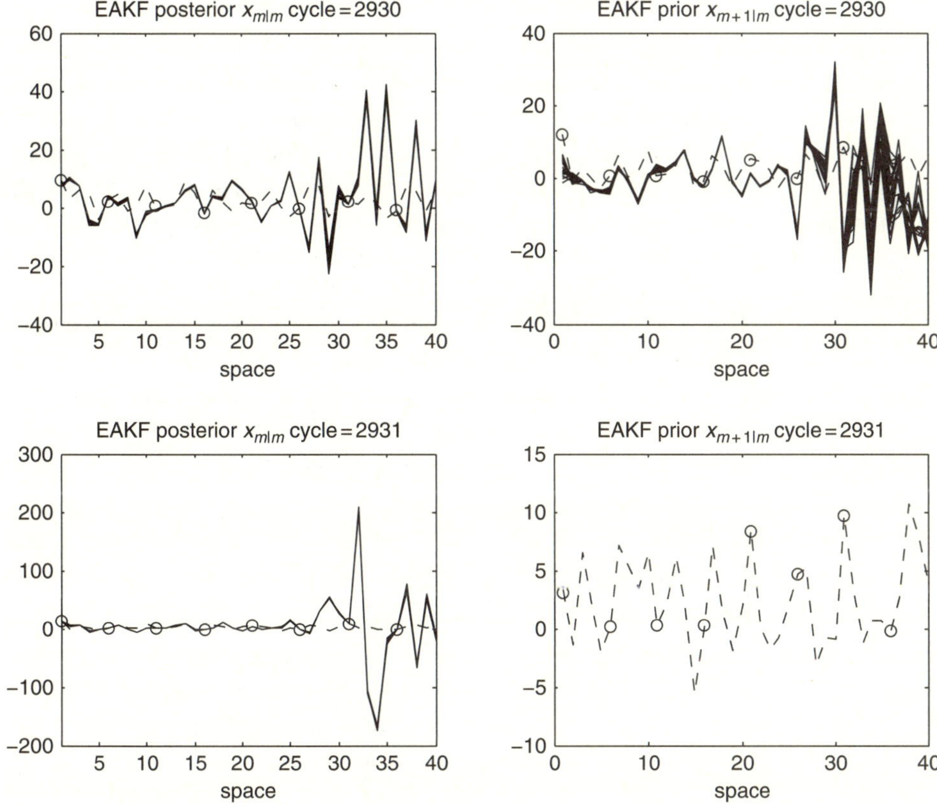

Figure 11.6 Snapshots of 27 members exhibiting catastrophic filter divergence from the 80-member ensemble in EAKF at assimilation cycle 2931. The parameters are $F = 8, P = 5, T_{\text{obs}} = 0.234, r^o = 0.01$. All the posterior ensemble members on the left panels are collapsing on the same exploding large-amplitude state. In each panel, we show the posterior or prior states (in solid black), the true signal (dashes) and observations (in circle). In the last panel, the prior state is not plotted since it diverges to ∞.

will show that we can avoid this catastrophic filter divergence with vastly cheaper computational cost and have significant filter skill by utilizing suitable linear stochastic models. Thus, we will apply the understanding gained in Chapters 6 and 7 for filtering turbulent signals of linear systems to filtering nonlinear turbulent dynamical systems.

From the viewpoint of nonlinear analysis, the catastrophic filter divergence of ETKF and EAKF in a chaotic dynamical system with the absorbing ball property clearly needs further mathematical theory. The catastrophic divergence in Figs 11.5 and 11.6 resembles classical nonlinear instability for finite difference schemes (see Majda and Timofeyev, 2002, p. 66) but new mechanisms must occur here. It is an important problem to understand this issue through more rigorous mathematical analysis.

11.4 The two-layer quasi-geostrophic model

Prototype studies of both mid-latitude (between 30° and 65° in both hemispheres) atmospheric and oceanic turbulent dynamics often involve the quasi-geostrophic approximation of the primitive equations (Salmon, 1998). This approximation, roughly speaking, assumes small Rossby number (significantly less than one), small variation of the Coriolis parameter, and comparable length-scale of motion relative to the radius of deformation. Physically, this approximation filters out the gravity waves and captures the low-frequency variability which has geostrophic balance. See Majda (2003) for a rigorous treatment of this filtering process.

As a simple paradigm, we consider the two-layer quasi-geostrophic (QG) model in a double periodic domain with instability induced by mean vertical shear (Salmon, 1998). The properties of the turbulent cascade have been extensively discussed in this setting, e.g. see Salmon (1998) and the citations in Smith *et al.* (2002). The governing equations for the two-layer QG model with a flat bottom, rigid lid and equal depth layers H can be written as

$$\begin{aligned} \frac{\partial q_1}{\partial t} + J(\psi_1, q_1) + U\frac{\partial q_1}{\partial x} + (\beta + k_d^2 U)\frac{\partial \psi_1}{\partial x} + \nu\nabla^s q_1 = 0, \\ \frac{\partial q_2}{\partial t} + J(\psi_2, q_2) - U\frac{\partial q_2}{\partial x} + (\beta - k_d^2 U)\frac{\partial \psi_2}{\partial x} + \kappa\nabla^2\psi_2 + \nu\nabla^s q_2 = 0, \end{aligned} \tag{11.5}$$

where subscript 1 denotes the top layer and 2 the bottom layer; ψ is the perturbed stream function; $J(\psi, q) = \psi_x q_y - \psi_y q_x$ is the Jacobian term representing nonlinear advection; U is the zonal mean shear; β is the meridional gradient of the Coriolis parameter; q is the perturbed quasi-geostropic potential vorticity, defined as follows

$$q_i = \beta y + \nabla^2\psi_i + \frac{k_d^2}{2}(\psi_{3-i} - \psi_i), \quad i = 1, 2, \tag{11.6}$$

where $k_d = \sqrt{8}/L_d$ is the wavenumber corresponding to the Rossby radius of deformation L_d; the coefficient κ is the Ekman bottom drag coefficient; and the dissipative operator $\nu\nabla^s q$ is added to filter out the energy buildup on the smaller scales. Here, the hyperviscosity coefficient $\nu \equiv \nu(s, \Delta t)$ is chosen such that it only damps the smaller scale. Note that Eqns (11.5) are the prognostic equations for perturbations around a uniform shear with stream function $\Psi_1 = -Uy$, $\Psi_2 = Uy$ as the background state.

In our numerical simulations, the true signal is generated by resolving Eqns (11.5) with 63 modes meridionally and zonally, which corresponds to $128 \times 128 \times 2$ grid points. With such resolution, the numerical integration of this turbulent system only needs slightly more than 30,000 state variables since one can always compute ψ from knowing q and vice versa via Eqn (11.6). This model has two important non-dimensionalized parameters: $b = \beta\,(L/2\pi)^2\,/U_o$, where $U_o = 1$ is the horizontal non-dimensionalized velocity scale and L is the horizontal domain size in both directions (we choose $L = 2\pi$), and $F = (L/2\pi)^2\,/L_d^2$, which is inversely proportional to the deformation radius. As in Kleeman and Majda (2005), we consider the same two cases: one with deformation radius $F = 4$ that roughly mimicks a turbulent jet in the mid-latitude atmosphere and another

Table 11.5 Parameter values for the numerical experiments; $(k, \ell)_{max}$ is the horizontal wavenumber accociated with the maximum growth rate, $\tilde{\sigma}_{max} = Im[c]k$, where c is the wave speed.

Regime	b	F	U	κ	$\tilde{\sigma}_{max}$	$(k, \ell)_{max}$
Atmosphere	2	4	$0.2U_o$	0.1	0.21	(3,0)
Ocean	2	40	$0.1U_o$	0.1	0.48	(9,0)
Atmosphere	2	4	$0.2U_o$	0.2	0.18	(3,0)

one with smaller deformation radius $F = 40$ that mimicks a turbulent jet in the ocean (see Table 11.5 for details). Numerically, Eqns (11.5) are very stiff in the ocean case since the term $k_d^2 U \partial_x \psi_i$, which is proportional to F, is large. The velocity magnitude U is chosen such that both cases exhibit baroclinic instability with a turbulent cascade. For the atmosphere case with $F = 4$, we will also consider model errors by doubling its bottom drag coefficient from $\kappa = 0.1$ to $\kappa = 0.2$ which still retains turbulent cascades from baroclinic instability (we will discuss this in Section 4). In Table 11.5, we also give the growth rate of the most unstable mode. For the atmosphere case, instability occurs on waveband $\{(k,\ell) : 1 \le k \le 7, 0 \le \ell \le 6, 2.8 < \sqrt{k^2+\ell^2} < 7.2\}$ for the stronger bottom drag case and on waveband $\{(k,\ell) : 1 \le k \le 6, 0 \le \ell \le 6, 2.8 < \sqrt{k^2+\ell^2} < 6.4\}$ for the weaker bottom drag case. In the ocean case, instability occurs on a wider waveband, $\{(k,\ell) : 1 \le k \le 15, 0 \le \ell \le 15, 8.5 < \sqrt{k^2+\ell^2} < 15.6\}$. Notice that the maximum growth rate decreases when the bottom drag is stronger.

The large-scale components of these turbulent systems are overwhelming barotropic (see Salmon (1998) and fig. 1 in Kleeman and Majda (2005)). In the atmosphere case, the barotropic mode, which provides the bulk depth-averaged component of the flow with the stream function $\psi_b = (\psi_1 + \psi_2)/2$, is dominated by a strong zonal jet (see the barotropic velocity vector field and stream function in Fig. 11.7). On the other hand, the baroclinic modes govern the transport of heat and have the stream function $\psi_c = (\psi_1 - \psi_2)/2$. Turbulence theory (Salmon, 1998) and citations in Smith *et al.* (2002) suggest on inverse cascade to the barotropic modes at large scales. This is confirmed in the simulations presented here. In the ocean case, there are transitions between blocked and unblocked flow with both large-scale Rossby waves and zonal jets (see Fig. 11.7).

For both the atmosphere and ocean cases, the most energetic modes are on the large-scale barotropic modes (see Fig. 11.8) in the statistical equilibrium state (see Kleeman and Majda (2005) for a discussion of the energy cascades in these two cases with topography). The horizontal axis in Fig. 11.8 corresponds to the barotropic modes, ordered according to empirical orthogonal functions (EOFs) using the barotropic stream function variance, i.e. from the largest to the smallest variance (averaged over a long period of time). In the atmosphere case, 95% of the energy or variability is represented by modes 1–5, which corresponds to the two-dimensional horizontal Fourier modes (0,1), (1,1), (−1,1), (0,2), and

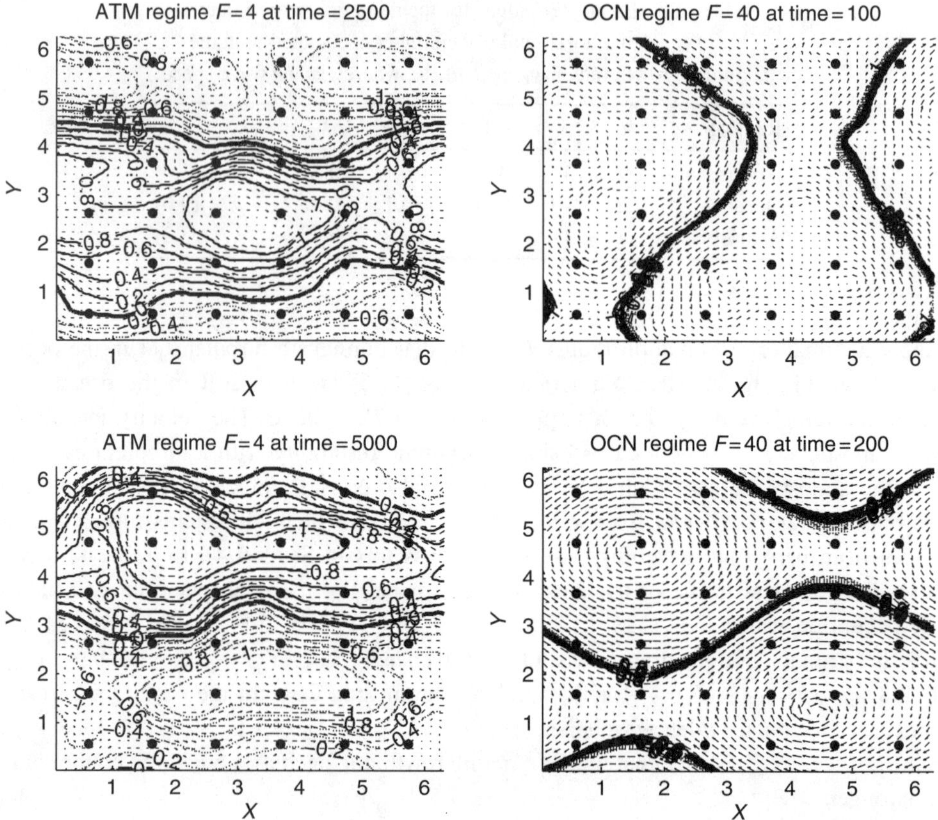

Figure 11.7 The barotropic velocity vector field (arrows), the stream function (contour) and the observation network (dots): first column for atmosphere case at times 2500 and 5000, the second column for ocean case at times 100 and 200.

(1,0), respectively. Notice that the large-scale zonal jet modes, (0,1) and (0,2), carry the largest and the fourth largest variances. In the ocean case, the first five modes are ordered as follows: (1,0), (0,1), (1,1), (−1,1), and (0,2). Here the large-scale zonal jet modes carry the second and fifth largest variances and the Rossby mode (1,0) has the largest variance. Secondly, the magnitudes of the variances of the first two modes are comparable, which indicates competition between two distinct regimes (Rossby waves and zonal jets, see Fig. 11.7). In the third case, the atmosphere with stronger bottom drag, the fifth most energetic mode is wavenumber (2,0). According to the EOF basis, wavenumber (1,0), which is the fifth mode when the bottom drag is weaker, becomes mode 8 when the bottom drag is stronger. Notice also that when the bottom drag coefficient is stronger, the fraction of the variance in the first zonal mode decreases but the variance increases for modes 2–10 (see Fig. 11.8).

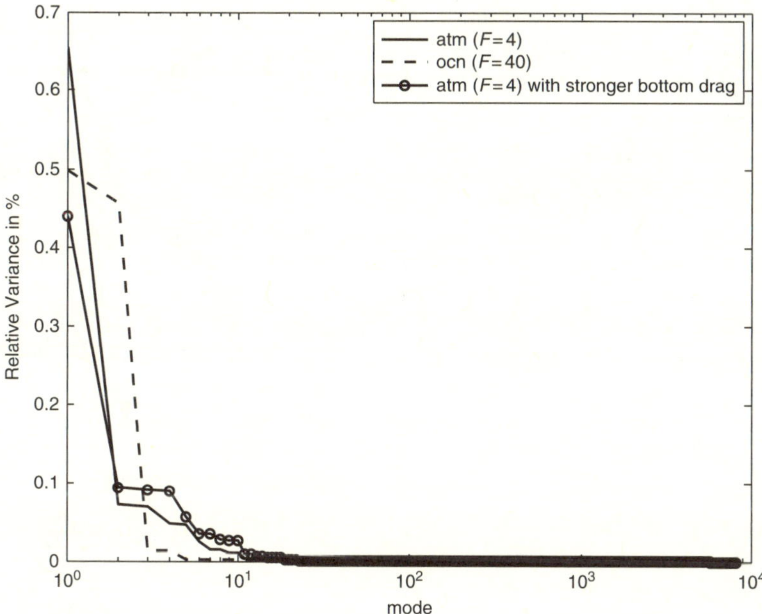

Figure 11.8 Percentage of variances of the barotropic stream function as functions of modes, ordered from the largest to the smallest.

In Fig. 11.9, we show the marginal pdfs of the first five and the eighth most energetic modes for both the atmosphere with weaker and stronger bottom drag and the ocean cases. These marginal pdfs are generated through bin-counting the barotropic stream function, centered at 0 (each panel in Fig. 11.9 shows a histogram of $d\psi = \hat{\psi}_b - \langle\hat{\psi}_b\rangle$) and they encompass solutions of Eqns (11.5) up to $T = 10,000$ time units at every 0.25 time interval in the atmosphere case and up to $T = 400$ time units at every 0.01 time interval in the ocean case. In the atmosphere case, we see that unlike the Rossby modes, the zonal modes are far from Gaussian. But when the bottom drag is doubled, the zonal mode 1 has a tighter marginal distribution and mode 4 looks more Gaussian. In Fig. 11.10, we show the correlation functions of each mode; note that the correlation time to be used in MSM1 and MSM2 in Chapters 12 and 13 is basically the integral of the correlation function (see Chapter 12 and Majda *et al.* (2005) for the exact formulas). In the atmosphere case, the zonal mode (0,1) is weakly mixed and it decays very slowly. In the ocean case, on the other hand, the same zonal mode (0,1) is far from Gaussian (see Fig. 11.9) and decays much faster (see Fig. 11.10). When the bottom drag is changed, there is a significant discrepancy in the correlation functions since the EOF-based modes are ordered differently. In particular, the oscillatory correlation on the fifth mode (or wavenumber (1,0)) when the bottom drag is weak appears as the eighth most energetic mode in the stronger bottom drag case and it replaces a non-oscillatory correlation function of wavenumber (2,1). These oscillatory correlations are associated with Rossby waves (Salmon, 1998).

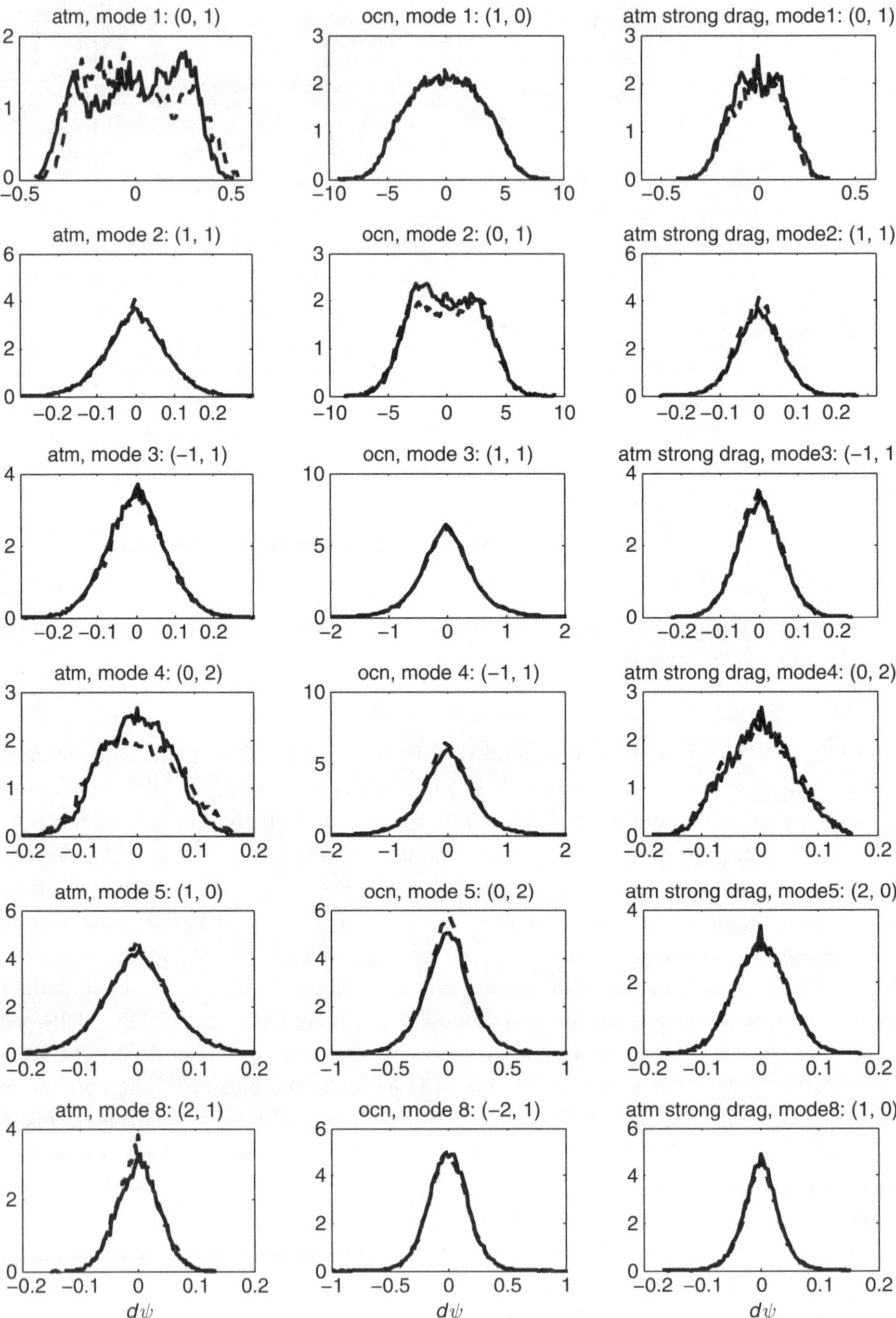

Figure 11.9 Marginal pdfs of the barotropic stream function (centered around its mean). Solid indicates real part and dashes indicates imaginary part. First column for atmosphere case, the second column for ocean case, and the third column for atmosphere case with stronger bottom drag.

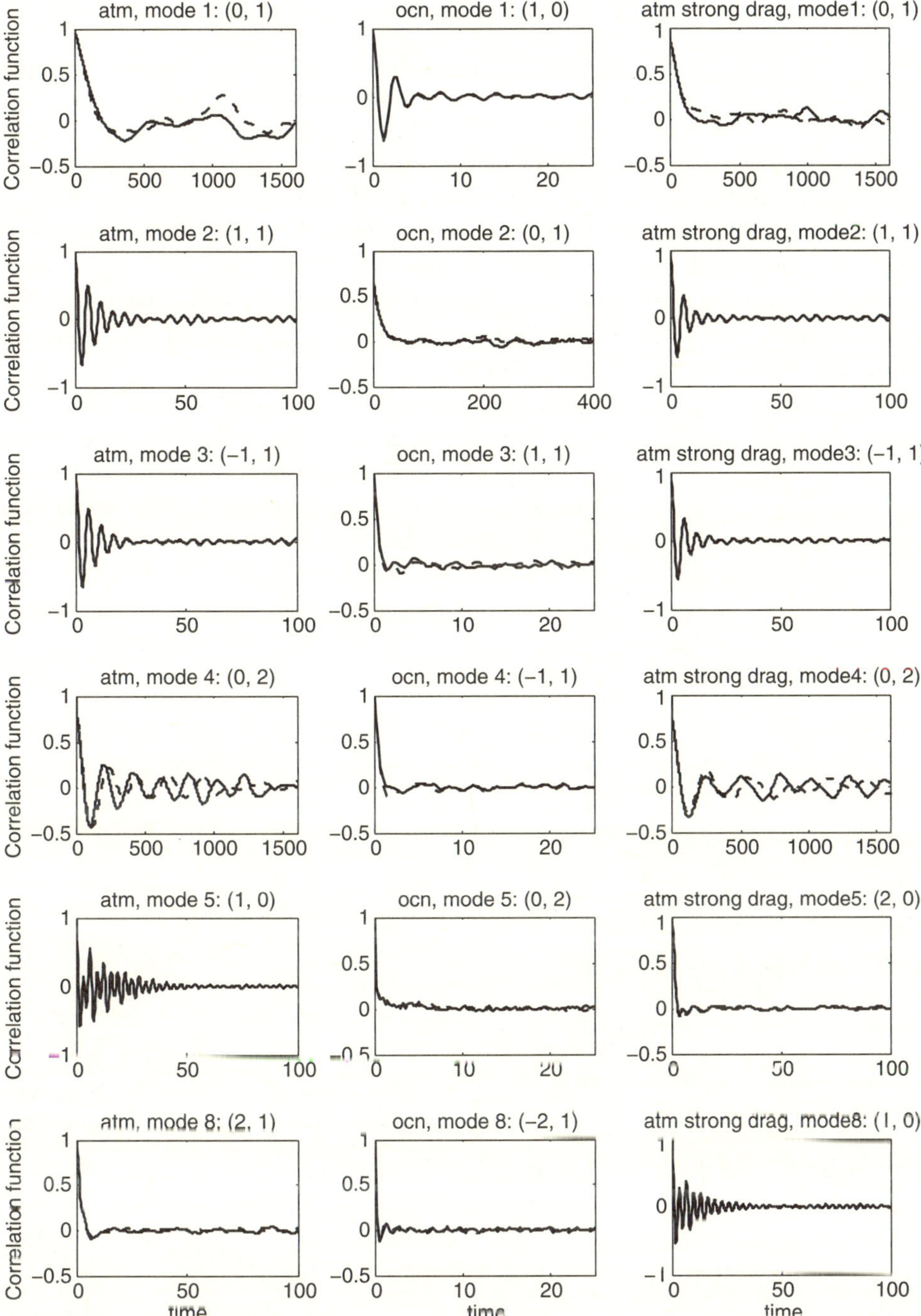

Figure 11.10 Correlation functions of the barotropic stream function as functions of time. Solid indicates real part and dashes indicates imaginary part. First column for atmosphere case, the second column for ocean case,and the third column for atmosphere case with stronger bottom drag.

11.5 Local least-square EAKF on the QG model

In this section, we describe filtering true signals from the two-layer QG model with a sequential data assimilation technique using the local-least-squares ensemble adjustment Kalman filter (LLS-EAKF) of Anderson (2003). Our motivation for choosing this scheme is because it produces very accurate solutions in many contexts and it does not require any singular value decomposition as in many other types of ensemble Kalman filters (see e.g. Anderson (2001); Bishop *et al.* (2001); Ott *et al.* (2004) and Chapter 9). The local least-squares framework is a numerical approximation of EAKF that assumes a least-squares relation between prior distributions of an observation and model state variables. Here, the posterior update consists of two steps: first, update the prior ensemble estimate of the observation variable with a scalar ensemble filter; second, perform a linear regression to the prior ensemble member of the state variable based on the increment on the observation variable. In our simulation, we implement the sequential data assimilation locally in space, that is, each observation corrects only variables within a two-dimensional rectangular box of size $(2D+1)\times(2D+1)$ centered at the corresponding observation location. The localization is used to suppress spurious long-distance correlations produced by a limited ensemble size (Anderson, 2001; Hamill *et al.*, 2001; Houtekamer and Mitchell, 2001).

A one-analysis step beginning with a prior ensemble of state variables within a rectangular box of radius D grid points, $u^i_{m+1|m} \in \mathbb{R}^{(2D+1)^2}$, where the index $i = 1,\ldots,K$ denotes the ensemble member, is given as follows:

1. We compute $v^i_{m+1|m} = g(u^i_{m+1|m})$; in our case $g \in \mathbb{R}^{1\times(2D+1)^2}$ is zero everywhere and one for the component that corresponds to the location of the observation; hence $v^i_{m+1|m}$ is scalar.
2. Compute the ensemble average
$$\bar{u}_{m+1|m} = \frac{1}{K}\sum_{i=1}^{K} u^i_{m+1|m}, \quad \bar{v}_{m+1|m} = \frac{1}{K}\sum_{i=1}^{K} v^i_{m+1|m}.$$
3. Compute the cross-covariances
$$\sigma_{UV} = \frac{1+r}{K-1}UV^T, \quad \sigma_{VV} = \frac{1+r}{K-1}VV^T,$$
where r is the variance inflation coefficient (in our numerical experiment, we will empirically find the optimal inflation coefficient). Each column of U is $u^i_{m+1|m} - \bar{u}_{m+1|m}$ and each column of V is $v^i_{m+1|m} - \bar{v}_{m+1|m}$. In our case, $\sigma_{UV} \in \mathbb{R}^{(2D+1)^2\times 1}$ and $\sigma_{VV} \in \mathbb{R}$.
4. Correct the mean observation state
$$\bar{v}_{m+1|m+1} = \bar{v}_{m+1|m} + (\sigma_{VV} + r^o)^{-1}\sigma_{VV}(v_{m+1} - \bar{v}_{m+1|m}),$$
where v_m is a scalar observation with noise variance r^o.
5. Correct each ensemble member of the observation state
$$v^i_{m+1|m+1} = \bar{v}^i_{m+1|m+1} + \sqrt{\frac{r^o}{r^o+\sigma_{VV}}}(v^i_{m+1|m} - \bar{v}_{m+1|m})$$

and compute $\Delta v^i_{m+1} = v^i_{m+1|m+1} - v^i_{m+1|m} \in \mathbb{R}$.

6. For each ensemble member, we update each ensemble member with the least-square formula

$$u^i_{m+1|m+1} = u^i_{m+1|m} + \frac{\sigma_{UV}}{\sigma_{VV}} \Delta v^i_{m+1}. \tag{11.7}$$

7. Repeat steps 1–6 for the remaining observations one-by-one. If there are overlaps between local boxes, use the updated variables from earlier analysis for the overlapped regions instead of the prior forecasts.

We consider sparse observations of the barotropic stream function at 36 grid points, distributed uniformly in a two-dimensional 2π-periodic domain (see the observation network in Fig. 11.7). Specifically, these observations are generated by adding white noise to a realization of the barotropic stream function, ψ_b, at every discrete observation time T_{obs},

$$\psi^o_m(x_i, y_j) = \psi^t_{b,m}(x_i, y_j) + \sigma^o_m(x_i, y_j), \quad i, j = 1, \ldots, 6, \tag{11.8}$$

where $x_i = 2\pi(12 + (i - 1)21)/128$ and $y_i = 2\pi(12 + (j - 1)21)/128$ are the zonal and meridional observation locations, respectively; $\sigma^o_m(x_i, y_j) \sim \mathcal{N}(0, r^o)$ is a white noise with observation noise variance r^o; subscript 'm' denotes discrete time, superscript 't' denotes true signal, and superscript 'o' denotes observations.

We choose such a sparse observation network to be consistent with a typical sparse ocean observation platform (Cane *et al.*, 1996). In fact, these observations are sufficient to explain more than 90% of the time averaged model variability. In particular, the 12 most energetic modes shown in Fig. 11.8 correspond to wavenumbers $|k| \leq 2, 0 \leq \ell \leq 2$, ignoring the complex conjugate modes (−2,0), (−1,0), and the zeroth mode (0,0). The Fourier representation of the observations in (11.8) is given as follows

$$\hat{\psi}^o_{k,\ell,m} = \sum_{(k_i,\ell_j)\in\mathcal{A}(k,\ell)} \hat{\psi}^t_{k_i,\ell_j,m} + \hat{\sigma}^o_{k,\ell,m}, \tag{11.9}$$

where $\mathcal{A}(k, \ell) = \{(k_i, \ell_j), k_i = k + (2M + 1)i, \ell_j = \ell + (2M + 1)j, |k_i|, |\ell_j| \leq N, i, j \in \mathbb{Z}\}$ is the aliasing set for wavenumber (k, ℓ), where $M = 2$ and $N = 63$ are the total observation and true signal wavenumbers, respectively, in each (zonal and meridional) horizontal direction (see Harlim and Majda (2008b, 2010a) for an explicit example in the one-dimensional context); $\hat{\psi}^o$ and $\hat{\psi}^t$ are the Fourier coefficients of the stream functions ψ^o and ψ^t, respectively; and $\hat{\sigma}^o_{k,\ell} \sim \mathcal{N}(0, \hat{r}^o)$, where $\hat{r}^o = r^o/36$. Theoretically, $(2M + 1) \times (2M + 1) = 5 \times 5$ observations are sufficient (Harlim and Majda, 2008b, 2010a) to recover the dominant modes $|k|, |\ell| \leq M = 2$ but we observe at 6×6 grid points since the practical FFT algorithm used for integrating our QG model is designed to transform an even number of grid points in each horizontal direction.

For the atmosphere case, the radius of deformation is $L_d \approx 0.7$ when the length of each side of the domain is 2π. To represent the radius of deformation, we choose a local box with radius $D = L_d/(2\pi) \times (2N + 1) \approx 14$ grid points. For the ocean case, the corresponding deformation radius is $D \approx 5$; however we numerically find that the solutions

Table 11.6 Average RMS errors (averaged over 10,000 assimilation cycles ignoring the first 200 cycles) with LLS-EAKF.

$F = 4$ Scheme	$T_{\text{obs}} = 0.5$, perfect	$r^o = 0.0856$ model error	$T_{\text{obs}} = 0.25$, perfect	$r^o = 0.1711$ model error
$\sqrt{r^o}$	0.2925	0.2925	0.4137	0.4137
LLS-EAKF	0.1582	0.1471	0.1790	0.1510
$F = 40$	$T_{\text{obs}} = 0.01$,	$r^o = 0.25E$	$T_{\text{obs}} = 0.02$,	$r^o = 0.5E$
$\sqrt{r^o}$	4.1672		4.1672	
LLS-EAKF $D = 14$	6.7656		7.0683	
LLS-EAKF $D = 25$	6.9412		–	

blow up beyond machine infinity with such a small radius D. We empirically find that the ocean case requires $D = 14$ to avoid this numerical blowup. For the atmosphere case, both the true signal and the data assimilation are performed by integrating (11.5) with time step $\Delta t = 0.001$. For the ocean case, unfortunately, we have to tune the forecast time step carefully to avoid numerical blowup. This is due to the numerical stiffness discussed earlier in Section 11.5 when F is large; in our simulations, we generate the true signal with $\Delta t = 0.00025$, but for the data assimilation we have to reduce the forecast time step by 1/5, i.e. $\Delta t = 0.00005$. This makes the LLS-EAKF or any ensemble filters that involve ensemble forecasting computationally very expensive because the major computational cost is from propagating each ensemble member. Besides the two empirical parameters above, the implementation requires tunings of ensemble size and variance inflation coefficient to avoid catastrophic filter divergence due to ensemble collapse as discussed in Section 11.3 and Harlim and Majda (2010a) (we use an ensemble of size $K = 48$ and variance inflation $r = 0.2$). Note that these parameters need to be retuned when observation times or noise variances are varied.

In the atmosphere regime, the perfect model simulation is performed with both the filter model and the true signal integrated with a bottom drag coefficient $\kappa = 0.1$. For this parameter regime, the model variability is $E = 0.34$. The RMS errors shown in Table 11.6 suggest that the filtered solutions are very accurate and even significantly below the observation errors. We also consider a model error case by filtering exactly the same set of observations and observation error variance but with a filter model that has stronger bottom drag coefficient, $\kappa = 0.2$, whereas the true signal has only $\kappa = 0.1$. For this case, we find that LLS-EAKF with the same parameters as in the weaker bottom drag case (local box radius $D = 14$, ensemble size $K = 48$, variance inflation coefficient $r = 0.2$, observation time $T_{\text{obs}} = 0.25$, and observation noise variance $r^o = 0.1711$) produces smaller RMS error (see Table 11.6) relative to the perfect simulation above. This extremely high filtering skill is because the variability of the RMS LLS-EAKF with stronger bottom drag ($E = 0.14$) is much smaller than that with weaker bottom drag ($E = 0.34$). This causes the former configuration to be more stable than the latter (see the maximum growth rate reduction when the bottom drag is stronger in Table 11.5); thus, an ensemble of 48 members provides sufficiently accurate prior statistics. Furthermore, there is always a possibility that we do not tune LLS-EAKF optimally.

The ocean case is an extremely hard test case because of the numerical stiffness in propagating the dynamics due to the smaller radius of deformation. We will only consider the perfect model simulation with bottom drag coefficient $\kappa = 0.1$. Numerically, we find that LLS-EAKF performs well on some periods of time but poorly in other periods of time (we will show this in more detail in Chapter 13). We also find that a larger local box radius $D = 25$ does not reduce the errors (see Table 11.6). In Chapter 13, we will compare the results in this section with an accurate and numerically fast filtering strategy that uses stochastic parametrization. There, we will check the filter performance not only through the RMS error but also through the spatial correlation as well as the spectrum recovery. We will see that the appropriate versions of the stochastic filters with judicious model errors discussed in Chapters 12 and 13 are not only vastly cheaper than LLS-EAKF but they also have comparable filter skill to LLS-EAKF in the atmospheric regime and have extremely high skill beyond LLS-EAKF in the oceanic regime.

12

Filtering turbulent nonlinear dynamical systems by linear stochastic models

In the previous chapter, we discussed the L-96 model and showed the skill of two ensemble square-root filters, ETKF and EAKF, on filtering this model in various turbulent regimes. As we discussed in Chapter 1, with the ensemble approach, there is an inherently difficult practical issue of small ensemble size in filtering statistical solutions of these complex problems due to the large computational overload in generating individual ensemble members through the forward dynamical operator (Haven *et al.*, 2005). Furthermore, as we have seen in Chapter 11, the ensemble square-root filters (ETKF, EAKF) on the L-96 model suffer from severe catastrophic filter divergence, where solutions diverge beyond machine infinity in finite time in many chaotic regimes, when the observations are partially available. For the two-layer QG model, we also found that extensive calibration is needed on the EAKF with the local least-squares framework to avoid catastrophic filter divergence (see chapter 11 and Harlim and Majda (2010b)); yet, the filtered solutions are not accurate in the numerically stiff "oceanic" regime. Naturally one would ask whether there is any skillful reduced filtering strategy that can overcome these challenges of computational overhead.

In this chapter, we discuss a radical approach for filtering nonlinear systems which has several desirable features including high computational efficiency and robust skill under variation of parameters. In particular, we implement an analogue of the Fourier diagonal filter as developed in Chapters 6–8 on the nonlinear L-96 model discussed in Chapter 11 with varying degrees of nonlinearity in the true dynamics ranging from weakly chaotic to fully turbulent. This approach introduces physical model error on top of the model errors due to the numerical discretization scheme since the diagonal linear model completely ignores the nonlinear interactions that occur in the true dynamics and this is why we call it a radical filtering strategy; we replace the nonlinear terms in the original model by a stochastic noise and a linear damping term (the combination of these two terms together is an Ornstein–Uhlenbeck process as discussed in Chapter 2) on each Fourier mode with the hope that these additional terms regenerate the turbulent nature of the true signal. This stochastic modeling is a "poor-man's" approach for modeling turbulent signals; its motivation is discussed thoroughly in Chapter 5 and by Delsole (2004).

Our goal in this chapter is to show that even with this "poor-man's" model, we can obtain a reasonably skillful filtered solution beyond trusting the observations with appropriate parametrization guided by the mathematical off-line test criteria developed in Chapters 6 and 7. In the case where the ensemble filter fails, this radical strategy provides accurate filtered solutions beyond simply trusting the observations. We do not advocate this "poor-man's" strategy as the end product but we hope that the encouraging results shown in this chapter serve as a motivation for filtering nonlinear turbulent dynamical systems in this non-standard fashion; in particular, we hope to convince readers that ignoring correlations between different Fourier modes of nonlinearly mixed signals is a judicious approach; different reduced stochastic models should be tested for more filtering skill beyond the results in this chapter. For example, in Chapter 13, we show how to improve these methods further through nonlinear stochastic parameter estimation of the damping and forcing coefficients "on the fly", from the observations of the dynamics alone.

Furthermore, we will show that this alternative strategy avoids catastrophic filter divergence on filtering regularly spaced sparse observations. We will show that although the filtering skill with these model errors is significant but inferior to the expensive EAKF with the perfect model when the system is weakly chaotic, this reduced filtering strategy produces more accurate filtered solutions in a fully turbulent regime. We will also see that the model error through this "poor-man's" approach restricts the prior distribution to be Gaussian, which prohibits catastrophic filter divergence.

12.1 Linear stochastic models for the L-96 model

Recall the L-96 model (11.1) discussed in Chapters 1 and 11,

$$\frac{du_j}{dt} = (u_{j+1} - u_{j-2})u_{j-1} - u_j + F \tag{12.1}$$

where $j = 0, ..., 2N - 1$ with periodic boundary conditions. We consider the following rescaling, discussed by Majda *et al.* (2005) and Majda and Wang (2006),

$$u_j = \bar{u} + E_p^{1/2}\tilde{u}_j \text{ and } t = E_p^{-1/2}\tilde{t}, \tag{12.2}$$

in order to non-dimensionalize the fluctuations to have unit energy independent of F. Here $\bar{u}$ represents the (temporal) mean state and

$$E_p = \frac{1}{2T}\sum_{j=0}^{2N-1}\int_{T_o}^{T_o+T}(u_j(t) - \bar{u})^2 dt$$

is the average variance in the energy fluctuation. In our numerical results below, both $\bar{u}$ and E_p are computed by integrating (12.1) with the four-step Runge–Kutta method (RK4) with time step $\Delta t = 1/64$ up to time $T = 10,000$ units (see Majda *et al.*, 2005, for values of E_p for various F). The rescaling in (12.2) is chosen such that the rescaled model has zero mean state and a unit energy perturbation.

Substituting (12.2) into (12.1) and after some algebra, we obtain the rescaled L-96 model

$$\frac{d\tilde{u}_j}{d\tilde{t}} = E_p^{-1}(F - \bar{u}) + E_p^{-1/2}((\tilde{u}_{j+1} - \tilde{u}_{j-2})\bar{u} - \tilde{u}_j) + (\tilde{u}_{j+1} - \tilde{u}_{j-2})\tilde{u}_{j-1}. \tag{12.3}$$

In Majda *et al.* (2005) and Majda and Wang (2006), a detailed linear stability analysis of (12.3) is studied with the following definition for the direct and inverse discrete Fourier transforms

$$\hat{u}_k = \frac{1}{J}\sum_{j=0}^{J-1} \tilde{u}_j e^{-2\pi i k j/J}, \quad \tilde{u}_j = \sum_{k=0}^{J-1} \hat{u}_k e^{2\pi i k j/J}. \tag{12.4}$$

Here, we follow the strategy of stochastic modeling of shear turbulence (Delsole, 2004). That is, we make the following approximation of the nonlinear terms

$$(\tilde{u}_{j+1} - \tilde{u}_{j-2})\tilde{u}_{j-1} \to \sum_{k=0}^{J-1} \Big((-d_k + i\omega_k)\hat{u}'_k + \sigma_k \dot{W}_k\Big) e^{2\pi i k j/J}, \tag{12.5}$$

where $\dot{W}_k$ is a complex white noise in time for each Fourier mode k. The additional damping term in (12.5) is important because it neutralizes the additional energy from the white noise and hence the two terms together (which is a well-known Ornstein–Uhlenbeck process) can be designed to statistically conserve the total energy (Delsole, 2004) on the attractor; physically they attempt to mimic the discrete advective nonlinear term in the full L-96 model (12.1). This turbulent approximation is a "poor-man's" strategy with no mathematically rigorous justification; however, numerical implementations of this approximation in more advanced turbulent climate modeling (Majda *et al.*, 2001; Majda and Timofeyev, 2004; Majda *et al.*, 2006; Franzke and Majda, 2006) produce reasonably accurate statistical estimates of the original model provided that the original signal from the full nonlinear model is strongly chaotic or fully turbulent. Such approximations were discussed in detail earlier in Chapter 5 and here we give an explicit concrete example of their implementation on a turbulent dynamical system.

Now, the linear SDE for each mode $\hat{u}'_k$ becomes

$$\frac{d\hat{u}'_k}{d\tilde{t}} = \big(-d_k + \omega_1(k) + i(\omega_2(k) + \omega_k)\big)\hat{u}'_k + \frac{1}{E_p}(F - \bar{u})\delta_k + \sigma_k \dot{W}_k, \tag{12.6}$$

where

$$\omega_1(k) = (\bar{u}(\cos(2\pi k/J) - \cos(4\pi k/J)) - 1)/\sqrt{E_p}, \text{ and} \tag{12.7}$$

$$\omega_2(k) = \bar{u}(\sin(2\pi k/J) + \sin(4\pi k/J))/\sqrt{E_p}. \tag{12.8}$$

Next, we will describe two simple strategies for parametrizing d_k, ω_k and σ_k such that the Fourier modes $\hat{u}'_k$ of the linear stochastic model in (12.6) have the same long-time statistical properties as the Fourier modes $\hat{u}_k$ from the L-96 model.

12.1.1 Mean stochastic model 1 (MSM1)

A simple strategy to determine the filter model parameters is to fit the variance and the integral of the time autocorrelation function for each mode to match the corresponding empirical statistics of the solutions of the L-96 model. The variance and the autocorrelation function of each mode $\hat{u}_k$ of the original system can be computed as follows

$$E_k \equiv \mathrm{Var}(\hat{u}_k) = \overline{|\hat{u}_k(\tilde{t}) - \bar{\hat{u}}_k|^2} \tag{12.9}$$

$$R_{\hat{u}_k}(\tau) = \frac{\overline{(\hat{u}_k(\tilde{t}) - \bar{\hat{u}}_k)(\hat{u}_k(\tilde{t}+\tau) - \bar{\hat{u}}_k)^*}}{\mathrm{Var}(\hat{u}_k)}. \tag{12.10}$$

We denote the integral of the correlation function as

$$\int_0^\infty R_{\hat{u}_k}(\tau)d\tau = T_k - \mathrm{i}\theta_k, \tag{12.11}$$

where T_k is a decorrelation time of the mode $\hat{u}_k$. In Fig. 12.1, we show E_k and T_k for various $F = 0, 6, 8, 16$. In the unforced case $F = 0$, we also set the linear damping $-u_j$ in (12.1) to be zero and therefore it is an energy-conserving system with equipartition equilibrium statistical mechanics (Majda *et al.*, 2005; Majda and Wang, 2006). From these

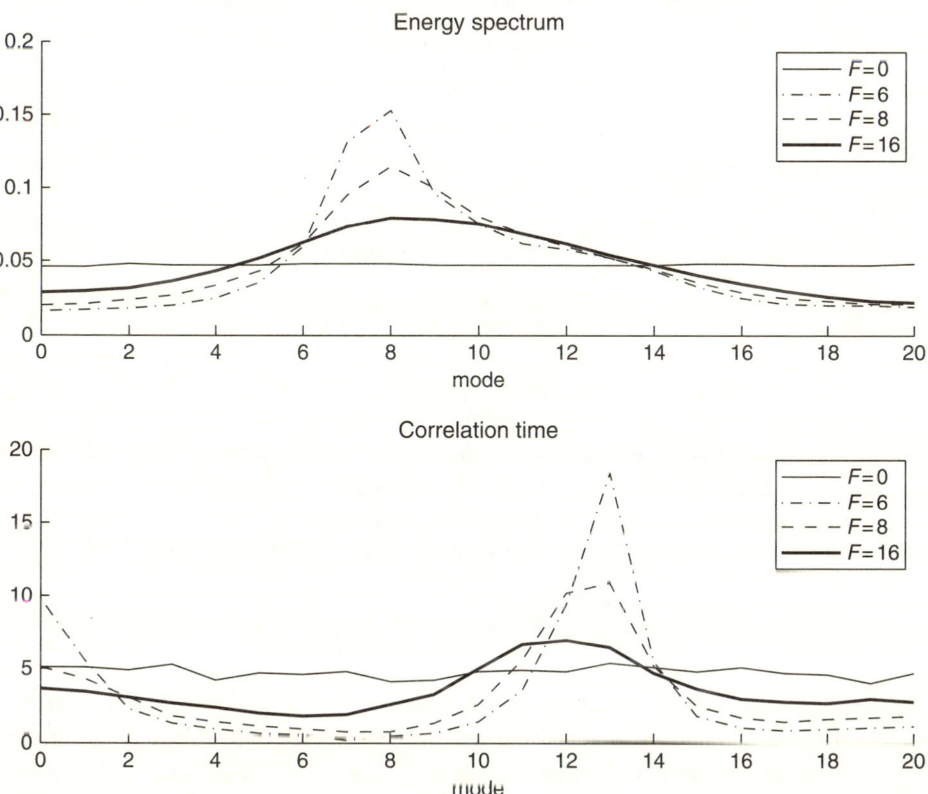

Figure 12.1 The top figure shows the equilibrium variances E_k for various F as a function of wavenumbers k. The bottom figure shows the correlation time T_k for various F as a function of wavenumbers.

statistical estimates, we notice that the equilibrium variance E_k, which represents the climatological energy, is more uniformly distributed in frequency space as F increases; there are no turbulent cascades for this model as in Chapter 5 and the spectrum approaches white noise for larger F.

We have three parameters, E_k, T_k and θ_k, which will be used to determine the three unknowns, d_k, ω_k and σ_k. For the linear process $\hat{u}'_k(\tilde{t})$ given by (12.6), the corresponding values of the variance and of the integral of the correlation function at equilibrium state are given by

$$\mathrm{Var}(\hat{u}'_k) = \frac{\sigma_k^2}{2(d_k - \omega_1(k))},$$
$$\int_0^\infty R_{\hat{u}'_k}(\tau)d\tau = \frac{1}{d_k - \omega_1(k) + \mathrm{i}(\omega_k + \omega_2(k))}.$$

See the detailed computations of these in Chapter 5. To find the unknown parameters, we solve the following equations

$$\frac{1}{d_k - \omega_1(k) + i(\omega_k + \omega_2(k))} = T_k - \mathrm{i}\theta_k,$$
$$\frac{\sigma_k^2}{2(d_k - \omega_1(k))} = E_k.$$

The solution is given by

$$d_k = \omega_1(k) + \frac{T_k}{T_k^2 + \theta_k^2}, \tag{12.12}$$
$$\omega_k = -\omega_2(k) + \frac{\theta_k}{T_k^2 + \theta_k^2},$$
$$\sigma_k^2 = 2E_k \frac{T_k}{T_k^2 + \theta_k^2}.$$

These formulas together with (12.6) determine the perfect regression strategy. Notice that the parameters in (12.12) are always realizable with the constraints in (12.11). This regression strategy was first developed by Majda *et al.* (2009) and utilized recently by Harlim and Majda (2010a) where it is called the **mean stochastic model 1 (MSM1).**

12.1.2 Mean stochastic model 2 (MSM2)

The standard way to build a linear stochastic model for shear turbulence (Delsole, 2004) is to keep the linear frequency of each mode, $\omega_2(k)$, fixed at the climatological background operator but change only the damping and white noise forcing to satisfy the variance constraint (12.11). In this case, the linear SDE for each mode $\hat{u}'_k$ becomes

$$\frac{d\hat{u}'_k}{dt} = \big(-d_k + \omega_1(k) + i\omega_2(k)\big)\hat{u}'_k + \frac{1}{E_p}(F - \bar{u})\delta_k + \sigma_k \dot{W}_k. \tag{12.13}$$

Now, we have only two unknown parameters, d_k and σ_k, to be determined and we utilize the real part of the integrated correlation function in (12.11). The regression solves

$$T_k = \frac{d_k - \omega_1(k)}{(d_k - \omega_1(k))^2 + \omega_2(k)^2},$$
$$E_k = \frac{\sigma_k^2}{2(d_k - \omega_1(k))}.$$

The first equation is quadratic in d_k and has the following solutions

$$d_k = \omega_1(k) + \frac{1 \pm \sqrt{1 - 4T_k^2\omega_2(k)^2}}{2T_k}. \tag{12.14}$$

In order to ensure that the damping is real and positive, the expression under the square root has to be positive. Due to this restriction, we have a realizability constraint on the correlation times T_k. If the correlation time T_k that we obtain from the original nonlinear model does not satisfy this constraint, then this value of T_k is unrealizable and we have to "tune" the dissipation. The second parameter, σ_k^2, is computed via

$$\sigma_k^2 = 2(d_k - \omega_1(k))E_k. \tag{12.15}$$

This second regression strategy is the one utilized by Harlim and Majda (2008a, 2010a) where it is called the **climate stochastic model (CSM)**; in Harlim and Majda (2010b), this model is called the **mean stochastic model 2 (MSM2)**. Note that a model error in recovering the variance of the climatological mean state is committed whenever we need to "time" the dissipation to satisfy realizability.

12.1.3 Observation time model error

Let $\vec{u}(t_i), i = 1, \ldots, T = 500$, be a trajectory solution of the L-96 model (12.1), sampled discretely at every observation time $t_{i+1} - t_i = T_{\text{obs}}$. At each observation time interval, we integrate differential equation (12.6) for MSM1 (or (12.13) for MSM2) with initial condition $\hat{u}'_k(t_i)$ and time step T_{obs} to obtain $\hat{u}'_k(t_i + T_{\text{obs}})$. We define a crude measure of the observation time model error as the RMS average difference between solutions of the original L-96 model (12.1) and the linear stochastic differential equation (12.6) (or (12.13)) in physical space after the time T_{obs}, following Tribbia and Baumhefner (1988). In our earlier discussion (in Section 12.1) we referred to this quantity as the (physical) model error, so for the rest of this chapter we will understand it as the model error after time T_{obs}.

In Fig. 12.2, we show the model errors for both approaches, MSM1 and MSM2, as functions of observation time $T_{\text{obs}} = n\Delta t$, where $n = 2, 5, 15$, and $\Delta t = 1/64$. From Fig. 12.2, we see that both model errors increase as functions of time and the external forcing F. The model error increment as a function of time is well understood since a linear approximation of a nonlinear function typically holds only for sufficiently short time. The model error increment as a function of F is also obvious since the dynamical system becomes more

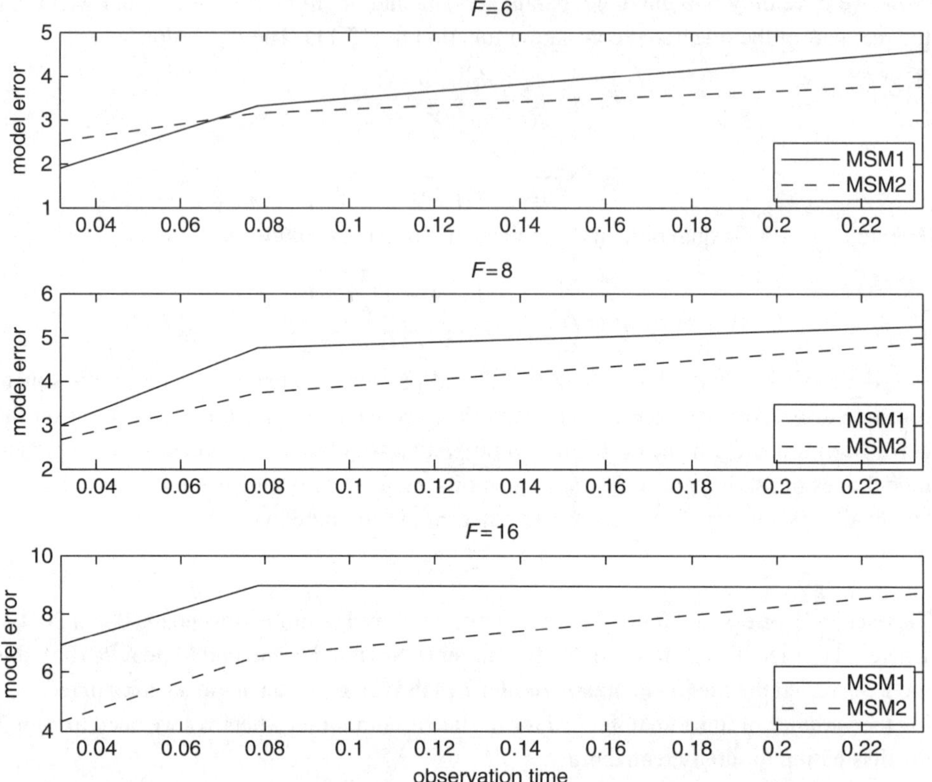

Figure 12.2 Observation time model errors as functions of observation time, T_{obs}: the first row is for $F = 6$, the second row is for $F = 8$, and the third row is for $F = 16$.

chaotic and the number of positive Lyapunov exponents increases as a function of F (see Chapter 1 and Table 11.1 in Chapter 11). Thus, the attractor size also increases with F; we will find below that the filtering skill with the stochastic models increases as solutions of L-96 become more turbulent ($F = 8, 16$) despite the larger model errors. We also notice that except for the case of $F = 6$ and $T_{\text{obs}} = 0.03$, the model error of MSM1 is almost always larger than that of MSM2. With these stochastic models we implement, for both plentiful or sparse regularly spaced observations, the Fourier filtering strategies in Chapters 6 and 7 which ignore correlation from different aliasing sets. We call these filtering methods FDKF MSM1,2.

12.2 Filter performance with plentiful observation

For plentiful observation, as discussed in Chapter 6, the Fourier domain approach reduces the filtering problem to independent scalar filters with observation model defined as follows

$$\hat{v}_{k,m} = \hat{u}_{k,m} + \hat{\sigma}^o_{k,m}, \quad |k| \le N, \tag{12.16}$$

where $\hat{v}_{k,m}$ is the Fourier coefficient of the rescaled observation, $\tilde{v} = (v - \bar{u})/\sqrt{E_p}$, and $\hat{\sigma}^o_{k,m} \sim \mathcal{N}(0, \hat{r}^o)$. In Fig. 12.3, we show the RMS average errors for the various observation times $T_{\rm obs} = n\Delta t$ (with $n = 2, 5, 15$ and $\Delta t = 1/64$) and observation error variance $\hat{r}^o = r^o/(2NE_p)$, where r^o is the observation noise variance in physical space. For benchmark purposes, we include simulations of EAKF (see Chapter 11) with the original nonlinear model in (12.1) as well as with imperfect models through MSM1 and MSM2. We refer to the former as EAKF true and the latter as EAKF MSM1 and EAKF MSM2, consecutively. Thus, EAKF MSM1, 2 use the stochastic models for filtering with model error with the finite ensemble method EAKF so correlations between different aliasing sets occur in the finite ensemble. In the numerical experiments below, EAKF is implemented with an ensemble of size $K = 40 = 2N$ and variance inflation $r = 0.05$.

From our numerical simulations (see Fig. 12.3), we find that FDKF almost always produces filtered solutions that are more accurate than simply trusting the observations. The only exception is when $F = 6$ and observation error $\sqrt{r^o} = 0.1$ in which the FDKF solutions are comparable to simply trusting the observations with RMS error that is only slightly worse than that of the perfect model simulation with EAKF. A reasonable explanation for this result is that the observation error is so small that both FDKF and EAKF weigh the posterior solutions toward the observations. The same explanation also justifies the small discrepancy between FDKF MSM1 and FDKF MSM2 when the observation error is small. When the observation error increases, the difference between FDKF MSM1 and FDKF MSM2 becomes more apparent (see the thick and thin solid lines in Fig. 12.3). In particular FDKF MSM2 is better when the observation time interval is shorter and FDKF MSM1 is more skillful when the observation time interval is longer. The reason for this is because the linear frequency in MSM2, $\omega_2(k)$, is an accurate approximation only for shorter times. When the observation time interval is longer, $n = 15$, the frequency coefficients from the long-time statistics as in MSM1, $\omega_2(k) + \omega_k$, produces better approximations. This fact, that FDKF MSM1 supersedes FDKF MSM2 for large observation times diminishes as F increases since the correlation time is short when the model is in fully turbulent regime (see Fig. 12.1). We also find that both EAKF MSM1 and EAKF MSM2 are not as good as FDKF MSM1 and FDKF MSM2. This confirms the advantage of ignoring correlations between different modes in filtering with model errors besides saving computational labor.

Finally, we find that EAKF with the perfect model is not robust; in some regimes such as $F = 6, n = 2, r^o = 4$, it produces the best filtered solutions (see Figs 12.3–12.5), however there are many regimes where the filtered solution diverges (e.g. see Figs 12.6 and 12.7 for regimes with $F = 6, n = 15, \sqrt{r^o} = 0.25$). The general trend is that when the system is weakly turbulent the perfect model simulation produces the most accurate solution better than FDKF. In the fully turbulent regime, FDKF retains its filtering skill beyond EAKF and trusting the observations (e.g. see Figs 12.3, 12.8, 12.9). This general trend agrees with results in turbulent climate modeling (Majda *et al.*, 2001; Majda and Timofeyev, 2004; Majda *et al.*, 2006; Franzke and Majda, 2006); the linear stochastic approximation in (12.5) produces reasonably accurate statistical estimation provided that the original signal

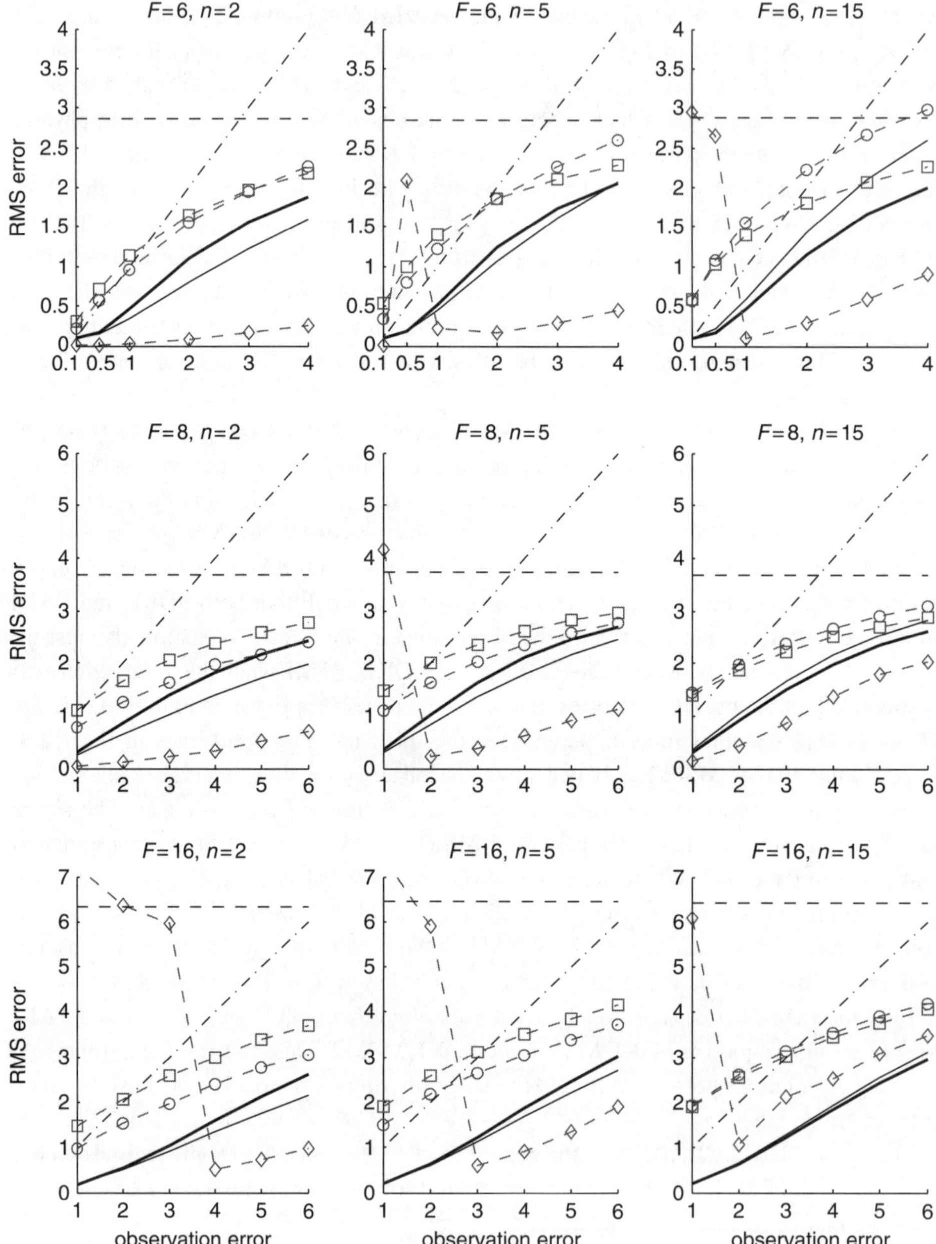

Figure 12.3 Plentiful observations: RMS errors as functions of observation errors ($\sqrt{r^o}$): the first row is for $F = 6$, the second row is for $F = 8$, and the third row is for $F = 16$. In each row, results for $n = 2, 5$, and 15 are shown. In each plot, EAKF true is denoted by dashes with diamonds, no filter by horizontal dashes, observation errors by black dashed-dotted line, FDKF MSM1 by thick solid line, FDKF MSM2 by thin solid line, EAKF MSM1 is denoted by dashes with squares, and EAKF MSM2 by dashes with circles.

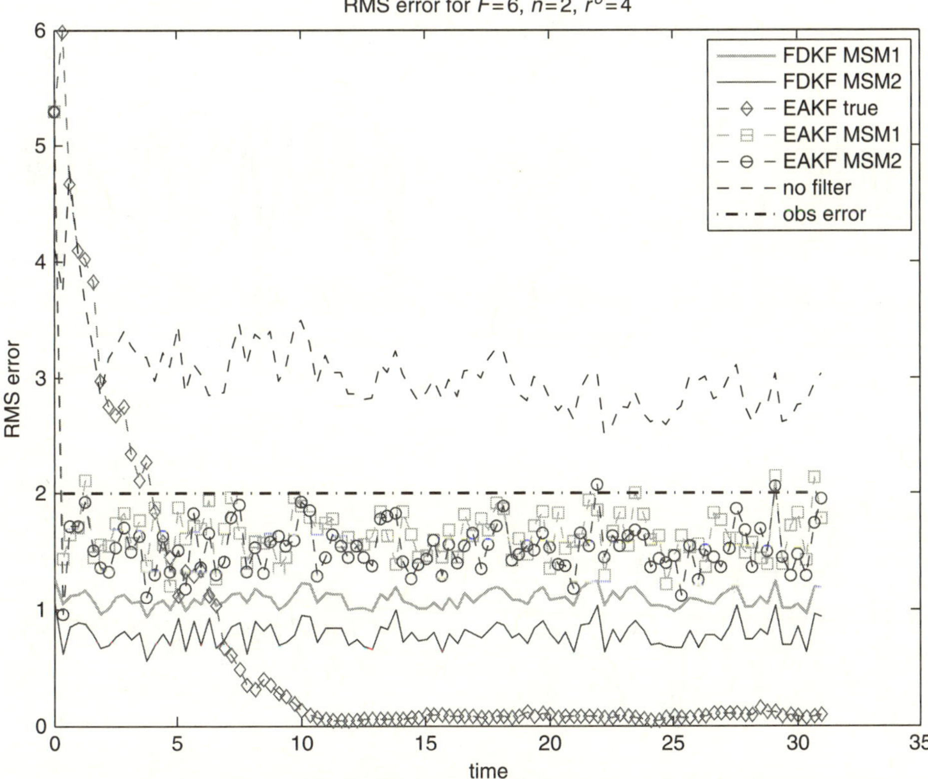

Figure 12.4 Plentiful observations: Note the superior skill of EAKF true. Regime $F = 6, n = 2, r^o = 4$, and RMS temporal average errors as functions of time.

from the full nonlinear model is strongly chaotic or fully turbulent. In addition to that, the skill of EAKF true is also sensitive for small r^o whereas FDKF is not (see Fig. 12.3). For a longer time series of training data, it is possible to use multi-level autoregressive linear models for each decoupled Fourier mode separately; there is an improvement in filtering skill over MSM1 for weakly chaotic regimes, $F = 5, 6$ (Kang and Harlim, 2011).

12.3 Filter performance with regularly spaced sparse observations

For the discussion in this section, as in Chapter 7, we assume that observations $\hat{v}_{k,m}$ are sparsely available at every $P > 1$ model grid point with observation errors reflected by a Gaussian distribution $\hat{\sigma}^o_{\ell,m} \sim \mathcal{N}(0, \hat{r}^o)$. In physical space, this variance corresponds to $r^o = 2NE_p\hat{r}^o/P$. By Theorem 7.1 in Chapter 7, the Fourier coefficients of these regularly spaced observations are given as follows

$$\hat{v}_{\ell,m} = \sum_{k\in\mathcal{A}(\ell)} \hat{u}_{k,m} + \hat{\sigma}^o_{\ell,m}, \quad |\ell| \leq N/P \tag{12.17}$$

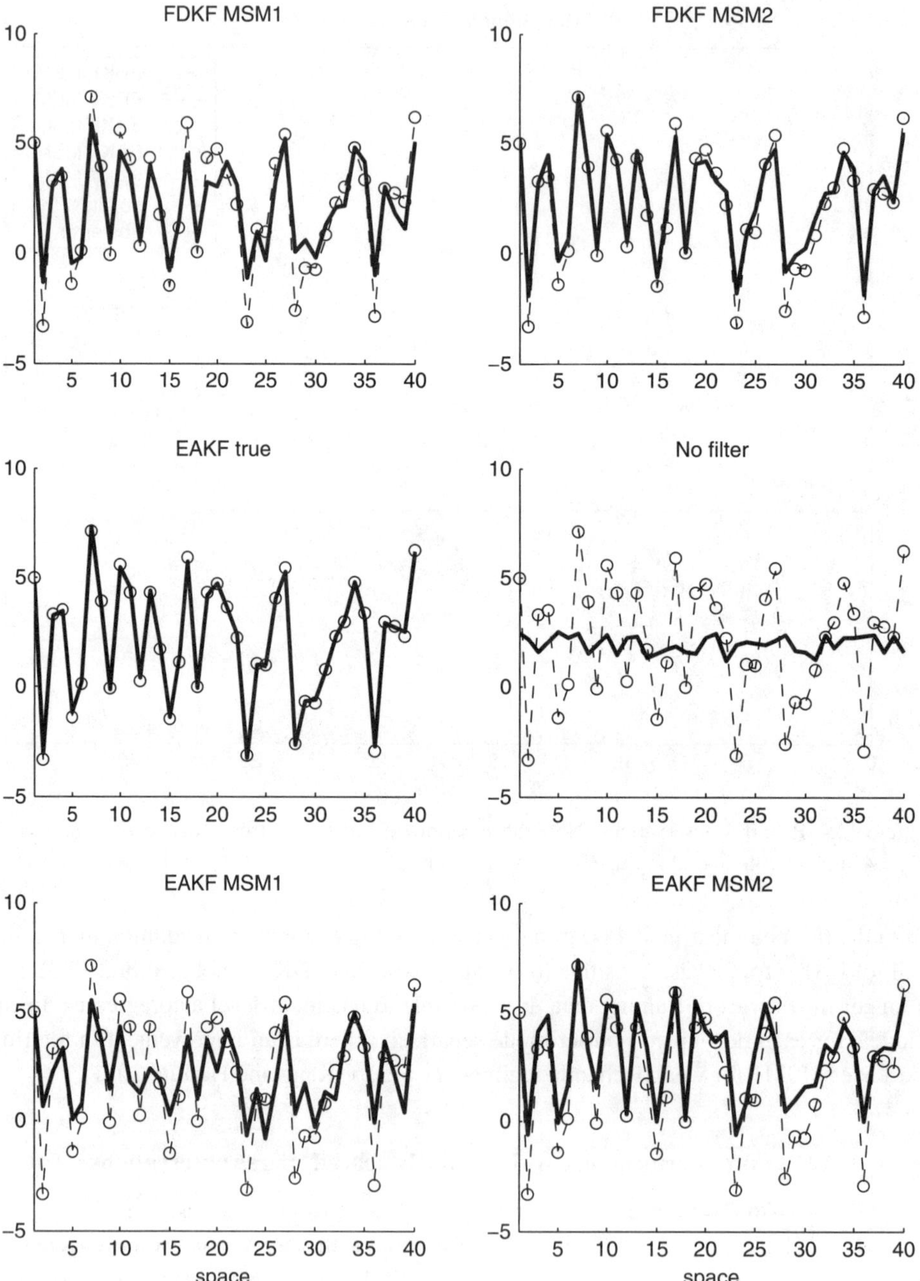

Figure 12.5 Plentiful observations: Regime $F = 6$, $n = 2$, $r^o = 4$, filtered solution (solid), true signal (dashes) and observations (circle) as functions of model space after 500 assimilation cycles.

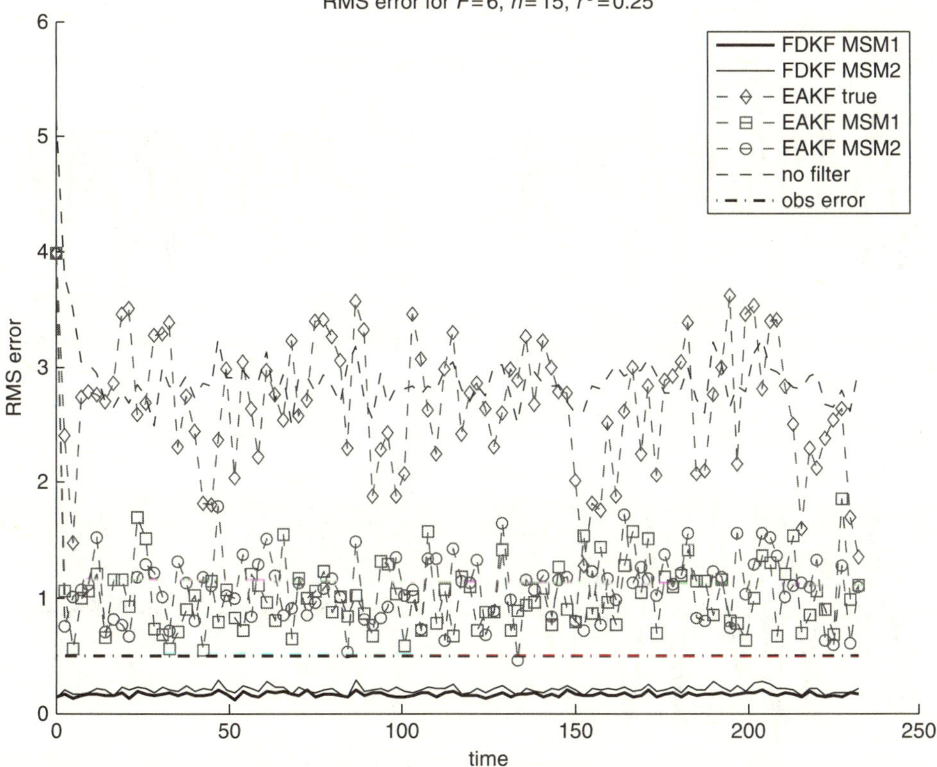

Figure 12.6 Plentiful observations: Note the superior skill of FDKF MSM1, MSM2. Regime $F = 6$, $n = 15$, $r^o = 0.25$, and RMS temporal average errors as functions of time.

where

$$\mathcal{A}(\ell) = \{k|k = \ell + q(2N/P), |k| \leq N, q \in \mathbb{Z}\} \tag{12.18}$$

is the aliasing set for mode ℓ. For simplicity, we assume that $2N/P$, which represents the total number of observations, is an integer value. As an example, consider $N = 20$ as in the L-96 model and $P = 2$. Then the model has 21 Fourier modes whereas the observations have 11 Fourier modes. According to formula (12.18), there are a total of 11 aliasing sets: $\mathcal{A}(0) = \{0, -20\}$, $\mathcal{A}(1) = \{1, -19\}, \ldots, \mathcal{A}(9) = \{9, -11\}$, and $\mathcal{A}(10) = \{10\}$. Thus, the filtering problem is decoupled into 10 two-dimensional filters with scalar observations for Fourier modes in aliasing sets 0 to 9 and a scalar filter on the tenth mode. This filtering reduction (FDKF) simply ignores the correlation between the Fourier coefficients in different aliasing sets as discussed in Chapter 7.

In the numerical simulations below, we compare our Fourier domain strategy to EAKF in the perfect model, a severe test. We implement EAKF with an ensemble size that doubles

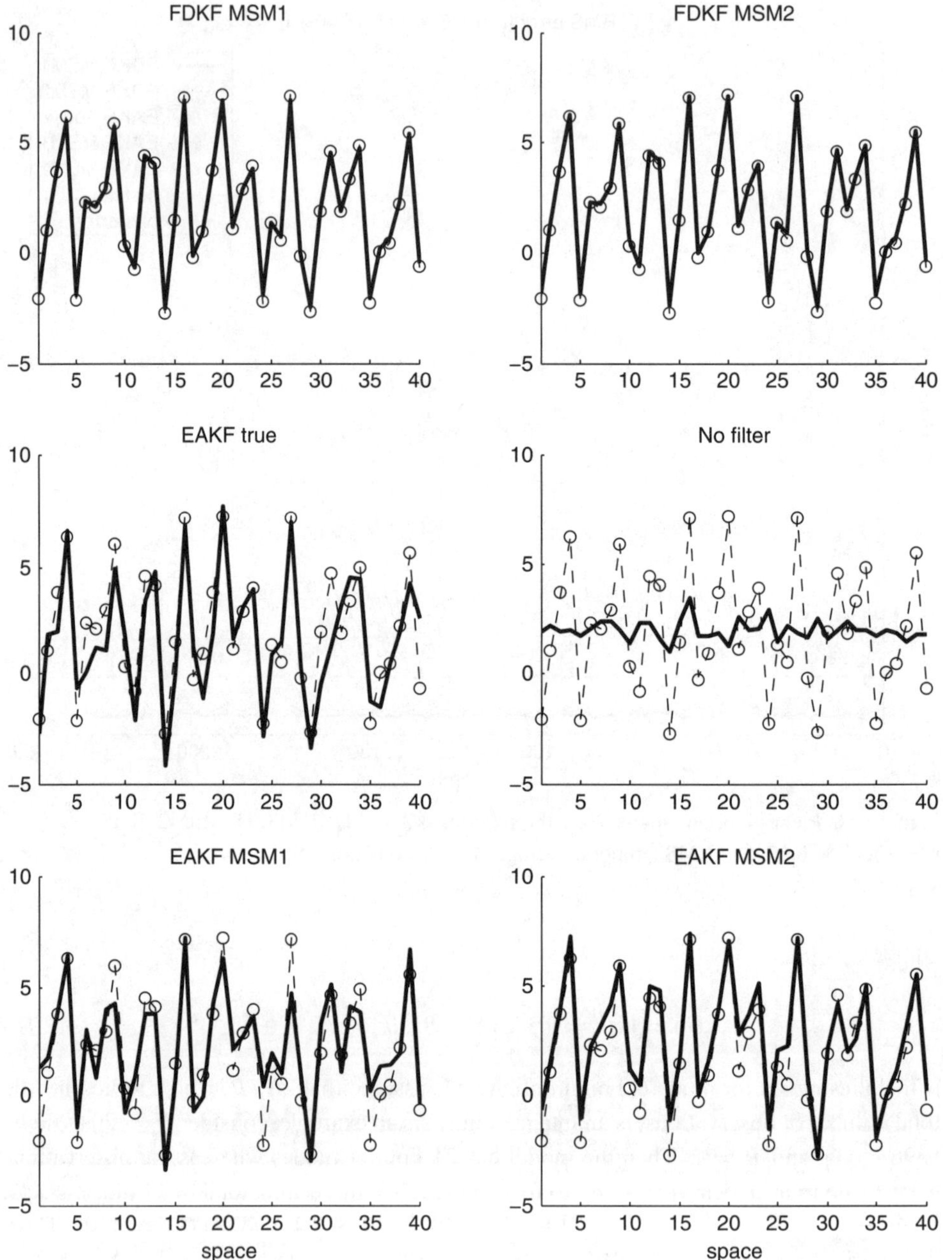

Figure 12.7 Plentiful observations: Regime $F = 6$, $n = 15$, $r^o = 0.25$, filtered solution (solid), true signal (dashes) and observations (circle) as functions of model space after 500 assimilation cycles.

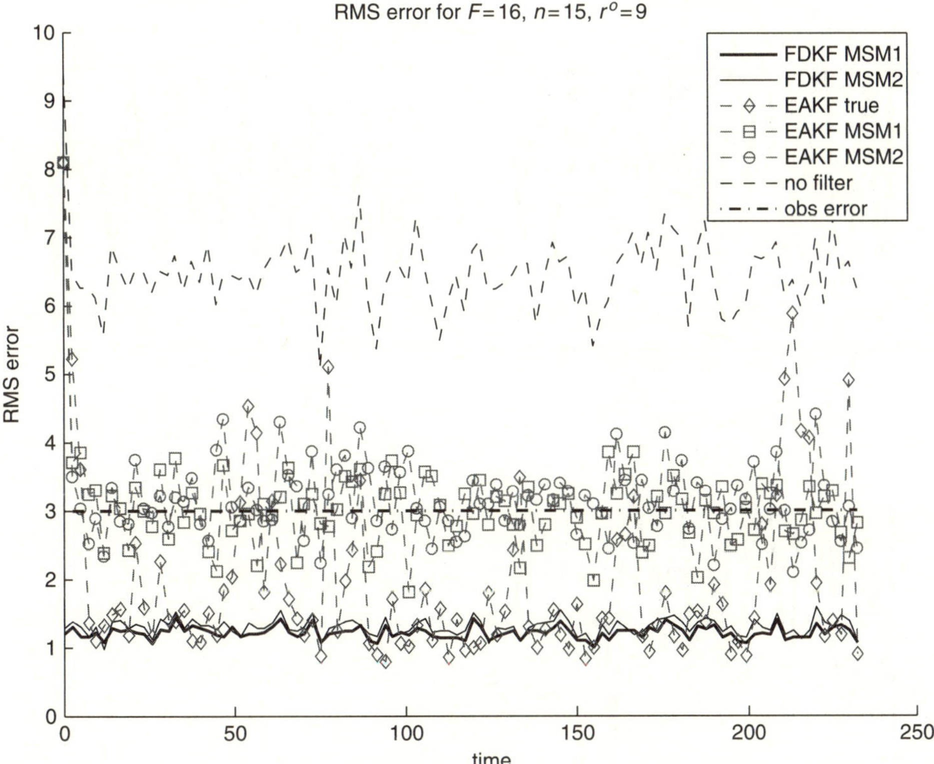

Figure 12.8 Plentiful observations: Note the superior skill of FDKF MSM1, MSM2. Regime $F = 16$, $n = 15$, $r^o = 9$, and RMS temporal average errors as functions of time.

the phase space dimension, $K = 80 = 4N$, and variance inflation coefficient $r = 0.05$. Each simulation is run for 5000 assimilation cycles.

12.3.1 Weakly chaotic regime

For the weakly chaotic regime, $F = 6$, we will also show a variant of FDKF which simply filters only the most energetic mode in each aliasing set. This approximate filter reduces to uncoupled scalar filters in which the unfiltered Fourier coefficients trust the dynamics fully. This approach is simply the reduced Fourier domain Kalman filter or RFDKF as in Section 7.3.2 of Chapter 7 or Section 8.3 of Chapter 8 with energetic primary mode. We should also note that both stochastic models MSM1 and MSM2 in this section are exactly the same as those used in the study for the plentiful observation case in Section 12.2.

In the first numerical simulation, we consider the weakly chaotic regime, $F = 6$, with a long observation time, $n = 15$ (i.e. $T_{\text{obs}} = 0.234$). We consider sparse observations with $P = 2$ and error variance $r^o = 1.96$, the square of one-half of our rough estimate of the

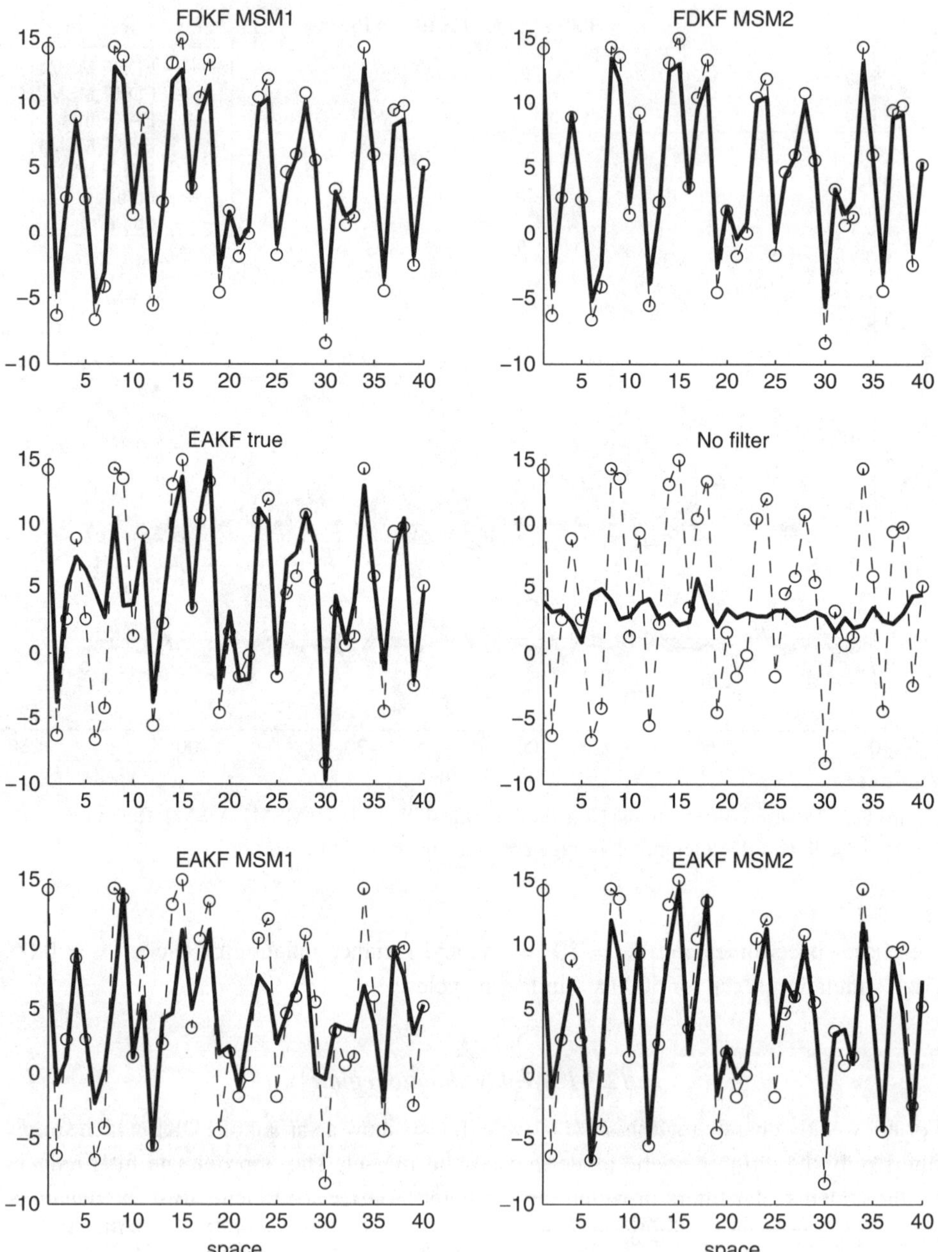

Figure 12.9 Plentiful observations: Regime $F = 16$, $n = 15$, $r^o = 9$, filtered solution (solid), true signal (dashes) and observations (circle) as functions of model space after 500 assimilation cycles.

Table 12.1 Average RMS errors and spatial correlations for simulations with $F = 6$, $P = 2$ $r^o = 1.96$, and $n = 15$ (correspond to $T_{\text{obs}} = 0.234$). This is a regime where EAKF true is superior and the RFDKF methods have comparable significant skill to FDKF for MSM1.

Spatial domain scheme	RMS	corr.	Fourier domain scheme	RMS	corr.
EAKF true	0.88	0.95	FDKF MSM1	2.27	0.60
EAKF MSM1	2.52	0.50	FDKF MSM2	2.63	0.66
EAKF MSM2	2.92	0.57	RFDKF MSM1	2.39	0.56
No filter	2.85	0	RFDKF MSM2	3.90	0.59

size of the chaotic attractor, which is 2.8 for this weakly chaotic regime (see table 11.2 of chapter 11 of Majda *et al.* (2005) and Majda and Wang (2006)). Figure 12.10 shows snapshots of the filtered solutions after 5000 assimilation cycles from various approaches including the perfect model simulation EAKF true, the three filtering strategies EAKF, FDKF and RFDKF in the presence of model errors through MSM1 and MSM2, and the unfiltered solutions. Notice that the filtered solutions of the Fourier domain filters with model errors are less skillful than those with the EAKF true; in the region with large-amplitude wave trains for L-96 model with $F = 6$, the presence of model errors hurts the filter when observations are not available. However, FDKF shows substantial skill (much better than the unskillful solutions) and notice that the model errors through MSM1 and MSM2 also hurt the ensemble method, EAKF, at similar unobserved locations (compare the snapshots in Fig. 12.10). In fact, the ensemble filter with model errors produces larger average RMS error and lower average spatial correlation compared to the exactly identical simulation with FDKF (see Table 12.1). We also notice that there is filtering skill even with the cheapest scheme RFDKF MSM1 with solutions comparable to those produced by EAKF MSM1. The surprising skill of RFDKF can be justified as follows: the strongly energetic modes in the weakly chaotic L-96 model are concentrated within modes 6–10 (see Fig. 12.1) and the structure of the aliasing set defined in Chapter 7 allows these energetic modes to be distributed in different aliasing sets so that the unfiltered modes (modes where RFDKF fully trusts the dynamics) are the weakly energetic modes. On the other hand, when the model is strongly chaotic $F = 8$ or fully turbulent $F = 16$, the energy spectrum becomes more homogeneous so that ignoring some of the energetic modes may hurt the filter substantially as shown in Chapter 7. In other experiments (not shown here), we confirm the anticipated declining skill with RFDKF for stronger chaotic regimes.

12.3.2 Strongly chaotic regime

In the second experiment, we consider sparser observations with $P = 4$ so there are only 10 regularly spaced observations. The results in Fig. 12.11 are from a simulation with this sparse observation network in the strongly chaotic regime of L-96 with $F = 8$ with observation noise size $r^o = 3.24$ (i.e. the square of one-half of the chaotic attractor size

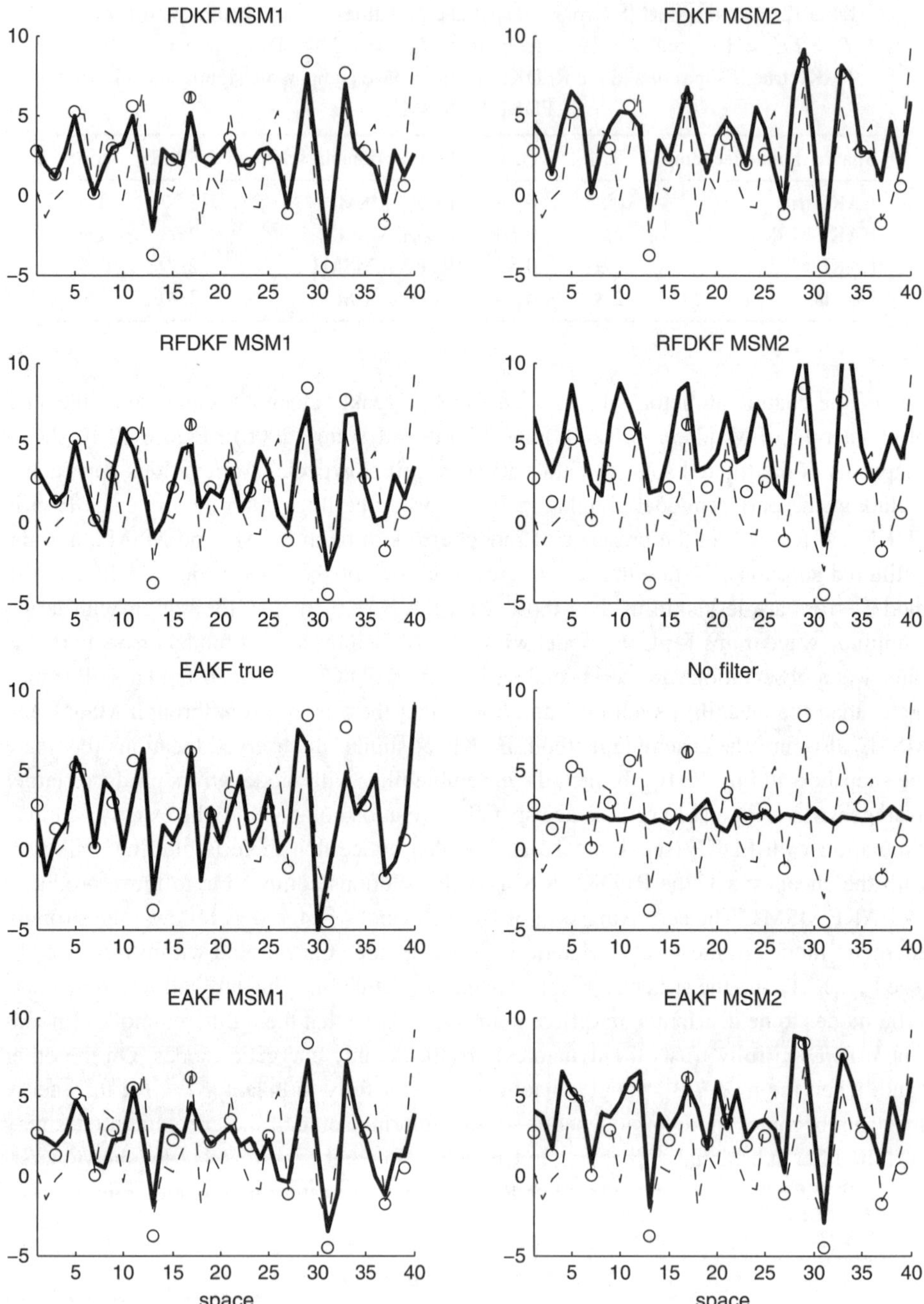

Figure 12.10 Snapshots of the filtered solutions at time 1172.1 (or right after 5000 assimilation cycles) for $F = 6$, $P = 2$, $r^o = 1.96$, and $n = 15$. In each panel, the true signal is denoted as dashes, the observations as circles, and the filtered solution as a solid line. Note the superior skill of EAKF true in this weakly chaotic regime with partial observations at a longer time.

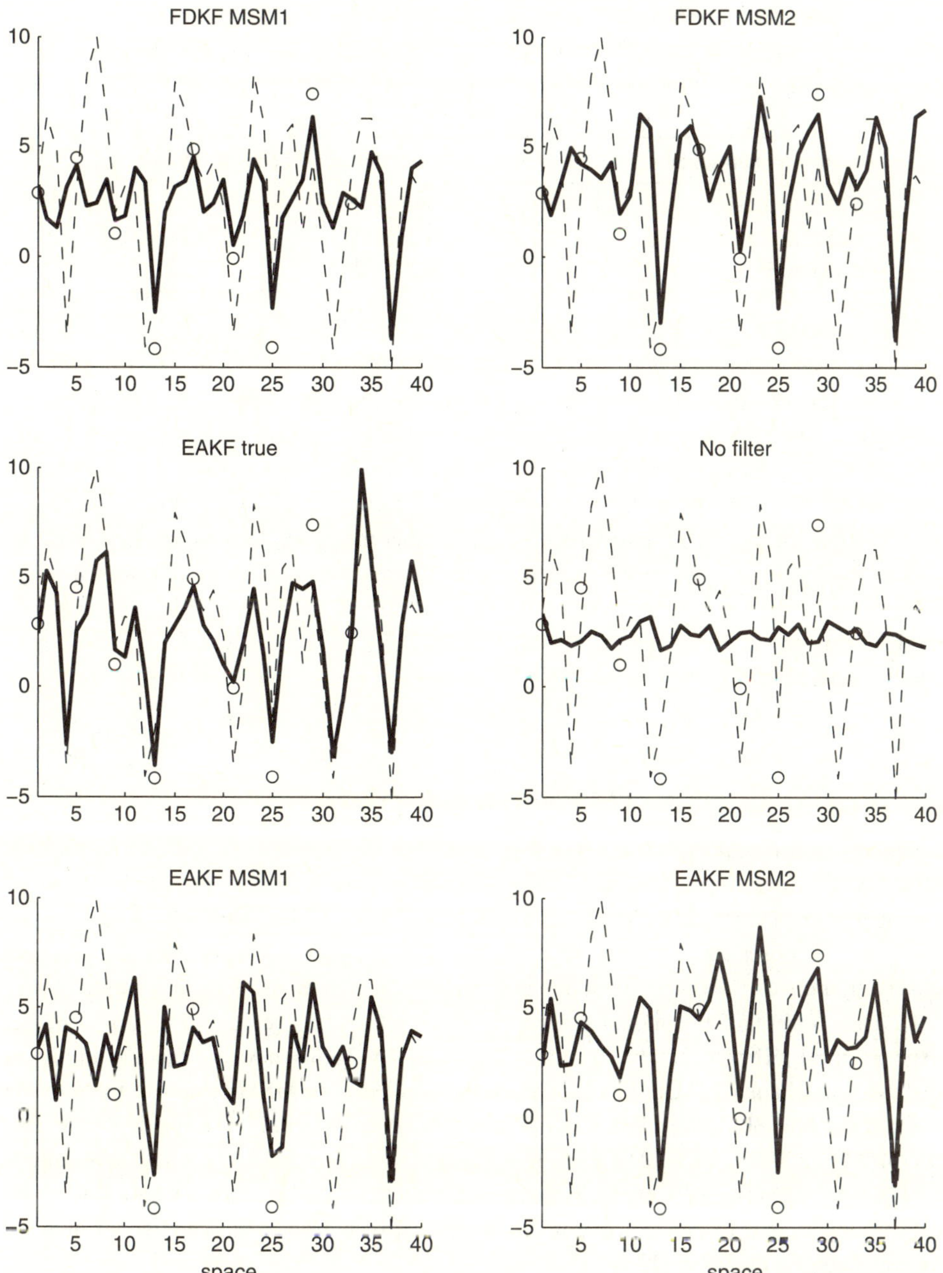

Figure 12.11 Snapshots of the filtered solutions at time 1172.1 (or right after 5000 assimilation cycles) for $F = 8$, $P = 4$, $r^o = 3.24$, and $n = 15$. In each panel, the true signal is denoted as dashes, the observations as circles, and the filtered solution as a solid line.

Table 12.2 Average RMS errors and spatial correlations for simulations with $F = 8$, $P = 4$, $r^o = 3.24$, and $n = 15$ (corresponding to $T_{\text{obs}} = 0.234$). This is a regime where EAKF true is mildly superior at a much larger computational cost than FDKF MSM1.

Spatial domain scheme	RMS	corr.	Fourier domain scheme	RMS	corr.
EAKF true	2.69	0.68	no filter	3.66	–
EAKF MSM1	3.33	0.43	FDKF MSM1	3.18	0.49
EAKF MSM2	3.71	0.44	FDKF MSM2	3.53	0.48

3.6) and observation time $n = 15$ ($T_{\text{obs}} = 0.234$). Qualitatively, we find that the filtered solutions underestimate the unobserved peaks in the true signal even with perfect model simulation. The presence of model errors through linear stochastic models does not magnify the RMS errors as significantly as we saw in the earlier experiments in the weakly chaotic regime; MSM1 increases errors by about 1.18 times in this experiment, compared to 2.57 times in the earlier experiment. In this experiment, the filtered solution with the perfect model simulation, EAKF true, produces filtered solutions with the lowest average RMS error of 2.69 and the highest average spatial correlation of 0.68. In the presence of model errors, the RMS error with MSM1 is slightly lower than that with EAKF true but its spatial correlation is only 0.5 (see Table 12.2).

12.3.3 Fully turbulent regime

In the third numerical experiment (see Fig. 12.12 for the snapshots of the filtered solution after 5000 assimilation cycles and Table 12.3 for its corresponding average RMS errors and spatial correlations), we consider the fully turbulent regime for L-96 with $F = 16$ and a shorter observation time $n = 5$ ($T_{\text{obs}} = 0.078$), observation density $P = 2$, and observation error $r^o = 0.81$ (which is the square of a quarter of its chaotic attractor size 6.4). Thus, only half the model points are observed. In this regime, the perfect model experiment, EAKF true, exhibits a catastrophic filter divergence beyond machine infinity as discussed in Chapter 11. On the other hand, FDKF MSM1 or MSM2 improve the filtering skill (again, see Table 12.3) with non-trivial spatial correlation 0.67. We also find that if we use MSM1 or MSM2 instead of the original nonlinear L-96 model to EAKF, the solutions do not exhibit catastrophic filter divergence. This result suggests that a way to avoid the catastrophic filter divergence is to confine the prior distribution to be Gaussian through the linear stochastic model in (12.6).

With such encouraging results, we will revisit MSM1 and MSM2 in Chapter 13 in which we will use them to filter sparsely observed quasi-geostrophic turbulent signals (see Section 11.4). There, we will compare MSM1 and MSM2 with an extended Kalman filtering strategy that uses an exactly solvable stochastic parametrization algorithm and with the LLS-EAKF discussed in Section 11.5.

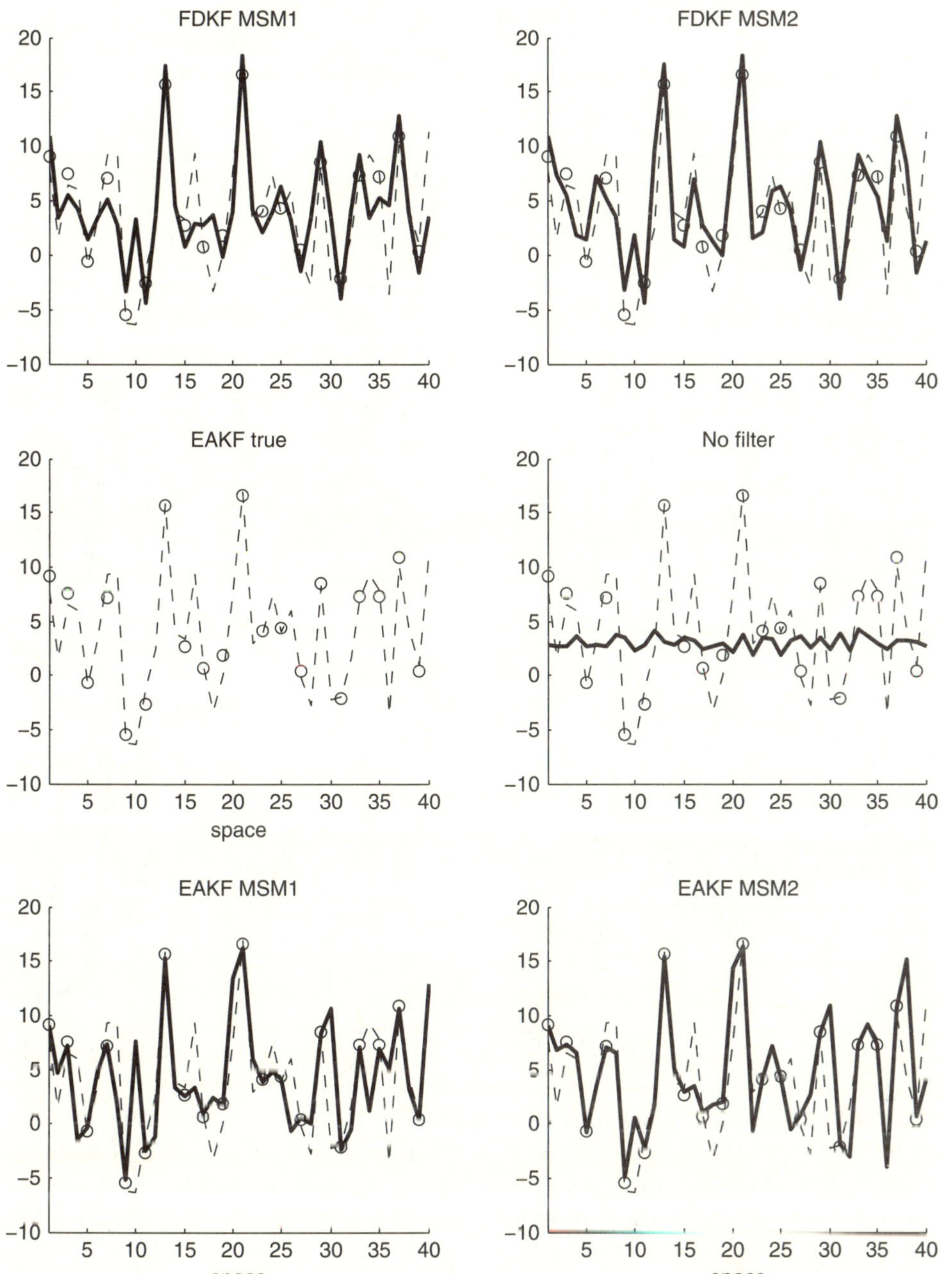

Figure 12.12 Snapshots of the filtered solutions at time 390.7 (or right after 5000 assimilation cycles) for $F = 16$, $P = 2$, $r^o = 0.81$, and $n = 5$. In each panel, the true signal is denoted as dashes, the observations as circles, and the filtered solution as a solid line.

Table 12.3 Average RMS errors and spatial correlations for simulations with $F = 16$, $P = 2$, $r^o = 0.81$, and $n = 5$ (corresponding to $T_{\text{obs}} = 0.078$). This is a regime where FDKF is superior.

Spatial domain scheme	RMS	corr.	Fourier domain scheme	RMS	corr.
EAKF true	∞	–	no filter	6.38	–
EAKF MSM1	5.31	0.59	FDKF MSM1	4.73	0.67
EAKF MSM2	5.15	0.63	FDKF MSM2	4.84	0.68

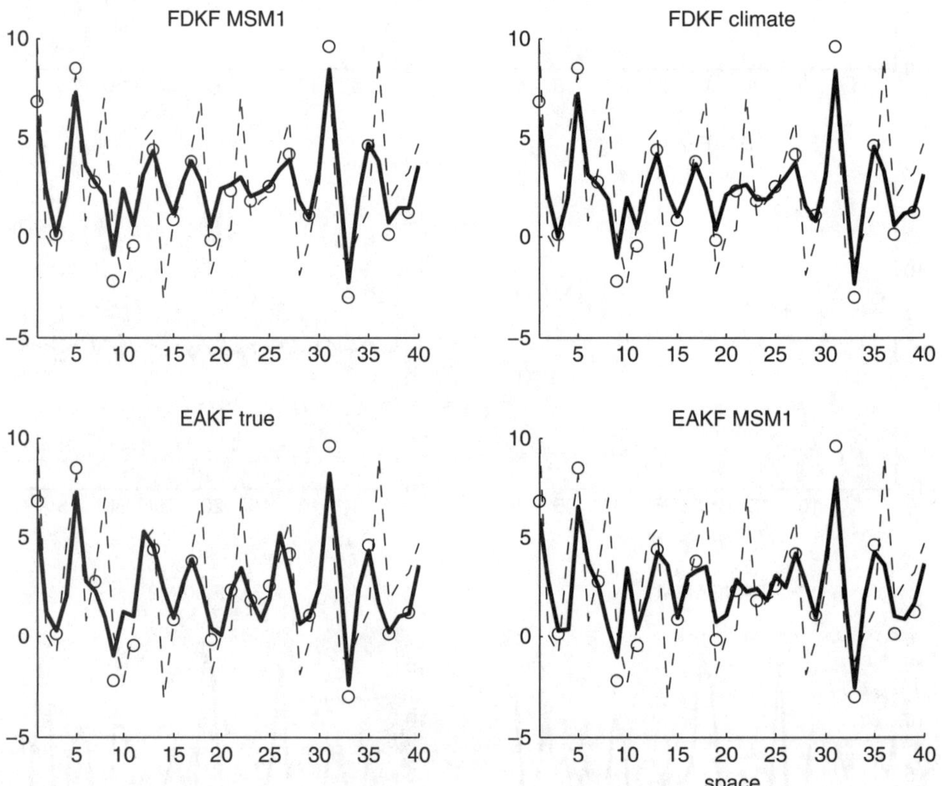

Figure 12.13 Snapshots of the filtered solutions at time 11,721.10 (or right after 5000 assimilation cycles) for $F = 8$, $P = 2$, $r^o = 3.24$, and $n = 150$. In each panel, the true signal is denoted as dashes, the observations as circles, and the filtered solution as a solid line. Note the lack of a filtered signal for EAKF true because catastrophic filter divergence has occured while FDKF MSM1, 2 have significant skill in this turbulent regime.

12.3.4 Super-long observation times

Finally, we consider a super-long observation time with $n = 150$ ($T_{\text{obs}} = 2.34$, which corresponds to 12 days which is far beyond the three-day decorrelation time of the L-96 model with $F = 8$ (Lorenz, 1996)). In this numerical experiment, we set all parameters as in the second experiment: $F = 8$, $r^o = 3.24$, but with $P = 2$ instead of $P = 4$. We find that (see also Fig. 12.13 for snapshots) the average RMS error and the average spatial correlation of the perfect model simulation EAKF true are, consecutively, 2.97 and 0.58, which are slightly worse than those of FDKF MSM1 (RMS = 2.83, corr = 0.64). For such a super-long observation time, we find that the filtered solutions with EAKF MSM1 (RMS = 3.04, corr = 0.57) is not very different from that of EAKF true. In fact, we find similar filtering skill as FDKF MSM1 if we simply use the climatological mean and covariance as the prior statistics (FDKF CLIMATE in Fig. 12.13 produces RMS = 2.81, corr = 0.64). Thus, for a very long observation time beyond the decorrelation time, a dynamic-less filter akin to the 3D-VAR (Lorenc, 1986) scheme is an alternative strategy.

13

Stochastic parametrized extended Kalman filter for filtering turbulent signals with model error

Throughout the book, we have stressed that a central issue in practical filtering of turbulent signals is model error. Naively, one might think that model errors always have a negative effect on filter performance and indeed this was illustrated in Chapters 2 and 3 in simple examples with simple time differencing methods like backward or forward Euler with associated natural time discrete noise. However, a central issue of this book is to emphasize that judicious model errors in the forward operator, guided by mathematical theory, can both ameliorate the effect of the curse of ensemble size for turbulent dynamical systems and retain high filtering skill, remarkably, often exceeding that with the perfect model! In particular, we have illustrated these principles with various model errors arising from using approximate numerical solvers (Chapter 2), reduced strategies for filtering dynamical systems with instability (Chapters 3 and 8), reduced strategies for filtering sparsely observed signals (Chapter 7) and simple linear stochastic models for filtering turbulent signals from nonlinear dynamical systems (Chapters 5, 10, and 12). In our earlier discussion (see Chapters 8, 10, and 12), we demonstrated that the off-line strategy accounting for model errors, the mean stochastic model (MSM), under some circumstances produces reasonably accurate filtered solutions. However, this off-line strategy often has limited skill for estimating real-time prediction problems with rapid fluctuations that are often observed in nature since the off-line strategy (MSM) relies heavily on a fixed parameter set that is extracted from long-time or climatological statistical quantities such as the energy spectrum and correlation time. An alternative strategy to deal with model errors in filtering is to learn these parameters adaptively from observations of the true signal. The main idea in this approach that is often practised in filtering with model errors is to augment the dynamical system for u with an unknown parameter λ,

$$\frac{du}{dt} = F(u, t, \lambda), \tag{13.1}$$

with an approximate dynamical equation for the parameters

$$\frac{d\lambda}{dt} = g(\lambda). \tag{13.2}$$

The right-hand side of (13.2) is often chosen on an ad hoc basis as $g(\lambda) \equiv 0$ or white noise forcing with a small variance (Friedland, 1969, 1982). The static model, $g = 0$, works well for systems with constant bias whereas for systems with time varying bias, the white noise model, $g = \sigma \dot{W}$, yields reasonably accurate filtered solutions provided that the variance of the noise is sufficiently small. Partial observations of the signal from nature are often processed by an extended Kalman filter (EKF, see Chapter 9 or Anderson and Moore (1979); Chui and Chen (1999)) applied to the augmented system in (13.1) and (13.2) where the parameters λ are estimated adaptively from these partial observations. Note that even if the original model in (13.1) is linear, it can readily have nonlinear dependence on the parameters λ through (13.2) so typically an EKF involving the linear tangent approximation as discussed in Chapter 9 is needed for parameter estimation in this standard case. Some recent applications of these and similar ideas to complex nonlinear dynamical systems can be found in Pitcher (1977); Dee and Silva (1998); Dee and Todling (2000); Baek *et al.* (2006); and Anderson (2007).

The prototype filter model for turbulent dynamical systems studied earlier in Chapters 8 and 12 involves linear stochastic models with fixed damping coefficient and forcing based on climatological regression fitting. As discussed earlier in Chapter 5, this type of model is the simplest stochastic model for turbulence (Delsole, 2004). More elaborate and prohibitively expensive stochastic models and direct closures for turbulence are available in the literature (Salmon, 1998) where the damping and forcing at each Fourier mode are determined from elaborate global space–time nonlinear interactions across scales; O'Kane and Frederiksen (2008) have implemented one of these elaborate closures as the prediction step operator in the filtering of atmospheric blocking events with plentiful observations with high skill. Motivated by this work, it is interesting to develop computationally cheap improved algorithms for filtering turbulent signals which learn the damping and forcing coefficients in each Fourier mode "on the fly" and mimic the effect of such prohibitively expensive closures.

With the input from the above two different directions, the authors and B. Gershgorin have recently developed a family of stochastic parametrized extended Kalman filters (SPEKF) for improving filtering and prediction skill with model errors (Gershgorin *et al.*, 2010a,b; Harlim and Majda, 2010b). Unlike the classical procedures for parameter augmentation using a linear tangent model in EKF, these algorithms utilize nonlinear exactly solvable statistics for estimating parameters of dissipation and forcing "on the fly". The exactly solvable feature here utilizes the mathematical tools described in Chapter 10. This is a major conceptual and practical advantage since algorithms for parameter estimation can often be ill-conditioned and the linear tangent approximation in EKF can be very inaccurate (Branicki *et al.*, 2011).

In Section 13.1, we describe SPEKF on a single Fourier mode (Gershgorin *et al.*, 2010b) and we discuss filtering the true signal with intermittent bursts of instability as described in Chapter 8. In Section 13.2, we test this algorithm on sparsely observed idealized spatially extended systems with instability (see Section 8.2) in which we will introduce additional model errors through hidden forcing (Gershgorin *et al.*, 2010a). Finally, in Section 13.3, we compare SPEKF with the finite ensemble local least-squares ensemble adjustment Kalman filter (LLS-EAKF) (see Chapter 11 and Anderson (2001)) and the mean stochastic model (MSM) in Chapter 12 on a realistic model for atmospheric or oceanic shear turbulence (Harlim and Majda, 2010b), the two-layer QG model introduced in Chapter 1 and discussed in Chapter 11.

13.1 Nonlinear filtering with additive and multiplicative biases: One-mode prototype test model

In the test model here, the signal from nature is assumed to be given by the solution of the complex scalar Langevin equation with time-dependent damping

$$\frac{du(t)}{dt} = -\gamma(t)u(t) + i\omega u(t) + \sigma\dot{W}(t) + f(t), \tag{13.3}$$

where $\dot{W}(t)$ is complex white noise and $f(t)$ is a prescribed external forcing. To generate significant model errors as well as to mimic intermittent chaotic instability as often occurs in nature, we allow $\gamma(t)$ to switch between stable ($\gamma > 0$) and unstable ($\gamma < 0$) regimes according to a two-state Markov jump process as discussed in Chapter 8. Here we regard $u(t)$ as representing one of the modes from nature in a turbulent signal as is often done in turbulence models (Majda *et al.*, 2005; Delsole, 2004; Salmon, 1998; Majda and Grote, 2007), and the switching process can mimic physical features such as intermittent baroclinic instability (Pedlosky, 1979). As often occurs in practice, we assume that the switching process details are not known and only averaged properties are modeled. In Chapter 8, the mean stochastic model (MSM) with significant model error given by

$$\frac{du(t)}{dt} = -\bar{\gamma}u(t) + i\omega u(t) + \sigma\dot{W}(t) + \tilde{f}(t) \tag{13.4}$$

is utilized for filtering; here $\bar{\gamma} > 0$ is an average damping constant and $\tilde{f}(t)$ is possibly an incorrectly specified forcing. Note that the unstable process is completely hidden from the MSM model so there are potentially large model errors in filtering with MSM.

To compensate for model error "on the fly", we consider a stochastic model for the evolution of the state variable $u(t)$ together with combined additive, $b(t)$, and multiplicative, $\gamma(t)$, bias correction terms,

$$\begin{aligned}
\frac{du(t)}{dt} &= (-\gamma(t) + i\omega)u(t) + b(t) + f(t) + \sigma\dot{W}(t),\\
\frac{db(t)}{dt} &= (-\gamma_b + i\omega_b)(b(t) - \hat{b}) + \sigma_b\dot{W}_b(t),\\
\frac{d\gamma(t)}{dt} &= -d_\gamma(\gamma(t) - \hat{\gamma}) + \sigma_\gamma\dot{W}_\gamma(t),
\end{aligned} \tag{13.5}$$

for improving filtering with model errors. We refer to system (13.5) as the **combined model**. Here, ω is the oscillation frequency of $u(t)$, $f(t)$ is an external forcing, and σ characterizes the strength of the white noise forcing $\dot{W}(t)$. Also, parameters γ_b and d_γ represent the damping and parameters σ_b and σ_γ represent the strength of the white noise forcing of the additive and multiplicative bias correction terms, respectively. The stationary mean bias correction values of $b(t)$ and $\gamma(t)$ are given by $\hat{b}$ and $\hat{\gamma}$, correspondingly, and the frequency of the additive noise is denoted as ω_b. Note that the white noise $\dot{W}_\gamma(t)$ is real-valued while the white noises $\dot{W}(t)$ and $\dot{W}_b(t)$ are complex-valued and their real and imaginary parts are independent real-valued white noises. It is important to realize that the parameters of the state variable $u(t)$ come from the characteristics of the physical system, which is modeled by the first equation in (13.5). On the other hand, the parameters of $b(t)$ and $\gamma(t)$, $\{\gamma_b, \omega_b, \sigma_b, d_\gamma, \sigma_\gamma\}$, are introduced in the model and, in principle, cannot be directly obtained from the characteristics of the physical system. One of the goals of this section is to establish "universality regimes" for these parameters, that is, to study how the results of the filtering are sensitive to wide variations of these parameters, and to obtain some insights into how to choose them in an efficient robust fashion. System (13.5) is considered with the initial values $u(t_0) = u_0, b(t_0) = b_0, \gamma(t_0) = \gamma_0$, which are independent Gaussian random variables with the known statistics: $\langle u_0\rangle$, $\langle \gamma_0\rangle$, $\langle b_0\rangle$, $\mathrm{Var}(u_0)$, $\mathrm{Var}(\gamma_0)$, $\mathrm{Var}(b_0)$, $\mathrm{Cov}(u_0, u_0^*)$, $\mathrm{Cov}(u_0, \gamma_0)$, $\mathrm{Cov}(u_0, b_0)$, $\mathrm{Cov}(u_0, b_0^*)$.

We also consider two special cases of the combined model (13.5): the **additive model** when we only have the additive bias correction with constant damping $\gamma(t) = \hat{\gamma}$; and the **multiplicative model** when we only have multiplicative bias correction, i.e. $b = 0$. Note that the additive model is linear and nearly similar to that considered earlier by Friedland (1982) except that we add a linear damping term $-\gamma_b$ to guarantee a non-divergent additive bias $b(t)$ term. These two special cases are designed for the purpose of comparison with the combined model. Moreover, studying the filter performance based on the combined model as well as on the multiplicative and additive models will help us understand which component of the bias corrections is more important depending on the physical properties of the system, such as its damping and external forcing. We will find out that there are exceptional situations when the multiplicative model performs slightly better than the combined model and there are situations when the additive model is as good as the combined model. The former situation occurs because of sampling error in the additive bias term, $b(t)$, when the external forcing is specified correctly. However, we will demonstrate via extensive numerical study that the combined model is the most robust to the variations of parameters and, therefore, it should be the method of choice when there is no additional information that indicates in advance that either the multiplicative or additive model is better in a particular situation.

13.1.1 Exact statistics for the nonlinear combined model

In this section, we show how to find the first- and second-order statistics of the combined model (13.5). These statistics are the main formulas for updating the prior statistics in the Kalman filter. Note that the first equation in (13.5) is nonlinear and, therefore, in general, Gaussian initial values at $t = t_0$ will not stay Gaussian for $t > t_0$. However, the special structure of the first equation in (13.5) allows us to find exact analytical formulas for the first and second (and, in principle, any) order statistics. The technique of obtaining the first- and second-order statistics for the kind of nonlinearity used in system (13.5) was studied by Gershgorin and Majda (2008, 2010) and discussed in Chapter 10. Here, we will follow the same procedure outlining the most interesting and important points. In principle, these formulas follow from general mathematical formulas for conditional Gaussian processes (Bensoussan, 2004); however, the detailed properties of the explicit formulas have central importance here.

The exact path-wise solution for system (13.5) is given as follows:

$$b(t) = \hat{b} + (b_0 - \hat{b})e^{\lambda_b(t-t_0)} + \sigma_b \int_{t_0}^{t} e^{\lambda_b(t-s)} dW_b(s), \tag{13.6}$$
$$\gamma(t) = \hat{\gamma} + (\gamma_0 - \hat{\gamma})e^{-d_\gamma(t-t_0)} + \sigma_\gamma \int_{t_0}^{t} e^{-d_\gamma(t-s)} dW_\gamma(s),$$

where $\lambda_b = -\gamma_b + \mathrm{i}\omega_b$, and $\hat{b}$ and $\hat{\gamma}$ are the stationary bias correction values of $b(t)$ and $\gamma(t)$. Now, we introduce the new notation

$$\hat{\lambda} = -\hat{\gamma} + \mathrm{i}\omega,$$
$$J(s,t) = \int_{s}^{t} (\gamma(s') - \hat{\gamma})ds'.$$

Then, using this notation, we find the solution for $u(t)$

$$u(t) = e^{-J(t_0,t)+\hat{\lambda}(t-t_0)}u_0 + \int_{t_0}^{t} \Big(b(s) + f(s)\Big)e^{-J(s,t)+\hat{\lambda}(s-t_0)}ds$$
$$+\sigma \int_{t_0}^{t} e^{-J(s,t)+\hat{\lambda}(s-t_0)} dW(s). \tag{13.7}$$

We start with the statistics of $b(t)$ and $\gamma(t)$. The second and third equations in (13.5) are linear SDEs and their solutions are Gaussian given Gaussian initial data. The mean and covariance of $b(t)$ and $\gamma(t)$ are the following

$$\begin{aligned}
\langle b(t)\rangle &= \hat{b} + (\langle b_0\rangle - \hat{b})e^{\lambda_b(t-t_0)},\\
\langle \gamma(t)\rangle &= \hat{\gamma} + (\langle \gamma_0\rangle - \hat{\gamma})e^{-d_\gamma(t-t_0)},\\
\mathrm{Var}(b(t)) &= \mathrm{Var}(b_0)e^{-2\gamma_b(t-t_0)} + \frac{\sigma_b^2}{2\gamma_b}\Big(1 - e^{-2\gamma_b(t-t_0)}\Big),\\
\mathrm{Var}(\gamma(t)) &= \mathrm{Var}(\gamma_0)e^{-2d_\gamma(t-t_0)} + \frac{\sigma_\gamma^2}{2d_\gamma}\Big(1 - e^{-2d_\gamma(t-t_0)}\Big),\\
\mathrm{Cov}(b(t), b(t)^*) &= \mathrm{Cov}(b_0, b_0^*)e^{2\lambda_b(t-t_0)},\\
\mathrm{Cov}(b(t), \gamma(t)) &= e^{(-d_\gamma+\lambda_b)(t-t_0)}\mathrm{Cov}(b_0, \gamma_0).
\end{aligned} \tag{13.8}$$

Next, we find the mean $\langle u(t)\rangle$ using Eqn (13.7)

$$\langle u(t)\rangle = e^{\hat{\lambda}(t-t_0)}\Big\langle u_0 e^{-J(t_0,t)}\Big\rangle + \int_{t_0}^{t} e^{\hat{\lambda}(t-s)}\Big\langle b(s)e^{-J(s,t)}\Big\rangle ds$$
$$+ \int_{t_0}^{t} e^{\hat{\lambda}(t-s)} f(s)\Big\langle e^{-J(s,t)}\Big\rangle ds, \tag{13.9}$$

Following the discussion in Chapter 10 (Gershgorin and Majda, 2008, 2010), the averages on the right-hand side of Eqn (13.9) can be computed using the characteristic function of the Gaussian random process $J(s,t)$. Recall that for given complex Gaussian z and real Gaussian x, we have

$$\langle ze^{bx}\rangle = \big(\langle z\rangle + b\mathrm{Cov}(z,x)\big)e^{b\langle x\rangle + \frac{b^2}{2}\mathrm{Var}(x)}. \tag{13.10}$$

Note that u_0 and $b(s)$ are Gaussian and so is $J(s,t)$ since it is an integral of Gaussian random process $\gamma(t)$. Applying Eqn (13.10) to Eqn (13.9) and then using Eqn (13.6), we find

$$\begin{aligned}\langle u(t)\rangle &= e^{\hat{\lambda}(t-t_0)}\Big(\langle u_0\rangle - \mathrm{Cov}(u_0, J(t_0,t))\Big)e^{-\langle J(t_0,t)\rangle + \frac{1}{2}\mathrm{Var}(J(t_0,t))}\\ &\quad + \int_{t_0}^{t} e^{\hat{\lambda}(t-s)}\Big(\hat{b} + e^{\lambda_b(s-t_0)}\Big(\langle b_0\rangle - \hat{b} - \mathrm{Cov}(b_0, J(s,t))\Big)\Big)\\ &\quad \times e^{-\langle J(s,t)\rangle + \frac{1}{2}\mathrm{Var}(J(s,t))}ds\\ &\quad + \int_{t_0}^{t} e^{\hat{\lambda}(t-s)} f(s) e^{-\langle J(s,t)\rangle + \frac{1}{2}\mathrm{Var}(J(s,t))}ds.\end{aligned} \tag{13.11}$$

Using the linear property of the covariance, we find

$$\begin{aligned}\mathrm{Cov}(u_0, J(s,t)) &= \frac{1}{d_\gamma}\Big(e^{-d_\gamma(s-t_0)} - e^{-d_\gamma(t-t_0)}\Big)\mathrm{Cov}(u_0,\gamma_0),\\ \mathrm{Cov}(b_0, J(s,t)) &= \frac{1}{d_\gamma}\Big(e^{-d_\gamma(s-t_0)} - e^{-d_\gamma(t-t_0)}\Big)\mathrm{Cov}(b_0,\gamma_0).\end{aligned}$$

In order to find $\langle J(s,t)\rangle$, we integrate Eqn (13.8) to obtain

$$\langle J(s,t)\rangle = \frac{1}{d_\gamma}\Big(e^{-d_\gamma(s-t_0)} - e^{-d_\gamma(t-t_0)}\Big)(\langle\gamma_0\rangle - \hat{\gamma}). \tag{13.12}$$

The variance of $J(s,t)$ can be found using the Itô isometry formula (Gardiner, 1997)

$$\Big\langle\Big(\int g(t)dW(t)\Big)^2\Big\rangle = \int g^2(t)dt,$$

for any deterministic $g(t)$. Then, we obtain

$$\begin{aligned}\mathrm{Var}(J(s,t)) &= \frac{1}{d_\gamma^2}\Big(e^{-d_\gamma(s-t_0)} - e^{-d_\gamma(t-t_0)}\Big)^2\mathrm{Var}(\gamma_0)\\ &\quad - \frac{\sigma_\gamma^2}{d_\gamma^3}\Big[1 + d_\gamma(s-t) + e^{-d_\gamma(s+t-2t_0)}\\ &\quad \times\Big(-1 - e^{2d_\gamma(s-t_0)} + \cosh(d_\gamma(s-t))\Big)\Big].\end{aligned} \tag{13.13}$$

Finally, we find that the mean of $u(t)$ at time t can be expressed in terms of the initial statistics of the system (13.5) at time t_0. The integrals on the right-hand side of Eqn (13.11) can be approximated using the trapezoidal rule,

$$\int_{t_0}^{t} g(s)ds \approx \frac{h}{2}\big(g(t_0) + g(t)\big) + h \sum_{j=1}^{N-1} g(t_0 + jh),$$

which gives second-order accuracy, where $h = (t - t_0)/N$ is a time step in the equidistant partition of the interval $[t_0, t]$ into N subintervals. In our numerical simulations, we empirically find that $h = 10^{-3}$ insures numerical accuracy. Similarly as in Chapter 10 (Gershgorin and Majda, 2008, 2010), we can find the second-order statistics of $u(t)$, $\gamma(t)$, and $b(t)$ such as $\text{Var}(u(t))$, $\text{Cov}(u(t), u^*(t))$, $\text{Cov}(u(t), \gamma(t))$, $\text{Cov}(u(t), b(t))$, and $\text{Cov}(u(t), b^*(t))$. However, since these computations are long, we present them in Appendix A to this chapter. In Fig. 13.1, we show the analytically derived statistics (solid line) of $u(t)$, $b(t)$ and $\gamma(t)$ with their respective estimate through Monte Carlo simulations (circles). We note the excellent agreement between the analytically obtained statistics and

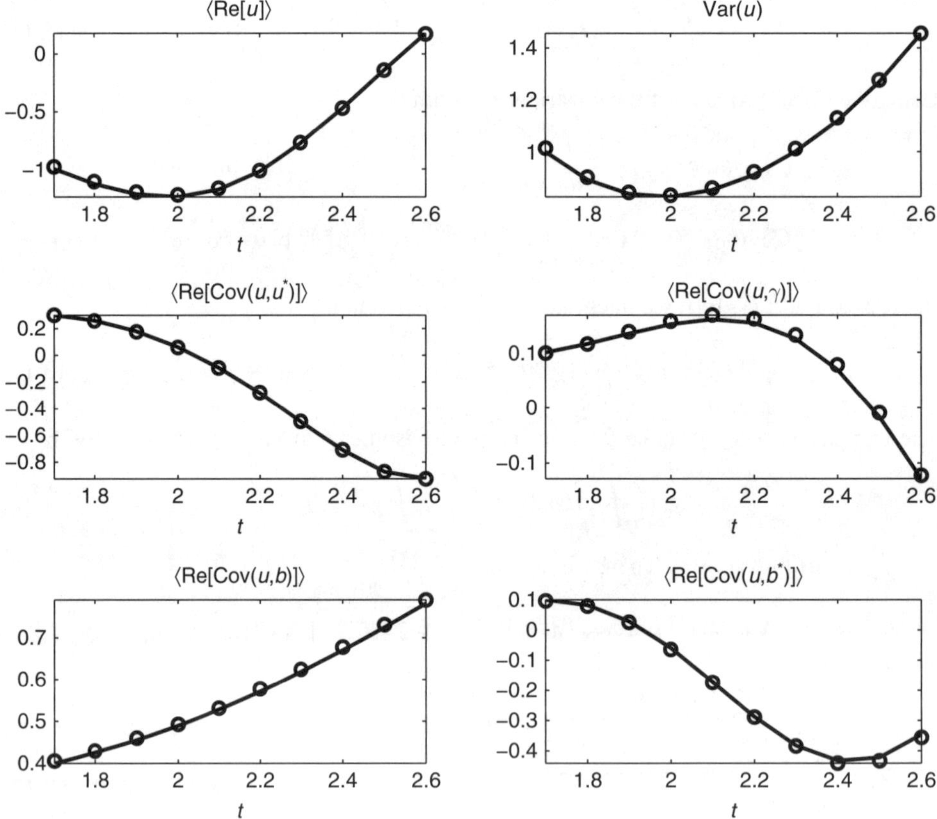

Figure 13.1 First- and second-order statistics of $u(t)$, $b(t)$ and $\gamma(t)$. The solid line corresponds to the analytical formulas for the statistics and circles correspond to Monte Carlo averaging of an emsemble of solutions.

their Monte Carlo approximation which gives us high confidence in the validity of the analytic formulas.

13.1.2 The stochastic parametrized extended Kalman filter (SPEKF)

Recall that we only have observations of $u(t)$ but not of $b(t)$ or $\gamma(t)$ since the last two variables are parameters artificially introduced in the model. Therefore, we consider the observation model

$$v = G(u, b, \gamma)^T + \sigma^o, \tag{13.14}$$

where σ^o is a Gaussian noise with zero mean and variance r^o and

$$G = \begin{pmatrix} 1 & 0 & 0 \end{pmatrix}$$

is the observation operator. The filtering problem with the combined model in (13.5) and observation model in (13.14) constitutes the **stochastic parametrized extended Kalman filter** (SPEKF). Different from the standard EKF (see Chapter 9) that linearizes the filtering problem, the SPEKF uses the exactly solvable nonlinear statistical values computed in the last section as its prior statistics in each assimilation step.

Now, as discussed and applied in Chapters 3, 7 and 10, we discuss the observability of the linearized SPEKF. That is, we linearize it around a mean state

$$\begin{aligned} u(t) &= \bar{u}(t) + \tilde{u}(t), \\ b(t) &= \bar{b}(t) + \tilde{b}(t), \\ \gamma(t) &= \bar{\gamma}(t) + \tilde{\gamma}(t). \end{aligned}$$

Then, the homogeneous part of the dynamics of the perturbation vector $\tilde{U} = (\tilde{u}, \tilde{b}, \tilde{\gamma})^T$ is given by $d\tilde{U}/dt = A\tilde{U}$, where the matrix A is given by

$$A = \begin{pmatrix} \lambda & 1 & -\bar{u} \\ 0 & \lambda_b & 0 \\ 0 & 0 & -d_\gamma \end{pmatrix},$$

with $\lambda = -\bar{\gamma} + i\omega$. The solution of this linearized perturbed dynamics is given by $\tilde{U}(t) = F\tilde{U}(t_0)$, where

$$F = e^{At} = \begin{pmatrix} e^{\lambda t} & \frac{e^{\lambda_b t} - e^{\lambda t}}{\lambda - \lambda_b} & -\bar{u}\frac{e^{\lambda t} - e^{-d_\gamma t}}{d_\gamma + \lambda} \\ 0 & e^{\lambda_b t} & 0 \\ 0 & 0 & e^{-d_\gamma t} \end{pmatrix}.$$

Recall that observability (see Chapter 3) of the linear system with operators F and G is characterized by the rank of the observability matrix $\mathcal{O} = (G^T, F^T G^T, (F^T)^2 G^T)$. In our filtering problem, the observability matrix is

$$\mathcal{O} = \begin{pmatrix} 1 & e^{\lambda t} & e^{2\lambda t} \\ 0 & \frac{e^{\lambda_b t} - e^{\lambda t}}{\lambda - \lambda_b} & \frac{\left(e^{\lambda_b t} - e^{\lambda t}\right)^2}{(\lambda - \lambda_b)^2} \\ 0 & -\bar{u}\frac{e^{\lambda t} - e^{-d_\gamma t}}{d_\gamma + \lambda} & -\bar{u}^2 \frac{\left(e^{\lambda t} - e^{-d_\gamma t}\right)^2}{(d_\gamma + \lambda)^2} \end{pmatrix}.$$

Here, the linearized system is not observable when

$$\begin{aligned} \det(\mathcal{O}) &= -\bar{u}\frac{\left(e^{\lambda t} - e^{-d_\gamma t}\right)\left(e^{\lambda_b t} - e^{\lambda t}\right)}{(d_\gamma + \lambda)(\lambda - \lambda_b)} \left(\frac{e^{\lambda_b t} - e^{\lambda t}}{\lambda_b - \lambda} + \bar{u}\frac{e^{\lambda t} - e^{-d_\gamma t}}{d_\gamma + \lambda} \right) \\ &= 0. \end{aligned} \tag{13.15}$$

This determinant vanishes for any t if $\bar{u} = 0$ in the case $\lambda \neq \lambda_b$ and $d_\gamma + \lambda \neq 0$. The first term, $\bar{u} = 0$, shows the loss of observability in the multiplicative bias correction term, while the vanishing of the last factor in (13.15) corresponds to the loss of observability in the additive bias correction term for specific times depending on $\bar{u}, \lambda_b, d_\gamma$. This fact can be easily verified by checking the observability for both the additive and multiplicative models, respectively. Here, the places where the observability is violated define the values of u where it will be difficult to observe either the multiplicative or additive parameter correction.

13.1.3 Filtering one mode of a turbulent signal with instability with SPEKF

Next, we discuss the numerical results for filtering the true signal generated from solutions of the Langevin equation (13.3) with the two-state Markov damping process discussed in Chapter 8. In our numerical simulation below, we will use exactly the same parameters as in Chapter 8 to generate the intermittent burst of instability; the damping term takes a value of $d^+ = 2.27$ in the stable regime and $d^- = -0.04$ in the unstable regime; we fix the switching rate $\nu = 0.1$ for switching from the stable to the unstable regime and $\mu = 0.2$ vice versa; with these parameters, we have an average damping of $\bar{d} = 1.5$; in order to connect with a specific concrete physical model discussed earlier in Chapters 5 and 8, we also fix the frequency $\omega = 8.91/k = 8.91/5 = 1.78$ to be that of the fifth mode of the barotropic Rossby waves in a one-dimensional periodic domain and the noise strength σ to satisfy the following constraint, $\sigma^2/\bar{d} = E = k^{-3} = 1/125$ (see Chapter 5). Figure 13.2 shows the real component of the true signal (solutions of (13.3) with the above parameters) and its corresponding two-state damping coefficient as functions of time for the unforced and the periodically forced equations.

We will consider the filtering problem with the additive, multiplicative and combined model denoted as **SPEKF-A, SPEKF-M** and **SPEKF-C**, respectively, and we will compare these approaches with the perfectly specified damping model and the mean stochastic model, discussed in Chapter 8, with only one Fourier mode. In this section, we consider the filter performance only for the periodically forced case, i.e.

$$f(t) = \mathrm{e}^{\mathrm{i}0.15t}.$$

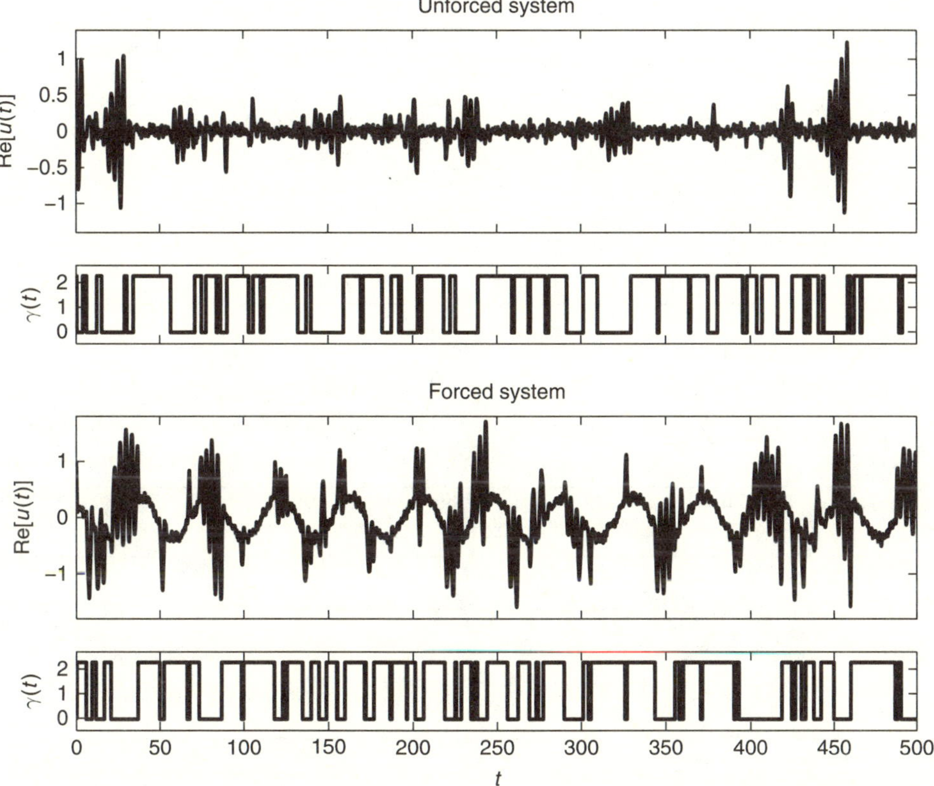

Figure 13.2 Unforced (first panel) and forced (third panel) versions of the true trajectory ($\mathrm{Re}[u(t)]$ shown) with switching damping, $\gamma(t)$, (second and fourth panels for the unforced and forced case, correspondingly).

Interested readers can consult Gershgorin *et al.* (2010b) for the unforced case. We set $\hat{b} = 0$ since the additive bias is supposed to be for perturbations around 0 and $\hat{\gamma} = \bar{d}$ since $\bar{d} = 1.5$ is the average equilibrium damping strength. We run the filter for 2000 assimilation cycles and we quantify the performance by computing the root mean square (RMS) difference between the true signal, u_m, and the posterior mean state, $\bar{u}_{m|m}$. In Figs 13.3–13.5, we show the trajectories of the posterior mean states $u(t)$, $b(t)$, $\gamma(t)$, and the Kalman weights with parameter sets: $\{\gamma_b = 0.1\bar{d}, \omega_b = \omega, \sigma_b = 5\sigma, d_\gamma = 0.01\bar{d}, \sigma_\gamma = 5\sigma\}$ for SPEKF-C, $\{d_\gamma = 0.01\bar{d}, \sigma_\gamma = 5\sigma\}$ for SPEKF-M, and $\{\gamma_b = 0.1\bar{d}, \sigma_b = 5\sigma\}$ for SPEKF-A, for simulations with observation time $\Delta t = 0.25$ and observation noise variance $r^o = E = 0.008$. We find that SPEKFs are almost as skillful as the perfectly specified model which has RMS error 0.042 and their RMS errors, 0.04–0.05, are smaller than the observation error, 0.06. On the other hand, the RMS error of MSM, 0.14, is approximately two times larger than the observation error. The choice of parameters in SPEKF-C suggests

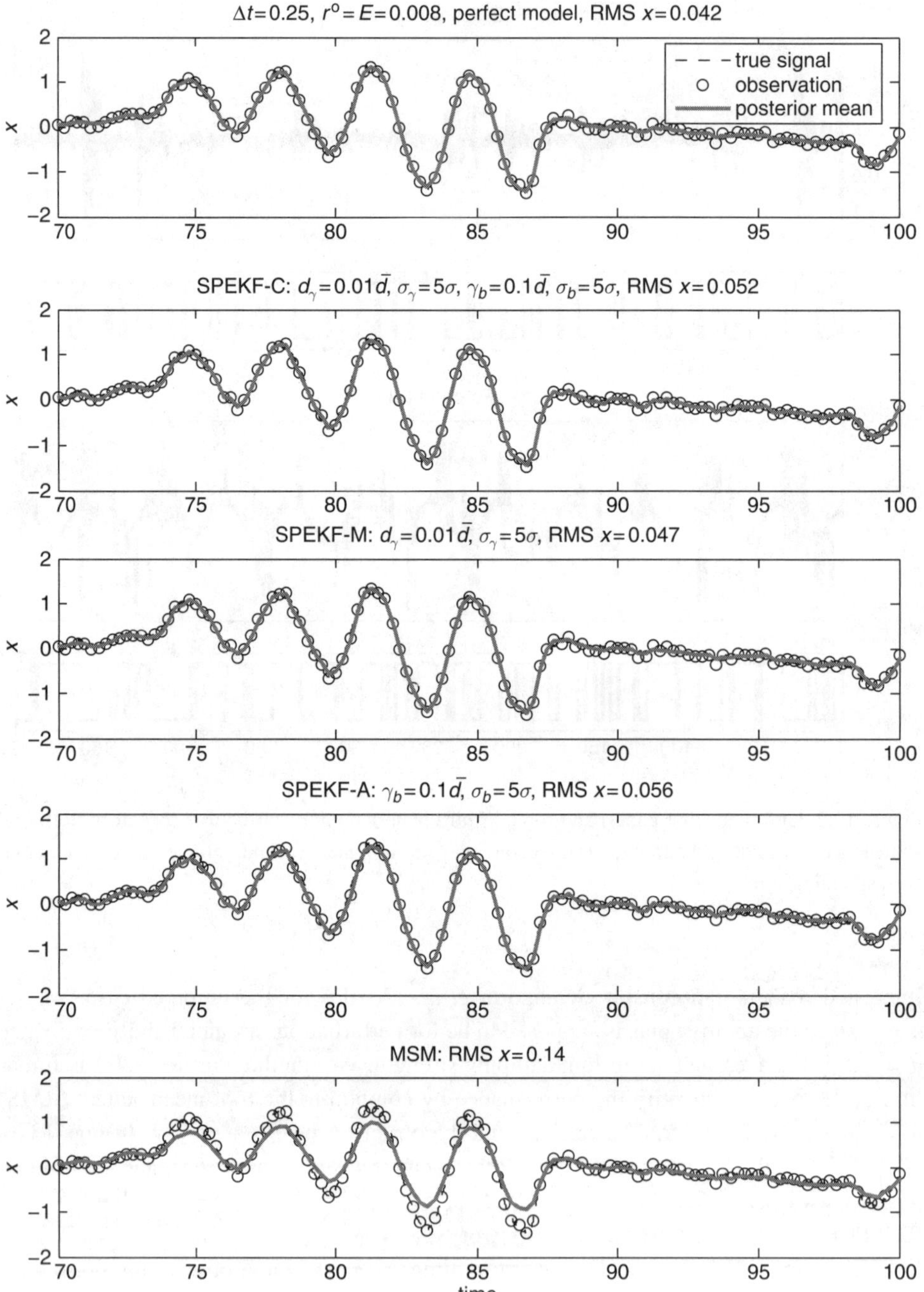

Figure 13.3 Posterior mean state $x(t) = \mathrm{Re}[u(t)]$ (thick solid line) as a function of time for simulations with $\Delta t = 0.25$, $r^o = E$ and stochastic parameters $\{\gamma_b = 0.1\bar{d}, \omega_b = \omega, \sigma_b = 5\sigma, d_\gamma = 0.01\bar{d}, \sigma_\gamma = 5\sigma\}$, compared with the true signals (dashes) and observations (circle).

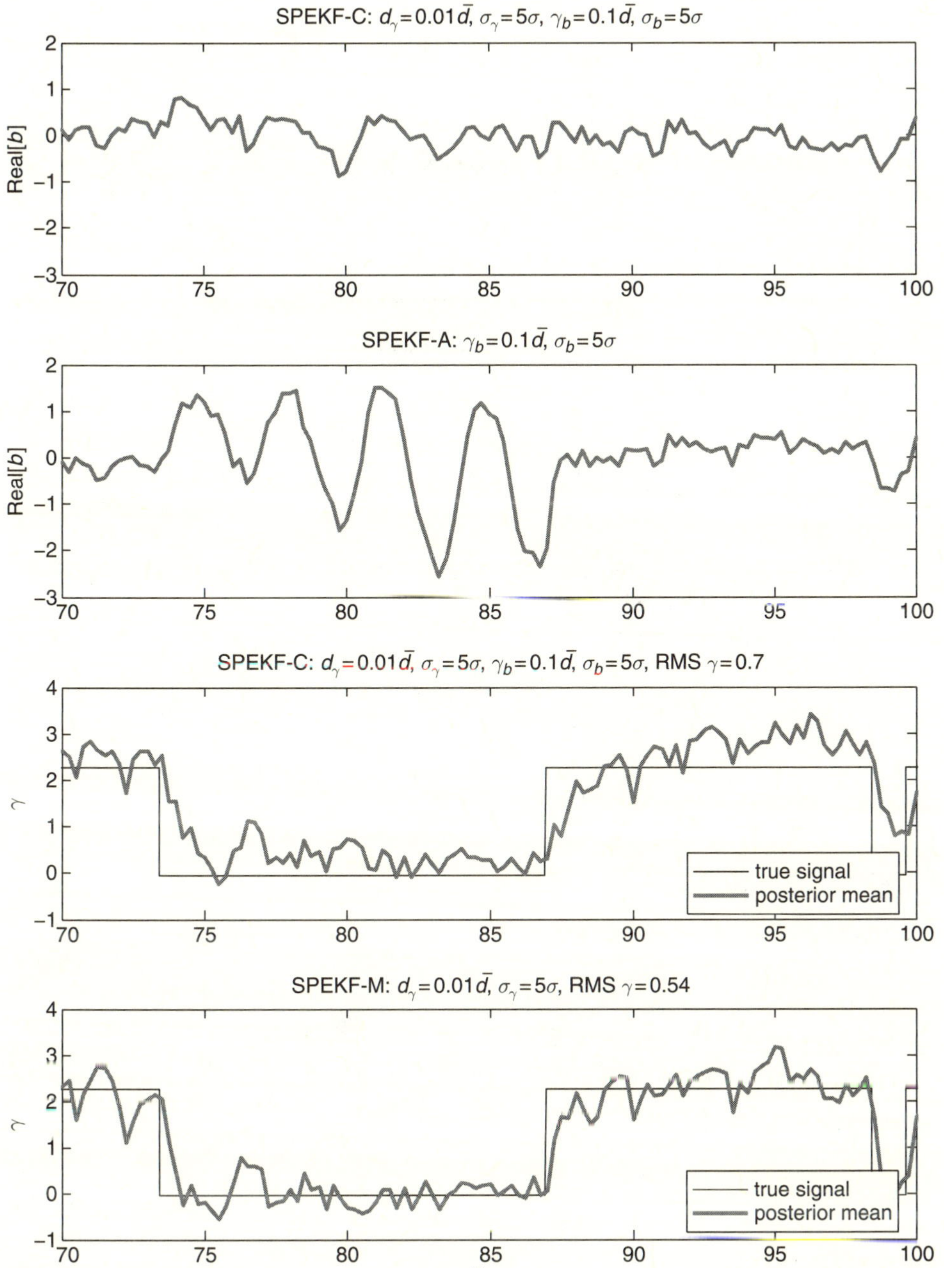

Figure 13.4 Posterior mean bias correction terms, $\mathrm{Re}[b(t)]$ and $\gamma(t)$, as functions of time of the corresponding simulations in Fig. 13.3.

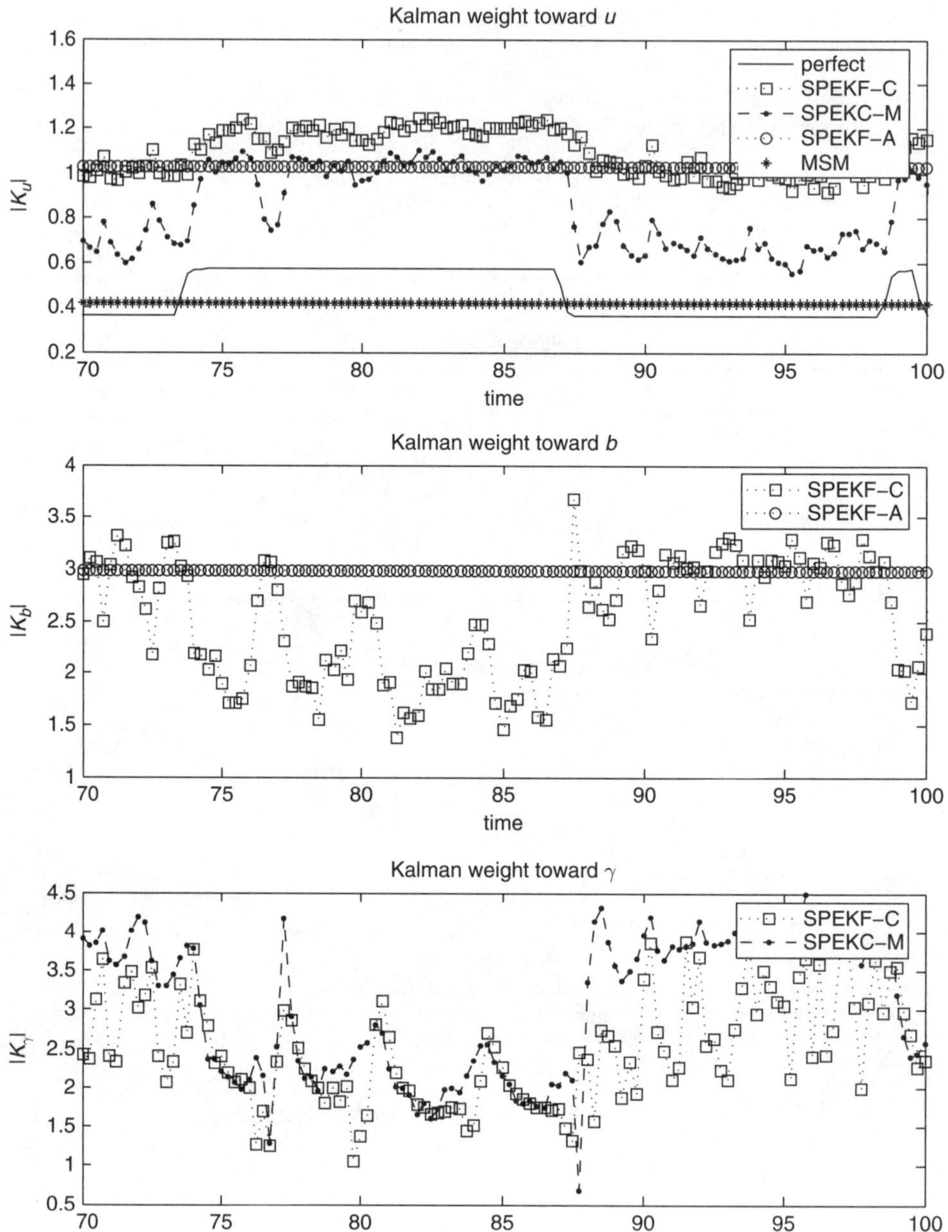

Figure 13.5 Kalman weights toward $u(t)$, $b(t)$ and $\gamma(t)$ as functions of time of the corresponding simulations in Fig. 13.3.

a relatively significant contribution of additive bias correction: $b(t) \approx \mathcal{O}(10^{-1})$ and the Kalman weight relative to the additive bias, $|K_b|$, fluctuates around 2.5 in Fig. 13.5. Later, we shall see that the additive bias correction, $b(t)$, becomes even more significant when the forcing is specified incorrectly. Here we find substantial skill in tracking $\gamma(t)$ with SPEKF-C (see Fig. 13.4) as well as SPEKF-M. This result suggests that the multiplicative bias correction term, $\gamma(t)$, in SPEKF-C becomes as important as the additive bias correction term, $b(t)$ in the present context.

Robustness and sensitivity toward variations of stochastic parameters

Notice that in the numerical experiment above, we choose a particular stochastic parameter set $\{\gamma_b, \omega_b, \sigma_b, d_\gamma, \sigma_\gamma\}$ to obtain accurate filtered solutions. In Gershgorin *et al.* (2010b), we find that the filtering skill of SPEKF-C is the least sensitive toward variation of stochastic parameters compared to SPEKF-A and SPEKF-M. There are some extreme parameter regimes where SPEKF-C behaves poorly. This includes the regime when damping γ_b is too large such that SPEKF-C behaves exactly like SPEKF-M. In this situation, both schemes produce poor filtering skill comparable to MSM when the multiplicative damping d_γ is too strong or when the noise strength σ_γ is too small since the multiplicative bias correction dynamics are nothing more than weak perturbations around the average equilibrium damping, $\bar{d}$, used in MSM. When the observation time is beyond the decorrelation time, both schemes also fail for weak damping d_γ (here the multiplicative bias correction term is roughly a Wiener process which has unbounded variance as a function of time) or strong noise σ_γ (too much fluctuation in the OU process overestimates the multiplicative bias correction term, $\gamma(t)$) (see Gershgorin *et al.*, 2010b, for extensive discussion and documentation).

Learning the forcing from the filtering

Consider the situation when the external forcing $f(t)$ is either partially known or unknown. If we regard the true signal as arising from a turbulent component of a more complex system, such a circumstance arises readily. In this situation, the additive bias correction term $b(t)$ is used as a learning tool for the external forcing. First, we consider the case when the external forcing is completely unknown. This is a severe test case; thus, we set $f(t) \equiv 0$ in Eqn (13.5) and apply the filters to recover a true signal that has the same forcing as in the previous section. In Fig. 13.6, we demonstrate the results of filtering using all three methods (SPEKF-C, SPEKF-M and SPEKF-A) with the incorrectly prescribed external forcing. We note that the combined and additive models retain most of their skill in filtering $u(t)$ (RMS errors of SPEKF-C is 0.055 and of SPEKF-A is 0.059); on the other hand, the performance of the multiplicative model (with RMS error 0.111) became much worse than in the earlier situation with the correctly specified forcing. The explanation of this behavior comes from observing the dynamics of the additive bias correction $b(t)$ (Fig. 13.6, second panel). For both the combined and additive models, $b(t)$ captures the external forcing $f(t)$ that is present in the true signal. Moreover, in the additive model, $b(t)$ also tries to recover the multiplicative part of the noise by adding the high-frequency

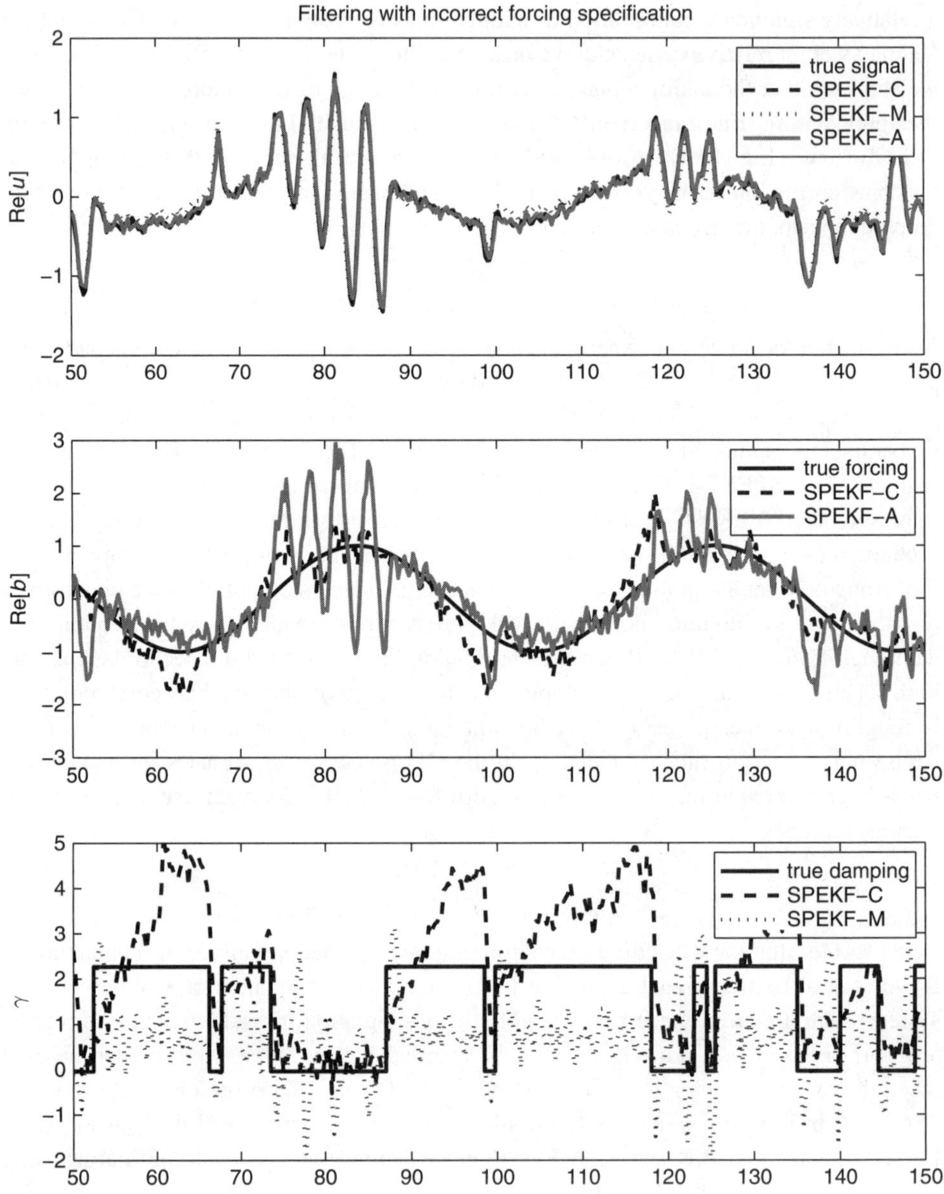

Figure 13.6 Filter performance with incorrect forcing specification for $r^o = E$, $\Delta t = 0.25$. The first panel shows $u(t)$ (the true signal is shown with the solid black line, the SPEKF-C posterior is shown with dashes, the SPEKF-M posterior is shown with dots, and the SPEKF-A posterior is shown with the solid gray line), the second panel shows the additive bias correction $b(t)$ for SPEKF-C and SPEKF-A together with the true forcing $f(t)$, and the third panel shows the multiplicative bias correction $\gamma(t)$ for SPEKF-C and SPEKF-M together with the true damping.

oscillations with frequency ω and large amplitude whenever $\gamma(t)$ takes negative values. On the other hand, the multiplicative model performs poorly – the RMS error is larger than the average observation error, $\sqrt{r^o/2}$ =0.063. There is no means to repair the incorrectly specified forcing by using the multiplicative model. Moreover, the multiplicative model gives the wrong estimation for the damping parameter $\gamma(t)$.

Next, we study how our three filtering models perform, when the external forcing is not specified exactly, as parameters in the incorrect forcing signal are varied systematically. Suppose the amplitude, or the frequency, or the phase of the forcing are unknown, while the remaining two parameters are known. Then, we can vary the unknown parameter around its value in the true signal and study how the filter skill changes. In Fig. 13.7, we show how the filter skill changes when we vary the amplitude of the forcing A_f (first panel), the frequency of the forcing ω_f (second panel) and the phase of the forcing ϕ_f (third panel), where the forcing has the form

$$f(t) = A_f e^{\mathrm{i}(\omega_f t + \phi_f)}.$$

The true signal was generated with the values $A_f = 1$, $\omega_f = 0.15$ and $\phi_f = 0$. We note that for all three filters and for all three parameters of the forcing, the filters produce the results with the minimum error at the true values of the forcing parameters. Moreover, the filters allow for the variations of the amplitude, A_f, and the phase, ϕ_f, around their respective values in the true signal. However, for SPEKF-M, even slight variations of the frequency, ω_f, lead to the deterioration of the filtered signal with the RMS error greater than the observation error. We also note again that the multiplicative model is not suitable for filtering with incorrect forcing specification. On the other hand, both the combined and additive models produce filtered solutions with skill comparable to the skill of the perfectly specified model for various frequencies.

There is a recent comprehensive study of the SPEKF models and various recent linear and nonlinear extended Kalman filters in a variety of difficult test regimes (Branicki *et al.*, 2011). While the SPEKF filter outperforms all the other filters with model error and linear tangent Kalman filters often exhibit divergence, the nonlinear Gaussian closure filters almost perform as well as SPEKF and provide an interesting alternative filtering strategy.

13.2 Filtering spatially extended turbulent systems with SPEKF

In this section, we discuss filtering the spatially extended turbulent system with intermittent bursts of instability given by the canonical model in Section 8.2 of Chapter 8. Similar to the canonical model for instability described in Chapter 8, we consider a stochastic PDE with time dependent damping Langevin equation (13.3) for the first five Fourier modes, i.e.

$$\frac{du_k(t)}{dt} = -\gamma_k(t)u_k(t) + \mathrm{i}\omega_k u_k(t) + \sigma_k \dot{W}_k(t) + f_k(t), \quad k = 1, \ldots, 5,$$

and linear Langevin equation with constant damping $\bar{d}$ for modes $k > 5$. The damping coefficient of the first five Fourier modes alternates between two states (see Table 13.1)

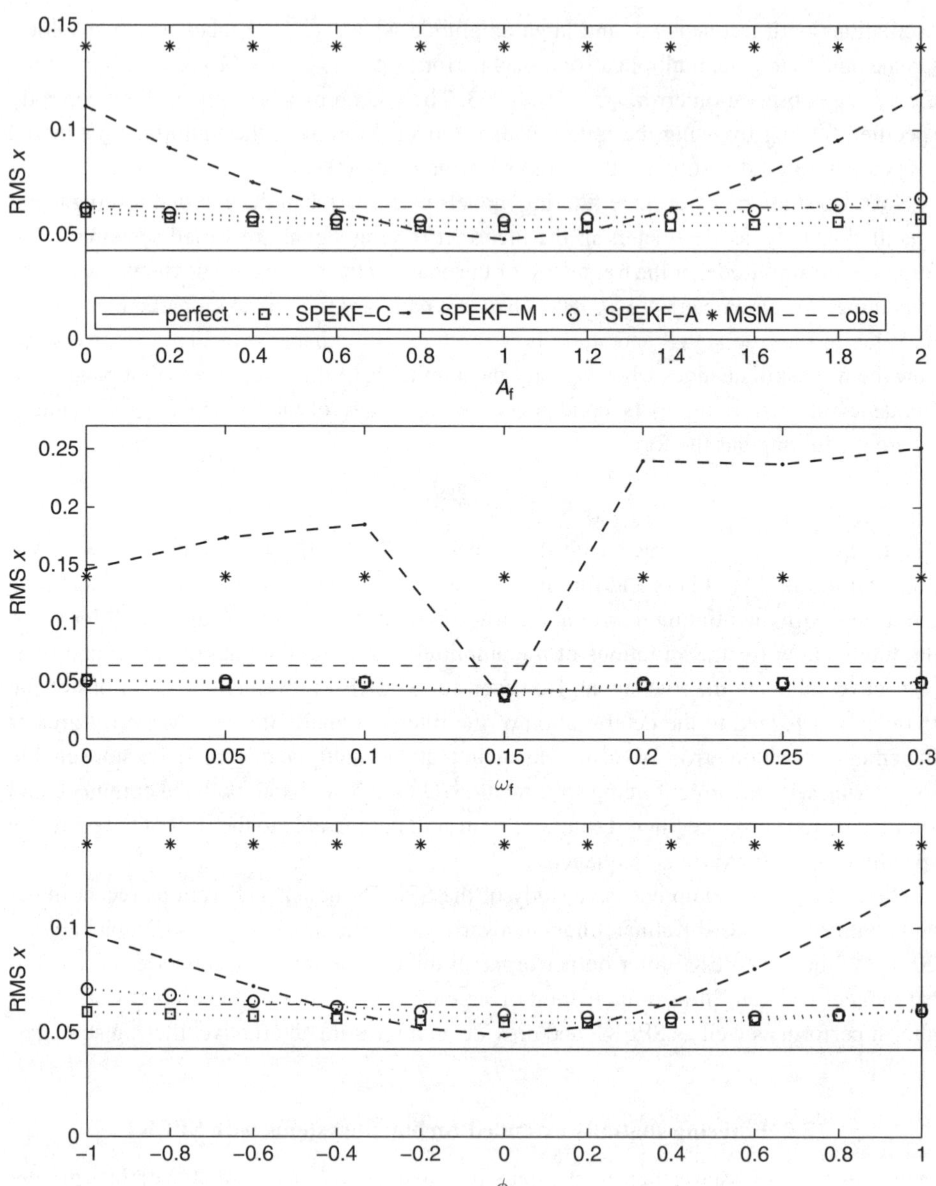

Figure 13.7 Filter performance with incorrect forcing specification. The first panel shows the dependence of the RMS error on the forcing amplitude A_f, the second panel on the forcing frequency ω_f, and the third panel on the forcing phase ϕ_f.

Table 13.1 Damping coefficients for simulating the true signal.

modes (k)	stable	unstable
1,2	$d_w = 1.3$	$d_s = 1.6$
3,4,5	$d^+ = 2.25$	$d^- = -0.04$
≥ 6	$\bar{d} = 1.5$	$\bar{d} = 1.5$

with exponentially distributed switching times. In our simulation below, we use the same switching rates as in Section 13.1.3 above; $\nu = 0.1$ for switching from the stable to the unstable regime and $\mu = 0.2$ vice versa. We also fix the remaining physical parameters to be exactly identical to those in Chapter 5; Rossby wave frequency $\omega_k = 8.91/k$ and energy spectrum $E_k = k^{-3}$; we use a finite discretization up to mode $N = 52$ so that we have a total of $2N + 1 = 105$ grid points. In the numerical simulations below, we only consider sparse regularly spaced observations at every $P = 2N + 1/2M + 1 = 7$ grid points; thus, the highest wavenumber of the observations is $M = 7$.

In Fig. 13.8(a), we show a space–time plot of the solutions of this stochastic toy model for the Rossby waves with the instability transitions (the "switching" SPDE) described above, solved with time step 0.001 and discretized with a total of $2N + 1 = 105$ grid points in a one-dimensional 2π-periodic domain. In Fig. 13.8(b), we also show the stability regime, i.e. the damping strength of modes 3–5, which is typically unknown in reality. In this snapshot, we see a strong coherent westward wind burst which begins two days after the unstable transition. The fact that the occurrence of this westward wind burst is not exactly right after regime transition makes this turbulent signal an extremely hard test problem. In the remainder of this section, we refer to this turbulent signal as the true signal that we want to filter and predict. In particular, we will observe this true signal for a period of 500 time units (or days) at every 0.25 time interval. We simulate the observations at sparse uniformly distributed locations as described earlier with error noise variance $r^o = 0.3$ in the correctly specified forcing case and $r^o = 0.5$ in the incorrectly specified forcing case. Both noise variances in k-space, $r^o/(2M + 1)$, exceed the equilibrium energy spectrum $E_k = k^{-3}$ of the unforced system for $k > 3$. When the turbulent signal is externally forced (with periodic forcing), its energy spectrum is given as follows

$$E_k = \lim_{T \to \infty} \frac{1}{2T} \int_0^T |\hat{u}_k(s)|^2 ds. \tag{13.16}$$

In Fig. 13.9(b), we find that the above choice of observation error variance, indeed, exceeds the energy spectrum (13.16) for wavenumbers $k > 7$. In this figure, the energy spectrum (13.16) is approximated by averaging over a finite amount of time $T = 450$ days after a transient period of 50 days. For the remainder of this section, the term "energy spectrum" refers to the spectrum of the forced system defined in (13.16), and not to that of the unforced case, k^{-3}. We will discuss results with different noise level in Section 13.2.3.

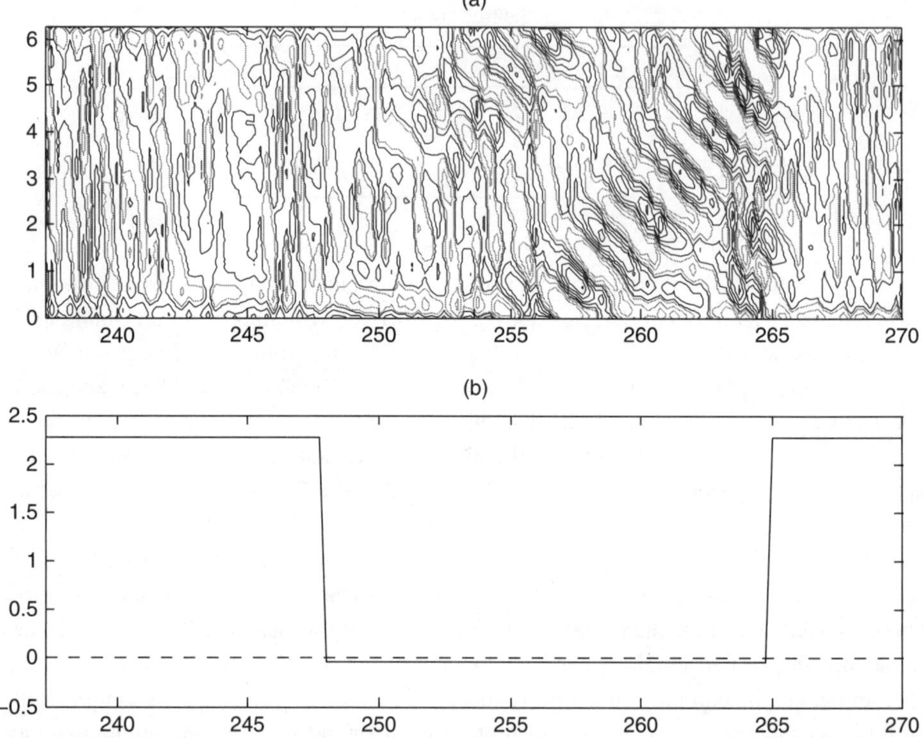

Figure 13.8 The true signal $u(x, t)$ for a period of time between $t = 240$ and $t = 270$. (a) Space–time plot with contour interval 2 units/day: dark contours for positive values and gray contours for negative values. (b) Damping $\gamma(t)$ for the solution $u(x, t)$ shown in panel (a).

In Chapter 8, we considered two filtering strategies including the perfectly specified model and the mean stochastic model (as in Section 13.1 above), where both strategies are implemented with the reduced Fourier domain Kalman filter (RFDKF, see Chapter 7) since the observations are sparse, regularly spaced, and the turbulent spectrum, k^{-3}, decays rapidly at small wavelengths. Here, we follow the same procedure for SPEKF. In particular, we filter the observed modes, $1 \leq k \leq M = 7$, with the following adjusted complex scalar coefficients,

$$\hat{v}'_{\ell,m} \equiv \hat{v}_{\ell,m} - \sum_{k \in \mathcal{A}(\ell), k \neq \ell} \hat{u}_{k,m} = \hat{u}_{\ell,m} + \hat{\sigma}^o_m, \tag{13.17}$$

where $\hat{v}_\ell$ denotes the original observed modes and it is adjusted by the total aliased modes. These aliased modes are unobserved and they are updated solely with the mean model (Langevin equation with constant damping coefficient $\bar{d}$ in (13.4)). In the remainder of this section, we show the filter performance for the following five strategies including the perfectly specified model, the mean stochastic model (MSM), and the stochastic

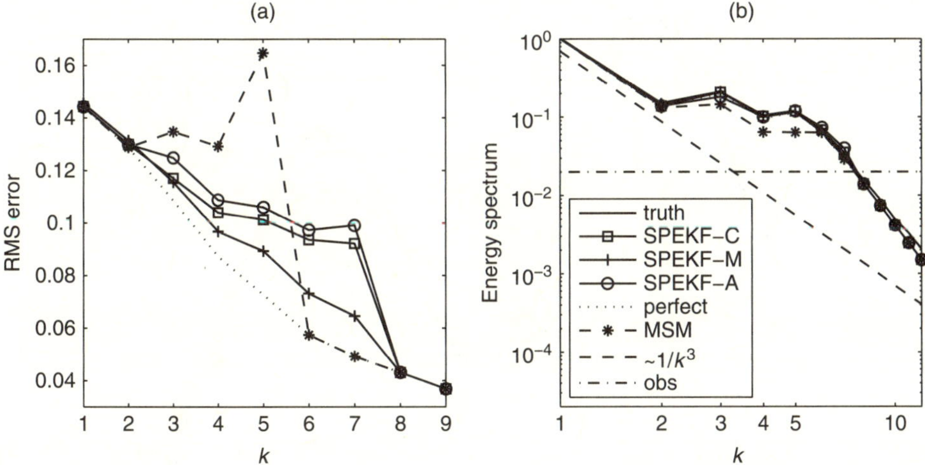

Figure 13.9 Filtering with correctly specified forcing: (a) RMS errors as functions of wavenumbers; (b) energy spectrum as a function of wavenumbers.

parametrized strategies, SPEKF-A, SPEKF-M and SPEKF-C. We will quantify the performance by measuring the root mean square (RMS) difference between the true signal, $u^t(x_j, t_m)$, and the filtered solution, $\bar{u}(x_j, t_m)$ and we will also compare the energy spectra generated from the true signal with those generated through various posterior filtered solutions. Note that recovery of a turbulent energy spectrum from sparse regular observations is both an interesting bulk test of filter performance and a quantity of great practical interest.

For SPEKFs, we choose stochastic parameters that belong to the robust set of parameters based on the comprehensive study of Gershgorin *et al.* (2010b); they are $\{\gamma_{b,k} = 0.1\bar{d}, \omega_{b,k} = \omega_k, \sigma_{b,k} = 4\sigma_5 = 0.6, d_{\gamma,k} = 0.1\bar{d}, \sigma_{\gamma,k} = 4\sigma_5 = 0.6\}$ for filtered wavenumbers $k = 1, \dots, 7$. In Section 13.2.3, we will check the robustness of the filter skill when these parameters are changed.

13.2.1 Correctly specified forcing

In Fig. 13.9(a), we compare the RMS errors of the nine most energetic modes. The perfectly specified filter provides the smallest error as expected since the perfect model utilizes the exact dynamics including the turbulent bursts of instability. The multiplicative model produces a filtered solution with error just slightly larger than the error of the perfectly specified filter. Note that the error of the multiplicative model is still very low for both stable and unstable regions of the spectrum (as described in Table 13.1 instability occurs in modes 3, 4 and 5). Next, the combined and additive models produce the filtered solutions with errors that are almost the same as the errors of the perfectly specified filter for modes 1 and 2 and then deviate from the corresponding values of the perfectly specified filter error

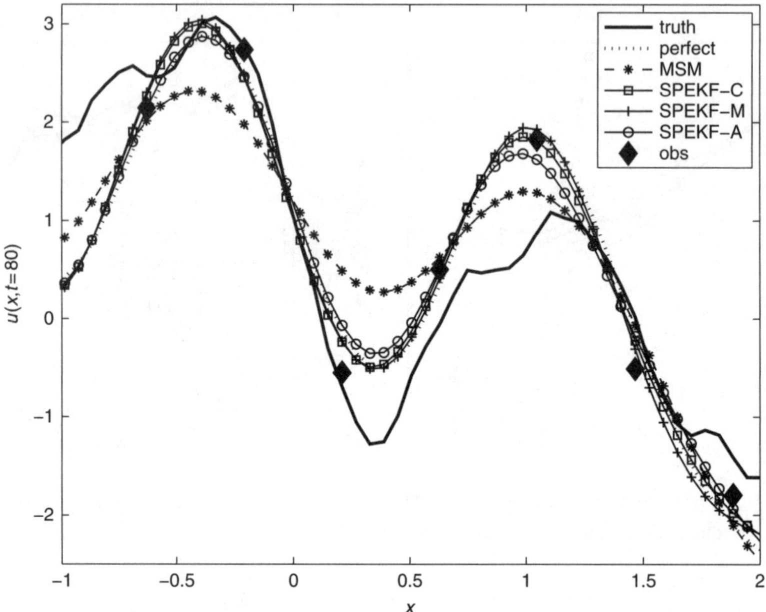

Figure 13.10 Filtering with correctly specified forcing: Snapshot of the true trajectory in the unstable regime at time $t = 80$ together with filtered signals as a function of space. We only show the region with $x \in [-1, 2]$ for clarity of presentation.

for higher modes as expected with perfectly specified forcing. Finally, the MSM has a very large error for the unstable modes 3, 4 and 5. On modes higher than 5, MSM is identical to the perfectly specified filter since the true damping of these modes is exactly equal to the average damping of the system $\bar{d} = 1.5$. One important practical bulk quantity to recover from filtering is the energy spectrum (13.16) of the true signal. In Fig. 13.9(b), we see that the energy spectrum of the posterior state of MSM underestimates the true energy spectrum of the unstable modes. On the other hand, the multiplicative and combined models produce filtered solutions with energy spectrum very close to the true energy spectrum for all seven filtered modes. The energy spectrum of the additive model is also close to the true energy spectrum with just slight deviations from it.

In Fig. 13.10, we show a snapshot of the true trajectory in the unstable regime at time $t = 80$ together with the posterior mean states of various filtered solutions. There, we observe that in the unstable regime all SPEKFs produce filtered solutions that are much closer to the true signal relative to that of the MSM. Of course, with such sparse observations at every seventh grid point, we cannot expect to recover the details of the spatial pattern. This is explained by the fact that the multiplicative bias correction $\gamma_k(t)$ and additive bias correction $b_k(t)$ help the SPEKF to recover the true dynamics better than the MSM with averaged damping, especially in the unstable modes 3–5, which we will discuss next.

Table 13.2 Spatial RMS errors for simulations with the unforced case, correctly forced case, and incorrectly forced case.

Forcing	Unforced case	Correct forcing	Incorrect forcing
r^o	0.2	0.3	0.5
RMSE of perfect filter	0.346	0.391	0.454
RMSE of MSM	0.394	0.477	0.728
RMSE of MSM$_{f=0}$	–	–	1.169
RMSE of SPEKF-C	0.380	0.444	0.588
RMSE of SPEKF-M	0.359	0.418	0.787
RMSE of SPEKF-A	0.398	0.457	0.601

In Fig. 13.11, we show the time series of the additive and multiplicative bias corrections, $b_k(t)$ and $\gamma_k(t)$, respectively, for wavenumbers $k = 3, 4, 5$ with intermittent unstable damping. In the panels for $\gamma_k(t)$, we also show the true value of the damping. We note that $\gamma_k(t)$ of both the combined and multiplicative models follow the trajectory of the true damping. However, the multiplicative model produces a better estimate of the true damping relative to the combined model in all three unstable modes. As a result, we expect the multiplicative model to produce a better approximation to the true signal $u(x, t)$, which is confirmed in Table 13.2. On the other hand, the panels of Fig. 13.11 that correspond to the additive bias correction $b_k(t)$ show that $b_k(t)$ do not deviate much from zero, except for the times when the damping is unstable. In the unstable regime, both the combined and additive models use the additive bias correction to recover the true signal. However, since the model error is multiplicative for the correctly specified forcing, we expect the multiplicative model to show the best performance among the three SPEKFs. This result is also confirmed in Table 13.2, in which we present the RMS errors in real space. In the same table, we also report results for the unforced case ($\hat{f}_k = 0$ both for the true signal and the filter) run with slightly smaller $r^o = 0.2$ (which is greater than k^{-3} for wavenumbers $k > 4$); here we find similar conclusions as in the forced case discussed above: the multiplicative model is the method of choice.

13.2.2 Unspecified forcing

Here, we consider a true signal with forcing given by

$$\hat{f}_k(t) = A_{f,k} \exp\Big(\mathrm{i}(\omega_{f,k} t + \phi_{f,k})\Big), \tag{13.18}$$

for $k = 1, \dots, 7$ with amplitude $A_{f,k}$, frequency $\omega_{f,k}$, and phase $\phi_{f,k}$ drawn randomly from uniform distributions,

$$\begin{aligned} A_{f,k} &\sim U(0.6, 1), \\ \omega_{f,k} &\sim U(0.1, 0.4), \\ \phi_{f,k} &\sim U(0, 2\pi), \\ \hat{f}_k &= \hat{f}^*_{-k}, \end{aligned}$$

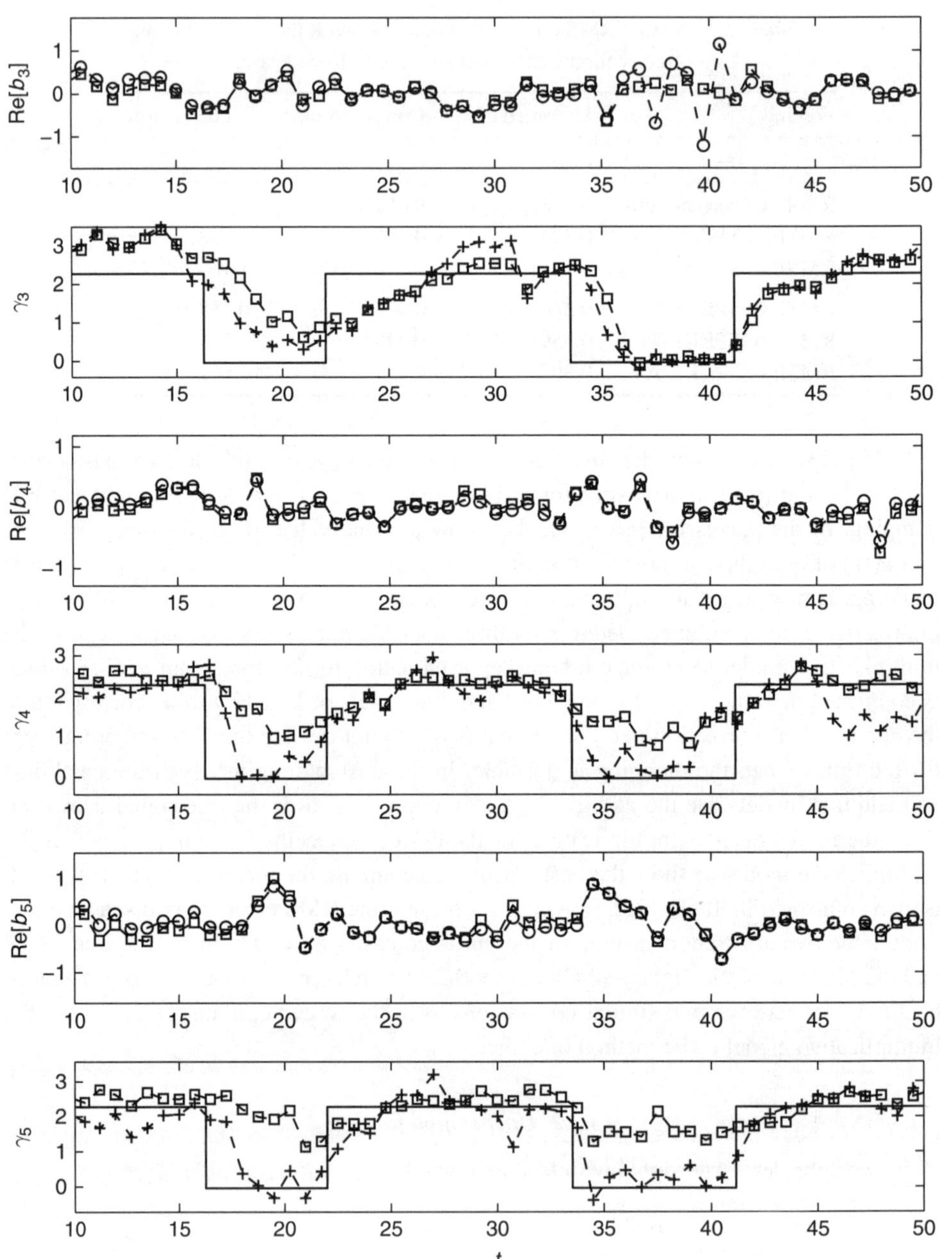

Figure 13.11 Filtering with correctly specified forcing: additive bias correction, $b_k(t)$, and multiplicative bias correction, $\gamma_k(t)$, for SPEKF-C (squares), SPEKF-A (circles) and SPEKF-M (pluses) for modes $k = 3, 4, 5$ with unstable damping. The solid line shows the true damping $\gamma(t)$.

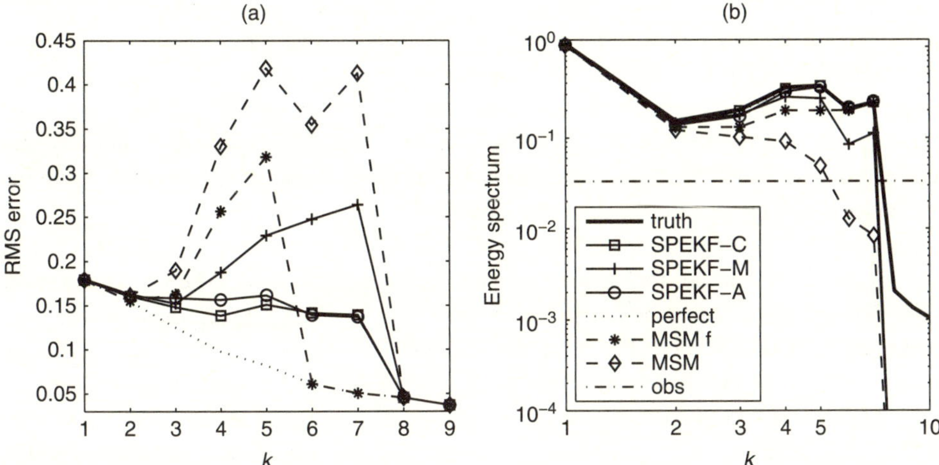

Figure 13.12 Filtering with unspecified forcing: (a) RMS errors as functions of wavenumbers; (b) energy spectrum as a function of wavenumbers.

and unforced, $\hat{f}_k(t) = 0$, for modes $k > 7$. However, we do not specify this true forcing to the filter model, i.e. we use $\tilde{f}_k = 0$ for all modes.

In Fig. 13.12, we compare the RMS errors and the energy spectra of the true signal and of the filtered solutions mode by mode. In terms of RMS errors, SPEKF-C is the best strategy, followed by SPEKF-A and both RMS errors are smaller than that of the MSM with perfectly specified forcing in the unstable modes 3–5. The high filtering skill with SPEKF-M in the perfectly specified forcing (see Table 13.2) case deteriorates as the forcing is incorrectly specified; in this case, the absence of the additive bias correction term, b_k, in the multiplicative model degrades the filtering skill significantly. In terms of the energy spectrum, we find that the first seven most energetic modes are filtered very well by the perfectly specified filter as well as by the combined and additive models if we compare with the spectrum of the true signal. The multiplicative model does not provide a good approximation of the energy spectrum for the wavenumbers $k > 3$. The MSM with correctly specified forcing misses the unstable modes 3, 4 and 5, while the rest of the modes are filtered well. The MSM with unspecified forcing only filters the first two modes well, while missing the rest of the modes. Similar conclusions hold when the spatial RMS error, reported in Table 13.2, is used for the performance measure.

13.2.3 Robustness and sensitivity to stochastic parameters and observation error variances

In this section, we study how sensitive the proposed SPEKFs are to variations of stochastic parameters and how the skill of the filters varies for different values of observation error variance, r^o, for a fixed set of stochastic parameters. In this study, we keep all but one of the parameters fixed and vary the remaining parameter in a broad range of values. The fixed parameters are $\{d_{\gamma,k} = 0.1\bar{d}, \gamma_{b,k} = 0.1\bar{d}, \sigma_{\gamma,k} = 4\sigma_5, \sigma_{b,k} = 4\sigma_5, \omega_{b,k} = 1\}$. It is very important to realize that we use the same set of stochastic parameters for all the switching modes, that is, wavenumbers $|k| \leq 7$. These modes have different energies and different correlations in time. Therefore, using the same set of stochastic parameters for a number of modes is a tough test for the robustness of the SPEKFs.

In Fig. 13.13, we demonstrate the dependence of the spatially averaged RMS errors of the various filters on the stochastic parameters and observation variance for the correctly specified forcing case. We note that both SPEKF-M and SPEKF-C are robust to the variations of $d_\gamma/\bar{d}$ and σ_γ but SPEKF-M has smaller RMS errors relative to SPEKF-C (see panels (a) and (b) in Fig. 13.13). The robustness in SPEKF-C toward these two parameters is very similar to the robustness of the fifth mode mentioned above in Section 13.1.3 and studied in detail by Gershgorin *et al.* (2010b). For the SPEKF-M, the sensitivity toward large damping and small noise on the fifth mode study (Gershgorin *et al.*, 2010b) disappears when all modes are accounted for and this is because the errors in the non-switching modes are smaller. The sensitivity of the filtering skill toward variations of the additive stochastic parameters, $\gamma_b, \omega_b, \sigma_b$ reflects the one-mode study on the fifth mode in Gershgorin *et al.* (2010b) (see panels (c) and (e) in Fig. 13.13); SPEKF-C is robust toward variations both in the damping and frequency phase. SPEKF-C produces rather high errors when the magnitude of σ_b approaches the observation error; this result exactly reflects what we found in the one-mode study in Gershgorin *et al.* (2010b), except there the observation error is rather small. From Fig. 13.13(e), we conclude that the RMS errors of all the filters increase as functions of r^o. Moreover, the multiplicative model (SPEKF-M) performs better than the combined model, SPEKF-C, which in turn is better than the additive model, SPEKF-A.

When forcing is incorrectly specified (see Section 13.2.2), the situation changes considerably. In this case, the skill of SPEKF-C and SPEKF-A is very robust with respect to variations of multiplicative stochastic parameters, d_γ, σ_γ (see panels (a) and (b) in Fig. 13.14). On the other hand, SPEKF-M is not skillful at all as we found and discussed in Section 13.2.2. When the additive stochastic parameters are chosen such that γ_b is small and σ_b is rather large, then the skill of the combined and additive models is good (see Fig. 13.14(c),(d)). There is one particular regime of parameters when both the combined and additive models fail, that is, the regime when the additive noise is too small. In this regime we naturally expect the combined and additive model to perform poorly since the additive bias correction is needed to recover the unspecified forcing. The performance of the combined and additive models is quite robust to the changes of ω_b unless this parameters takes values of order 10 or larger (see Fig. 13.14(f)). So we conclude that in our model the frequency of the additive bias correction $b_k(t)$ should not exceed the maximum frequency of the system which is equal $\omega_k = 8.91$ for $k = 1$. In Fig. 13.14(e), we demonstrate the performance of the filters as a function of observation variance r^o. Here we see that the combined model is the best for the whole range of observation variance. As we

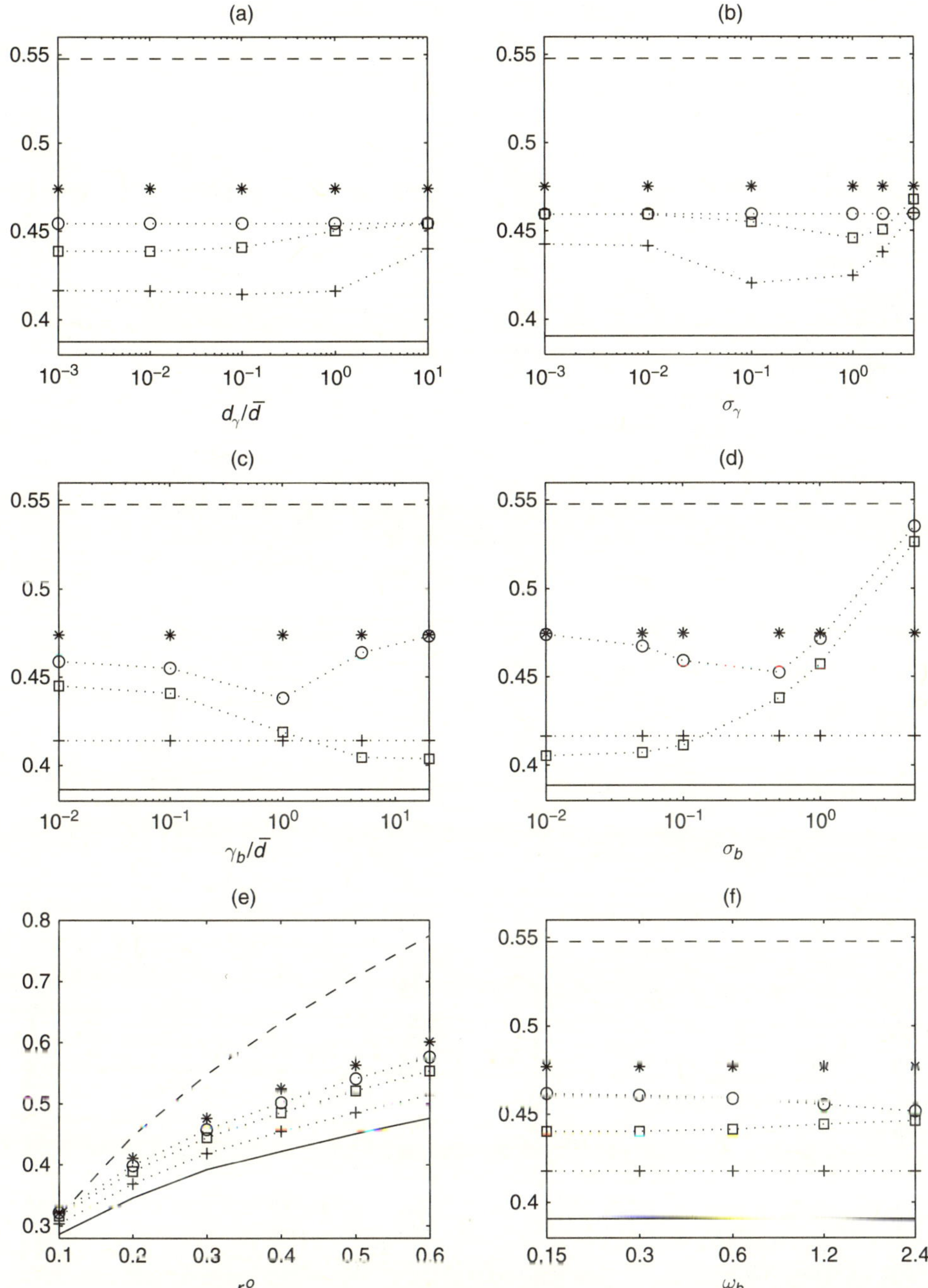

Figure 13.13 Filtering with correctly specified forcing: spatially averaged RMS errors of the perfectly specified filter (solid line), MSM (asterisks), SPEKF-C (squares), SPEKF-M (pluses), SPEKF-A (circles) and observation error (dashes). The fixed parameters had the values $d_\gamma/\bar{d} = 0.1$, $\gamma_b/\bar{d} = 0.1$, $\sigma_\gamma = 0.6$, $\sigma_b = 0.6$, $\omega_b = 1$, $r^o = 0.3$.

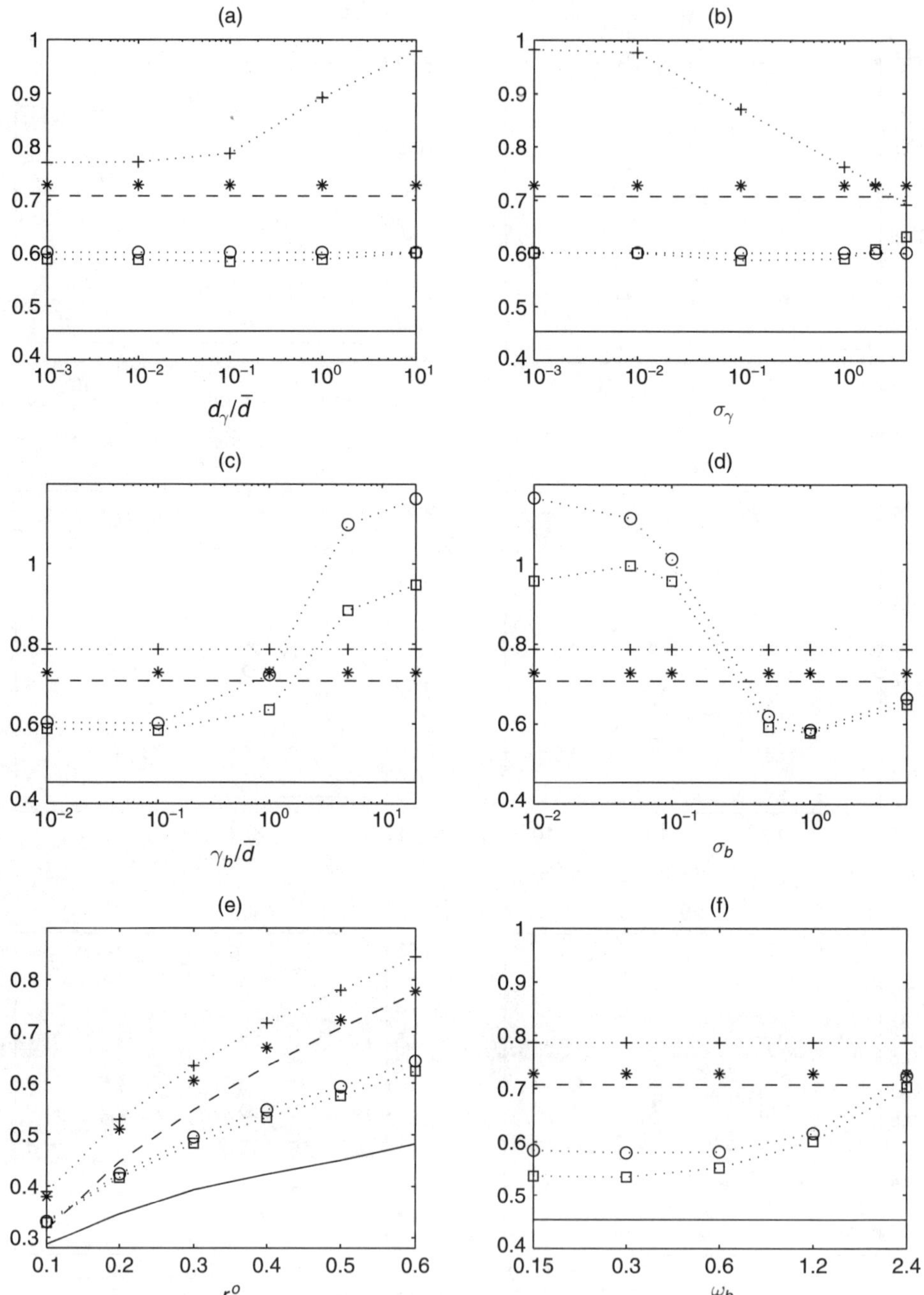

Figure 13.14 Filtering with unspecified forcing: Spatially averaged RMS errors of the perfectly specified filter (solid line), MSM (asterisks), SPEKF-C (squares), SPEKF-M (pluses), SPEKF-A (circles) and observation error (dashes). The fixed parameters had the values $d_\gamma/\bar{d} = 0.1$, $\gamma_b/\bar{d} = 0.1$, $\sigma_\gamma = 0.6$, $\sigma_b = 0.6$, $\omega_b = 1$, $r^o = 0.5$.

already pointed out earlier in this section, it is very important to realize that the same set of stochastic parameters is used for all wavenumbers with $|k| < 7$ even though initially this set was chosen for the fifth mode of our model as in Gershgorin *et al.* (2010b). Here, we have shown that this set of stochastic parameters belongs to a broad range of robust stochastic parameters which produce similar high skill filtered solutions with SPEKFs, and, therefore, one should not tune the stochastic parameters for a specific mode; this attractive feature makes the SPEKF algorithm useful in a practical context as discussed below in Section 13.3.

In Gershgorin *et al.* (2010a), we also discuss the predictive skill of the three stochastic parametrized models above: the combined, additive, and multiplicative models. Here, we omit that discussion and we will instead discuss the recent success of utilizing the SPEKF algorithm to filter the sparsely observed two-layer QG model (see Chapter 11 for the details of the model).

13.3 Application of SPEKF to the two-layer QG model

With the encouraging results on the spatially extended test models above, we test the SPEKF algorithm to filter more realistic turbulent signals defined by the two-layer model in (11.5) of the quasi-geostrophic potential vorticity defined in (11.6) in Chapter 11 with over 30,000 degrees of freedom (Harlim and Majda, 2010b). Thirty-six very sparse noisy regularly spaced observations were utilized in filtering these turbulent solutions (see Fig. 11.7). Two regimes of turbulent signals were considered corresponding to the atmosphere with $F = 4$ and the ocean with $F = 40$ (see Table 11.5) where $F = L_d^{-2}$ and L_d is the Rossby radius defined below (11.6) in Chapter 11; the situation $F = 4$ mimics atmospheric turbulence while $F = 40$ mimics oceanic turbulence with realistic sparse noisy observations. The mean stochastic models in Chapter 12, MSM1, MSM2, were utilized in the projected RFDKF filtering algorithms (see Section 13.2) for 36 large-scale Fourier modes; recall from Sections 12.1.1 and 12.1.2 that MSM1 is based on a perfect regression strategy for the climate statistics while MSM2 is based on a regression strategy from shear turbulence theory (Delsole, 2004) with larger model errors in the climate statistics. The SPEKF algorithm to learn the damping and forcing was utilized with the MSM1 model. These SPEKF and MSM1, MSM2 algorithms are remarkably cheap computational algorithms for filtering and tens of thousands of times faster than LLS-EAKF (see Section 11.5) for the perfect model.

In Table 13.3, we show the filtering skill for all of these filters in the atmospheric case, $F = 4$, with and without model errors in damping. Remarkably, for these sparsely observed turbulent geophysical flows, as shown in Table 13.3, the very cheap SPEKF filters have comparable excellent filtering skill to LLS-EAKF both with and without model errors. The mean model MSM1 with perfect regression strategy has high filtering skill while MSM2 has the worse filter performance. A snapshot of the filtering skill for the stream function in the case with model error is depicted in Fig. 13.15. The true signal is depicted in the upper left panel and the true signal corrupted by observational noise, which is what is actually

Table 13.3 Average RMS errors (averaged over 10,000 assimilation cycles ignoring the first 200 cycles) for $F = 4$.

$F = 4$	$T_{\text{obs}} = 0.5, r^o = 0.0856$		$T_{\text{obs}} = 0.25, r^o = 0.1711$	
Scheme	Perfect	Model error	Perfect	Model error
SPEKF	0.1486	0.1522	0.1567	0.1618
MSM1	0.1921	0.2018	0.2010	0.2101
MSM2	0.2731	0.2962	0.2705	0.2933
$\sqrt{r^o}$	0.2925	0.2925	0.4137	0.4137
LLS-EAKF	0.1582	0.1471	0.1790	0.1510

Table 13.4 Average RMS errors (averaged over 10,000 assimilation cycles ignoring the first 200 cycles) for $F = 40$.

$F = 40$	$T_{\text{obs}} = 0.01, r^o = 0.25E$	$T_{\text{obs}} = 0.02, r^o = 0.5E$
SPEKF	1.0087	1.2193
MSM-1	0.9964	1.2341
MSM-2	5.2059	6.2890
$\sqrt{r^o}$	4.1672	4.1672
LLS-EAKF $D = 14$	6.7656	7.0683
LLS-EAKF $D = 25$	6.9412	–

measured by the filtering algorithms, is shown in the upper right panel. The middle panels show the result of filtering by SPEKF (left) and LLS-EAKF (right) while the bottom panels show the filtering results by MSM1 and MSM2. The excellent skill of the SPEKF filter and LLS-EAKF is evident from the small RMS errors and high spatial correlation in Fig. 13.16.

The oceanic case with $F = 40$ is extremely challenging for using the perfect model for filtering because the two-layer equations in (11.6) become stiff with multiple time-scales. Even after extensive parameter tuning in the LLS-EAKF algorithm as discussed in Chapter 11, as reported in Table 13.4, there is very little skill in filtering by the LLS-EAKF algorithm or MSM2 in this regime. On the other hand, both the cheap SPEKF and even cheaper MSM1 filtering algorithms have very high skill in filtering these turbulent signals. These results are confirmed by Fig. 13.17 which gives a snapshot of the stream function as described earlier in Fig. 13.15. In Fig. 13.18, we see that LLS-EAKF performs well during the periods of time 0–10 and 110–135 but poorly in the remaining periods. Here, both SPEKF and MSM1 are significantly accurate (low RMS errors and high spatial correlation). The high filtering skill with MSM and SPEKF1 is also signified by accurate spectra recovery for modes 1–5 with SPEKF and 1–4 for MSM1. For more discussion, see Harlim and Majda (2010b). An important filtering problem in contemporary oceanography is to recover the turbulent heat fluxes from sparse satellite altimetry measurements at the surface. A stochastic super-resolution algorithm which utilizes both a Monte Carlo version

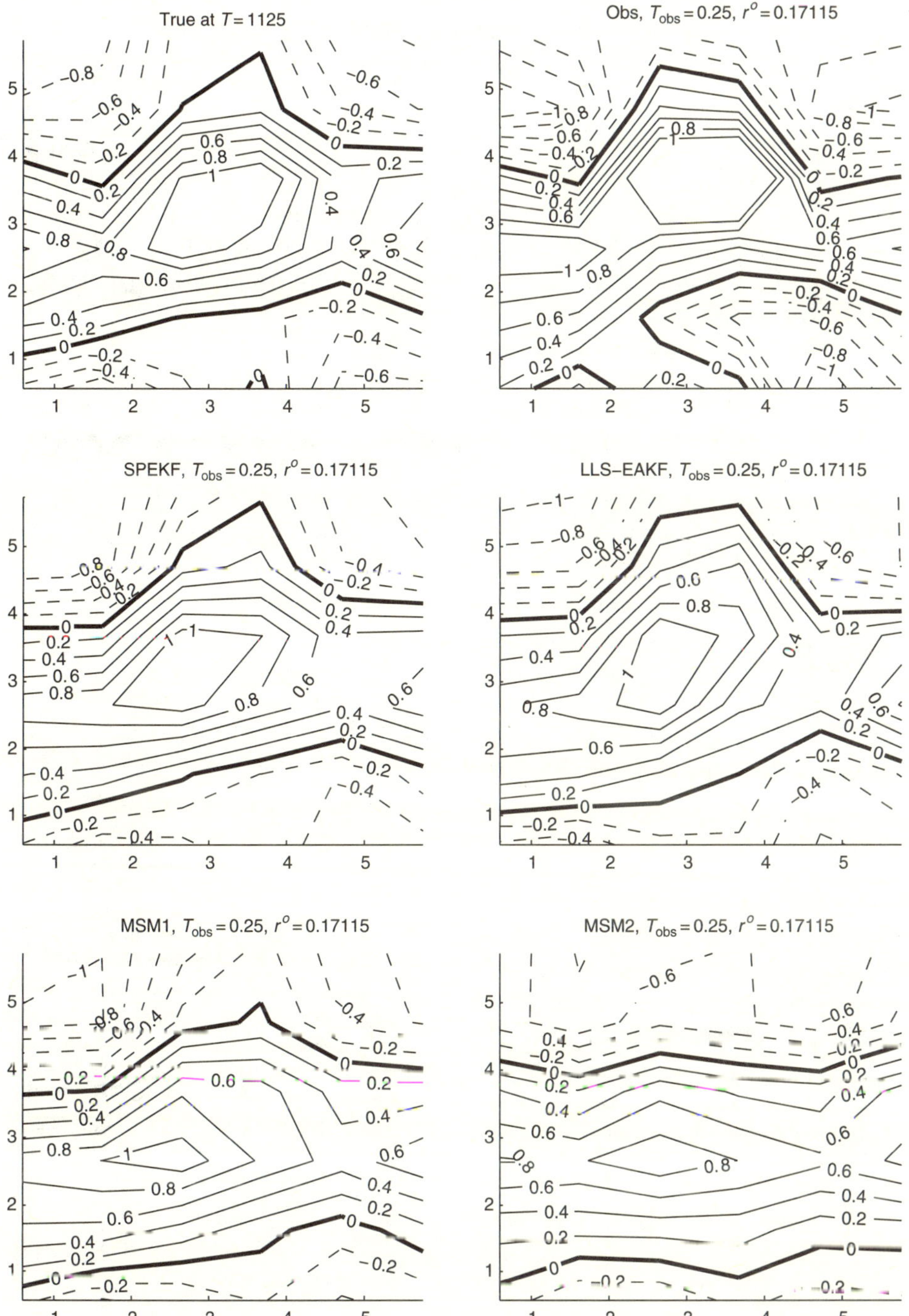

Figure 13.15 Model error case: Snapshots at time 1125 for filtering with $T_{\mathrm{obs}} = 0.25$ and $r^o = 0.1711$.

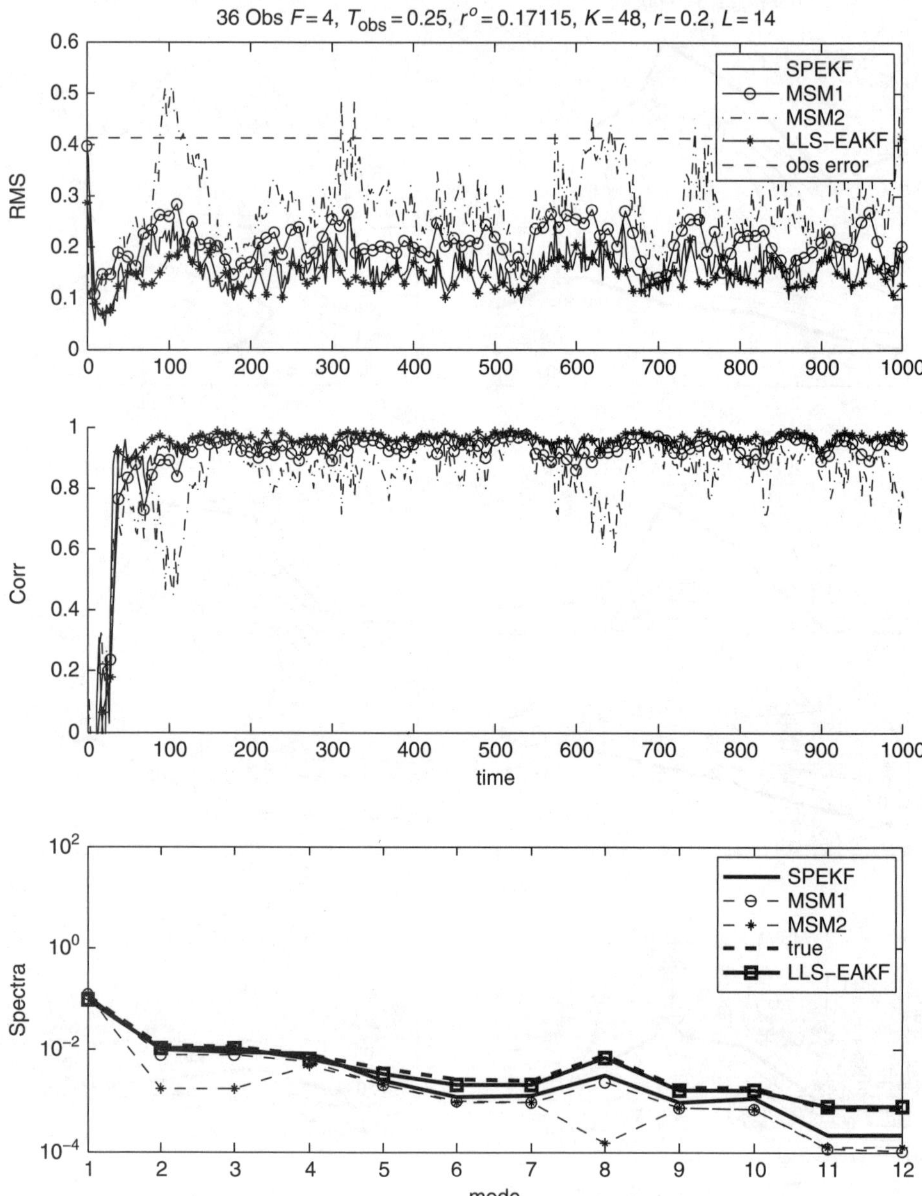

Figure 13.16 Model error case: Filtering skill for $T_{\text{obs}} = 0.25$ and $r^o = 0.1711$.

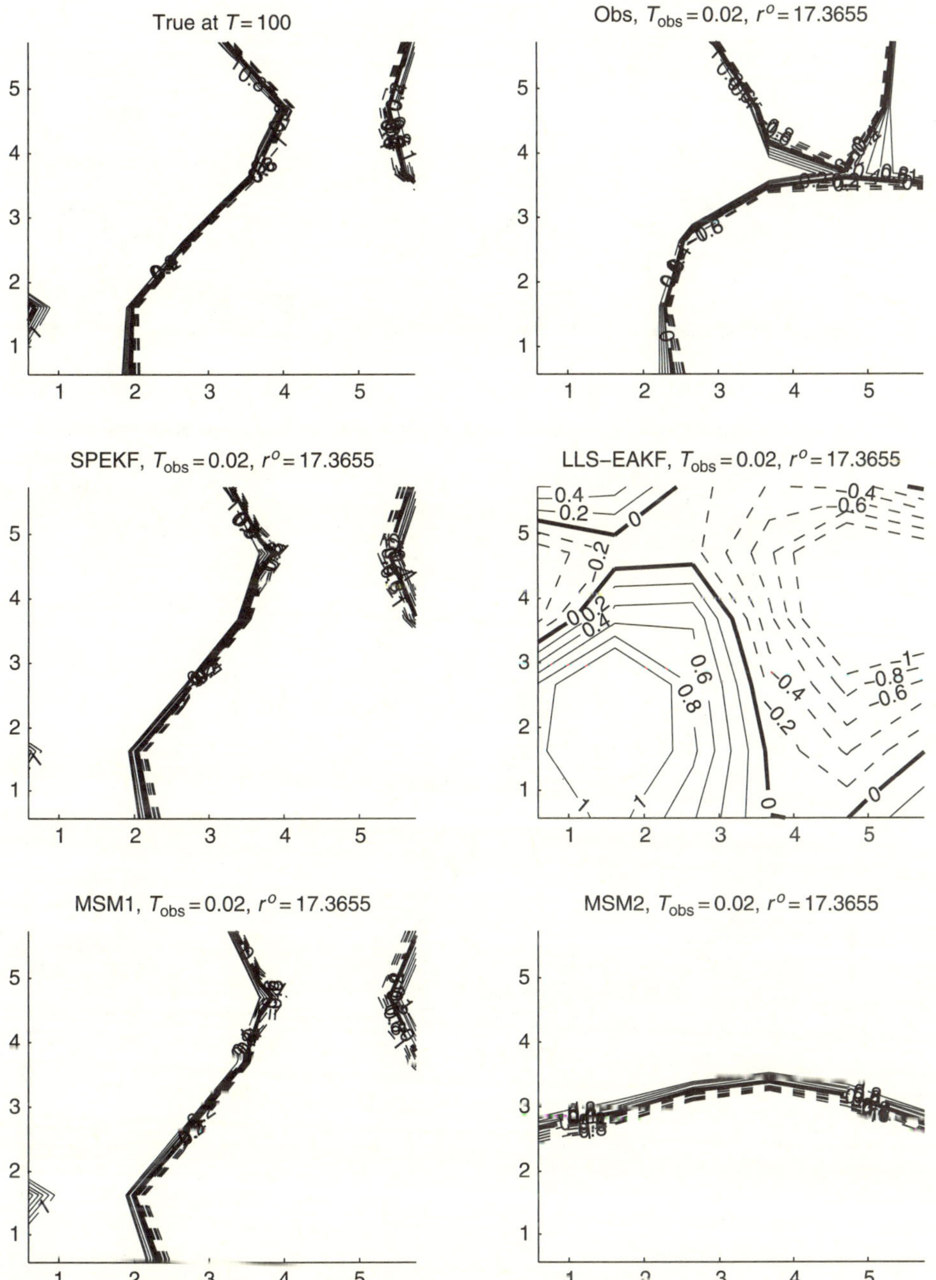

Figure 13.17 Ocean case ($F = 40$): Snapshots at time 100 for filtering with $T_{obs} = 0.02$ and $r^o = 17.36$.

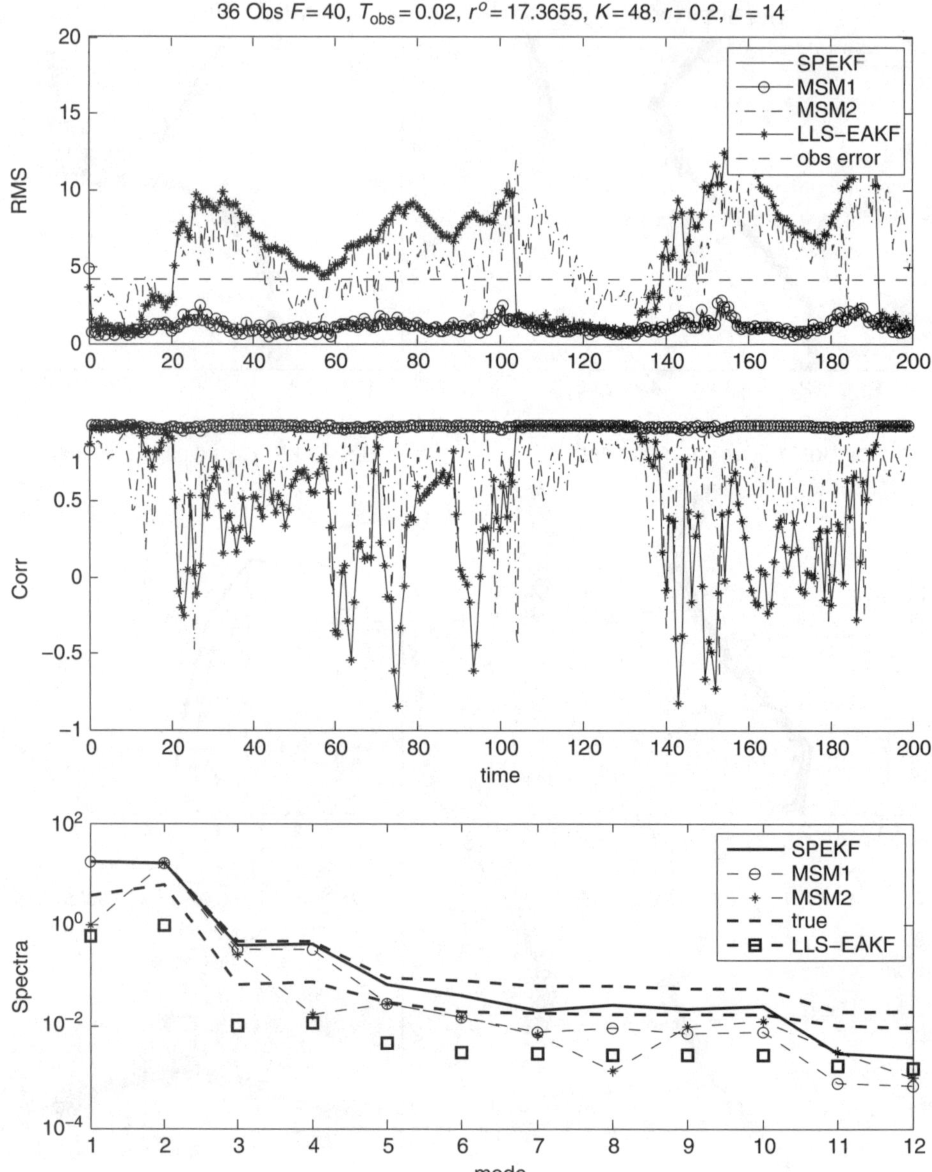

Figure 13.18 Ocean case ($F = 40$): Filtering skill for $T_{\text{obs}} = 0.02$ and $r^o = 17.36$.

of SPEKF and the remarkable fact developed in Chapter 7 where aliasing helps filtering skill yields significant filtering skill in this difficult application (Keating *et al.*, 2011).

Appendix

In this appendix, we compute the second-order statistics of $u(t)$, $\gamma(t)$ and $b(t)$. These analytical formulas are used in the Kalman filter formulation to propagate the posterior covariance from previous analysis to obtain a prior covariance for the next analysis.

A.1. $\mathrm{Var}(u(t))$

We compute the variance using

$$\mathrm{Var}(u(t)) = \langle |u(t)|^2 \rangle - |\langle u(t) \rangle|^2.$$

In terms of the following notation

$$u(t) = A + B + C,$$

where

$$A = e^{-J(t_0,t)+\hat{\lambda}(t-t_0)} u_0, \tag{13.19}$$

$$B = \int_{t_0}^{t} \Big(b(s) + f(s) \Big) e^{-J(s,t)+\hat{\lambda}(t-s)} ds, \tag{13.20}$$

$$C = \sigma \int_{t_0}^{t} e^{-J(s,t)+\hat{\lambda}(t-s)} dW(s), \tag{13.21}$$

we rewrite

$$\langle |u(t)|^2 \rangle = \langle |A^2| \rangle + \langle |B^2| \rangle + \langle |C^2| \rangle + 2\mathrm{Re}[(A^* B)]. \tag{13.22}$$

We find the right-hand side of Eqn (13.22) term by term

$$\langle |A|^2 \rangle = e^{-2\hat{\gamma}(t-t_0)} \Big(|\langle u_0 \rangle|^2 + \mathrm{Var}(u_0) - 4\mathrm{Re}[\langle u_0 \rangle^* \mathrm{Cov}(u_0, J(t_0,t))] + 4|\mathrm{Cov}(u_0, J(t_0,t))|^2 \Big) e^{-2\langle J(t_0,t) \rangle + 2\mathrm{Var}(J(t_0,t))},$$

where we used the following property of Gaussian random variables (Gershgorin and Majda, 2008, 2010)

$$\langle z w e^{bx} \rangle = \Big[\langle z \rangle \langle w \rangle + \mathrm{Cov}(z, w^*) + b\big(\langle z \rangle \mathrm{Cov}(w, x) + \langle w \rangle \mathrm{Cov}(z, x) \big) + b^2 \mathrm{Cov}(z, x) \mathrm{Cov}(w, x) \Big] e^{b\langle x \rangle + \frac{b^2}{2} \mathrm{Var}(x)}, \tag{13.23}$$

for complex Gaussian z and w and real Gaussian x.

Next, we have

$$\langle |B|^2 \rangle = \int_{t_0}^{t} ds \int_{t_0}^{t} dr b_{\mathrm{Var}}(s, r),$$

where

$$\begin{aligned}
b_{\mathrm{Var}}(s,r) = {} & e^{-\hat{\gamma}(2t-s-r)+\mathrm{i}\omega(r-s)} \\
& \times e^{-\langle J(s,t)\rangle - \langle J(r,t)\rangle + \frac{1}{2}\mathrm{Var}(J(s,t)) + \frac{1}{2}\mathrm{Var}(J(r,t)) + \mathrm{Cov}(J(s,t),J(r,t))} \\
& \times \Bigg[\Big(\langle b(s) b^*(r) \rangle - \langle b(s) \rangle \big[\mathrm{Cov}(b^*(r), J(s,t)) \\
& + \mathrm{Cov}(b^*(r), J(r,t)) \big] - \langle b^*(r) \rangle \big[\mathrm{Cov}(b(s), J(s,t)) \\
& + \mathrm{Cov}(b(s), J(r,t)) \big] + \big[\mathrm{Cov}(b^*(r), J(s,t)) \\
& + \mathrm{Cov}(b^*(r), J(r,t)) \big] \big[\mathrm{Cov}(b(s), J(s,t)) \\
& + \mathrm{Cov}(b(s), J(r,t)) \big] \Big) + f^*(r) \Big(\langle b(s) \rangle \\
& - \mathrm{Cov}(b(s), J(s,t) - \mathrm{Cov}(b(s), J(r,t)) \Big) \\
& + f(s) \Big(\langle b(r) \rangle^* - \mathrm{Cov}(b(r), J(s,t)^* \\
& - \mathrm{Cov}(b(r), J(r,t))^* \Big) + f(s) f^*(r) \Bigg],
\end{aligned}$$

with

$$\begin{aligned}
\Big\langle b(s) b(r)^* \Big\rangle = {} & (1 - e^{\lambda_b (s-t_0)})(1 - e^{\lambda_b^*(r-t_0)}) |\hat{b}|^2 \\
& + e^{\lambda_b (s-t_0)} (1 - e^{\lambda_b^*(r-t_0)}) \hat{b}^* \langle b_0 \rangle \\
& + e^{\lambda_b^*(r-t_0)} (1 - e^{\lambda_b (s-t_0)}) \hat{b} \langle b_0 \rangle^* \\
& + e^{\lambda_b (s-t_0)} e^{\lambda_b^*(r-t_0)} (\mathrm{Var}(b_0) + |\langle b_0 \rangle|^2), \\
& + \frac{\sigma_b^2}{2\gamma_b} \Big(e^{-\gamma_b (s+r-2\min(s,r))} - e^{-\gamma_b (s+r-2t_0)} \Big) e^{\mathrm{i}\omega_b (s-r)}, \\
\mathrm{Cov}(b(r), J(s,t)) = {} & \frac{1}{d_\gamma} \Big(e^{-d_\gamma (s-t_0)} - e^{-d_\gamma (t-t_0)} \Big) e^{\lambda_b (r-t_0)} \mathrm{Cov}(b_0, \gamma_0).
\end{aligned}$$

We compute the covariance of $J(s,t)$ and $J(r,t)$ for $t_0 \le r \le s \le t$ in the following way

$$\mathrm{Cov}(J(s,t), J(r,t)) = \mathrm{Var}(J(s,t)) + \mathrm{Cov}(J(s,t), J(r,s)),$$

where

$$\begin{aligned}
\mathrm{Cov}(J(s,t), J(r,s)) = {} & \frac{\mathrm{Var}(\gamma_0)}{d_\gamma^2} \Big(e^{-d_\gamma (t-t_0)} - e^{-d_\gamma (s-t_0)} \Big) \\
& \times \Big(e^{-d_\gamma (s-t_0)} - e^{-d_\gamma (r-t_0)} \Big) \\
& - \frac{\sigma_\gamma^2}{2 d_\gamma^3} \Big(e^{-d_\gamma (t-s)} - e^{-d_\gamma (t-r)} + e^{-d_\gamma (t+s-2t_0)} - e^{-d_\gamma (t+r-2t_0)} - 1 \\
& + e^{-d_\gamma (s-r)} - e^{-2d_\gamma (s-t_0)} + e^{-d_\gamma (s+r-2t_0)} \Big). \qquad (13.24)
\end{aligned}$$

In order to find this covariance for $t_0 \le s \le r \le t$, we use the fact that $\mathrm{Cov}(J(s,t), J(r,t)) = \mathrm{Cov}(J(r,t), J(s,t))$ and follow the same procedure but for s and r switched with each other. Next, we find

$$\begin{aligned}\langle |C|^2\rangle &= \sigma^2 \int_{t_0}^{t} e^{-2\hat{\gamma}(t-s)} \Big\langle e^{-2J(s,t)} \Big\rangle ds \\ &= \sigma^2 \int_{t_0}^{t} e^{-2\hat{\gamma}(t-s)} e^{-2\langle J(s,t)\rangle + 2\mathrm{Var}(J(s,t))} ds.\end{aligned}$$

Finally, we find

$$\begin{aligned}\langle A^* B\rangle &= e^{-\hat{\gamma}(t-t_0)} \int_{t_0}^{t} e^{(\lambda_b - \mathrm{i}\omega)(s-t_0) - \hat{\gamma}(t-s)} \Big\langle u_0^* b_0 e^{-J(t_0,t)-J(s,t)} \Big\rangle ds \\ &\quad + e^{-\hat{\gamma}(t-t_0)} \int_{t_0}^{t} e^{\mathrm{i}\omega(t_0-s) - \hat{\gamma}(t-s)} \\ &\quad \Big(\hat{b}\big(1 - e^{\lambda_b(s-t_0)}\big) + f(s)\Big) \Big\langle u_0 e^{-J(t_0,t)-J(s,t)} \Big\rangle^* ds.\end{aligned}$$

where

$$\begin{aligned}\Big\langle u_0^* b_0 e^{-J(t_0,t)-J(s,t)} \Big\rangle &= \Big(\mathrm{Cov}(u_0^*, b_0^*) + \langle u_0\rangle^* \langle b_0\rangle \\ &\quad - \langle u_0\rangle^* \Big[\mathrm{Cov}(b_0, J(t_0,t)) + \mathrm{Cov}(b_0, J(s,t))\Big] \\ &\quad - \langle b_0\rangle \Big[\mathrm{Cov}(u_0, J(t_0,t)) + \mathrm{Cov}(u_0, J(s,t))\Big]^* \\ &\quad + \Big[\mathrm{Cov}(b_0, J(t_0,t)) + \mathrm{Cov}(b_0, J(s,t)\Big] \\ &\quad \times \Big[\mathrm{Cov}(u_0, J(t_0,t)) + \mathrm{Cov}(u_0, J(s,t))\Big]^* \Big) \\ &\quad \times e^{-\langle J(t_0,t)\rangle - \langle J(s,t)\rangle + \frac{1}{2}\mathrm{Var}(J(t_0,t)) + \frac{1}{2}\mathrm{Var}(J(s,t)) + \mathrm{Cov}(J(t_0,t) J(s,t))} \\ \Big\langle u_0 e^{-J(t_0,t)-J(s,t)} \Big\rangle &= \Big(\langle u_0\rangle - \mathrm{Cov}(u_0, J(t_0,t)) - \mathrm{Cov}(u_0, J(s,t)) \\ &\quad \times e^{\langle J(t_0,t)\rangle - \langle J(s,t)\rangle + \frac{1}{2}\mathrm{Var}(J(t_0,t)) + \frac{1}{2}\mathrm{Var}(J(s,t)) + \mathrm{Cov}(J(t_0,t) J(s,t))}.\end{aligned}$$

A.2. $\mathrm{Cov}(u(t), u^*(t))$

We use the definition of the covariance to find

$$\mathrm{Cov}(u(t), u^*(t)) = \langle u(t)^2\rangle - \langle u(t)\rangle^2.$$

We find that

$$\langle u(t)^2\rangle = \langle A^2\rangle + \langle B^2\rangle + 2\langle AB\rangle, \tag{13.25}$$

where we have used the independence of $W(t)$ of other random variables. We find the right-hand side of Eqn (13.25) term by term:

$$\langle A^2\rangle = e^{2\hat{\lambda}(t-t_0)}\Big(\langle u_0\rangle^2 + \mathrm{Cov}(u_0, u_0^*) - 4\langle u_0\rangle \mathrm{Cov}(u_0, J(t_0,t))$$
$$+4\mathrm{Cov}(u_0, J(t_0,t))^2\Big)e^{-2\langle J(t_0,t)\rangle + 2\mathrm{Var}(J(t_0,t))}.$$

Next, we have

$$\langle B^2\rangle = \int_{t_0}^{t} ds \int_{t_0}^{t} dr b_{\mathrm{Cov}}(s,r),$$

where

$$b_{\mathrm{Cov}}(s,r) = e^{\hat{\lambda}(2t-s-r)}e^{-\langle J(s,t)\rangle - \langle J(r,t\rangle + \frac{1}{2}\mathrm{Var}(J(s,t)) + \frac{1}{2}\mathrm{Var}(J(r,t)) + \mathrm{Cov}(J(s,t),J(r,t))}$$
$$\times\Bigg[\Big(\langle b(s)b(r)\rangle - \langle b(s)\rangle$$
$$\times\big[\mathrm{Cov}(b(r), J(s,t)) + \mathrm{Cov}(b(r), J(r,t))\big]$$
$$-\langle b(r)\rangle\big[\mathrm{Cov}(b(s), J(s,t)) + \mathrm{Cov}(b(s), J(r,t))\big]$$
$$+\big[\mathrm{Cov}(b(r), J(s,t)) + \mathrm{Cov}(b(r), J(r,t))\big]$$
$$\times\big[\mathrm{Cov}(b(s), J(s,t)) + \mathrm{Cov}(b(s), J(r,t))\big]\Big)$$
$$+ f(r)\Big(\langle b(s)\rangle - \mathrm{Cov}(b(s), J(s,t) - \mathrm{Cov}(b(s), J(r,t))\Big)$$
$$+ f(s)\Big(\langle b(r)\rangle - \mathrm{Cov}(b(r), J(s,t) - \mathrm{Cov}(b(r), J(r,t))\Big)$$
$$+ f(s)f(r)\Bigg],$$

with

$$\Big\langle b(s)b(r)\Big\rangle = (1-e^{\lambda_b(s-t_0)})(1-e^{\lambda_b(r-t_0)})\hat{b}^2 + e^{\lambda_b(s-t_0)}(1-e^{\lambda_b(r-t_0)})\hat{b}\langle b_0\rangle$$
$$+ e^{\lambda_b(r-t_0)}(1-e^{\lambda_b(s-t_0)})\hat{b}\langle b_0\rangle + e^{\lambda_b(s-t_0)}e^{\lambda_b(r-t_0)}$$
$$\times(\mathrm{Var}(b_0) + |\langle b_0\rangle|^2).$$

Finally, we find

$$\langle AB\rangle = \int_{t_0}^{t} e^{\hat{\lambda}(2t-s-t_0)+\lambda_b(s-t_0)}\Big\langle u_0 b_0 e^{-J(t_0,t)-J(s,t)}\Big\rangle ds$$
$$+\int_{t_0}^{t} e^{\hat{\lambda}(2t-s-t_0)}\Big(\hat{b}(1-e^{\lambda_b(s-t_0)}) + f(s)\Big)\Big\langle u_0 e^{-J(t_0,t)-J(s,t)}\Big\rangle ds.$$

where

$$\left\langle u_0 b_0 e^{-J(t_0,t)-J(s,t)} \right\rangle = \Big(\mathrm{Cov}(u_0, b_0^*) + \langle u_0 \rangle \langle b_0 \rangle$$
$$- \langle u_0 \rangle \Big[\mathrm{Cov}(b_0, J(t_0,t)) + \mathrm{Cov}(b_0, J(s,t)) \Big]$$
$$- \langle b_0 \rangle \Big[\mathrm{Cov}(u_0, J(t_0,t)) + \mathrm{Cov}(u_0, J(s,t)) \Big]$$
$$+ \Big[\mathrm{Cov}(b_0, J(t_0,t)) + \mathrm{Cov}(b_0, J(s,t) \Big]$$
$$\times \Big[\mathrm{Cov}(u_0, J(t_0,t)) + \mathrm{Cov}(u_0, J(s,t)) \Big] \Big)$$
$$\times e^{-\langle J(t_0,t)\rangle - \langle J(s,t)\rangle + \frac{1}{2}\mathrm{Var}(J(t_0,t)) + \frac{1}{2}\mathrm{Var}(J(s,t)) + \mathrm{Cov}(J(t_0,t)J(s,t))},$$

A.3. $\mathrm{Cov}(u(t), \gamma(t))$

The covariance of $u(t)$ and $\gamma(t)$ is found as follows

$$\mathrm{Cov}(u(t), \gamma(t)) = \langle u(t)\gamma(t) \rangle - \langle u(t) \rangle \langle \gamma(t) \rangle = \langle u(t)(\gamma(t) - \hat{\gamma}) \rangle$$
$$+ \langle u(t) \rangle (\hat{\gamma} - \langle \gamma(t) \rangle).$$

We compute the first term

$$\langle u(t)(\gamma(t) - \hat{\gamma}) \rangle = e^{\hat{\lambda}(t-t_0)} \left\langle u_0 e^{-J(t_0,t)} (\gamma(t) - \hat{\gamma}) \right\rangle$$
$$+ \int_{t_0}^{t} e^{\hat{\lambda}(t-s)} \left\langle (b(s) + f(s)) e^{-J(s,t)} (\gamma(t) - \hat{\gamma}) \right\rangle ds$$
$$= -e^{\hat{\lambda}(t-t_0)} \frac{\partial}{\partial t} \left\langle u_0 e^{-J(t_0,t)} \right\rangle$$
$$- \int_{t_0}^{t} e^{\hat{\lambda}(t-s)} \frac{\partial}{\partial t} \left\langle (b(s) + f(s)) e^{-J(s,t)} \right\rangle ds,$$

where

$$\frac{\partial}{\partial t} \left\langle u_0 e^{-J(t_0,t)} \right\rangle = \Big[-\mathrm{Cov}(u_0, \gamma(t)) + \Big(\langle u_0 \rangle - \mathrm{Cov}(u_0, J(t_0,t)) \Big)$$
$$\times \Big(\hat{\gamma} - \langle \gamma(t) \rangle + \frac{1}{2} \frac{\partial}{\partial t} \mathrm{Var}(J(t_0,t)) \Big) \Big]$$
$$\times e^{-\langle J(t_0,t)\rangle + \frac{1}{2}\mathrm{Var}(J(t_0,t))},$$

and

$$\frac{\partial}{\partial t} \left\langle (b(s) + f(s)) e^{-J(s,t)} \right\rangle = \Big[-\mathrm{Cov}(b(s), \gamma(t)) + \Big(\langle b(s) \rangle + f(s) - \mathrm{Cov}(b(s), J(s,t)) \Big)$$
$$\times \Big(\hat{\gamma} - \langle \gamma(t) \rangle + \frac{1}{2} \frac{\partial}{\partial t} \mathrm{Var}(J(s,t)) \Big) \Big] e^{-\langle J(s,t)\rangle + \frac{1}{2}\mathrm{Var}(J(s,t))}.$$

The derivative of $\mathrm{Var}(J(s,t))$ with respect to t has the following form

$$\frac{\partial}{\partial t}\mathrm{Var}(J(s,t) = -\frac{1}{d_\gamma^2}\Big(\sigma_\gamma^2\big(e^{-d_\gamma(t-s)}-1\big)+\big(\sigma_\gamma^2-2d_\gamma\mathrm{Var}(\gamma_0)\big)\\ \times\big(e^{-d_\gamma(t+s-2t_0)}-e^{-2d_\gamma(t-t_0)}\big)\Big).$$

A.4. Cov$(u(t), b(t))$

We have

$$\mathrm{Cov}(u(t), b(t)) = \langle u(t)b^*(t)\rangle - \langle u(t)\rangle\langle b(t)\rangle^*.$$

We find

$$\langle u(t)b^*(t)\rangle = \langle u(t)\rangle\hat{b}^*\big(1-e^{\lambda_b^*(t-t_0)}\big)+e^{(\hat{\lambda}+\lambda_b^*)(t-t_0)}\Big\langle u_0b_0^*e^{-J(t_0,t)}\Big\rangle\\ +e^{\lambda_b^*(t-t_0)}\int_{t_0}^t e^{\hat{\lambda}(t-s)}\Big\langle b_0^*b(s)e^{-J(s,t)}\Big\rangle ds+e^{\lambda_b^*(t-t_0)}\\ \times\int_{t_0}^t e^{\hat{\lambda}(t-s)}f(s)\Big\langle b_0e^{-J(s,t)}\Big\rangle^* ds\\ +\frac{\sigma_b^2}{2\gamma_b}\int_{t_0}^t e^{-\langle J(s,t)\rangle+\frac{1}{2}Var(J(s,t))}e^{(-\hat{\gamma}+i(\omega-\omega_b))(t-s)}\\ \times\big[e^{-\gamma_b(t-s)}-e^{-\gamma_b(s+t-2t_0)}\big]ds,$$

where

$$\Big\langle b_0e^{-J(s,t)}\Big\rangle = \Big(\langle b_0\rangle-\mathrm{Cov}(b_0, J(s,t))\Big)e^{-\langle J(s,t)\rangle+\frac{1}{2}\mathrm{Var}(J(s,t))}$$

$$\Big\langle u_0b_0^*e^{-J(t_0,t)}\Big\rangle = \Big(\mathrm{Cov}(u_0,b_0)+\langle u_0\rangle\langle b_0\rangle^*-\langle b_0\rangle^*\mathrm{Cov}(u_0,J(t_0,t))\\ -\langle u_0\rangle\mathrm{Cov}(b_0,J(t_0,t))^*+\mathrm{Cov}(u_0,J(t_0,t)\mathrm{Cov}(b_0,J(t_0,t))^*\Big)\\ \times e^{-\langle J(t_0,t)\rangle+\frac{1}{2}\mathrm{Var}(J(t_0,t))}$$

$$\Big\langle b_0^*b(s)e^{-J(s,t)}\Big\rangle = \Big(e^{\lambda_b(s-t_0)}\mathrm{Var}(b_0)+\langle b(s)\rangle\langle b_0\rangle^*-\langle b_0\rangle^*\mathrm{Cov}(b(s),J(s,t))\\ -\langle b(s)\rangle\mathrm{Cov}(b_0,J(s,t))^*+\mathrm{Cov}(b(s),J(s,t)\mathrm{Cov}(b_0,J(s,t))^*\Big)\\ \times e^{-\langle J(s,t)\rangle+\frac{1}{2}\mathrm{Var}(J(s,t))}$$

A.5. Cov$(u(t), b^*(t))$

We have

$$\mathrm{Cov}(u(t), b^*(t)) = \langle u(t)b(t)\rangle - \langle u(t)\rangle\langle b(t)\rangle.$$

We find

$$
\begin{aligned}
\langle u(t)b(t)\rangle = \langle u(t)\rangle\hat{b}\big(1 - e^{\lambda_b(t-t_0)}\big) + e^{(\hat{\lambda}+\lambda_b)(t-t_0)}\Big\langle u_0 b_0 e^{-J(t_0,t)}\Big\rangle \\
+e^{\lambda_b(t-t_0)}\int_{t_0}^{t} e^{\hat{\lambda}(t-s)}\Big\langle b_0 b(s) e^{-J(s,t)}\Big\rangle ds \\
+e^{\lambda_b(t-t_0)}\int_{t_0}^{t} e^{\hat{\lambda}(t-s)} f(s)\Big\langle b_0 e^{-J(s,t)}\Big\rangle ds,
\end{aligned}
$$

where

$$
\begin{aligned}
\Big\langle u_0 b_0 e^{-J(t_0,t)}\Big\rangle = \Big(&\mathrm{Cov}(u_0, b_0^*) + \langle u_0\rangle\langle b_0\rangle - \langle b_0\rangle\mathrm{Cov}(u_0, J(t_0,t)) \\
&-\langle u_0\rangle\mathrm{Cov}(b_0, J(t_0,t)) + \mathrm{Cov}(u_0, J(t_0,t)\mathrm{Cov}(b_0, J(t_0,t))\Big) \\
&\times e^{-\langle J(t_0,t)\rangle + \frac{1}{2}\mathrm{Var}(J(t_0,t))}
\end{aligned}
$$

$$
\begin{aligned}
\Big\langle b_0 b(s) e^{-J(s,t)}\Big\rangle = \Big(&e^{\lambda_b(s-t_0)}\mathrm{Cov}(b_0, b_0^*) + \langle b(s)\rangle\langle b_0\rangle - \langle b_0\rangle\mathrm{Cov}(b(s), J(s,t)) \\
&-\langle b(s)\rangle\mathrm{Cov}(b_0, J(s,t)) + \mathrm{Cov}(b(s), J(s,t)\mathrm{Cov}(b_0, J(s,t)\Big) \\
&\times e^{-\langle J(s,t)\rangle + \frac{1}{2}\mathrm{Var}(J(s,t))}.
\end{aligned}
$$

14

Filtering turbulent tracers from partial observations: An exactly solvable test model

Turbulent diffusion is a physical process that describes the transport of a tracer in a turbulent velocity field. Very often the tracer itself has very little or no influence on the background flow, in which case it is referred to as a passive tracer. Practically important examples include engineering problems such as the spread of hazardous plumes or pollutants in the atmosphere and contaminants in the ocean. Another class of problems that involve turbulent diffusion are climate science problems concerning the transport of greenhouse gases such as carbon dioxide and others. One of the characteristics of these systems is their complex multi-scale structure in both time and space. For example, the spatial scales of atmospheric flows span from planetary-scale Rossby waves to local weather patterns with the size of kilometers. Similarly, temporal scales involve both slow dynamics on the scales of decades as well as the fast dynamics on the scales of hours. Another remarkable property of many tracers in the atmosphere is their highly intermittent probability distributions with long exponential tails (Neelin *et al.*, 2011). Many contemporary applications in science and engineering involve real-time filtering of such turbulent non-Gaussian signals from nature with multiple scales in time and space.

Real-time tracking of a chemical plume released into the atmosphere or a contaminant injected into the ocean is another extremely important and practical example where real-time data assimilation plays a crucial role. As discussed throughout this book, major difficulties in accurate filtering and prediction of noisy turbulent spatially extended signals are typically caused by imperfect partial and sparse observations and model error due to inadequate numerical resolution or incomplete physical understanding. One way to evaluate the consequences of such complications and to build accurate and efficient filtering algorithms is to design a mathematically stringent test model that has the features of the realistic physical system but, on the other hand, allows for exact statistical solution. Such a model would not only help us to understand the underlying physics but could also be used for studying the role of observations and model error such as the use of eddy diffusivity models in data assimilation. We described and developed such a model for slow–fast systems in detail earlier in Chapter 10.

In this chapter, we introduce a test model for filtering turbulent tracers which has direct relevance to actual physical systems with the additional attractive feature of exactly solvable statistics for the mean and covariance despite the inherent statistical nonlinearity as we have done earlier in Chapters 10 and 13. The model for the velocity field is incompressible two-dimensional with a time-dependent cross-sweep, $U(t)$, in one direction and time and spatially dependent shear flow, $v(x,t)$, in the transverse direction. In the atmosphere it would correspond to the zonal jet (cross-sweep in the east–west direction) and the transverse Rossby waves (shear flow in the north–south direction), however different interpretations would arise in other physical contexts. We model the velocity field by a combination of the prescribed deterministic and Gaussian stochastic components. Such a simplified representation of the velocity field exhibits some empirical features of turbulent flows which are given by the Navier–Stokes equation, however, they are more manageable mathematically (Majda and Kramer, 1999). Transport of a passive tracer, T, by the fluid flow with velocity $\vec{v} = (U, v)^T$ can be modeled by the advection–diffusion equation (Majda and Kramer, 1999; Bourlioux and Majda, 2002).

$$\frac{\partial T}{\partial t} + \vec{v} \cdot \vec{\nabla}_{\vec{x}} T = \kappa \Delta_{\vec{x}} T - \gamma T + S(\vec{x}, t), \tag{14.1}$$

where κ is the molecular or eddy diffusivity, γ is uniform damping, and S is a source for the tracer. Typically the statistical solution for (14.1) cannot be obtained in closed form because of the nonlinear advection term $\vec{v} \cdot \vec{\nabla}_{\vec{x}} T$. This is a manifestation of the "turbulence moment closure problem". However, in certain simplified situations an exact analytical treatment is possible. For example, in this chapter we consider a one-dimensional version of the system when the tracer has a prescribed mean gradient in one direction and the one-dimensional fluctuations of the tracer around this mean gradient. The atmospheric analogy would include the north–south mean gradient of a tracer gas and zonal fluctuations restricted to a narrow band in the mid-latitudes. Despite all the simplifications, this test model carries such important characteristics of a realistic physical system as the turbulent energy cascade of the tracer and strongly intermittent probability distributions for the tracer with long exponential tails that are observed for tracers in the real atmosphere (Neelin *et al.*, 2011; Bourlioux and Majda, 2002).

As in Chapters 10 and 13 (Gershgorin and Majda, 2008, 2010), a nonlinear extended Kalman filter (NEKF) is designed for this model utilizing exact statistics. We test the performance of the NEKF on system (14.1) in various dynamical regimes. In particular, we are interested in the case of non-dispersive waves, which are often relevant in engineering applications, and dispersive waves such as Rossby waves that mimic the turbulent field in the atmosphere. We study how the type and sparseness of observations impact the filtering skill and how one can improve the performance of the NEKF by adding just one observation on the spatially extended grid. We find that even in very tough regimes when only few observations are available, the NEKF has excellent skill in recovering the spatial structure of the original truth signal as well as its spectral properties and the fat tailed tracer probability distribution. We also advocate here that the model in (14.1) has a lot of potential in studying the subtle issues of model error due to eddy diffusivity parametrization and other approximations.

14.1 Model description

We consider a model for a passive tracer $T(\vec{x}, t)$ which is advected by a velocity field $\vec{v}(\vec{x}, t)$. In general, the dynamics of such a tracer can be described by the following advection–diffusion equation (Majda and Kramer, 1999)

$$\frac{\partial T}{\partial t} + \vec{v} \cdot \vec{\nabla}_{\vec{x}} T = \kappa \Delta_{\vec{x}} T - \gamma T + S(\vec{x}, t), \tag{14.2}$$

where κ is the molecular diffusivity, γ represents the uniform damping, $S(\vec{x}, t)$ is an external source for the tracer, and the velocity field is considered to be incompressible, $\nabla_{\vec{x}} \vec{v} = 0$. We assume that the model for the velocity field is two dimensional and periodic in space

$$\vec{v}(\vec{x}, t) = (U(t), v(x, t))^T, \tag{14.3}$$

which automatically satisfies the incompressibility condition. Moreover, each component, the cross-sweep, $U(t)$, and the shear flow, $v(x, t)$, are Gaussian fields with prescribed dynamics. Suppose that the tracer has a known mean gradient

$$T(\vec{x}, t) = \alpha_1 x + \alpha_2 y + T'(x, y, t),$$

where $T'(x, y, t)$ denotes fluctuations around the mean gradient. Then, from (14.2), we find an equation for $T'(x, y, t)$:

$$\begin{aligned} &\frac{\partial T'(x, y, t)}{\partial t} + U(t) \frac{\partial T'(x, y, t)}{\partial x} + v(x, t) \frac{\partial T'(x, y, t)}{\partial y} \\ &= \kappa \left(\frac{\partial^2 T'(x, y, t)}{\partial x^2} + \frac{\partial^2 T'(x, y, t)}{\partial y^2} \right) - \gamma(t) T'(x, y, t) \\ &-\alpha_1 U(t) - \alpha_2 v(x, t) + S(x, y, t). \end{aligned} \tag{14.4}$$

It often appears in relevant physical systems that the fluctuations T' of the tracer only depend on the x spatial variable. In this situation (14.4) simplifies to a one-dimensional PDE

$$\frac{\partial T'(x, t)}{\partial t} + U(t) \frac{\partial T'(x, t)}{\partial x} = -\alpha v(x, t) + \kappa \frac{\partial^2 T'(x, t)}{\partial x^2} - \gamma T'(x, t), \tag{14.5}$$

where for simplicity we also assumed that $\alpha_1 \equiv 0$, $S(x, y, t) \equiv 0$ and $\alpha \equiv \alpha_2$. In the remainder of the chapter, we drop the prime superscript in (14.5). For a given velocity field, (14.5) is solved using Fourier series as utilized throughout this book. Moreover, as we also show below the statistics of the system $(U(t), v(x, t), T(x, t))^T$ can be computed exactly for a Gaussian velocity field despite the nonlinearity from the advection term, $U(t) \frac{\partial T'(x,t)}{\partial x}$. This remarkable feature makes the model extremely attractive for filtering via NEKF, which uses exact statistics to predict the next state of the system. Next, we show how to compute the first- and second-order statistics of this one-dimensional model for a general linear Gaussian velocity field, followed by a specific example that has direct relevance to the atmospheric flows.

14.2 System statistics

14.2.1 Tracer statistics for a general linear Gaussian velocity field

Here we show how to find the statistics of the tracer for a general Gaussian velocity field (14.3). It is convenient to rewrite (14.5) in Fourier space

$$\frac{\partial T_k(t)}{\partial t} + \mathrm{i}kU(t)T_k(t) = -\kappa k^2 T_k(t) - \gamma T_k(t) - \alpha v_k(t), \qquad T_{-k} = T_k^*, \quad \text{for } k \in [0, K] \tag{14.6}$$

where the subscript k denotes the kth Fourier mode of the corresponding variable. Moreover, unless stated otherwise, we will refer to the fluctuations of the tracer around the mean gradient as the "tracer". From (14.6), we can find that the tracer $T_k(t)$ is nonlinearly dependent on the Gaussian cross-sweep, $U(t)$. Therefore, the statistics of the tracer are non-Gaussian. Nevertheless, the special structure of the governing equation in (14.6) allows for exact analytical formulas for the first and second (and, in principle, any order) moments. Each Fourier mode $T_k(t)$ acts as a nonlinear forced-damped oscillator with a random frequency. Thus, while the physics is different here, the mathematical analysis is similar to that in Chapter 10. The path-wise solution of (14.6) becomes

$$T_k(t) = D_k(t_0, t)T_k(t_0)e^{-\mathrm{i}kJ_U(t_0,t)} - \alpha \int_{t_0}^{t} D_k(s, t)v_k(s)e^{-\mathrm{i}kJ_U(s,t)}ds, \tag{14.7}$$

where

$$D_k(s, t) = e^{-(\gamma+\kappa k^2)(t-s)}, \qquad J_U(s, t) = \int_s^t U(s')ds'. \tag{14.8}$$

Note that $D_k(s, t)$ is deterministic while $J_U(s, t)$ is Gaussian. We also assume a Gaussian initial condition $T_k(t_0)$ for the tracer. Then, we compute the mean $\langle T_k(t)\rangle$

$$\langle T_k(t)\rangle = D_k(t_0, t)\langle T_k(t_0)e^{-\mathrm{i}kJ_U(t_0,t)}\rangle - \alpha \int_{t_0}^{t} D_k(s, t)\langle v_k(s)e^{-\mathrm{i}kJ_U(s,t)}\rangle ds. \tag{14.9}$$

Using properties of the characteristic function of a Gaussian random variable, we obtain the following equality for any complex Gaussian z and any real Gaussian x (Gardiner, 1997; Bensoussan, 2004; Gershgorin and Majda, 2008, 2010)

$$\langle ze^{\mathrm{i}x}\rangle = \Big(\langle z\rangle + \mathrm{i}\mathrm{Cov}(z, x)\Big)e^{\mathrm{i}\langle x\rangle - \frac{1}{2}\mathrm{Var}(x)}. \tag{14.10}$$

Applying (14.10) to (14.9), we find

$$\begin{aligned}\langle T_k(t)\rangle &= D_k(t_0,t)\Big(\langle T_k(t_0)\rangle - ik\mathrm{Cov}(T_k(t_0), J_U(t_0,t))\Big)\\&\times e^{-ik\langle J_U(t_0,t)\rangle - \frac{k^2}{2}\mathrm{Var}(J_U(t_0,t))}\\&-\alpha\int_{t_0}^{t} D_k(s,t)\Big(\langle v_k(s)\rangle - ik\mathrm{Cov}(v_k(s), J_U(s,t))\Big)\\&\times e^{-ik\langle J_U(s,t)\rangle - \frac{k^2}{2}\mathrm{Var}(J_U(s,t))}ds.\end{aligned} \tag{14.11}$$

Note that the right-hand side of (14.11) only depends on the statistics of the prescribed Gaussian velocity field (14.3) and the initial value of the tracer at time t_0. Moreover, as we will demonstrate on a particular example below, the statistics of the velocity field at any time are also given by their initial values. Therefore, (14.11) gives an exact analytical formula for the mean $\langle T_k(t)\rangle$ as a functional of the Gaussian initial statistics of the velocity field and the tracer $(U(t_0), v_k(t_0), T_k(t_0))^T$. In appendix A of Gershgorin and Majda (2011), we derive similar formulas for the second-order statistics of the tracer. The reader is referred to that paper as well as Majda and Gershgorin (2011) for more details.

14.2.2 A particular choice of the Gaussian velocity field and its statistics

Suppose that the cross-sweep, $U(t)$, is given by the real part of a complex Ornstein–Uhlenbeck (OU) process

$$\begin{aligned}U(t) &= \mathrm{Re}[V(t)],\\ \frac{dV(t)}{dt} &= \lambda_U V(t) + f_U(t) + \sigma_U \dot{W}_U(t),\end{aligned} \tag{14.12}$$

where $\lambda_U = -\gamma_U + \mathrm{i}\omega_U$, $f_U(t)$ is the deterministic forcing which for simplicity is considered to be periodic

$$f_U(t) = A_f e^{\mathrm{i}(\eta t + \phi_f)} + B_f, \tag{14.13}$$

and $\dot{W}_U$ is a complex white noise process as discussed in Chapter 2. Note that by choosing the cross-sweep to be the real part of the complex OU process as opposed to the real OU process, we allow it to have highly oscillatory correlation functions, which provides a very interesting test case with negative effective diffusivity (Majda and Kramer, 1999). Next, the dynamical equation for the velocity component $v(x,t)$ is

$$\frac{dv(x,t)}{dt} = P\Big(\frac{\partial}{\partial x}\Big)v(x,t) + f_v(x,t) + \sigma_v(x)\dot{W}_v, \tag{14.14}$$

which is the simplest model for turbulence which we introduced in Chapter 5. Here again the function $f_v(x,t)$ is a known deterministic function, while $\dot{W}_v$ represents random white noise fluctuations in forcing arising from hidden nonlinear interactions and other processes. In Fourier space, the operator $P\left(\frac{\partial}{\partial x}\right)$ is given by the symbol $\hat{P}_k$ which has the form

$$\hat{P}_k = -\gamma_k + \mathrm{i}\omega_k \equiv \lambda_k.$$

We are looking for the solution of our model that consists of a finite number of Fourier modes $k \in [0, 2K]$ which is equivalent to the discretization of the 2π-periodic spatial domain into $2K+1$ points, $x_j = jh$, with $h = 2\pi/(2K+1)$, $j \in [1, 2K+1]$. Then, each Fourier component $v_k(t)$ is governed by the equation

$$\frac{dv_k(t)}{dt} = \lambda_k v_k(t) + f_k(t) + \sigma_k \dot{W}_k(t) \quad \text{for } k \in [0, K], \tag{14.15}$$

where $f_k(t)$ is the Fourier decomposition of $f_v(x,t)$, and the white noise of each Fourier mode $v_k(t)$ is independent complex white noise. The reality condition is expressed by

$$v_{-k} = v_k^*,$$

where the asterisk "*" denotes complex conjugate. Here, we discuss and compare two prototype models for the waves $v_k(t)$:

- non-dispersive waves with

$$\omega_k = -ck, \quad \gamma_k = d_v + \mu k^2, \tag{14.16}$$

  where d_v represents the uniform part of the dissipation and μ is the strength of the selective part of the dissipation;
- dispersive Rossby waves with

$$\omega_k = \frac{\beta k}{k^2 + F_s}, \quad \gamma_k = \nu(k^2 + F_s), \tag{14.17}$$

  where β represents rotation and F_s represents stratification.

Note that the case of non-dispersive waves may be more interesting for the engineering community while the case of Rossby waves has more implications for atmosphere–ocean science modeling. The strength of the white noise forcing for each mode v_k is given by σ_k which can be found if the energy spectrum E_k is prescribed as discussed earlier in Chapter 5. The energy of a wave $v_k(t)$ is given by its equilibrium variance

$$E_k \equiv \mathrm{Var}_{\mathrm{eq}}(v_k) = \frac{\sigma_k^2}{2\gamma_k}.$$

Therefore, we find

$$\sigma_k = \sqrt{2\gamma_k E_k}. \tag{14.18}$$

Here, we consider the energy spectrum for v_k with the following form

$$E_k = \begin{cases} E, & |k| \le k_0, \\ E\left(\frac{k}{k_0}\right)^{-5/3}, & |k| > k_0, \end{cases} \tag{14.19}$$

where we chose the equipartition of the energy spectrum to mimic energetic large-scale waves and a Kolmogorov spectrum for smaller scales. In Fig. 14.1(a), we illustrate such an

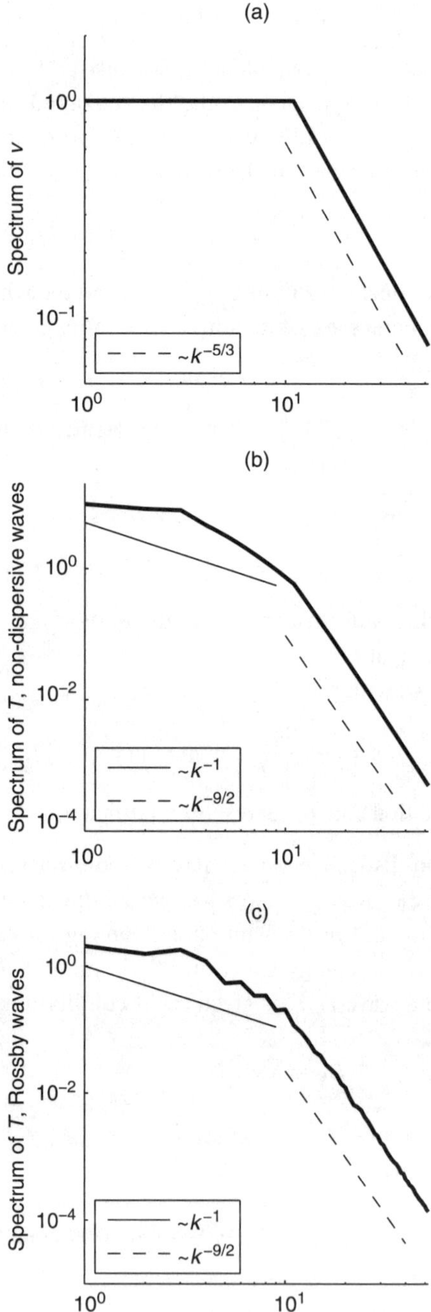

Figure 14.1 Panel (a): spectrum of the waves, v_k, described by (14.19) with $k_0 = 10$; panel (b): spectrum of the tracer, T_k, for the case of non-dispersive waves; panel (c): spectrum of the tracer, T_k, for the case of Rossby waves.

energy spectrum for v_k. Note that we can as well choose any other form of a spectrum for the waves $v_k(t)$ since it is an input parameter in our model and this flexibility is useful for other applications. For the external forcing $f_k(t)$, we choose an oscillating function

$$f_k(t) = A_k e^{\mathrm{i}(\xi_k t+\phi_k)}. \tag{14.20}$$

Now, we demonstrate the statistical solution for the specific Gaussian velocity field in (14.12) and (14.14). The velocity field is given by a Gaussian random processes (14.12) which is uniquely defined by its mean and covariance. Note that these statistics together with the first- and second-order statistics of the tracer given by (14.11) will be used below to build the NEKF for this system. Therefore, we will assume that all three components of the system $(U(t), v_k(t), T_k(t))^T$ are correlated at some initial time t_0 which always happens after a data assimilation time step, unlike in the statistically steady state when some of the cross-covariances vanish. We start with the cross-sweep $U(t)$. From (14.12), we find

$$\begin{aligned} V(t) &= e^{\lambda_U(t-t_0)}V(t_0) + \int_{t_0}^{t} e^{\lambda_U(t-s)} f_U(s)ds + \sigma_U \int_{t_0}^{t} e^{\lambda_U(t-s)} dW_U(s), \\ U(t) &= \mathrm{Re}[V(t)]. \end{aligned} \tag{14.21}$$

It is a simple exercise to show that the path-wise solution in (14.21) with the special form of forcing in (14.13) is equivalent to

$$\begin{aligned} U(t) &= \bar{U}(t) + \mathrm{Re}[V'(t)], \\ \bar{U}(t) &= U_0 + A_U \sin(\eta t), \end{aligned} \tag{14.22}$$

where U_0 and A_U are some constants and $V'(t)$ solves the unforced Langevin equation

$$\frac{dV'(t)}{dt} = \lambda_U V'(t) + \sigma_U \dot{W}_U(t),$$

with the same parameters as in (14.12). The form of the cross-sweep, $U(t)$, that is given in (14.22), is much simpler to use in our further calculations, therefore, we will assume that the constants U_0, A_U and η are given parameters of our model and (14.22) will be used as the definition of the cross-sweep although a more physically relevant but equivalent definition is in (14.12). The first and second-order statistics of $V'(t)$ become

$$\langle V'(t)\rangle = e^{\lambda_U(t-t_0)}\langle V'(t_0)\rangle, \tag{14.23}$$

$$\mathrm{Var}(V'(t)) = e^{-2\gamma_U(t-t_0)}\mathrm{Var}(V'(t_0)) + \frac{\sigma_U^2}{2\gamma_U}(1-e^{-2\gamma_U(t-t_0)}), \tag{14.24}$$

$$\mathrm{Cov}(V'(t), V'(t)^*) = e^{2\lambda_U(t-t_0)}\mathrm{Cov}(V'(t_0), V'(t_0)^*), \tag{14.25}$$

where for the cross-covariance, we used the definition $\mathrm{Cov}(a,b) = \langle ab^*\rangle - \langle a\rangle\langle b\rangle^*$.

Next, for Gaussian waves $v_k(t)$, we find the path-wise solution

$$v_k(t) = e^{\lambda_k(t-t_0)}v_k(t_0) + F_k(t_0,t) + \sigma_k \int_{t_0}^{t} e^{\lambda_k(t-s)} dW_k(s), \tag{14.26}$$

where

$$F_k(t_0,t)=\int_{t_0}^{t} e^{\lambda_k(t-s)} f_k(s)ds.$$

The first- and second-order statistics of $v_k(t)$ become

$$\langle v_k(t)\rangle = e^{\lambda_k(t-t_0)}\langle v_k(t_0)\rangle + F_k(t_0,t), \tag{14.27}$$

$$\begin{aligned}\mathrm{Cov}(v_k(t), v_l(t)) &= e^{(\lambda_k+\lambda_l^*)(t-t_0)} \\ &\times\left(\mathrm{Cov}(v_k(t_0), v_l(t_0)) + \tfrac{\sigma_k^2}{2\gamma_k}\delta_l^k\left(e^{2\gamma_k(t-t_0)}-1\right)\right).\end{aligned} \tag{14.28}$$

We note that the components of the velocity field $U(t)$ and $v_k(t)$ are governed by independent equations, however, they can be correlated provided that there is correlation between them at the initial moment t_0:

$$\mathrm{Cov}(V'(t), v_k(t)^*) = e^{(\lambda_U+\lambda_k)(t-t_0)}\mathrm{Cov}(V'(t_0), v_k(t_0)^*).$$

Now, with a specific form of the Gaussian velocity field, we can specify the mean and variance of J_U from (14.8) as well as other statistics from (14.11) for the mean of the tracer, $\langle T_k(t)\rangle$. Using (14.22), we decompose J_U into deterministic and stochastic parts

$$\begin{aligned}J_U(s,t) &= \bar{J}(s,t) + J(s,t),\\ \bar{J}(s,t) &= \int_s^t \bar{U}(s')ds' = (t-s)U_0 - \frac{A_U}{\eta}(\cos(\eta t)-\cos(\eta s)),\\ J(s,t) &= \int_s^t \mathrm{Re}[V'(s')]ds'.\end{aligned}$$

Then the mean of the tracer (14.11) becomes

$$\begin{aligned}\langle T_k(t)\rangle &= D_k(t_0,t)e^{-\mathrm{i}k\bar{J}(t_0,t)}\Big(\langle T_k(t_0)\rangle - \mathrm{i}k\mathrm{Cov}(T_k(t_0), J(t_0,t))\Big)\\ &\quad\times e^{-\mathrm{i}k\langle J(t_0,t)\rangle - \frac{k^2}{2}\mathrm{Var}(J(t_0,t))}\\ &\quad-\alpha\int_{t_0}^{t} D_k(s,t)e^{-\mathrm{i}k\bar{J}(s,t)}e^{\lambda_k(s-t_0)}\Big(\langle v_k(t_0)\rangle - \mathrm{i}k\mathrm{Cov}(v_k(t_0), J(s,t))\Big)\\ &\quad\times e^{-\mathrm{i}k\langle J(s,t)\rangle - \frac{k^2}{2}\mathrm{Var}(J(s,t))}ds\\ &\quad-\alpha\int_{t_0}^{t} D_k(s,t)e^{-\mathrm{i}k\bar{J}(s,t)}F_k(t_0,s)e^{-\mathrm{i}k\langle J(s,t)\rangle - \frac{k^2}{2}\mathrm{Var}(J(s,t))}ds,\end{aligned} \tag{14.29}$$

where

$$\begin{aligned}\langle J(s,t)\rangle &= \mathrm{Re}\left[\langle V'(t_0)\rangle\frac{1}{\lambda_U}\left(e^{\lambda_U(t-t_0)} - e^{\lambda_U(s-t_0)}\right)\right]\\ \mathrm{Var}(J(s,t)) &= \frac{\mathrm{Var}(V'(t_0))}{\gamma_U^2+\omega_U^2}e^{-\gamma_U(s+t-2t_0)}\Big(\cosh(\gamma_U(s-t)) - \cos(\omega_U(s-t))\Big)\\ &\quad+\frac{1}{2}\mathrm{Re}\left[\mathrm{Cov}(V'(t_0))\Big(\frac{e^{\lambda_U(s-t_0)} - e^{\lambda_U(t-t_0)}}{\lambda_U}\Big)^2\right]\end{aligned}$$

$$
+\frac{\sigma_U^2}{4\gamma_U}\left(2\mathrm{Re}\left[\frac{e^{\lambda_U(t-s)}-1}{\lambda_U^2}\right]\right.
$$
$$
\left.+\frac{e^{-2\gamma_U(t-t_0)}\left(e^{\lambda_U(s-t)}-1\right)+e^{-2\gamma_U(s-t_0)}\left(e^{\lambda_U(t-s)}-1\right)+2\gamma_U(t-s)}{\gamma_U^2+\omega_U^2}\right),
$$
$$
\begin{aligned}
\mathrm{Cov}(T_k(t_0),J(s,t)) &= \frac{1}{2}\left(\mathrm{Cov}(V'(t_0),T_k(t_0))^*\frac{e^{\lambda_U^*(t-t_0)}-e^{\lambda_U^*(s-t_0)}}{\lambda_U^*}\right.\\
&\quad\left.+\mathrm{Cov}(V'(t_0)^*,T_k(t_0))^*\frac{e^{\lambda_U(t-t_0)}-e^{\lambda_U(s-t_0)}}{\lambda_U}\right),\\
\mathrm{Cov}(v_k(t_0),J(s,t)) &= \frac{1}{2}\left(\mathrm{Cov}(V'(t_0),v_k(t_0))^*\frac{e^{\lambda_U^*(t-t_0)}-e^{\lambda_U^*(s-t_0)}}{\lambda_U^*}\right.\\
&\quad\left.+\mathrm{Cov}(V'(t_0)^*,v_k(t_0))^*\frac{e^{\lambda_U(t-t_0)}-e^{\lambda_U(s-t_0)}}{\lambda_U}\right).
\end{aligned}
\tag{14.30}
$$

14.2.3 Closed equation for the eddy diffusivity

Parametrization by eddy diffusivity is often used in practice, in engineering and atmosphere–ocean science to account for unresolved scales (Majda and Kramer, 1999; Majda and Gershgorin, 2011). In this situation, one parametrizes the collective effect of the nonlinear interaction of the turbulent velocity field on smaller scales on the tracer as effective or eddy diffusivity. Here, we unambiguously demonstrate the accuracy of such an eddy diffusivity approximation in the simplified model.

We obtain a differential equation for the mean $\langle T_k(t)\rangle$ by differentiating (14.11).

$$
\begin{aligned}
\frac{\partial\langle T_k(t)\rangle}{\partial t}+\mathrm{i}k\langle U(t)\rangle\langle T_k(t)\rangle &= -\kappa k^2\langle T_k(t)\rangle-\gamma\langle T_k(t)\rangle\\
&\quad -\alpha\langle v_k(t)\rangle+G_k(t_0,t),
\end{aligned}
\tag{14.31}
$$

where

$$
\begin{aligned}
G_k(t_0,t) &= -\frac{k^2}{2}D_k(t_0,t)\langle T_k(t_0)\rangle\frac{\partial}{\partial t}\Big(\mathrm{Var}(J_U(t_0,t))\Big)\\
&\quad\times e^{-\mathrm{i}k\langle J_U(t_0,t)\rangle-\frac{k^2}{2}\mathrm{Var}(J_U(t_0,t))}\\
&\quad-\alpha\frac{k^2}{2}\int_{t_0}^{t}\frac{\partial}{\partial t}\Big(\mathrm{Var}(J_U(s,t))\Big)D_k(s,t)\langle v_k(s)\rangle\\
&\quad\times e^{-\mathrm{i}k\langle J_U(s,t)\rangle-\frac{k^2}{2}\mathrm{Var}(J_U(s,t))}\,ds
\end{aligned}
$$

and vanishing initial correlations between the tracer and the velocity field were assumed at the initial time t_0 for simplicity. Note that (14.31) is an *exact* ODE for the mean tracer for the kth Fourier mode. The last term in (14.31), $G_k(t_0,t)$, is actually an enhancement to the diffusion term with a diffusion coefficient given by the kernel

$$\kappa(s,t) = \frac{1}{2}\frac{\partial}{\partial t}\Big(\mathrm{Var}(J_U(s,t))\Big).$$

Unlike standard diffusion, this kernel has memory in time. An extremely interesting question is how well can this non-local-in-time diffusion be approximated by a standard local-in-time diffusivity? To answer this question, we approximate the first term in the integrand by some constant

$$\kappa_{\mathrm{eddy}} \approx \frac{1}{2}\frac{\partial}{\partial t}\Big(\mathrm{Var}(J_U(s,t))\Big);$$

then (14.31) becomes

$$\begin{aligned}\frac{\partial\langle T_k(t)\rangle}{\partial t} + \mathrm{i}k\langle U(t)\rangle\langle T_k(t)\rangle &\approx -(\kappa+\kappa_{\mathrm{eddy}})k^2\langle T_k(t)\rangle \\ &\quad -\gamma\langle T_k(t)\rangle - \alpha\langle v_k(t)\rangle. \qquad (14.32)\end{aligned}$$

Here, this constant, κ_{eddy}, exactly represents the eddy diffusivity that enhances the diffusivity of the system due to smaller-scale nonlinear interactions. The model error due to the eddy diffusivity approximation can be quantified by comparing the exact mean given by (14.29) and its approximation given by the solution of the linear ODE (14.32). The important practical issue of model error due to the eddy diffusivity approximation in the filtering context can be addressed unambiguously using this test case.

14.2.4 Properties of the model

In this section, we describe the general statistical properties of the model (14.12), (14.14), (14.5), such as the energy spectrum, correlation time in equilibrium and probability distributions. We focus on two different examples of the waves $v_k(t)$, i.e. non-dispersive waves given by (14.16) and dispersive Rossby waves given by (14.17). We use the following physical intuition to choose the parameters for the model:

- the zonal jet $U(t)$ has positive mean and small enough standard deviation that ensures an eastward direction of the jet;
- the correlation time of the jet $U(t)$ is of the same order as the correlation time of the large-scale waves $v_k(t)$ and larger than the correlation time of the tracer $T_k(t)$;
- the dissipation parameters of both the waves and the tracer are chosen relatively small to mimic a realistic fully turbulent system;
- the mean gradient is large enough to make the energy spectrum of the tracer be of the same order as the energy spectrum of the waves, i.e. the tracer is strongly forced by the waves through the mean gradient; moreover, in the strong mean gradient regime, the tracer becomes strongly non-Gaussian with long exponential tails in the pdfs which very well mimics the observed statistics of various tracers in the atmosphere (Neelin *et al.*, 2011).

In Table 14.1, we give most of the parameters for the model (14.12), (14.14), (14.5). Note that for the forcing $f_k(t)$ in (14.20), we used the following parameters: A_k =

Table 14.1 Parameters for the model (14.12), (14.14), (14.5).

γ_U	ω_U	σ_U	U_0	A_U	η	c	β	F_s	ν, μ	d_v	γ	κ	α
0.04	1	0.3	1.7	0.5	2	1	8.91	16	0.002	0.03	0.04	0.001	1

$0.5 + 0.2\mathrm{U}(-0.5, 0.5)$, $\xi_k = 1.5 + 0.5\mathrm{U}(-0.5, 0.5)$, $\phi_k = \mathrm{U}(0, 2\pi)$, where $\mathrm{U}(a, b)$ is the uniform distribution in the interval (a, b). It is important to emphasize that although these parameters are chosen randomly from these distributions, we fix them for each test trajectory and consider constants. Moreover, in the numerical simulations the number of different Fourier modes for the waves, $v_k(t)$, and the tracer, $T_k(t)$, was either $K = 52$ which corresponds to $2K + 1 = 105$ spatial locations or $K = 10$ which corresponds to $2K + 1 = 21$ spatial locations. In the former case we define the energy spectrum of $v_k(t)$ in (14.19) with $k_0 = 10$ and in the latter case, we use $k_0 = 5$.

In Fig. 14.1, we show the spectra of the waves, v_k, and the spectra of the tracer for non-dispersive and Rossby waves. The spectrum is given by the equilibrium variance of the corresponding variable. The spectrum of the waves is prescribed by (14.19) and is independent of time. On the other hand, the spectrum of the tracer is set by the parameters of the model and the spectrum of the waves. Here, the spectrum of the tracer is time dependent because of the nonlinear dependence of the tracer on the time-dependent cross-sweep, $U(t)$. We compute a long-time average of the spectrum of the tracer, $T_k(t)$, in equilibrium and Gershgorin and Majda (2011) give analytical formulas for $\mathrm{Var}(T_k(t))$ in equilibrium. The spectrum of the tracer for the case of non-dispersive waves (Fig. 14.1(b)) turns out to have three distinct parts where it has approximately power-law behavior with different exponents. The first two parts correspond to the first 10 modes, where the velocity field has constant spectrum (Fig. 14.1(a)). The third part has a steep slope in the log-log scale with exponent approximately -4.5. The spectrum of the tracer for the case of Rossby waves (Fig. 14.1(c)) appears to have two distinct parts. The first part only spans over the 10 Fourier modes and has an exponent of approximately -1 which is the Batchelor spectrum for a tracer slowly forced at large scales (Majda and Kramer, 1999). The second part of the spectrum of the tracer is much steeper with the same exponent -4.5. One of the important tests and practical uses for a filter for turbulent systems is to recover the spectrum of the original truth signal. Below we will use this test to study how well the filter performs in various situations.

In Fig. 14.2, we demonstrate the equilibrium autocorrelation functions of the cross-sweep, $U(t)$, and the first six Fourier modes of the waves, $v_k(t)$, and the tracer, $T_k(t)$, for the case of non-dispersive waves given by (14.16). We used the following definition of the equilibrium autocorrelation function of a variable u

$$C_u(\tau) \equiv \frac{1}{\mathrm{Var}(u(t))} \langle \tilde{u}(t+\tau)\tilde{u}^*(t) \rangle,$$

where the averaging is taken over an ensemble of trajectories in equilibrium, and the tilde denotes the variable with its mean removed. The autocorrelation functions of the Gaussian variables $U(t)$ and $v_k(t)$ were computed analytically from (14.12) and (14.15):

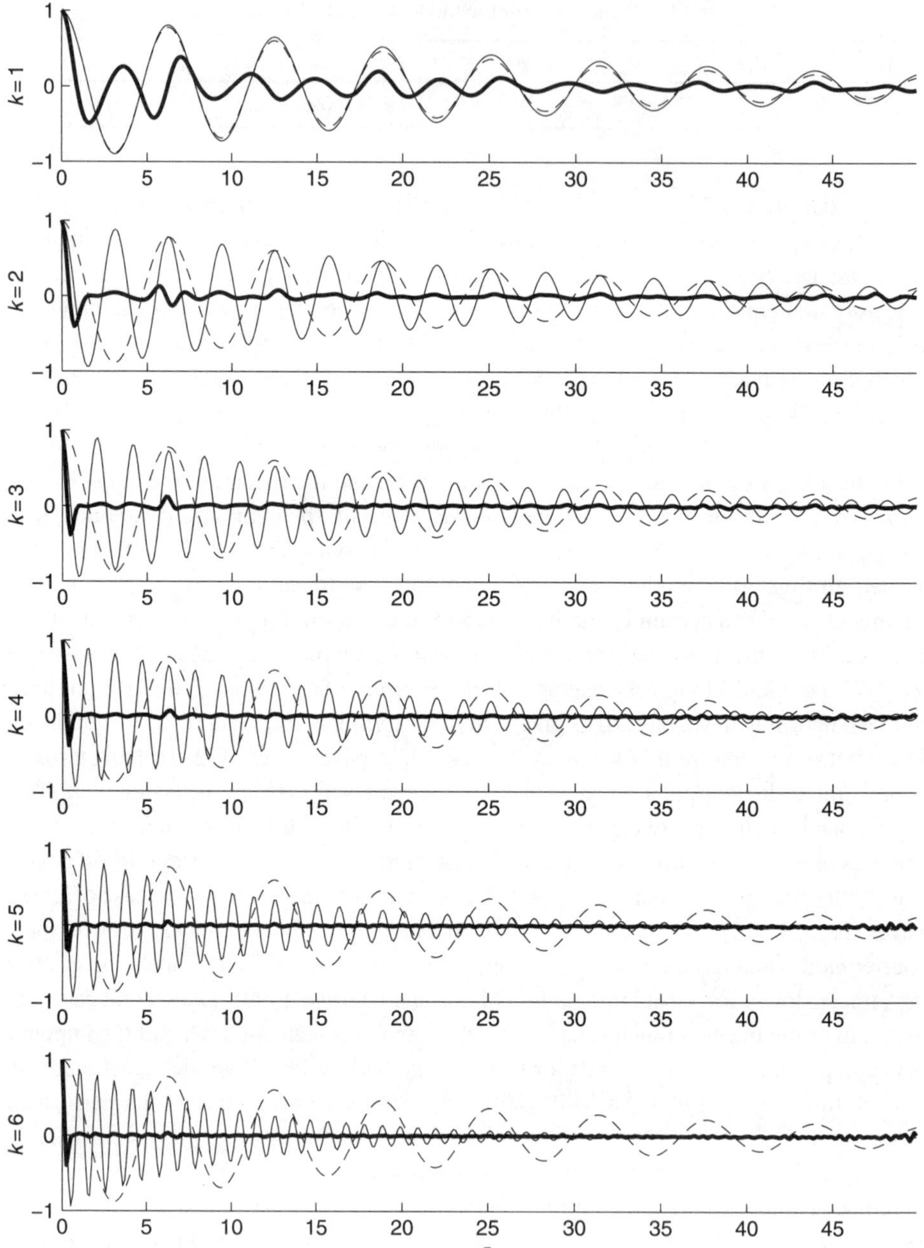

Figure 14.2 Real part of the autocorrelation functions of $U(t)$ (dashed line), $v_k(t)$ (thin solid line) and $T_k(t)$ (thick solid line) for the first six Fourier modes for the case of non-dispersive waves given by (14.16)

$$C_U(\tau) = \cos(\omega_U \tau) e^{-\gamma_U \tau},$$
$$C_{v_k} = e^{(-\gamma_k + i\omega_k)\tau},$$

For the tracer, T_k, we used an ensemble of $L = 400$ trajectories generated via (14.7) to compute the correlation function. We note that for $k = 3$ and higher the corresponding Fourier mode $T_k(t)$ decorrelates very fast while for the first three modes, the correlation time is significantly longer. It is also interesting to note that there are noticeable peaks in the correlation functions of the tracer for the first few modes at the times that are multiples of 2π time units. We attribute these peaks to the strongly non-Gaussian statistics of the tracer due to nonlinear advection of the tracer by the cross-sweep, $U(t)$, and strong forcing by the waves through the mean gradient.

In Fig. 14.3, we demonstrate the correlation time of both the velocity field and the tracer for the case of non-dispersive waves. We compute the correlation time of a variable $u(t)$ as an integral of the absolute value of the correlation function $C_u(\tau)$

$$\tau_{\text{corr},u} = \int_0^\infty |C_u(\tau)| d\tau. \tag{14.33}$$

For the Gaussian velocity field, we have

$$\tau_{\text{corr},U} = \frac{1}{\gamma_U},$$
$$\tau_{\text{corr},v_k} = \frac{1}{\gamma_k}.$$

On the other hand, for the tracer we use the correlation function from Fig. 14.2 to find the correlation time. For our choice of parameters from Table 14.1, the correlation time of the tracer is significantly shorter than the correlation time of the velocity field for each Fourier mode for the case of non-dispersive waves. Similarly, in Figs 14.4 and 14.5, we show the correlation functions and correlation times of the velocity field and the tracer for the case of Rossby waves given in (14.17). We note that a different dispersion relation, ω_k, leads to different oscillations of C_{v_k} for non-dispersive and Rossby waves. Moreover, the correlation functions of the tracer have a similar pattern but different detailed structure especially for the first few Fourier modes. Again, we note the presence of the peaks at the times that are multiples of 2π, which are even more pronounced for this case of Rossby waves. On the other hand, the correlation times for both cases of the linear dispersion ω_k are almost the same for the tracer and exactly the same for the velocity field since the linear dispersion does not affect the correlation time of a linear wave because of the absolute value in definition (14.33).

In Figs 14.6 and 14.7 we demonstrate the probability density functions (pdfs) of the first four Fourier modes of the tracer, T_k, and the pdfs of the tracer in physical space, $T(x)$, along with the corresponding Gaussian fits to these pdfs with the same mean and variance. In order to obtain these pdfs, we used a trajectory of length $t = 30,000$ time units in the equilibrium to construct the histograms of the corresponding Fourier modes and normalized them so that they integrate to unity. We note that both non-dispersive and Rossby

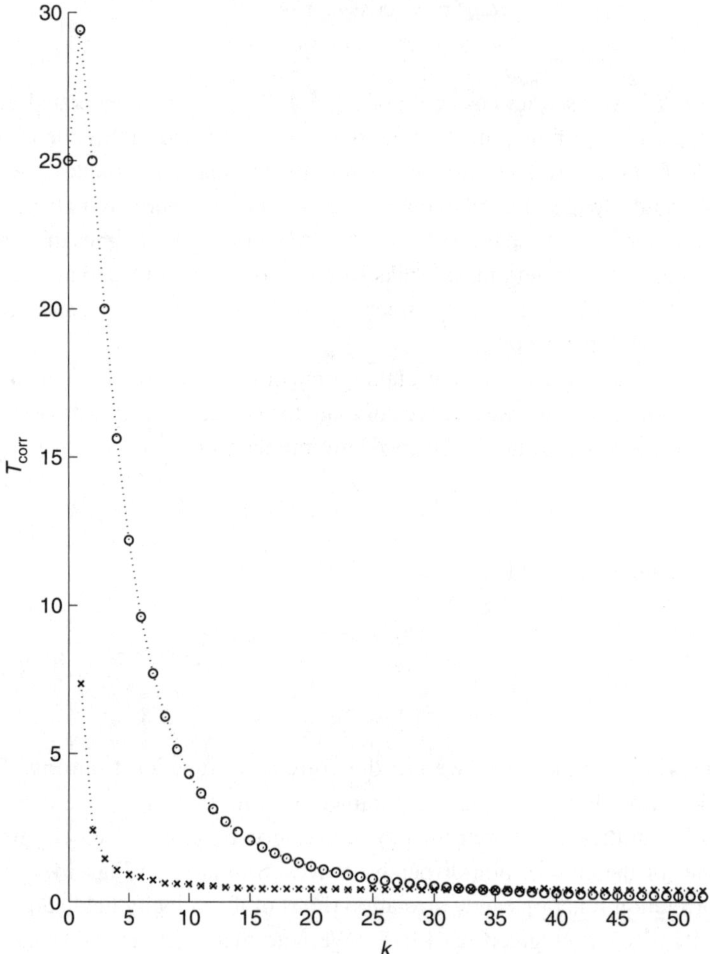

Figure 14.3 Correlation time of the velocity field $U(t)$, $v_k(t)$ (circles) and the tracer $T_k(t)$ (crosses) for different Fourier modes for the case of non-dispersive waves given by (14.16). Note that the correlation time for the cross-sweep $U(t)$ is shown at $k = 0$.

waves lead to strong intermittency that is expressed in long exponential tails. However, in the case of Rossby waves (Fig. 14.7) not only are the tails of the pdfs non-Gaussian, but also the peaks of the pdfs are narrow. These non-Gaussian strongly intermittent pdfs are very similar to the practically observed pdfs for tracers in the atmosphere (Neelin *et al.*, 2011; Gershgorin and Majda, 2011; Majda and Gershgorin, 2011) which makes this test model extremely interesting and physically relevant. A similar phenomenon was observed and explained by Bourlioux and Majda (2002) for the case of purely deterministic cross-sweep $U(t)$ through a different intermittency mechanism involving zeros of $U(t)$ as a source of intermittency.

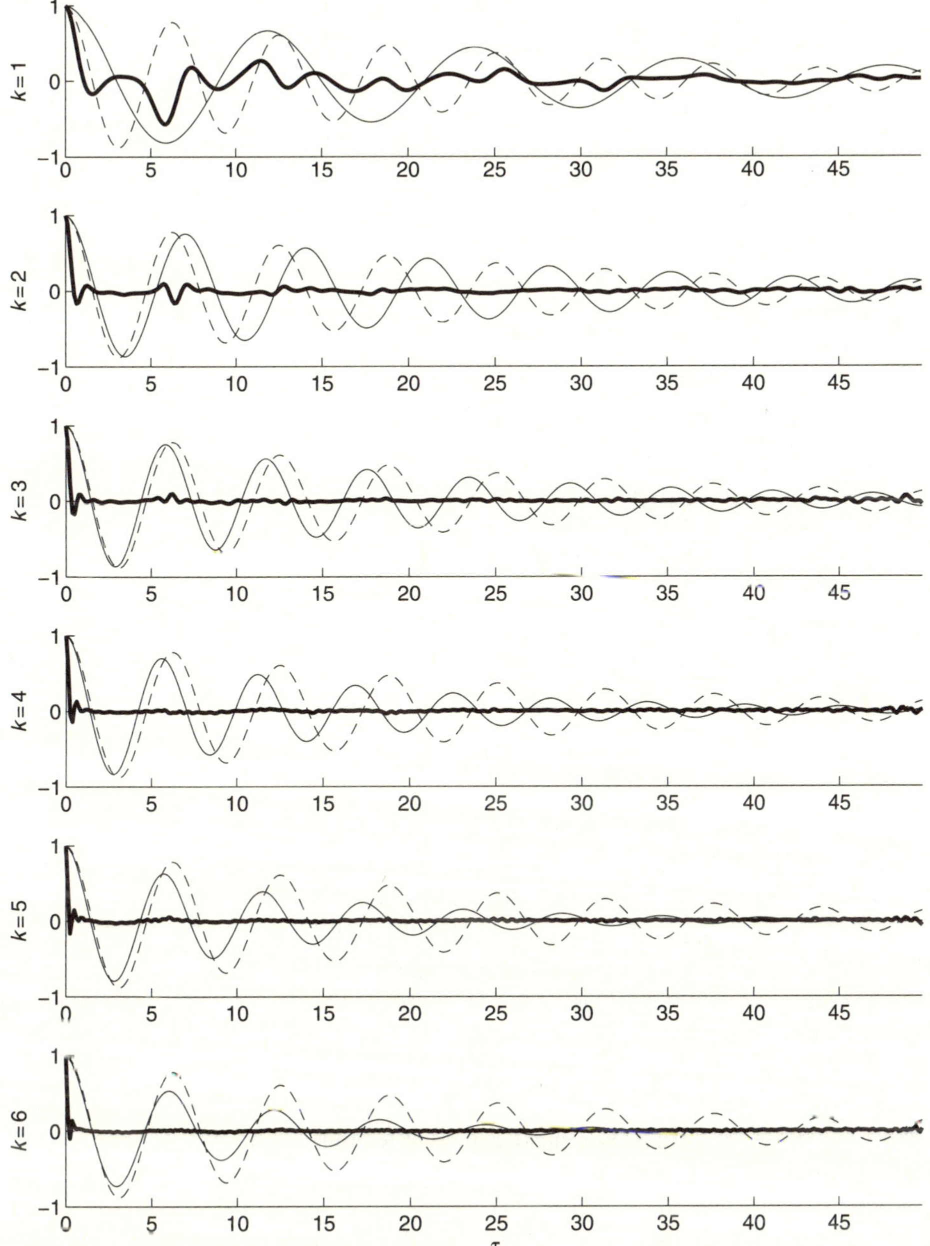

Figure 14.4 Real part of the autocorrelation functions of $U(t)$ (dashed line), $v_k(t)$ (thin solid line) and $T_k(t)$ (thick solid line) for the first six Fourier modes for the case of Rossby waves given by (14.17)

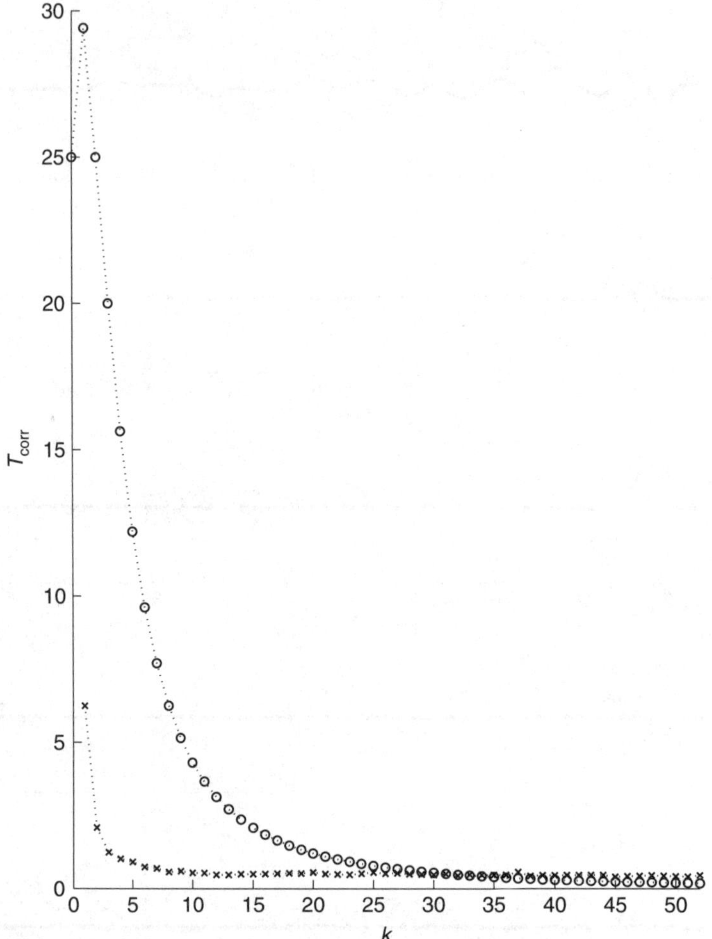

Figure 14.5 Correlation time of the velocity field $U(t)$, $v_k(t)$ (circles) and the tracer $T_k(t)$ (crosses) for different Fourier modes for the case of Rossby waves given by (14.17). Note that the correlation time for the cross-sweep $U(t)$ is shown at $k = 0$.

14.3 Nonlinear extended Kalman filter

14.3.1 Classical Kalman filter

First, we review the classical Kalman filter algorithm. Suppose, the evolution model for a vector $\vec{u}_m \equiv \vec{u}(t_m) \in \mathbb{C}^N$ as a function of discrete time $t_m = m\Delta t$ is given by

$$\vec{u}_{m+1} = \vec{F}_{m+1}(\vec{u}_m) + \vec{f}_{m+1} + \vec{\sigma}_{m+1}, \tag{14.34}$$

where $\vec{F} \in \mathbb{C}^N$ is a linear function, $\vec{f} \in \mathbb{C}^N$ is a deterministic forcing, and $\vec{\sigma} \in \mathbb{C}^N$ is a complex Gaussian noise. Suppose also that at each time t_m we obtain an observation, $\vec{z}_m \in \mathbb{C}^M$, of the true value of the signal, $\vec{u}_m$

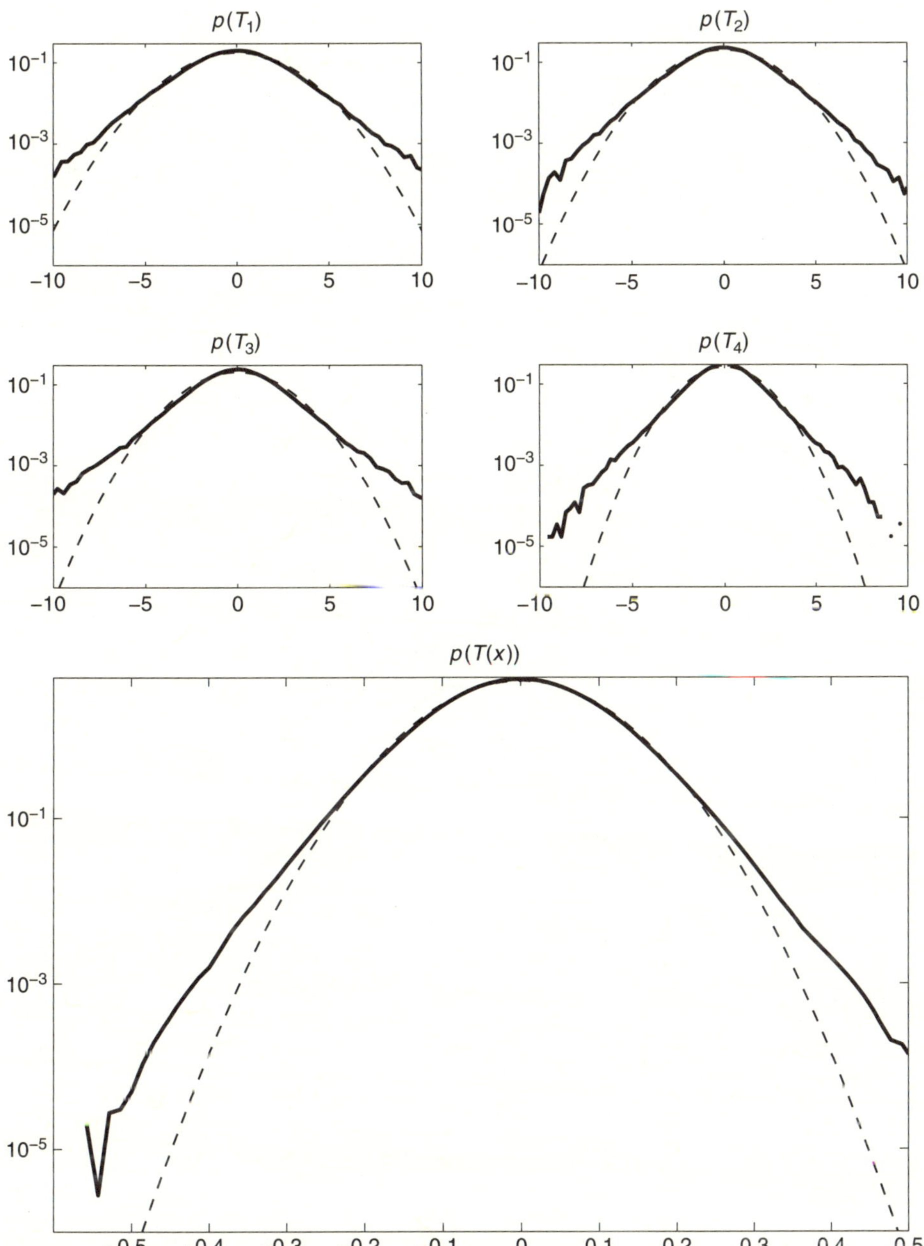

Figure 14.6 The four panels on the top demonstrate the pdfs with long exponential tails of the first four Fourier modes of the tracer T_k (solid line) and the Gaussian pdfs with the same mean and variance (dashed line) as the original pdfs. The lower panel shows the pdf of the tracer in physical space. The case of non-dispersive waves given by (14.16) is shown. Note the logarithmic scale of the y-axis. The pdfs of the real parts only are shown.

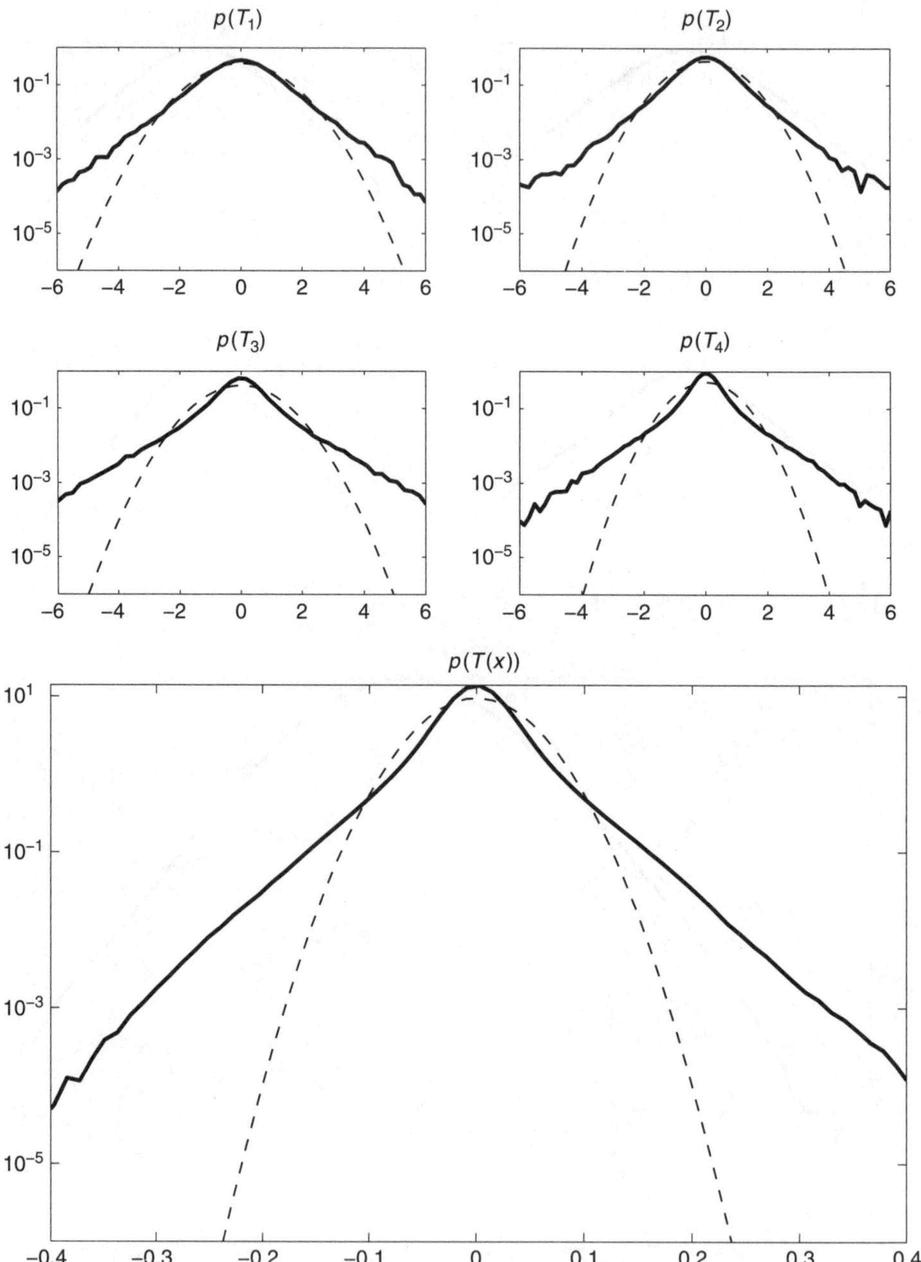

Figure 14.7 The four panels on the top demonstrate the pdfs with long exponential tails of the first four Fourier modes of the tracer T_k (solid line) and the Gaussian pdfs with the same mean and variance (dashed line) as the original pdfs. The lower panel shows the pdf of the tracer in physical space. The case of Rossby waves given by (14.17) is shown. Note the logarithmic scale of the y-axis. The pdfs of the real parts only are shown.

$$\vec{z}_{m+1} = G\vec{u}_{m+1} + \vec{\sigma}^o_{m+1}, \tag{14.35}$$

where $G \in \mathbb{C}^{M\times N}$ is a linear operator (a matrix) and $\vec{\sigma}^o \in \mathbb{C}^M$ is a complex Gaussian noise with mean zero and covariance $r^o \in \mathbb{C}^{M\times M}$. We refer to the time period between two successive observations, Δt, as the observation time. The classical Kalman filter for (14.34) and (14.35) produces the mean and covariance of $\vec{u}_{m+1}$ prior and posterior to knowing observations $\vec{z}_{m+1}$. The prior mean and covariance are denoted as $\langle\vec{u}\rangle_{m+1|m}$ and $r_{m+1|m}$ while the posterior mean and covariance are $\langle\vec{u}\rangle_{m+1|m+1}$ and $r_{m+1|m+1}$.

The prior update of the mean and covariance is done using (14.34)

$$\begin{aligned}\langle\vec{u}\rangle_{m+1|m} &\to \langle\vec{u}\rangle_{m+1|m+1},\\ r_{m+1|m} &\to r_{m+1|m+1},\end{aligned} \tag{14.36}$$

where the model error can also be introduced through the incorrect specification of the model in (14.34).

The posterior update is performed using the Kalman formulas

$$\begin{aligned}\langle\vec{u}\rangle_{m+1|m+1} &= \langle\vec{u}\rangle_{m+1|m} + K_{m+1}(\vec{z}_{m+1} - G\langle\vec{u}\rangle_{m+1|m}),\\ r_{m+1|m+1} &= (\mathcal{I} - K_{m+1}G)r_{m+1|m},\\ K_{m+1} &= r_{m+1|m}G^*(Gr_{m+1|m}G^* + r^o)^{-1},\end{aligned} \tag{14.37}$$

where $K_m \in \mathbb{C}^{N\times M}$ is the Kalman gain matrix and $\mathcal{I}$ is the identity matrix of size $N\times N$. The two-step procedure outlined in (14.36) and (14.37) provides the classical Kalman filter algorithm that produces the best possible approximation in the least-squares sense to the true signal when the model for the signal is linear Gaussian and the observations are also linear with Gaussian noise.

14.3.2 Nonlinear extended Kalman filter

Suppose that the governing model for the true dynamics of the system (14.34) is nonlinear and non-Gaussian. Then, the nonlinear extended Kalman filter as utilized earlier in Chapters 10 and 13 provides a way to extend the classical Kalman filter algorithm to nonlinear models. If the first- and second-order statistics of the nonlinear model (14.34) can be computed exactly, then they are used in the prior update step of the NEKF as in (14.36). The advantage of having exactly solvable statistics avoids linearizations as in the standard extended Kalman filter which is important in strongly non-Gaussian regimes.

As shown in Section 14.1, the system (14.12), (14.14), (14.5) has exactly solvable first- and second-order statistics despite the nonlinear and non-Gaussian tracer statistics due to the advection of the tracer by the Gaussian cross-sweep in Eqn (14.6). Therefore, this system provides an extremely interesting and physically relevant test case for the NEKF. In this situation, the state vector $\vec{u}$ has $4K+4$ components

$$\vec{u}(t) = \Big(\mathrm{Re}[V'(t)], \mathrm{Im}[V'(t)], v(x_1,t), \ldots,\\ v(x_{2K+1},t), T(x_1,t), \ldots, T(x_{2K+1},t)\Big)^T.$$

Note that we need to keep the imaginary part of V', $\text{Im}[V'(t)]$, to update the statistics of the cross-sweep, even though the tracer only depends on the real part of V'. We use (14.23), (14.27) and (14.11) to update $\langle \vec{u} \rangle$ from time $t_0 \equiv t_m$ to time $t \equiv t_{m+1}$. Similarly, we use (14.24), (14.25), (14.28) and the remaining cross covariances (see the appendix of Gershgorin and Majda (2011)). This procedure becomes a prior update step of the NEKF for our model. The posterior update is given by the same equation (14.37) as in the classical KF.

14.3.3 Observations

In order to generate the observations, we first generate a realization of the truth trajectory using Eqns. (14.7), (14.21) and (14.26). Then, we apply Eqn (14.35) to obtain the observations of various types depending on the matrix G. The spatially extended structure of the model (14.12), (14.14), (14.5) provides an opportunity to consider virtually any configuration of observations ranging from plentiful observations of each variable at every grid point to very sparse observations at only one grid point. In particular, we will focus our attention on the following observations

- plentiful observations at every grid point;
- partial observations, when only one or two variables out of the triplet V', v, T are observed;
- sparse observations, when only every pth point of the total $2K + 1$ points are observed;
- extremely sparse observations, with only one grid point where either v or T or both are observed.

Next, we demonstrate the typical forms of the observation matrix G for various types of observations. Note that for the cross-sweep U, we only observe the real part of V'. For the case of plentiful observations we have $G \in \mathbb{R}^{(4K+3)\times(4K+4)}$

$$G = \begin{pmatrix} 1 & 0 & 0 & 0 & \dots & 0 \\ 0 & 0 & 1 & 0 & \dots & 0 \\ 0 & 0 & 0 & 1 & \dots & 0 \\ \vdots & \vdots & \vdots & \vdots & \ddots & \vdots \\ 0 & 0 & 0 & 0 & 0 & 1 \end{pmatrix}.$$

Next, for the case of observation of the cross-sweep and sparse observation for both v and T with $p = 3$, the observation matrix becomes $G \in \mathbb{R}^{(2Q+1)\times(4K+4)}$ for $2K + 1 = pQ$

$$G = \begin{pmatrix} 1 & 0 & 0 & 0 & 0 & 0 & \dots & 0 \\ 0 & 0 & 1 & 0 & 0 & 0 & \dots & 0 \\ 0 & 0 & 0 & 0 & 0 & 1 & \dots & 0 \\ \vdots & \vdots & \vdots & \vdots & \vdots & \vdots & \ddots & \vdots \\ 0 & 0 & 0 & 0 & 0 & 0 & 0 & 1 \end{pmatrix}. \tag{14.38}$$

Similarly, one can write the observation matrix for any other configuration of observations. The observation noise covariance matrix r^o is diagonal with the entries on the diagonal equal to r_U^o for the observations of the cross-sweep, r_v^o for the observations of the waves and r_T^o for the observations of the tracer. The scalars r_U^o, r_v^o and r_T^o are chosen with standard deviation based on the attractor size of the corresponding variable, e.g 10% of the attractor size. These numbers will be specified below in the next section. To quantify the attractor size, we compute the statistically steady-state standard deviation of the corresponding signal. We note that in the current setup, the observations could be taken on an irregular grid as well, which can be practically relevant for the situations when there are more measuring devices in some areas of the domain of interest than in others. However, in this chapter we only consider uniform observational grids or the radical extreme limit of observations at a single point.

14.4 Filter performance

In this section, we discuss the performance of the NEKF filter described in Section 14.3.2 on the model (14.12), (14.14), (14.5). We first show the behavior of the filter on a typical trajectory of this system in physical space. Then, we study how well the filter performs in Fourier space by comparing the spectra of a truth signal with the spectra of the corresponding filtered signal. We consider a suite of various observational schemes as discussed in Section 14.3.3. Then, we study the performance of the NEKF as a function of observation noise strength and observation time. Finally, we consider a situation of extremely sparse observations and discuss how much of the truth signal can be recovered if the observations are only taken at one spatial location. The results of our study are compared for the cases of dispersive Rossby waves and non-dispersive waves.

We describe the performance of the filter by the filter skill, which measures the proximity of the filtered signal to the truth signal. We use the root mean square error (RMSE) to measure the filter skill

$$\mathrm{RMSE}(\vec{z}-\vec{w}) = \sqrt{\frac{1}{N}\sum_{j=1}^{N}|z_j - w_j|^2},$$

where $\vec{z}$ and $\vec{w}$ are the complex vectors to be compared and N is the length of each vector. The ratio of the RMSE and the typical magnitude of the signal gives the normalized percentage error. Moreover, to quantify how well the filtered signal recovers the pattern of the truth signal, we use the normalized cross-correlation function for two complex functions given by

$$\mathrm{XC}_C(\vec{z},\vec{w}) = \frac{1}{2}\Big(\mathrm{XC}_R(\mathrm{Re}[\vec{z}],\mathrm{Re}[\vec{w}]) + \mathrm{XC}_R(\mathrm{Im}[\vec{z}],\mathrm{Im}[\vec{w}])\Big), \tag{14.39}$$

where the cross-correlation between real vectors $\vec{x}$ and $\vec{y}$ is given by

$$\mathrm{XC}_R(\vec{x},\vec{y}) = \frac{\vec{x}\cdot\vec{y}}{\sqrt{|\vec{x}|^2|\vec{y}|^2}}.$$

14.4.1 Filtering individual trajectories in physical space

Figure 14.8 demonstrates a segment of a typical trajectory as a function of time. The first panel in Fig. 14.8 shows the dynamics of the cross-sweep, the second and third panels show the dynamics of the waves and the tracer at a fixed spatial location, x_{ob}, where observations are available. On the other hand, the last two panels show the dynamics of the waves and the tracer at a different spatial location, $x_{\mathrm{no\ ob}}$, where observations are not available. To generate this trajectory, we used the parameters from Table 14.1 for the Rossby wave case with $K = 52$ Fourier modes and $2K + 1 = 105$ spatial locations. For filtering, we use the observations of the cross-sweep (the real part of the fluctuations of the cross-sweep, $V'(t)$, around its mean state, $\bar{U}(t)$, as in (14.22) were actually observed and filtered), and sparse observations of the waves and the tracer with every third spatial location observed, i.e. with $p = 3$ and the observation matrix G given by (14.38). We use the observation time $\Delta t = 0.125$, which is shorter than the correlation times of all modes of the system that are shown in Fig. 14.5. The trajectory is $L = 2000$ assimilation cycles long. We set the standard deviation of the observation noise to be equal to 10% of the attractor size of the corresponding variable. In Fig. 14.8, the observations are shown wherever they are available. Moreover, we show both the prior and the posterior filtered signals. We note the exceptionally good skill of the NEKF: the filtered dashed line is almost always on top of the truth solid line. Note that in the first three panel of Fig. 14.8, where observations are available, the filtered signal tends to be between the prior forecast and the observations, which is in accord with the general theory of Kalman filters discussed throughout this book. By comparing the second and the fourth panels of Fig. 14.8 for the waves, we note that having observations at a particular location, x_{ob}, improves the filter skill, however, the general shape of the truth signal is recovered even at the location, $x_{\mathrm{no\ ob}}$, where observations are not available. Similar conclusions can be drawn from comparing the third and the fifth panels of Fig. 14.8 for the tracer. However, here the skill of the filter is excellent even at the non-observable location, $x_{\mathrm{no\ ob}}$. This is explained below by the fact that observation of the cross-sweep improve the filter skill for the tracer.

In Fig. 14.9(a),(b), we show a snapshot of the same trajectory at the time $t = 33.375$ (see Fig. 14.8 for the temporal evolution). We note that the observations are only given at every third spatial location as discussed above. By comparing the truth and the filtered signals by sight, we note the very high skill of the NEKF. Next, in Fig. 14.9 (c),(d), we show the cross-correlation given by (14.39) for both the waves and the tracer as a function of spatial location. The oscillating pattern of the cross-correlation in Fig. 14.9(c),(d) reflects the presence and the absence of the observations. At the locations where observations are available, the pattern correlation is close to 1, which is a sign of almost perfect recovery of the truth signal by the filtered signal. On the other hand, at the locations where observations are not available, the cross-correlation has values around 0.85 for the waves, $v(x, t)$, which is still significantly high. For the tracer, $T(x, t)$, the cross-correlation is always very high, of the order 0.98, with slightly higher skill at the locations with observations than at the locations without observations.

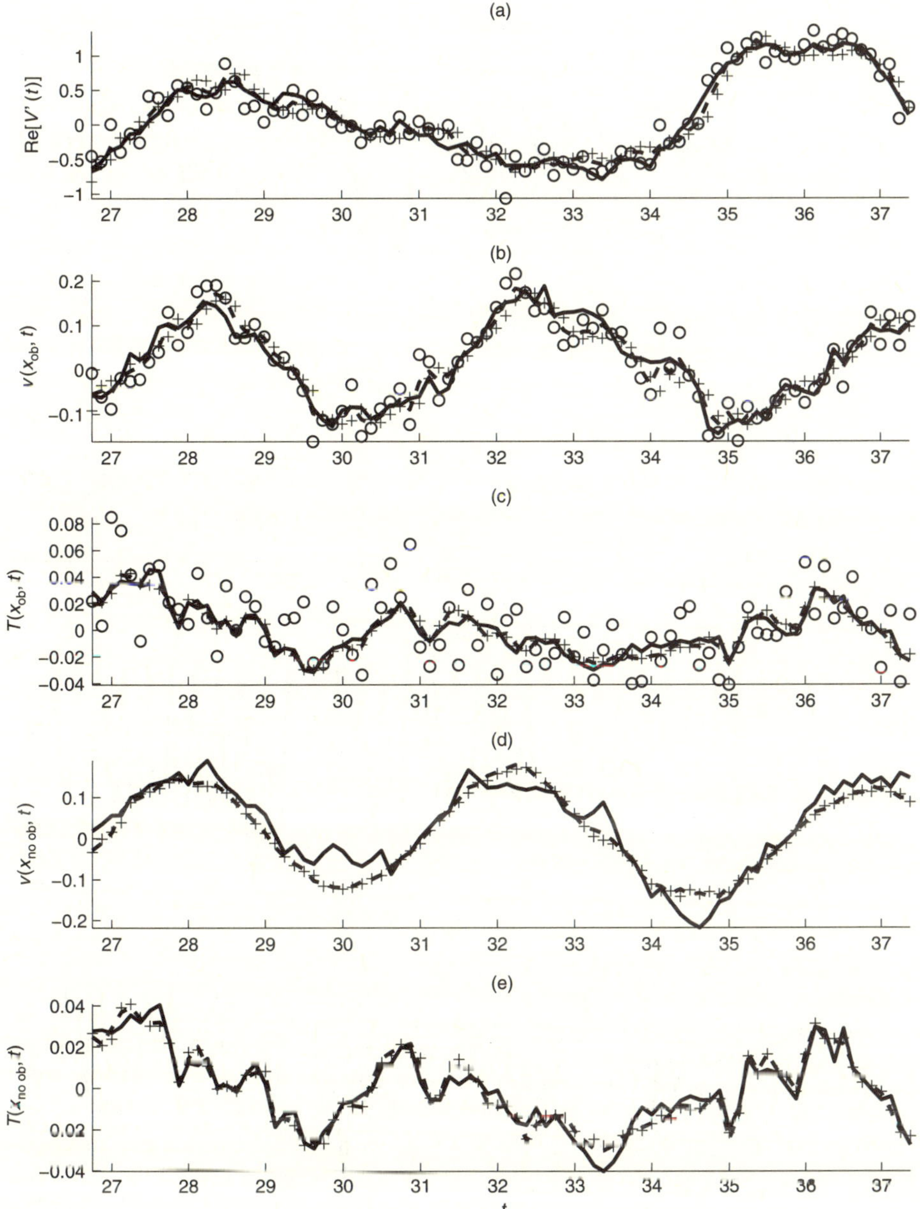

Figure 14.8 A segment of a typical trajectory for the Rossby wave case as a function of time: the truth signal (solid line), the observations (circles), the prior values (pluses) and the posterior values (dashed line). Panel (a) shows the cross-sweep (the real part of the fluctuations, $V'(t)$, is shown), panel (b) shows the waves, $v(x_{\text{ob}}, t)$, at a fixed spatial location x_{ob} where observations are available, panel (c) is the same as panel (b) but for the tracer $T(x_{\text{ob}}, t)$, panel (d) shows the waves, $v(x_{\text{no ob}}, t)$, at a fixed spatial location x_{ob} where observations are not available, and panel (c) is the same as panel (d) but for the tracer $T(x_{\text{no ob}}, t)$. Observations of the cross-sweep and sparse observations of both $v(x, t)$ and $T(x, t)$ with $p = 3$ were used.

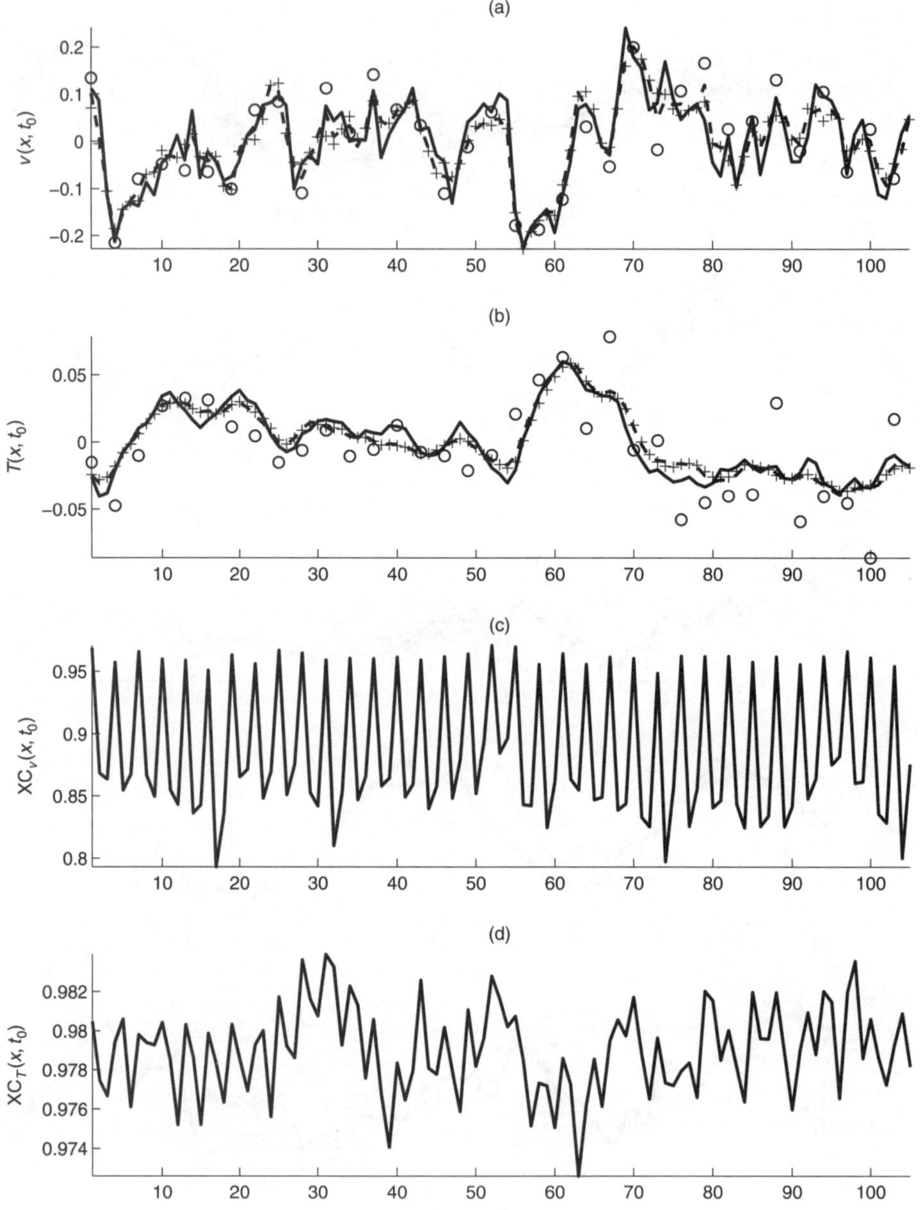

Figure 14.9 Panel (a): a segment of a typical trajectory, $v(x, t_0)$, for the Rossby wave case as a function of space for a fixed time $t_0 = 33.375$: the truth signal (solid line), the observations (circles), the prior values (pluses) and the posterior values (dashed line). Panel (b) is the same as panel (a) but for the tracer, $T(x, t_0)$. Panel (c) shows the pattern correlation as a function of space between the truth signal, $v(x, t)$, and the corresponding posterior values. Panel (d) is the same as panel (c) but for the tracer, $T(x, t_0)$. Observations of the cross-sweep and sparse observations of both $v(x, t)$ and $T(x, t)$ with $p = 3$ were used.

14.4.2 Recovery of turbulent spectra

Here, we study how well the NEKF recovers the spectrum of the original truth signal with various observation schemes. We test the NEKF on a sample trajectory that is $L = 2000$ assimilation cycles long, with each cycle $\Delta t = 0.125$. We use very small observation noise with a standard deviation of approximately 1% of the attractor size of the signal. We consider six different observation schemes:

1. observations of v only;
2. observations of T only;
3. observations of both U and v;
4. observations of both U and T;
5. observations of both v and T;
6. observations of U, v, and T.

Within each of these schemes, we take observations either at each or at every third or at every seventh spatial location to study how well sparse observations are filtered. We note that the NEKF uses the exact nonlinear statistics to build the prior forecast; moreover, we use very small observation time and observation noise. Therefore, most of the errors in the filtered solution can be attributed to the imperfect partial and sparse observations. This setup provides a transparent and unambiguous test case for the role of observations in the NEKF.

In Figs 14.10 and 14.11 we compare the truth spectra and the filtered spectra for the case of non-dispersive waves for both v and T. Since the equilibrium statistics are time dependent and we only have one sample trajectory, as opposed to having an ensemble of trajectories, we consider time averaged squared amplitudes, $\langle |v_k|^2 \rangle$ and $\langle |T_k|^2 \rangle$, instead of the time averaged variances which require knowledge of the time-dependent values of the means of v_k and T_k, information that may not be available. Here, we refer to these time averaged squared amplitudes as the spectrum, although this definition is technically different from the one used in Section 14.2.4; therefore, the truth spectra in Figs 14.10 and 14.11 are a little different from the "ideal" spectra in Fig. 14.1. In Figs 14.12 and 14.13, we compare the truth spectra and the filtered spectra for the case of Rossby waves for both v and T. After studying Figs 14.10–14.13, we draw the following conclusions:

- By decreasing the number of observed points on the grid, we notice a naturally expected decrease in the skill of the filtering. Moreover, this deterioration of the skill appears at high wavenumbers which is also not surprising since the high wavenumbers correspond to the small spatial scales in physical space.
- If only the waves, v, are observed then the filtered spectrum of waves is very close to the truth spectrum of waves, while the filtered spectrum of the tracer has a large error when compared to the truth spectrum of the tracer. However, the forms of the filtered and the truth spectra of the tracer are similar by sight.
- Observations of both components of the velocity field, i.e. the cross-sweep and the waves, significantly improve the skill in filtering the spectrum of the tracer even though the tracer is not directly observed.

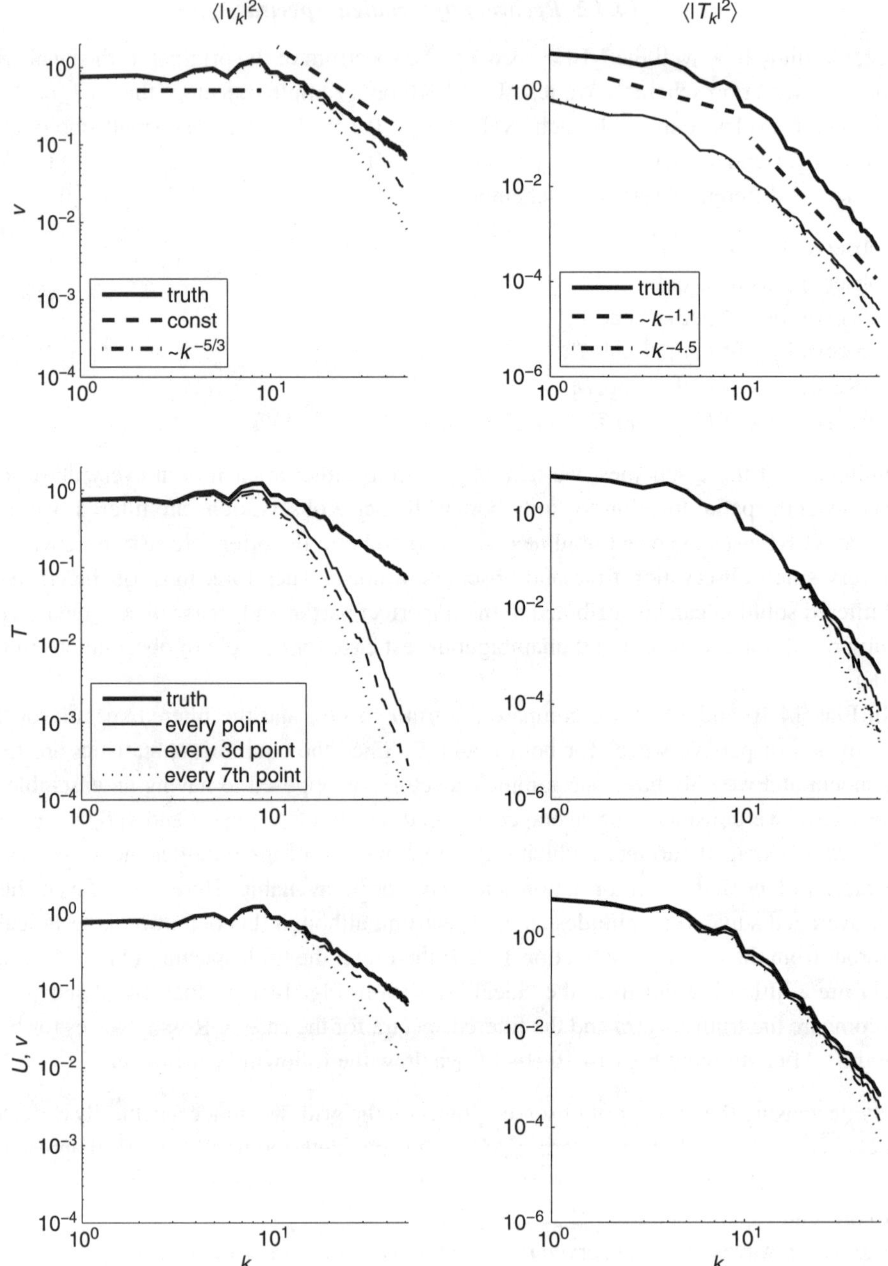

Figure 14.10 Spectrum recovery via the NEKF for the non-dispersive wave case and observations of v only (first row), T only (second row), and U and v (third row): the truth (thick solid line), the filtered signal with plentiful observations, i.e. $p = 1$ (thin solid line), the filtered signal with every third point observed (dashed line) and the filtered signal with every seventh point observed (dotted line). The left column shows the spectrum of the waves, v_k, and the right column shows the spectrum of the tracer, T_k.

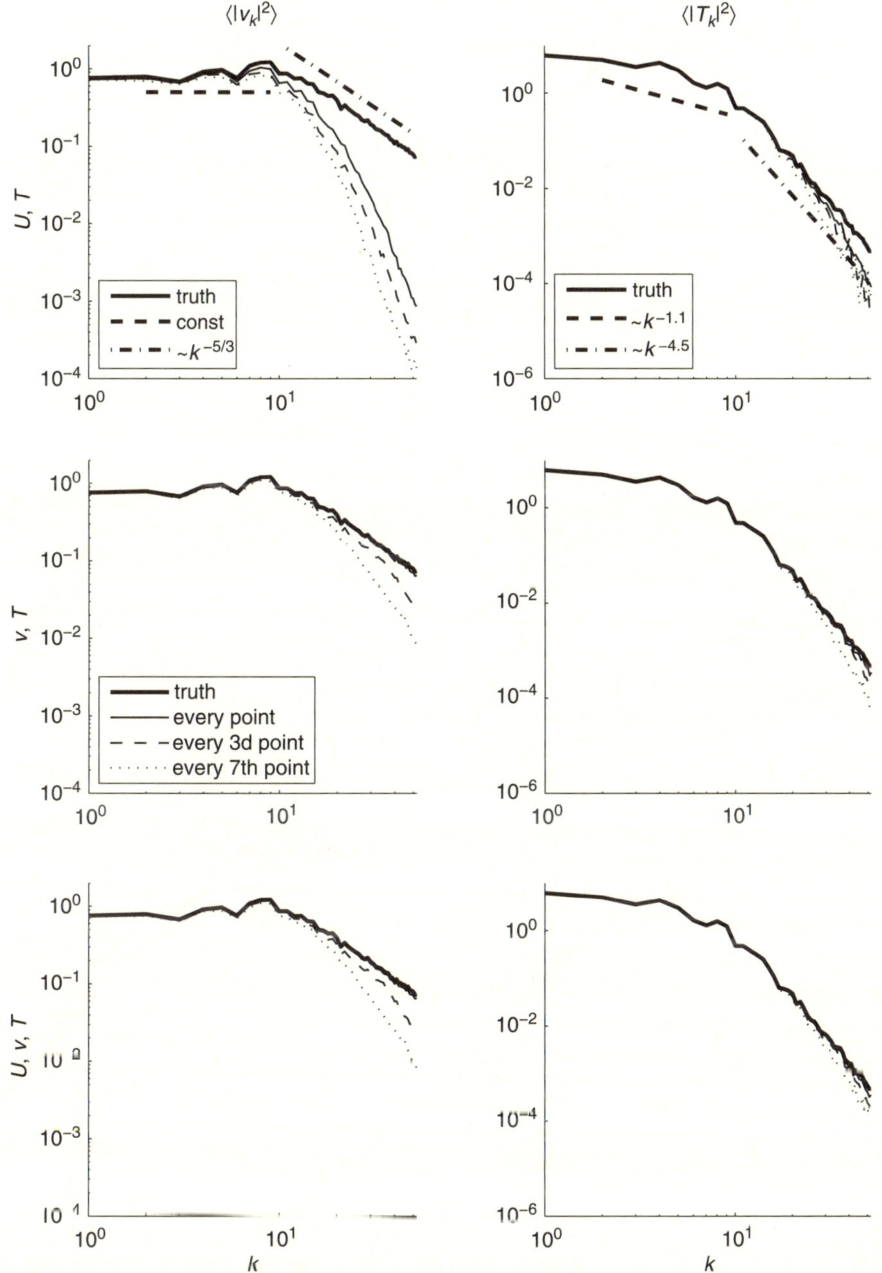

Figure 14.11 Spectrum recovery via the NEKF for the non-dispersive wave case and observations of U and T (first row), v and T (second row), and U, v and T (third row): the truth (thick solid line), the filtered signal with plentiful observations, i.e. $p = 1$ (thin solid line), the filtered signal with every third point observed (dashed line) and the filtered signal with every seventh point observed (dotted line). The left column shows the spectrum of the waves, v_k, and the right column shows the spectrum of the tracer, T_k.

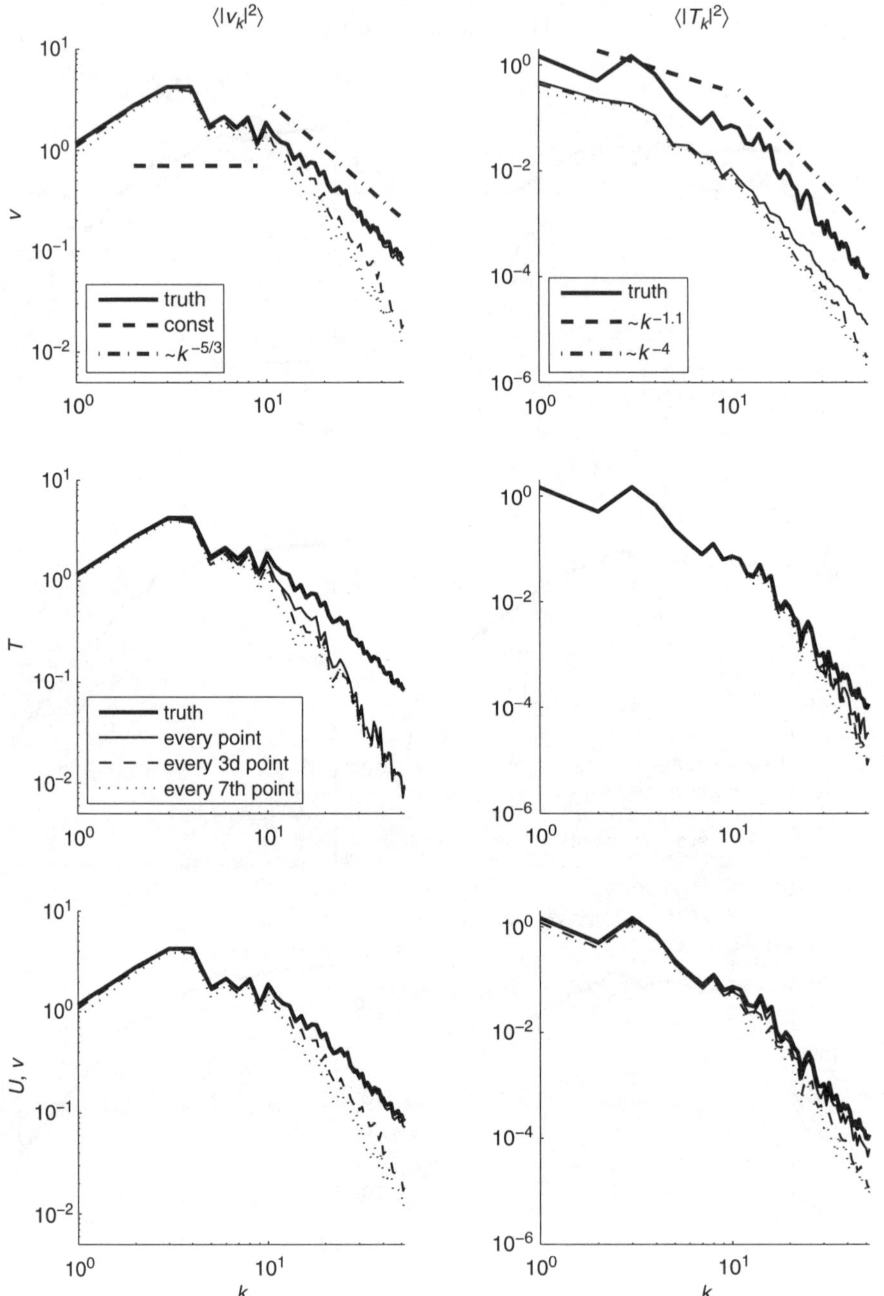

Figure 14.12 Spectrum recovery via the NEKF for the Rossby wave case and observations of v only (first row), T only (second row), and U and v (third row): the truth (thick solid line), the filtered signal with plentiful observations, i.e. $p = 1$ (thin solid line), the filtered signal with every third point observed (dashed line) and the filtered signal with every seventh point observed (dotted line). The left column shows the spectrum of the waves, v_k, and the right column shows the spectrum of the tracer, T_k.

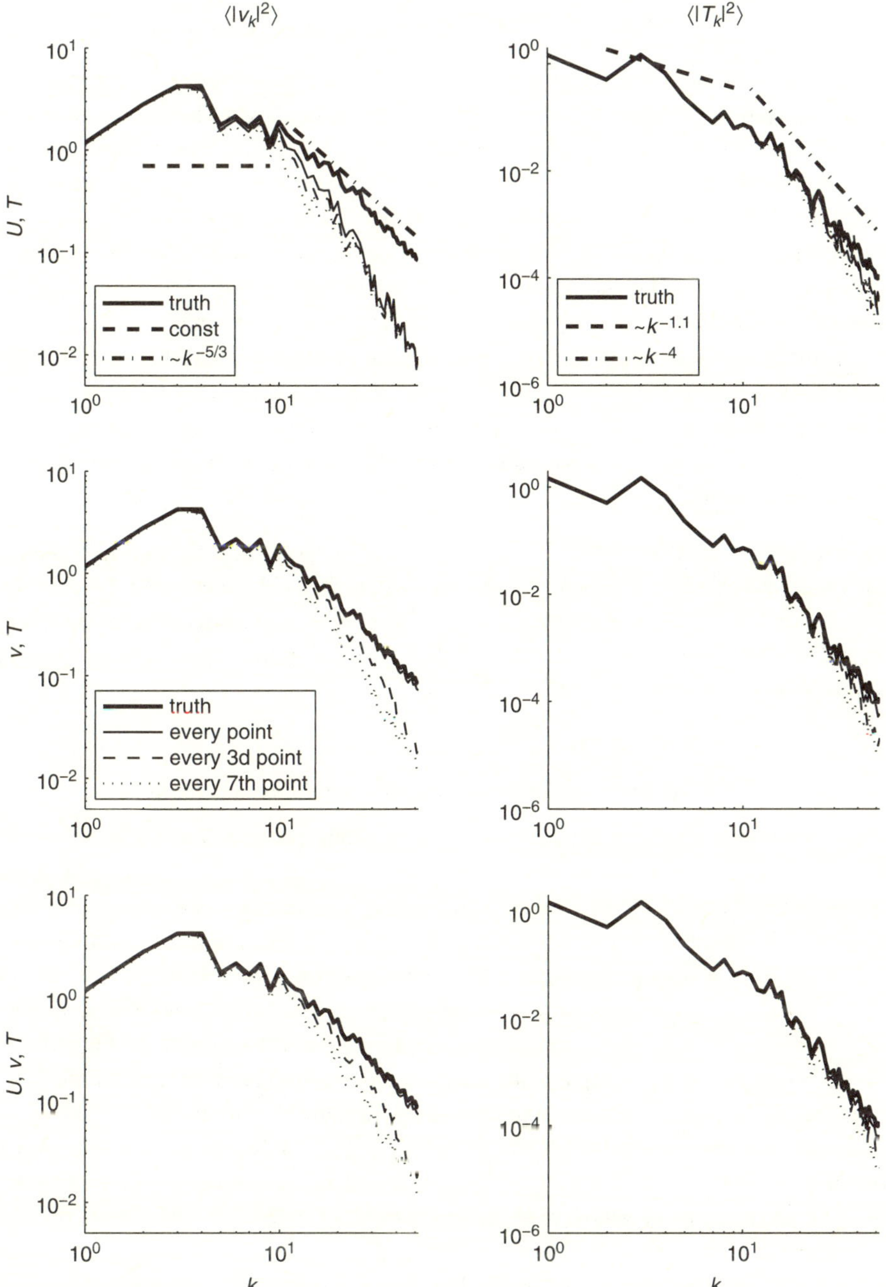

Figure 14.13 Spectrum recovery via the NEKF for the Rossby wave case and observations of U and T (first row), v and T (second row), and U, v, and T (third row): the truth (thick solid line), the filtered signal with plentiful observations, i.e. $p = 1$ (thin solid line), the filtered signal with every third point observed (dashed line) and the filtered signal with every seventh point observed (dotted line). The left column shows the spectrum of the waves, v_k, and the right column shows the spectrum of the tracer, T_k.

- Observations of the tracer alone provide very good skill in filtering the spectrum of the tracer. Moreover, the filtered spectrum of the waves at small wavenumbers is close to the truth spectrum of the waves, i.e. observations of the tracer alone carry information about the large-scale spatial structure of the waves. The recovery of the small-scale spatial structure of the waves requires additional observations of the waves.
- Additional observations of the cross-sweep when the tracer is observed do not help to improve the recovery of the spectrum of the waves at large wavenumbers which is a consequence of the fact that the cross-sweep and the waves have independent dynamics.
- Plentiful observations of all variables provide the best filtered skill because maximum possible information is utilized in the recovery of the spectra.
- The NEKF performs equally well in recovering the turbulent spectra for both the non-dispersive and dispersive Rossby wave systems for the case of sparse but not extremely sparse observations.

Note that here we concentrate on the filtering of the waves, v, and the tracer, T. The role of the observations of the cross-sweep will be discussed next in Section 14.4.4. Extensive discussion of filter skill in recovering the cross-sweep as well as detailed dependence of filter skill on the observation noise strength and observation time are given in Gershgorin and Majda (2011).

14.4.3 Recovery of the fat tail tracer probability distributions

Another important statistical characteristics of a stochastic turbulent signal is its equilibrium pdf. As we discussed in the introduction and in Section 14.2.4, one of the key properties of the model for the tracer with a mean gradient is the strongly intermittent pdfs with long exponential tails as in the tracers observed in the atmosphere (Neelin *et al.*, 2011). Therefore, it is important that a filtering algorithm recovers the pdf of the original true signal. Here, we study how well the NEKF recovers the truth pdf depending on the observations. To make the testing fair, we compare the histogram of the filtered signal with the histogram of the sample truth trajectory. Note that due to under-sampling with the shorter time series of the true signal, the histogram of the sample truth trajectory may be different from the ideal pdf of the truth signal. In Fig. 14.14, we demonstrate how the probability distribution of the tracer is recovered by the filtered signal for the Rossby wave case for different types of sparse observations with $p = 7$. We used the whole trajectory for the tracer in space and time ($2K + 1 = 105$ spatial points $\times$ 2000 assimilation cycles) and 80 bins to construct the histograms in Fig. 14.14. Note that the normalized histogram of the sample truth trajectory, which is filtered, is different from the ideal pdf of the tracer. However, the sample trajectory still has long exponential tails, which are less fat than in the ideal pdf. We note that the filtered histogram almost perfectly recovers the histogram of the sample truth trajectory for all types of observations except for the case of observations of the waves only. As we have seen earlier in Section 14.4.2, the NEKF also has poor skill in recovering the spectrum of the tracer when only the waves are observed, otherwise the

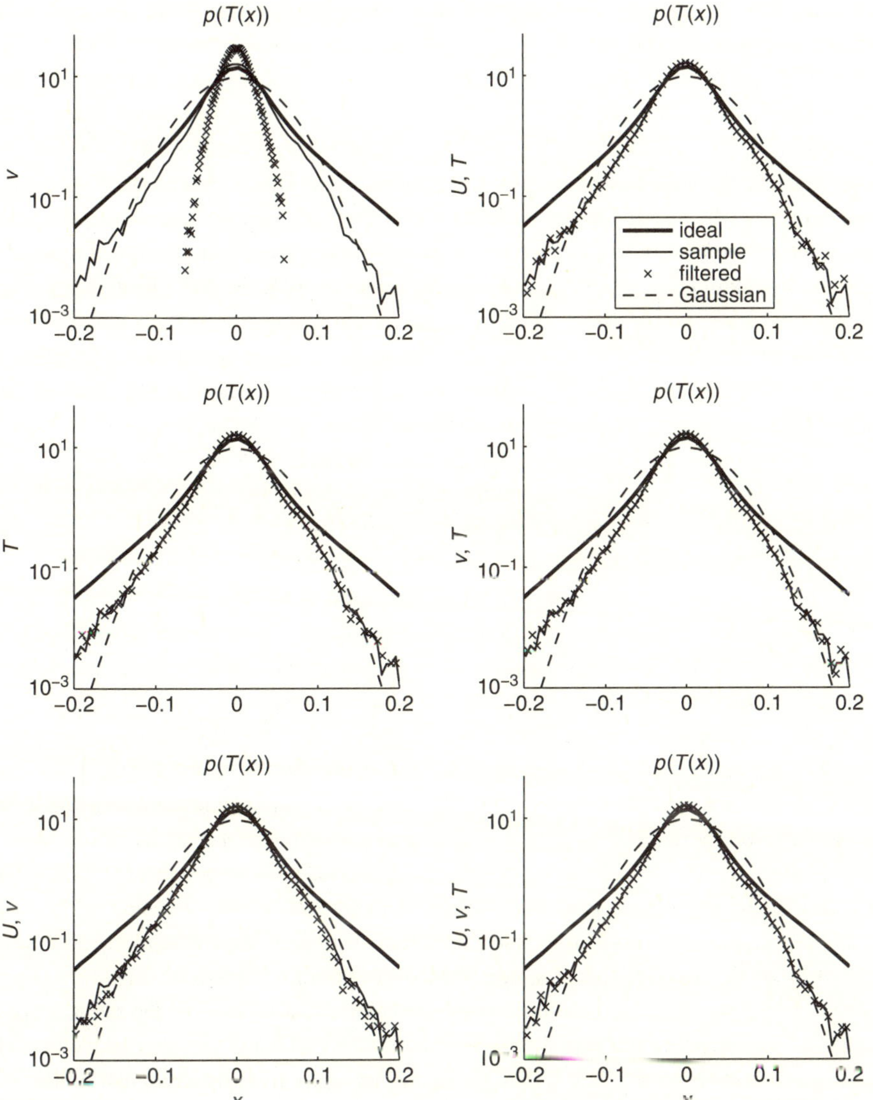

Figure 14.14 Recovery of the pdf of the tracer in physical space via NEKF for the Rossby wave case. The thick solid line shows the ideal pdf of the tracer in physical space as in Fig. 14.7, the dashed line shows the Gaussian with the same mean and variance, the thin solid line shows the normalized histogram of the sample trajectory which is filtered, and crosses show the normalized histogram of the filtered signal. The observed variables are shown to the right of the corresponding panel, sparse observations with $p = 7$ were used. Note the logarithmic scale of the y-axis.

spectrum of the tracer was recovered with high skill. Similar results were found for the case of non-dispersive waves.

14.4.4 Improvement of the filtering skill by adding just one observation

Let us consider the situation when only very sparse observations are available. How much can one improve the filtering skill by adding just one observation at a fixed spatial location? Will an observation of the velocity field or of the tracer provide better filtering skill? To mimic such situations, we carry out the following simulation. Suppose that in our model only the cross-sweep is observed. Then we fix some spatial location (which can be arbitrary because of the translational symmetry of the system) at which observations of either the waves, v, or the tracer, T, or both are taken. By comparing the filtering skill in all of these four filtering setups, we find answers to the questions asked at the beginning of this section.

We start with the case of non-dispersive waves. In Fig. 14.15, we show the skill of the NEKF in recovering the spatial structure of the waves and of the tracer using both the pattern cross-correlation, XC, and the root mean square error, RMSE. If only the cross-sweep U is observed, then the NEKF has no skill with the values of XC ≈ 0.1 and RMSE ≈ 1. If in addition to the cross-sweep, there is one observation of the waves, v, available at each assimilation cycle, we improve the skill of NEKF in filtering waves at the observation point and to the right of the observation point, i.e. in the direction of the zonal cross-sweep. At the observation point, we have almost perfect recovery of the waves with XC ≈ 1 and RMSE ≈ 0.1. Then, as we move further away to the right of the observation point, the filtering skill gradually drops to the value of XC ≈ 0.3 and RMSE ≈ 0.9. On the other hand, the skill in filtering the tracer improves a lot by adding only one observation of the waves: cross-correlation increases from XC ≈ 0.2 with observations of the cross-sweep alone to values as high as XC ≈ 0.75 at *each* point of the grid, not just around the observation point, and the RMSE decreases from RMSE ≈ 1 to RMSE ≈ 0.65 also uniformly around the spatial domain. Next, we study the improvement of the filtering skill by adding one observation of the tracer. In this situation, the skill in filtering waves increases very slightly up to XC ≈ 0.4 and RMSE ≈ 0.9 uniformly around the spatial domain. On the other hand, the skill of the NEKF in filtering the tracer increases significantly: at the observation site, XC ≈ 1 and RMSE ≈ 0.2, and as we move further away to the right of the observation point, i.e. in the direction of the zonal cross-sweep, the skill decreases to XC ≈ 0.85 and RMSE ≈ 0.65, which still indicate significant skill. Finally, if we add observations of both the waves and the tracer at the same spatial location in addition to the observations of the cross-sweep, the filtering skill for T improves just slightly over the situation when only the cross-sweep and the tracer at one spatial location are observed.

In Fig. 14.16, we compare the truth spectra of both v and T with the filtered spectra when in addition to the observations of the cross-sweep, the observations of either waves, or tracer, or both become available at one spatial location. We note that with the observations of the cross-sweep alone, the filtered spectra have very large errors. However, if the observations of either v, or T, or both, at one spatial location become available, the NEKF

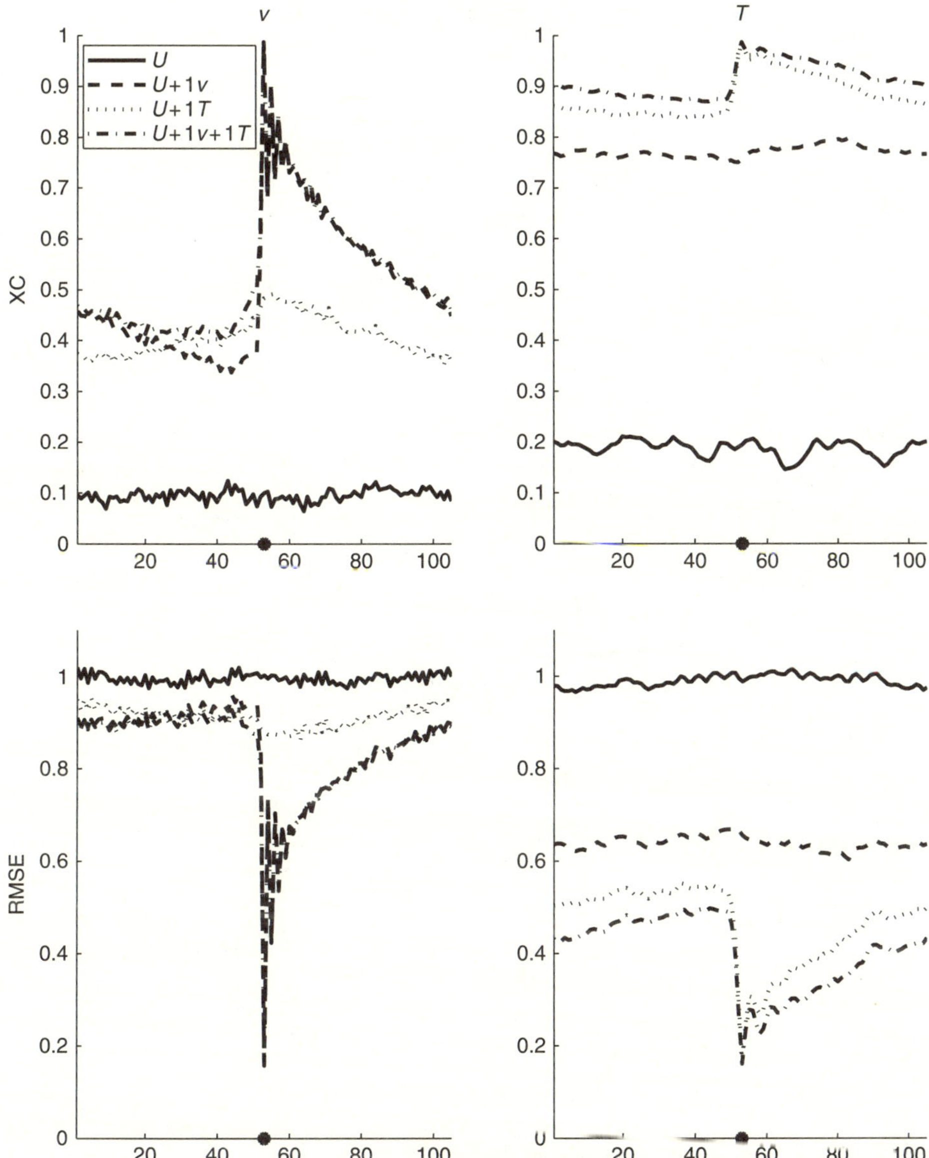

Figure 14.15 Improvement of the filter skill in physical space by adding an observation at just one spatial location for the case of non-dispersive waves. The top row shows the cross-correlation between the truth and the filtered signals, the bottom row shows the percentage RMS error of the filtered signal, the left column corresponds to the waves and the right column corresponds to the tracer. The solid line corresponds to the skill with the observations of the cross-sweep only, the dashed line corresponds to the skill with observations of the cross-sweep and the waves at one spatial location (denoted by a thick dot), the dotted line corresponds to the skill with observations of the cross-sweep and the tracer at one spatial location, and the dash-dotted line corresponds to the skill with observations of the cross-sweep and the waves and the tracer at one spatial location.

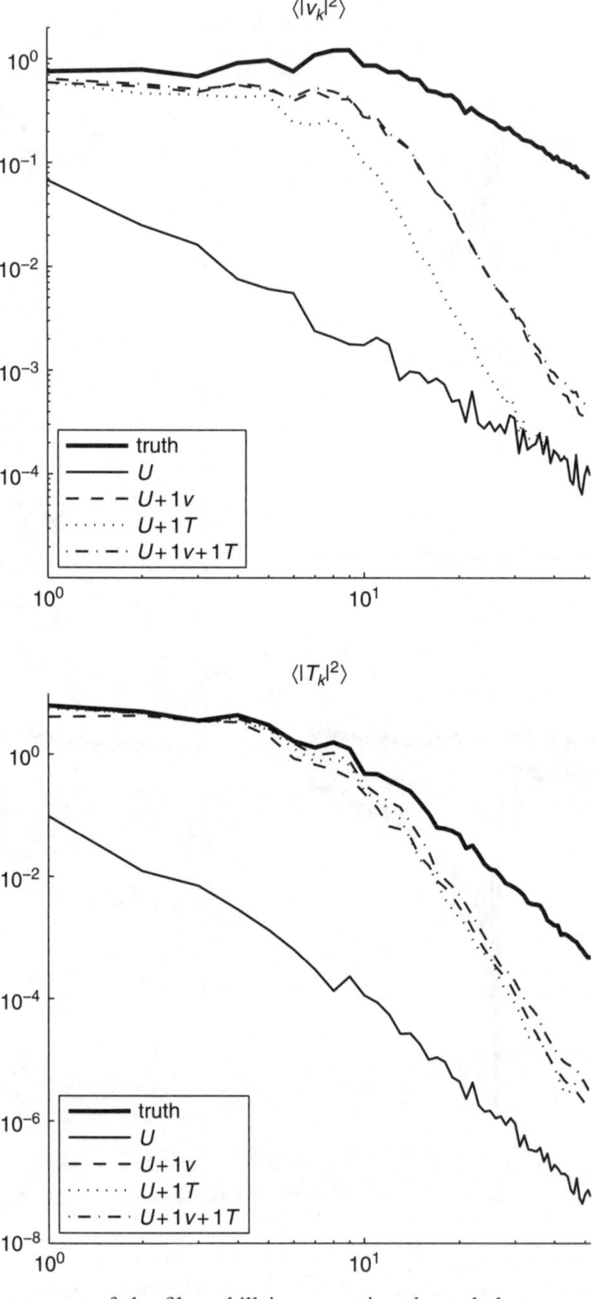

Figure 14.16 Improvement of the filter skill in recovering the turbulent spectrum by adding an observation at just one spatial location for the case of non-dispersive waves. The top panel shows the spectrum of the waves, and the bottom panel shows the spectrum of the tracer. The thick solid line corresponds to the truth spectrum, the thin solid line corresponds to the filtered spectrum with observations of the cross-sweep only, the dashed line corresponds to the filtered spectrum with observations of the cross-sweep and the waves at one spatial location only, the dotted line corresponds to the filtered signal with observations of the cross-sweep and the tracer at one spatial location only, and the dashed-dotted line corresponds to the observations of the cross-sweep and the waves and the tracer at one spatial location only.

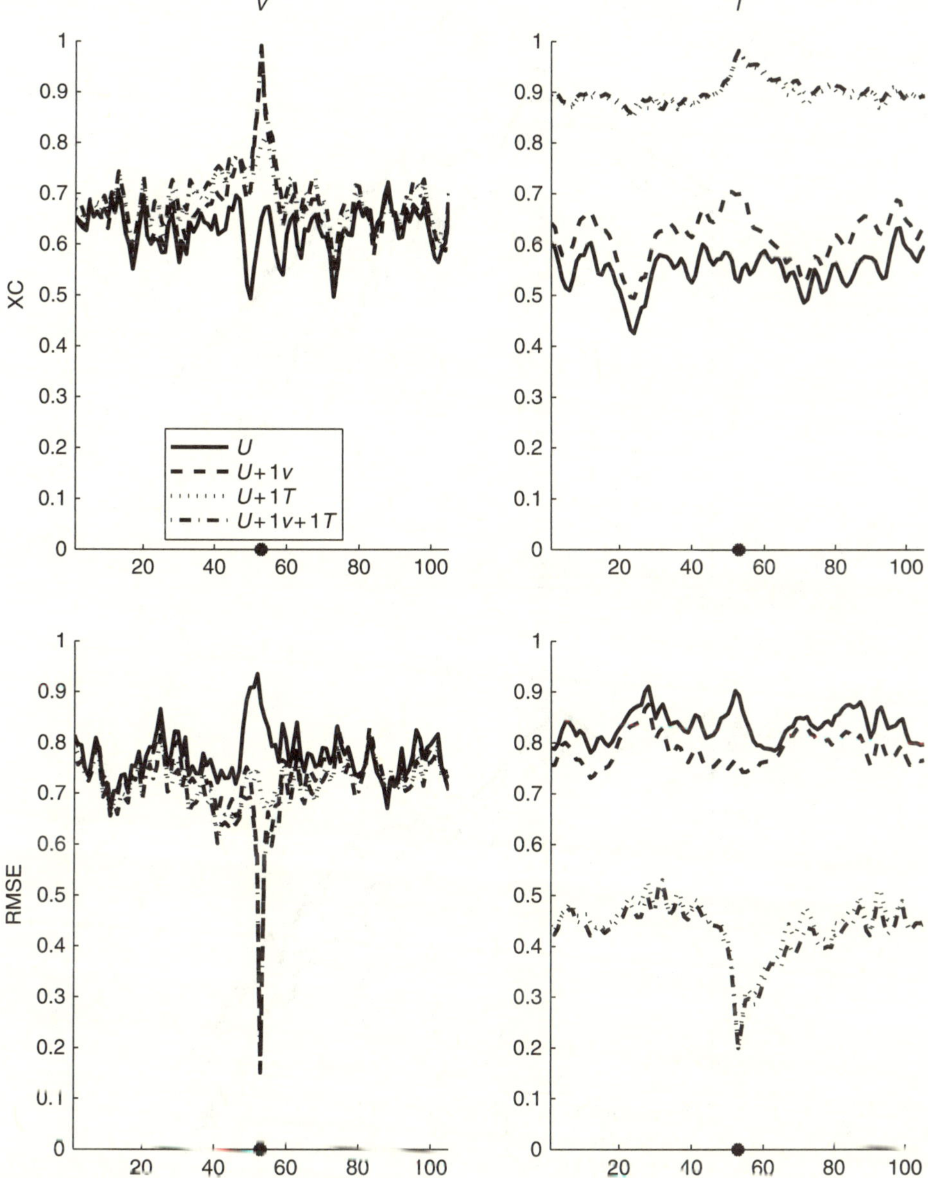

Figure 14.17 Improvement of the filter skill in physical space by adding an observation at just one spatial location for the case of Rossby waves. The top row shows the cross-correlation between the truth and the filtered signals, the bottom row shows the percentage RMS error of the filtered signal, the left column corresponds to the waves, and the right column corresponds to the tracer. The solid line corresponds to the skill with the observations of the cross-sweep only, the dashed line corresponds to the skill with observations of the cross-sweep and the waves at one spatial location (denoted by a thick dot), the dotted line corresponds to the skill with observations of the cross-sweep and the tracer at one spatial location, and the dash-dotted line corresponds to the skill with observations of the cross-sweep and the waves and the tracer at one spatial location.

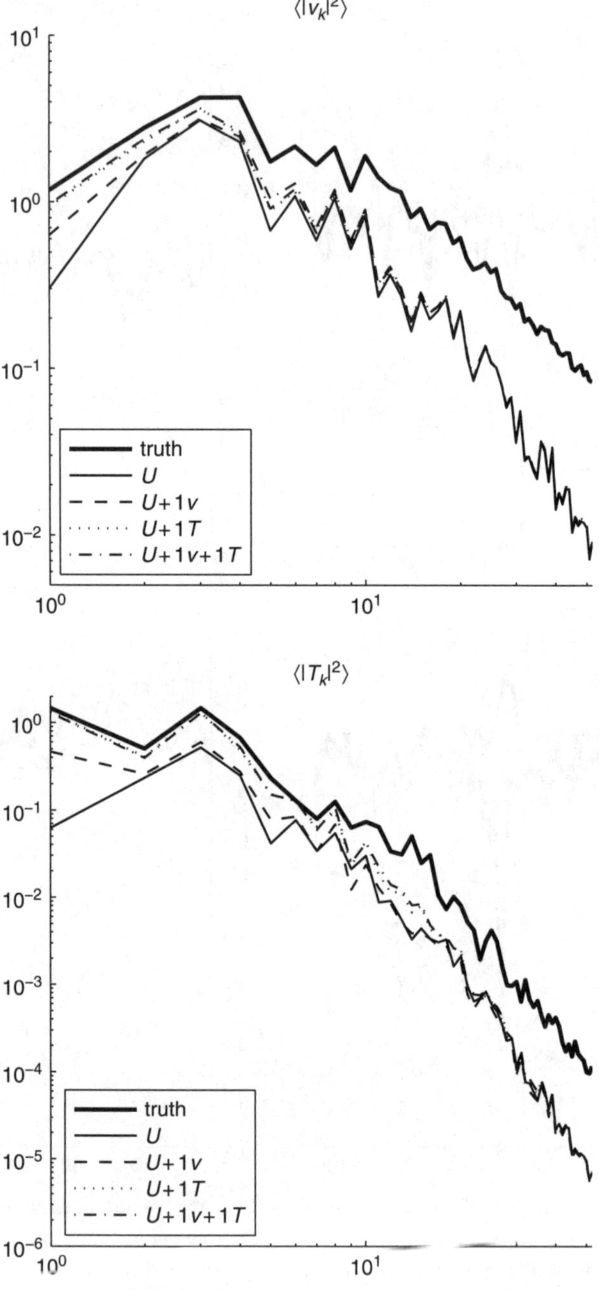

Figure 14.18 Improvement of the filter skill in recovering the turbulent spectrum by adding an observation at just one spatial location for the case of Rossby waves. The top panel shows the spectrum of the waves, and the bottom panel shows the spectrum of the tracer. The thick solid line corresponds to the truth spectrum, the thin solid line corresponds to the filtered spectrum with observations of the cross-sweep only, the dashed line corresponds to the filtered spectrum with observations of the cross-sweep and the waves at one spatial location only, the dotted line corresponds to the filtered signal with observations of the cross-sweep and the tracer at one spatial location only, and the dashed-dotted line corresponds to the observations of the cross-sweep and the waves and the tracer at one spatial location only.

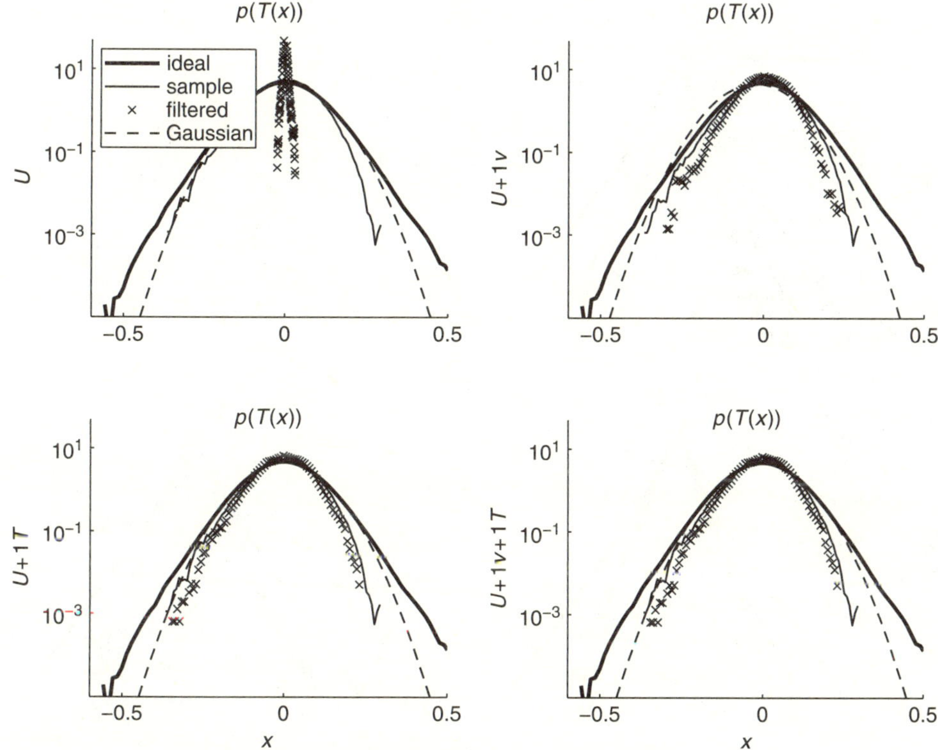

Figure 14.19 Improving recovery of the pdf of the tracer in physical space via NEKF for the non-dispersive wave case by adding an observation at just one spatial location. The thick solid line shows the ideal pdf of the tracer in physical space as in Fig. 14.7, the dashed line shows the Gaussian with the same mean and variance, the thin solid line shows the normalized histogram of the sample trajectory which is filtered, and crosses show the normalized histogram of the filtered signal. The top left panel corresponds to observations of the cross-sweep only, the top right panel corresponds to observations of the cross-sweep and the waves at one spatial location, the bottom left panel corresponds to observations of the cross-sweep and the tracer at one spatial location only, and the bottom right panel corresponds to observations of the cross-sweep and the waves and the tracer at one spatial location only. Note the logarithmic scale of the y axis.

recovers the forms of the spectra of both v and T, although the exponents of the turbulent cascades of the filtered signals are different from the corresponding exponents for the truth spectra. We note that for the first part of the spectrum that involve the first 10 modes, the recovery for both waves and tracer is excellent regardless of which variable is observed at the fixed spatial location. However, the high wavenumber parts of the filtered spectra have steeper decay than in the true spectra. It is natural to expect more skill in the low wavenumber part of the spectrum with sparse observations than in the high wavenumber part since sparse observations provide information about the large-scale spatial structure of the truth

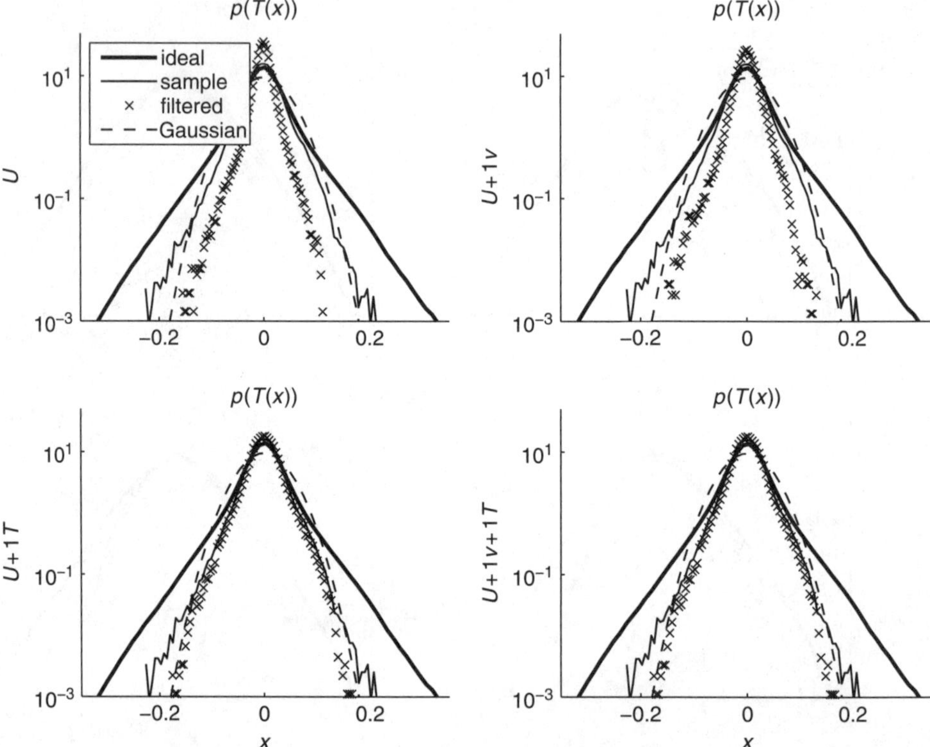

Figure 14.20 Improving recovery of the pdf of the tracer in physical space via NEKF for the Rossby wave case by adding an observation at just one spatial location. The thick solid line shows the ideal pdf of the tracer in physical space as in Fig. 14.7, the dashed line shows the Gaussian with the same mean and variance, the thin solid line shows the normalized histogram of the sample trajectory which is filtered, and crosses show the normalized histogram of the filtered signal. The top left panel corresponds to observations of the cross-sweep only, the top right panel corresponds to observations of the cross-sweep and the waves at one spatial location, the bottom left panel corresponds to observations of the cross-sweep and the tracer at one spatial location only, and the bottom right panel corresponds to observations of the cross-sweep and the waves and the tracer at one spatial location only. Note the logarithmic scale of the y-axis.

signal in the situation when the large scales carry most of the system variance. However, it is surprising to find that just one observation can improve the filtered spectrum so much.

We continue with the case of Rossby waves. In Fig. 14.17, we show the skill of the NEKF in recovering the spatial structure of the waves and of the tracer using both the pattern cross-correlation, XC, and the root mean square error, RMSE. We note that in this case even with the observations of the cross-sweep only, the filtered signal has some nontrivial skill with XC $\approx$ 0.65 and RMSE $\approx$ 0.8 for the waves and XC $\approx$ 0.55 and RMSE $\approx$ 0.8 for the tracer. Here, similar to the case of non-dispersive waves, the skill in filtering the tracer improves significantly if one observation of the tracer becomes available:

XC increases up to 0.9 and RMSE drops to 0.4. On the other hand, unlike in the case of non-dispersive waves, here, the skill of filtering the waves does not improve significantly when an observation of either v, or T, or both, become available at one point, except for the skill in the close proximity of this observation point. Moreover, the observations of the waves do not help to recover the tracer. Similar conclusions can be drawn from Fig. 14.18, for the spectra of the waves and the tracer.

In Fig. 14.19, we study the improvement of recovery of the pdf in physical space by the NEKF for the case of non-dispersive waves when only one observation of the waves and/or of the tracer becomes available. We note that with observations of the cross-sweep only, the filtered pdf is much more peaked than the truth pdf. However, when only one observation of either the waves or the tracer or both become available, the filtered pdf almost perfectly recovers the sample truth pdf. For the case of Rossby waves shown in Fig. 14.20, we note that if only the cross-sweep is observed, the filtered histogram has a higher central peak and much narrower tails that the pdf of the truth trajectory. Adding one observation of the waves does not improve the filtered pdf. However, adding one observation of the tracer makes the filtered pdf very close to the sample truth pdf.

To summarize this section, we have found that adding one observation of either the waves, or the tracer, or both, can significantly improve the filtering skill in both physical space and Fourier space. In the case of non-dispersive waves, the filtered signal is closer to the truth signal in the direction of the cross-sweep, i.e. the information that the observation carries, propagates in the direction of the cross-sweep. On the other hand, for Rossby waves the filtered signal for the tracer improves if an observation of the tracer becomes available and this improvement is only slightly better in the direction of the cross-sweep.

15

The search for efficient skillful particle filters for high-dimensional turbulent dynamical systems

In Chapters 9 and 11 we discussed finite ensemble filtering of nonlinear dynamical systems with Kalman-based (or linear) techniques and we demonstrated their advantages in the less turbulent regime and their limitation in the sparsely observed fully turbulent regime. We also showed their sensitivity toward variations of filtering parameters such as the variance inflation and ensemble size. Subsequently in Chapters 12–14, we discussed alternative strategies based on cheap reduced stochastic models to avoid all the disadvantages of ensemble Kalman filtering strategies and demonstrated their skill for filtering various turbulent nonlinear systems ranging from the idealized toy models like L-96 to quasi-geostrophic turbulence to turbulent diffusion models with non-Gaussian statistics.

Recently, there is an emerging need to develop nonlinear filtering methods beyond the ensemble Kalman filters since most physical problems are nonlinear and sometimes have highly non-Gaussian distributions. One of the most difficult problems is to devise a practically useful ensemble method based on the theoretically well-established particle filtering (or sequential Monte Carlo) methods (Del Moral, 1996; Del Moral and Jacod, 2001; Doucet *et al.*, 2008). The major challenge is to have a particle filter that is practically skillful for high-dimensional complex turbulent systems with only small ensemble size (small number of particles); this is very important because of the large overload in generating individual ensemble members through the forward dynamical operator (Haven *et al.*, 2005). Moreover, we require the particle filter to be skillful in fully turbulent regimes with spatially and temporally sparse observations with low noise (this is the toughest regime for the ensemble Kalman filter, see Chapter 12).

In this final chapter of the book, we present our attempt to devise such a particle filter. To help the reader, we present a short summary of particle filtering in Section 15.1 (more detailed discussion can be found in many particle filtering books such as Doucet *et al.* (2008)). In Section 15.2, we will discuss the main idea of our particle filter, the maximum entropy particle filter (MEPF), which is motivated by Jeff Anderson's rank histogram filter (RHF) (Anderson, 2010). In the same section, we will also discuss the potential dynamic range issues that one may face (with MEPF as well as RHF) in many situations and propose remedies. Subsequently in Sections 15.3–15.4, we demonstrate the filter performance on the L-63 model as well as the L-96 model. Finally, we close this chapter with a short discussion in Section 15.4. While the L-63 model is a low (three) dimensional weakly chaotic dynamical systems, it is highly non-Gaussian statistically. Thus, we note here that it is a major challenge to devise a particle method which successfully filters this system with very small ensemble sizes such as involving only 3–10 particles, especially for temporally sparse partial observations with low noise.

15.1 The basic idea of a particle filter

The main idea in particle filtering is to apply the Bayes theorem to update an ensemble of solutions (or particles) without assuming any Gaussianity on the prior or posterior distributions as well as on the observation likelihood function; thus, no explicit formulation like the Kalman filter equations is used. To be more precise, consider an ensemble posterior state $\{\vec{u}^k_{m|m} \in \mathbb{R}^N\}_{k=1,\ldots,K}$ of size K at time t_m with distribution function,

$$p_{m|m}(\vec{u}) = \sum_{k=1}^{K} w^k_m \delta(\vec{u} - \vec{u}^k_{m|m}), \tag{15.1}$$

where $\delta(x - x_0) = 1$ if $x = x_0$ and zero otherwise. In (15.1), the term w^k_m determines the weight (or importance) of each ensemble member $\vec{u}^k_{m|m}$ and for $p_{m|m}$ to be a probability density function, these weights satisfy $\sum_k w^k_m = 1$ and $w^k_m \geq 0$. Notice that with this density representation, we commit sampling errors in approximating the posterior distribution; such errors are unavoidable in a high-dimensional problem and this is the so-called "curse of dimensionality" issue in random sampling related problems.

In this setup, the goal of each step in particle filtering is to produce posterior particles with a distribution $p_{m+1|m+1}$ at the future time t_{m+1} that accounts for the statistical information about observations

$$\vec{v}_{m+1} = \vec{g}(\vec{u}_{m+1}) + \vec{\sigma}^o_{m+1}, \quad \vec{\sigma}^o_{m+1} \sim \mathcal{N}(0, r^o \mathcal{I}),$$

where the observation operator $\vec{g}$ can be nonlinear.

The first step in particle filtering is to produce a prior distribution at the observation time, $p_{m+1|m}$. Theoretically, for a dynamical system governed by the following stochastic differential equations,

$$d\vec{u} = \vec{f}(\vec{u}, t)dt + \vec{\sigma}(\vec{u}, t)d\vec{W}(t), \tag{15.2}$$

where $\vec{W}(t)$ is a vector of independent Wiener processes, as noted in Chapter 1, the exact prior density is a solution of the Fokker–Planck equation (or Liouville equation when the dynamical system is deterministic, $\vec{\sigma} = 0$) at finite time $t_{m+1} > t_m$ with initial condition $p(\vec{u}, t_m) = p_{m|m}(\vec{u})$, i.e.

$$\begin{aligned} p_t &= -\nabla \cdot (\vec{f} p) + \frac{1}{2}\nabla^2(\vec{\sigma} \otimes \vec{\sigma}^T p) \\ p(\vec{u}, t_m) &= p_{m|m}(\vec{u}). \end{aligned}$$

In practical particle filtering, one does not solve this Fokker–Planck equation since it is unfeasible for high-dimensional problems and we only have a finite sample of the initial distribution $p_{m|m}$ through particles $\{\vec{u}^k_{m|m}\}_{k=1,\dots,K}$. Typically, one propagates these particles forward in time by solving the SDE in (15.2) with initial conditions $\{\vec{u}^k_{m|m}\}_{k=1,\dots,K}$ to obtain a prior ensemble $\{\vec{u}^k_{m+1|m}\}_{k=1,\dots,K}$ and assumes that each prior particle is equally important, i.e.

$$p_{m+1|m}(\vec{u}) = \frac{1}{K}\sum_{k=1}^{K} \delta(\vec{u} - \vec{u}_{m+1|m}). \tag{15.3}$$

Given an observation likelihoood function $p(\vec{v}_{m+1}|\vec{u})$, the posterior density is produced through applying Bayes' formula as follows,

$$\begin{aligned} p_{m+1|m+1}(\vec{u}) = p(\vec{u}|\vec{v}_{m+1}) &= \frac{p(\vec{v}_{m+1}|\vec{u})\,p_{m+1|m}(\vec{u})}{\int p(\vec{v}_{m+1}|\vec{u})\,p_{m+1|m}(\vec{u})\,d\vec{u}} \\ &= \sum_{k=1}^{K} w^k_{m+1}\delta(\vec{u} - \vec{u}_{m+1|m}), \end{aligned}$$

where the first equality is true in the L^2-sense (see, e.g. Kallianpur, 1980; Oksendal, 2003; Bensoussan, 2004), the second equality is simply Bayes' theorem, and the third equality is the finite sample approximation through particle filtering with new weights

$$w^k_{m+1} = \frac{p(\vec{v}_{m+1}|\vec{u}^k_{m+1|m})}{\sum_{k=1}^{K} p(\vec{v}_{m+1}|\vec{u}^k_{m+1|m})}.$$

Notice here that the Bayes formula simply re-weights the particles. The main issue in practice is that many particles tend to have low weights in high-dimensional systems or in systems with a moderate dimensional chaotic attractor. If this two-step process is repeated over and over, then typically only one ensemble member will remain to have a large weight and the remaining become negligible (Bickel *et al.*, 2008; Bengtsson *et al.*, 2008; Snyder *et al.*, 2008; van Leeuwen, 2009). One way to avoid this ensemble collapse is to resample the particles (Doucet *et al.*, 2008) by duplicating those particles with large weights to replace those with small weights. The advantage of this resampling (known as sequential important resampling or SIR) is that we do not modify the particles so each ensemble member is dynamically balanced but the disadvantage is that we can only use this technique for a stochastic system as in (15.2) and a more serious problem is that the uncertainty of the prior particles is mostly due to the model uncertainty. There are various clever resampling strategies (Chorin and Krause, 2004; Del Moral, 1996; Del Moral and Jacod, 2001; Rossi and Vila, 2006, Chorin and Tu, 2009) but none of these are applicable for high-dimensional systems (Bickel *et al.*, 2008; Bengtsson *et al.*, 2008; Snyder *et al.*, 2008; van Leeuwen, 2009).

15.2 Innovative particle filter algorithms

In this section, we describe a particle filtering strategy with small ensemble size. The main idea (Anderson, 2010) is to use the particles to construct order statistics for marginal distributions, to update with the Bayes formula, and then resample the posterior particles in an orderly fashion such that they are equally important marginally. There are various ways to construct these marginal densities and in the next two sections we will discuss one method that uses discontinuous marginal densities: the rank histogram filter (due to Anderson, 2010) and our method that applies the maximum entropy principle to construct continuous marginal densities (maximum entropy particle filter).

15.2.1 Rank histogram particle filter (RHF)

Consider equally important prior particles $\{\vec{u}^k_{m+1|m}\}_{k=1,\dots,K}$. Instead of using the N-dimensional approximate prior density function in (15.3), let us construct an approximate marginal density function on each particle component $u^k_{m+1|m,j} \in \mathbb{R}$, $j = 1, \dots, N$ (different coordinate expansions may be explored in the future). In each projected coordinate j, the RHF assumes the particles partition the real line into $K+1$ subintervals with equal weight $1/(K+1)$. First, let us reorder particles $\{u^k_{m+1|m,j}\}$ into $u^{(1)} < u^{(2)} < \cdots < u^{(K)}$. Then we can realize the equipartition constraint with the following prior marginal density function,

$$
\begin{aligned}
p_{m+1|m,j}(u) = {} & \frac{1}{K+1}\sum_{k=1}^{K-1} \chi_{[u^{(k)},u^{(k+1)}]}(u)(u^{(k+1)} - u^{(k)})^{-1} \\
& + p_G(u|\mu_1, \sigma)\chi_{(-\infty,u^{(1)}]}(u) + p_G(u|\mu_N, \sigma)\chi_{[u^{(K)},\infty)}(u),
\end{aligned}
\tag{15.4}
$$

where $\chi_{[a,b]}(u)$ is a characteristic function on the interval $[a, b]$ and $p_G(u|\mu, \sigma)$ denotes a Gaussian density with mean μ and standard deviation σ. The last two terms in (15.4) suggest that we use two Gaussian tails on the two unbounded intervals $(-\infty, u^{(1)}]$ and $[u^{(K)}, \infty)$ with variance

$$
\sigma^2 = \langle (u_{m+1|m,j} - \langle u_{m+1|m,j}\rangle)^2 \rangle,
$$

where $\langle\cdot\rangle$ denotes the empirical variance obtained from ensemble averaging, and means, μ_1, μ_K, obtained by solving the equal weight constraints,

$$
\int_{-\infty}^{u^{(1)}} p_G(u|\mu_1, \sigma)du = \frac{1}{K+1} \quad \text{and} \quad \int_{u^{(K)}}^{\infty} p_G(u|\mu_K, \sigma)du = \frac{1}{K+1}.
$$

Notice that the approximate prior density in (15.4) is a piecewise continuous density function.

The second step in RHF is to approximate the likelihood function between each pair of particles as a constant equal to the average of the likelihood evaluated at the bounding particles and as a constant equal to the value of the likelihood function evaluated at the outermost particle member, that is,

$$\tilde{p}(v_{m+1,j}|u) \approx \sum_{k=1}^{K-1} \frac{1}{2}(p(v_{m+1,j}|u^{(k)}) + p(v_{m+1,j}|u^{(k+1)}))\chi_{[u^{(k)},u^{(k+1)}]}(u)$$
$$+p(v_{m+1,j}|u^{(1)})\chi_{(-\infty,u^{(1)}]}(x) + p(v_{m+1,j}|u^{(K)})\chi_{[u^{(K)},\infty)}(u). \tag{15.5}$$

The third step is to apply the Bayes theorem to update the approximate marginal density (15.4) to obtain an approximate posterior density

$$p(u|v_{m+1,j}) \approx \frac{\tilde{p}(v_{m+1,j}|u)p_{m+1|m,j}(u)}{\int_{\mathbb{R}} \tilde{p}(v_{m+1,j}|u)p_{m+1|m,j}(u)\,du}. \tag{15.6}$$

Notice that this approximate posterior density is also piecewise continuous with discontinuity points at the prior particles, $u^{(j)}$. The new particles, $\tilde{u}^{(1)} < \tilde{u}^{(2)} < \cdots < \tilde{u}^{(K)}$, are obtained through solving the nonlinear equations,

$$\int_{-\infty}^{\tilde{u}^{(1)}} p(u|v_{m+1,j})du = \frac{1}{K+1}, \tag{15.7}$$

$$\int_{u^{(k)}}^{\tilde{u}^{(k+1)}} p(u|v_{m+1,j})du = \frac{1}{K+1}, \quad k = 1, \ldots, K-1. \tag{15.8}$$

Thus, the jth component of the posterior particles, $u^k_{m+1|m+1,j}$, is obtained by reversing the ordered particles $\tilde{u}^{(k)}$ with respect to the ordering obtained from constructing the marginal prior distribution in (15.4). To find $\tilde{u}^{(1)}$, one inverts a cumulative distribution function of the normal distribution in Eqn (15.7). Subsequently, one can find $\tilde{u}^{(k+1)}$ incrementally from $k = 1$ to $K-1$ by solving the linear equation in (15.8). Here, Eqn (15.8) is linear because the approximate prior marginal density in (15.4) and likelihood function (15.5) are constant in the interior partitions. Note that one can also use a linear function to approximate the likelihood function as noted by Anderson (2010); in that case, Eqn (15.8) is a quadratic function.

This particle filtering strategy is computationally cheap because we sample marginally in a way that will not lead to ensemble collapse and each marginal sample can be done in parallel. Furthermore, this strategy is not restricted to stochastic dynamical models because we don't have multiple copies of identical particles as in the importance sampling particle filter mentioned in Section 15.1. One potential drawback is that the posterior particles may not be dynamical balanced especially since the RHF resamples from coarse approximate marginal distributions. Another technical problem that we will face is when all of the prior particles are in the tail of the observation likelihood function, then we will have difficulty in computing the normalizing constant of the posterior distribution (the denominator in (15.6)). We will address this dynamic range issue in Section 15.2.3.

15.2.2 Maximum entropy particle filter (MEPF)

In this section, we introduce one way to construct a smooth prior density function which matches the empirical statistics from the particle distributions. In particular, we will use the maximum entropy principle (Majda *et al.*, 2005; Majda and Wang, 2006) to construct a one-dimensional prior marginal density. Given the jth component of the prior particles $\{u^k_{j,m+1|m}\}_{k=1,\ldots,K}$, our goal is to find the least biased one-dimensional prior marginal density $p^L_{j,m+1|m}$ by maximizing the Shannon entropy,

$$\mathcal{S}(p) = -\int_{\mathbb{R}} p(u)\log(p(u))du, \tag{15.9}$$

subject to the following empirical moments constraints,

$$\begin{aligned}
\int_{\mathbb{R}} p^L_{j,m+1|m}(u)du &= 1,\\
\int_{\mathbb{R}} (u-\bar{u}_j)p^L_{j,m+1|m}(u)du &= 0\\
\int_{\mathbb{R}} (u-\bar{u}_j)^p p^L_{j,m+1|m}(u)du &= M_{j,p}, \quad 2 \le p \le 2L,
\end{aligned} \tag{15.10}$$

where $\bar{u}_j$ denotes the jth component of the prior ensemble mean,

$$\bar{u}_j = \frac{1}{K}\sum_{k=1}^{K} u^k_{j,m+1|m},$$

and $M_{j,p}$ denotes the jth component of the prior ensemble centered p-th moment,

$$M_{j,p} = \frac{1}{K}\sum_{k=1}^{K}(u^k_{j,m+1|m} - \bar{u}_j)^p, \quad 2 \le p \le 2L.$$

Here, we use an even number of moments to ensure a decaying solution for the constraint optimization problem above of the following form

$$p^L_{j,m+1|m}(u) \propto \exp\Big(\sum_{\ell=0}^{2L} \alpha_\ell (u-\bar{u}_j)^\ell\Big), \tag{15.11}$$

where α_ℓ are the Lagrange multipliers for the constraints (15.10). For a one dimensional marginal distribution with L sufficiently low (in our application we only consider up to fourth moments with $L = 2$), these Lagrange multipliers can be obtained instantly (in our numerical implementation, we use the routine developed by Abramov (2006, 2007, 2009, 2010)). See also the books by Majda *et al.* (2005) and Majda and Wang (2006), for many applications of such maximum entropy principles to fluid dynamics, and Haven *et al.* (2005) which applies these principles to small ensemble size prediction.

Assume that observations are available at discrete times $m\Delta t, m = 1, 2, 3, \ldots$, and they involve separate coordinates; specifically, we consider nonlinear observations of the following form

$$v_{j,m} = \sum_{\ell=0}^{L'} g_{j,\ell} u_{j,m}^{\ell} + \sigma_{j,m}^{o}, \quad 1 \leq j \leq N, \tag{15.12}$$

where the noise $\sigma_{j,m}^{o}$ is assumed to be spatially and temporally independent. Here, index j denotes the observation coordinate and plentiful (full) observations are available when $g_{j,\ell} \neq 0$ for every j in (15.12). For simplicity, we assume that the noise $\sigma_{j,m}^{o}$ is Gaussian with mean zero and covariance $r_j^o > 0$; the associated marginal likelihood function is

$$p(v_{j,m}|u) \propto \exp\Big(- \frac{(v_{j,m} - \sum_{\ell=0}^{L'} g_{j,\ell} u^{\ell})^2}{2r_j^o} \Big). \tag{15.13}$$

The form in (15.12) is the typical one when a turbulent dynamical system is partially observed at isolated spatial locations.

Given the prior marginal density in (15.11) and the marginal likelihood function in (15.13), compute the posterior marginal distribution by taking into account the observation of $v_{j,m+1}$ with the Bayes formula,

$$\begin{aligned} p_{j,m+1|m+1}^{L}(u) &\equiv p(u|v_{j,m+1}) \propto p(u)p(v_{j,m+1}|u) \\ &= p_{j,m+1|m}^{L}(u)p(v_{j,m+1}|u) \\ &= \exp\Big(\sum_{i=0}^{2L} \alpha_i (u - \bar{u}_j)^i - \frac{(v_{j,m+1} - \sum_{\ell=0}^{L'} g_{j,\ell} u^{\ell})^2}{2r_j^o} \Big). \end{aligned} \tag{15.14}$$

Assume $L' \leq L$; then $p_{j,m+1|m+1}^{L}$ is a one-dimensional exponential family of polynomials of order $2L$ (exactly like the prior marginal density in (15.11)), i.e.

$$p_{j,m+1|m+1}^{L}(u) = \frac{1}{Z} \exp\Big(\sum_{i=0}^{2L} \lambda_i (u - \bar{u}_j)^i \Big),$$

where Z is a normalization constant such that $\int_{\mathbb{R}} p_{j,m+1|m+1}^{L} = 1$. In our implementation with $L = L' = 2$, we perform the following algebra to avoid operations that simultaneously involve large and small numbers. First, we rewrite

$$\begin{aligned} v_{j,m+1} - \sum_{\ell=0}^{2} g_{j,\ell} u^{\ell} &= v_{j,m+1} - g_{j,0} + g_{j,2}(\bar{u}_j)^2 - (g_{j,1} + 2g_{j,2}\bar{u}_j)u \\ &\quad - g_{j,2}(u - \bar{u}_j)^2 \\ &= v_{j,m+1} - g_{j,0} + g_{j,2}(\bar{u}_j)^2 - (g_{j,1} + 2g_{j,2}\bar{u}_j)\bar{u}_j \\ &\quad - (g_{j,1} + 2g_{j,2}\bar{u}_j)(u - \bar{u}_j) - g_{j,2}(u - \bar{u}_j)^2 \\ &= \beta_0 - \beta_1(u - \bar{u}_j) - \beta_2(u - \bar{u}_j)^2, \end{aligned} \tag{15.15}$$

where

$$\beta_0 = v_{j,m+1} - g_{j,0} - g_{j,1}\bar{u}_j - g_{j,2}(\bar{u}_j)^2$$
$$\beta_1 = g_{j,1} + 2g_{j,2}\bar{u}_j$$
$$\beta_2 = g_{j,2}.$$

Taking the square of (15.15), we obtain

$$\beta_2^2(u-\bar{u}_j)^4 + 2\beta_1\beta_2(u-\bar{u}_j)^3 + (\beta_1^2 - 2\beta_0\beta_2)(u-\bar{u}_j)^2 - 2\beta_0\beta_1(u-\bar{u}_j) + \beta_0^2,$$

and thus

$$\lambda_0 = \alpha_0 - \frac{\beta_0^2}{2r_j^o}, \quad \lambda_1 = \alpha_1 + \frac{\beta_0\beta_1}{r_j^o}, \quad \lambda_2 = \alpha_2 - \frac{1}{2r_j^o}(\beta_1^2 - 2\beta_0\beta_2),$$
$$\lambda_3 = \alpha_3 - \frac{\beta_1\beta_2}{r_j^o}, \quad \lambda_4 = \alpha_4 - \frac{\beta_2^2}{2r_j^o}. \tag{15.16}$$

We compute the normalization factor Z with the Gauss–Hermite quadrature rule. In our implementation, we only use 20 mesh points $\{u_i\}$ as follows

$$Z = \int_{\mathbb{R}} \exp(-|\lambda_2|(u-\bar{u}_j)^2) f(u)du \approx \sum_{i=1}^{20} w_i f(u_i), \tag{15.17}$$

where

$$f(u) = \exp\Big(\sum_{j=0}^{2L} \lambda_j (u-\bar{u}_j)^j + |\lambda_2|(u-\bar{u}_j)^2\Big),$$

and w_i is the corresponding weight at u_i.

Then, we generate the posterior particles as in RHF in (15.7)–(15.8). In our implementation, however, we choose to solve the following independent nonlinear equations:

$$\int_{-\infty}^{\tilde{u}^{(k)}} p_{j,m+1|m+1}^L = \frac{k}{q+1}, \quad k = 1, \ldots, K, \tag{15.18}$$

to avoid potential numerical errors from the integral approximation that may occur if one computes the particles sequentially as in (15.7)–(15.8). This particle resampling technique involves standard algorithms for solving nonlinear equations such as the bisection line search or the Newton-based methods. In each optimization step, we use the standard composite trapezoidal rule with 100 mesh points to approximate the cumulative distribution functions in (15.18). Here, the domain of integration is determined by the smallest and largest roots of a Hermite polynomial of degree 20 that are available to us from the previous step in (15.17). Finally, we reorder the particles based on the ordering of the prior particles and obtain the jth component of the posterior particles $\{u_{j,m+1|m+1}^k\}_{k=1,\ldots,K}$. The other components of the posterior particles can be computed similarly in parallel fashion whenever their corresponding observations are available.

15.2.3 Dynamic range issues for implementing MEPF

Practically, we encounter several dynamic range issues when we implement the MEPF algorithm in various tough regimes. Some of these issues are also found in our implementation of RHF but they weren't reported by Anderson (2010) because only large observation noise variance was considered in that article. Below, we list all the issues we found from implementing the MEPF and RHF on the three-dimensional L-63 model for linear observation models $\{g_{j,0} = 0, g_{j,1} = 1, g_{j,2} = 0\}$, analyze the particular assimilation step when these dynamic range issues are encountered, and provide remedies for these issues. We will report the full numerical results for the enhanced MEPF, accounting for all these remedies, in Section 15.3 for filtering the L-63 model and in Section 15.4 for filtering the L-96 model.

15.2.4 Dynamic range A

When prior distributions are in the tail of the observation likelihood function, the normalization constant, Z, is too small, and essentially the prior particles do not provide any skill. We call this scenario the dynamic range A. As a simple remedy, we will set up a criterion such that the posterior density is exactly the observation likelihood function.

As an example, see Fig. 15.1 for assimilation step 207 on variable x of the L-63 model. In the top panel, we show the prior density (in dashes) obtained from maximizing Shannon's entropy in (15.9) subject to the first four empirical moments (15.10); to check the accuracy, we compare the cumulative distribution functions (cdf) obtained from the continuous prior density function through the analytical formula

$$\mathrm{cdf}(x) = \int_{-\infty}^{x} \exp\Big(\sum_{j=0}^{2L} \alpha_j (y-a)^j\Big) dy, \tag{15.19}$$

and from the equal weight prior particles x_j (see the middle panel) with empirical formula

$$\text{empirical cdf}(x) = \frac{1}{K+1} \sum_{j=1}^{K} \delta(x_j \leq x). \tag{15.20}$$

In this particular case, the dynamic range A occurs; all of the prior particles (diamonds) are in the tail of the likelihood function (the dash-dotted line; here we cannot see it because it coincides with the solid line that denotes the posterior density). In our implementation, we set the posterior to be identically the likelihood function whenever Z is less than some tolerance (we choose it to be 10^{-4}). Subsequently, we compute the posterior particles (see the squares) as prescribed in (15.18); to double-check the accuracy, we also compare the cdf obtained from the continuous posterior density function and the equal weight posterior particles (see the bottom panel).

We also found this dynamic range issue in our implementation of RHF of Section 15.2.1 and we essentially proceed with the same remedy in our implementation. A similar issue

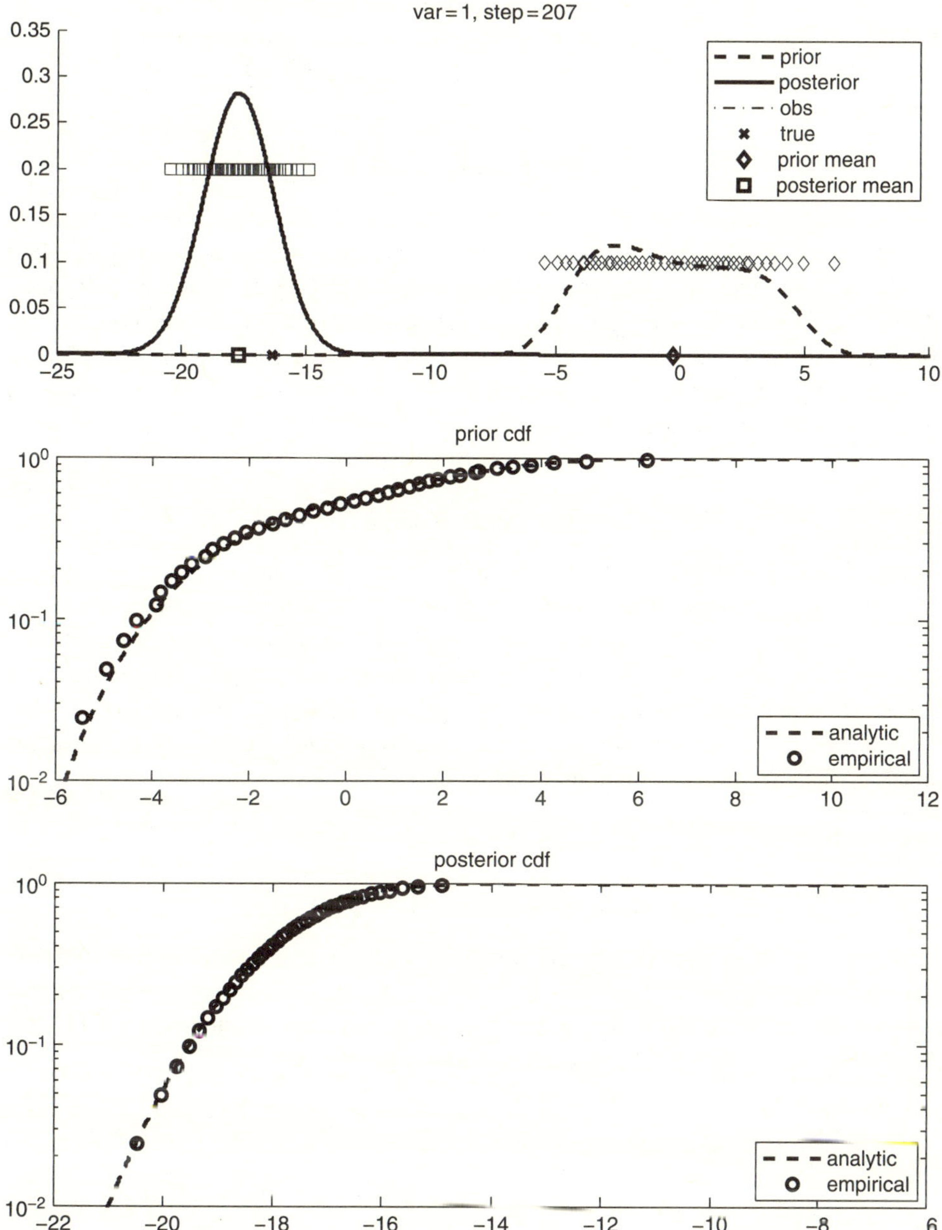

Figure 15.1 Dynamic range A: The top panel shows the prior and posterior densities and the observation likelihood function. Here the posterior is identical to the observation likelihood function since we fully trust the observation. We also show the true (×), posterior particles (square) and prior particles (diamond). In the two lower panels, we show the continuous and empirical cumulative distribution functions; see Eqns (15.19), (15.20), respectively.

was also reported by van Leeuwen (2009); indeed, the key point of his paper is to avoid this dynamic range issue by constraining the prior particles to be sufficiently close to the observations using an ad hoc nudging technique which he designed.

15.2.5 Dynamic range B

We define the dynamic range B to be a scenario where the normalization constant Z is large; such an issue occurs whenever the density function has unbounded tails (see Fig. 15.2). Notice that the prior particles cluster around 17.5 (see the top and middle panels of Fig. 15.2) but the cdf is infinite due to an unbounded left tail. When this issue occurs, we propose to use a Gaussian approximation, that is, set $\alpha_3 = \alpha_4 = 0$ and proceed. See for example in Fig. 15.3 that the posterior particles encompass the hidden true signal (denoted by the "$\times$" sign) after such a correction.

15.2.6 Dynamic range C

We found another dynamic range issue that occurs in the computations of the prior density function. In this extremely stiff regime, the particles are clustered into two groups such that the maximum entropy scheme diverges since the optimization problem has infeasible constraints (Junk, 2004). We call this scenario the dynamic range C. For example (see Fig. 15.4), there is a mismatch between the analytical and empirical prior cdf (see the middle panel). In our implementation, we set the posterior to equal the observation likelihood function whenever the maximum entropy minimization does not produce accurate prior distributions.

15.3 Filter performance on the L-63 model

In this section, we compare numerical results of MEPF with the RHF described in Section 15.2.1 and the EAKF described in Chapter 9 on the L-63 model. For the reader's convenience, we repeat this system of three equations here and refer to Section 9.4 for the detailed parameters and chaotic properties of this system:

$$\begin{aligned}\frac{dx}{dt} &= \sigma(y-x),\\ \frac{dy}{dt} &= \rho x - y - xz,\\ \frac{dz}{dt} &= xy - bz.\end{aligned} \tag{15.21}$$

In each of these numerical simulations, we do not apply the local-least squares framework described in Chapter 11 on the EAKF and we implement the EAKF with multiplicative variance inflation $r = 5\%$ (see Chapter 9). Below, we present numerical results for various regimes with different ensemble size, observation time, observation noise size,

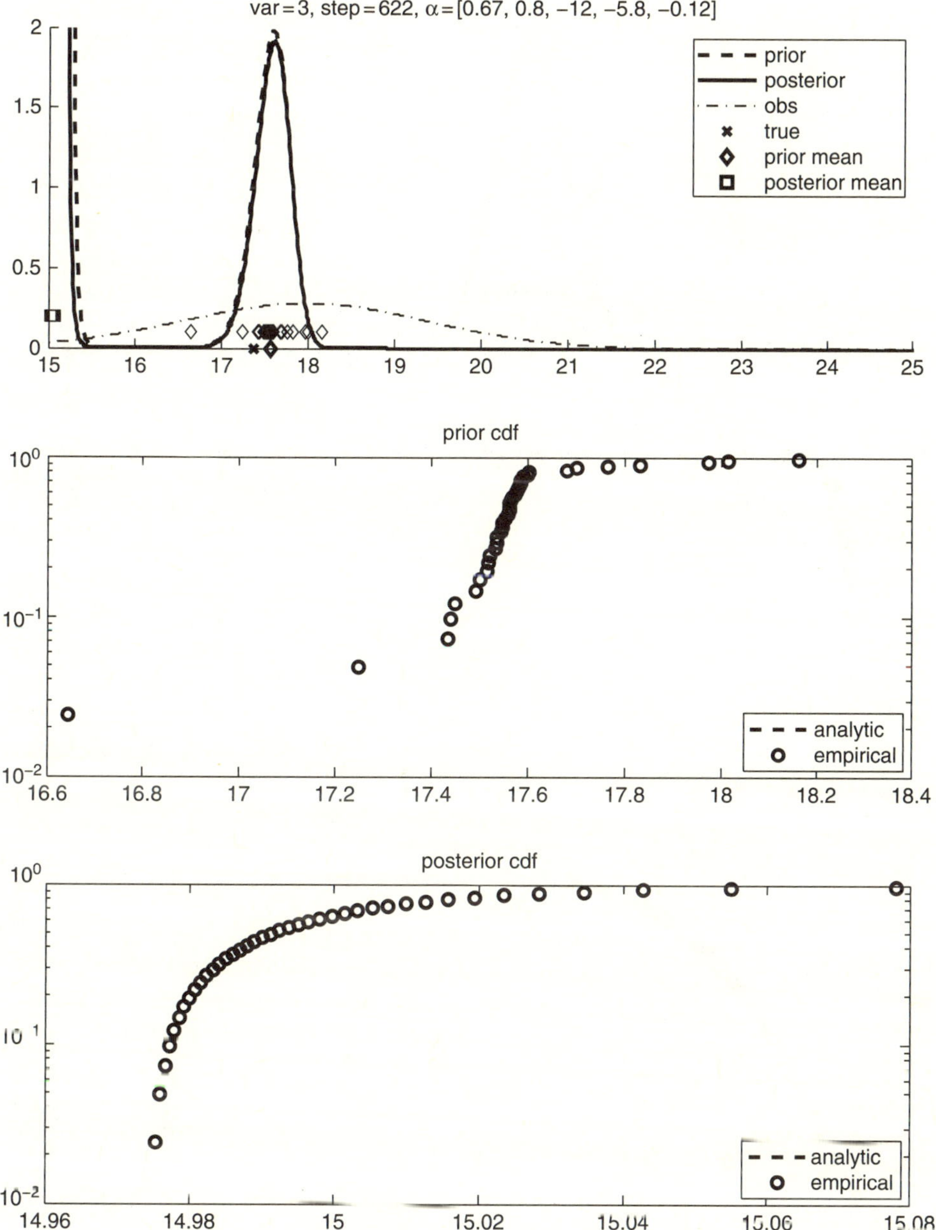

Figure 15.2 Dynamic range B: The top pdfs are not normalized for illustration purpose since $Z \gg 10^4$.

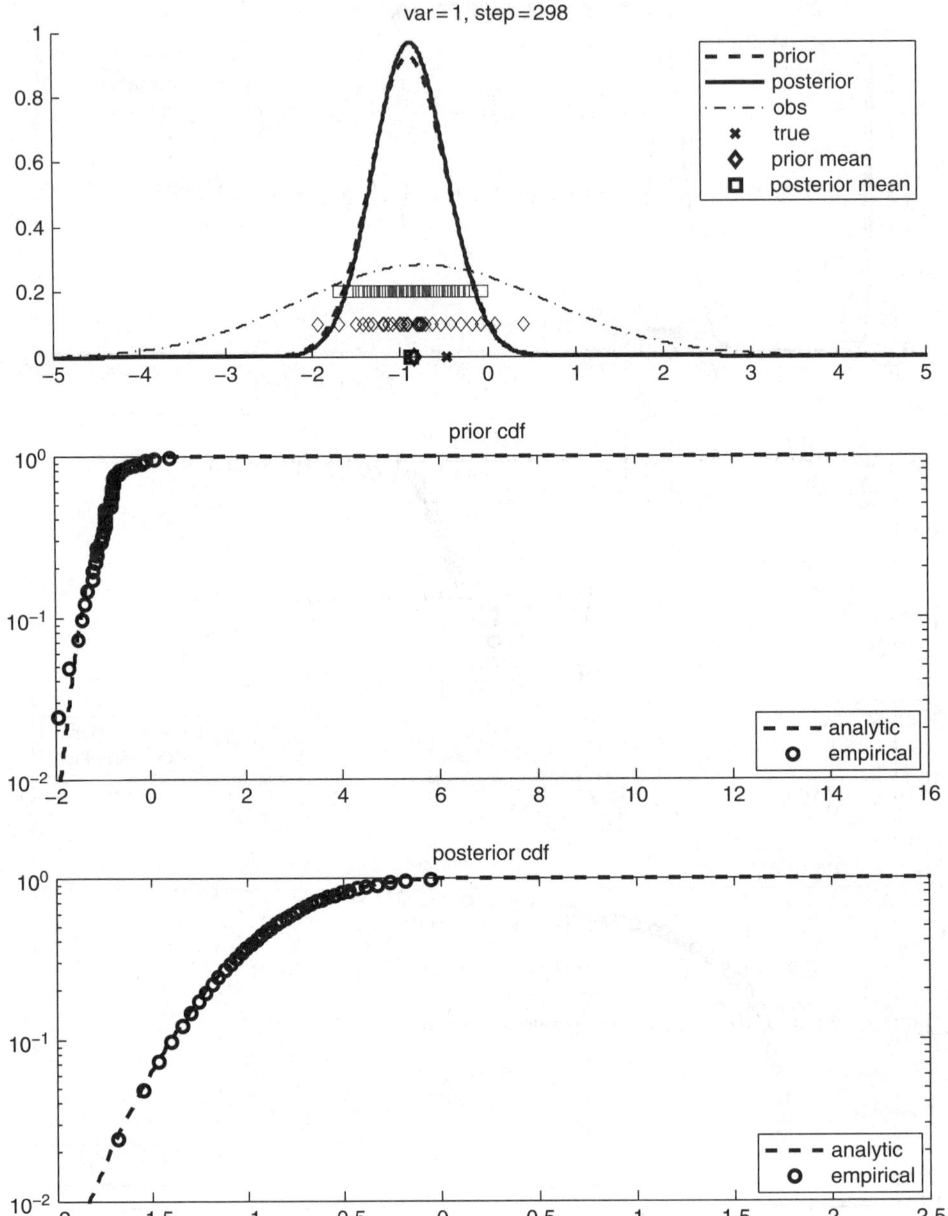

Figure 15.3 Here, we fix the dynamic range B issue by forcing the prior to be Gaussian by setting $\alpha_3 = \alpha_4 = 0$ whenever unbounded tails occur.

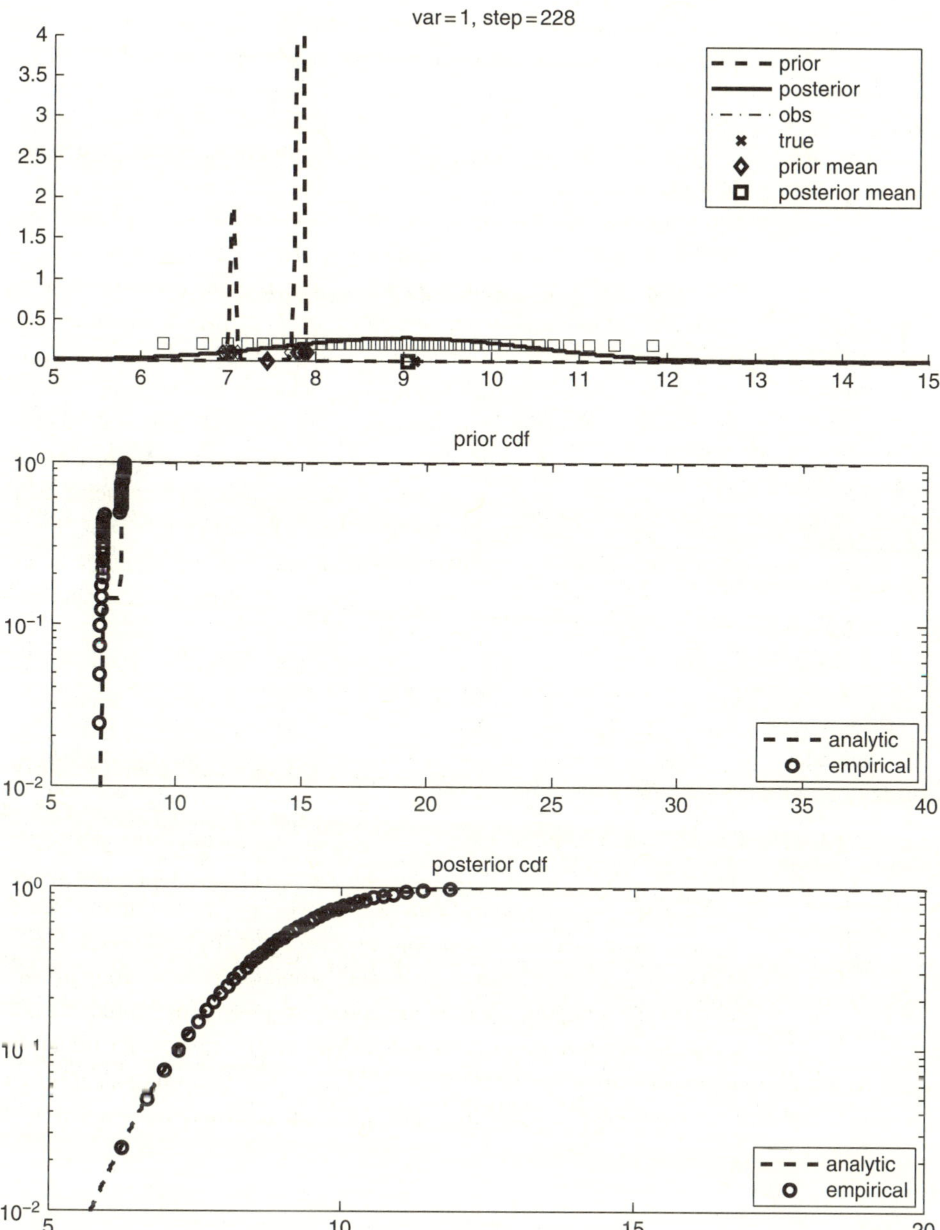

Figure 15.4 Dynamic range C: The entropy optimization is so stiff since the prior particle consists of two clustered.

observation network, and linear and nonlinear observation operators. As a reminder (see also Section 9.4), this model is chaotic with one positive Lyapunov exponent 0.906 that corresponds to a doubling time of 0.78 units. The shortest correlation time is 0.16 units and the natural variabilities (i.e. temporally averaged standard deviations) for x, y and z, are 7.91, 8.98, 8.6, respectively. In the following numerical experiments, we use these quantities as guidelines for choosing observation times and noises.

15.3.1 Regime where EAKF is superior

In this regime, we consider a large ensemble size $K = 40$, plentiful observations with relatively small noise variance $r^o = 0.25$ (compare to the square of the natural variability above), and a relatively short observation time $T_{\text{obs}} = 0.08$ (this is one-half of the correlation time). In this regime, we expect the Kalman-based EAKF to be superior since the linear and Gaussian approximations are relatively accurate here. Also, recall the results for EAKF in Chapter 9 for L-63 with this observation time and small ensemble size. To see the filter performance, we show the average RMS error as a performance measure on this short run (only 1000 assimilation cycles) in Table 15.1 and the corresponding time series of the average RMS error in Fig. 15.5. Based on the average RMS error in Table 15.1, EAKF seems to perform the best. However, when we observe carefully the time series of this RMS measure in Fig. 15.5 notice that the three schemes are sometimes accurate below the observation noise error (horizontal dashes) but occasionally inaccurate with large errors.

To understand why these large erratic errors occur, we display the prior and posterior distributions of each scheme at the times when the filtered solution is not accurate. In particular, EAKF and RHF fail whenever the particles are in the tail of the likelihood function (or they exhibit the dynamic range A scenario discussed in Section 15.2.3). See for example Fig. 15.6 for an assimilation at step 803 where EAKF fails, and notice that the prior ensembles of the RHF (black diamonds) are all on the tail of the likelihood function (dash-dotted line). In this case, the prior particles from the MEPF and RHF are on top of this likelihood function and more importantly, they are closer to the true hidden state (see "$\times$" on the x-axis).

Now, let us discuss a case where the MEPF is inadequate (see Fig. 15.7). In this particular instance, the MEPF prior particles (diamonds) exhibit the dynamic range A scenario; the

Table 15.1 Average RMS errors for 1000 assimilation cycles with an ensemble size of 40, plentiful observations with short observation time $T_{\text{obs}} = 0.08$, and observation noise variance $r^o = 0.25$.

Scheme	prior (x)	prior (y)	prior (z)	post (x)	post (y)	post (z)
EAKF	0.36	0.57	0.49	0.25	0.40	0.34
RHF	0.52	0.80	0.74	0.46	0.57	0.52
MEPF	0.39	0.60	0.57	0.38	0.46	0.39

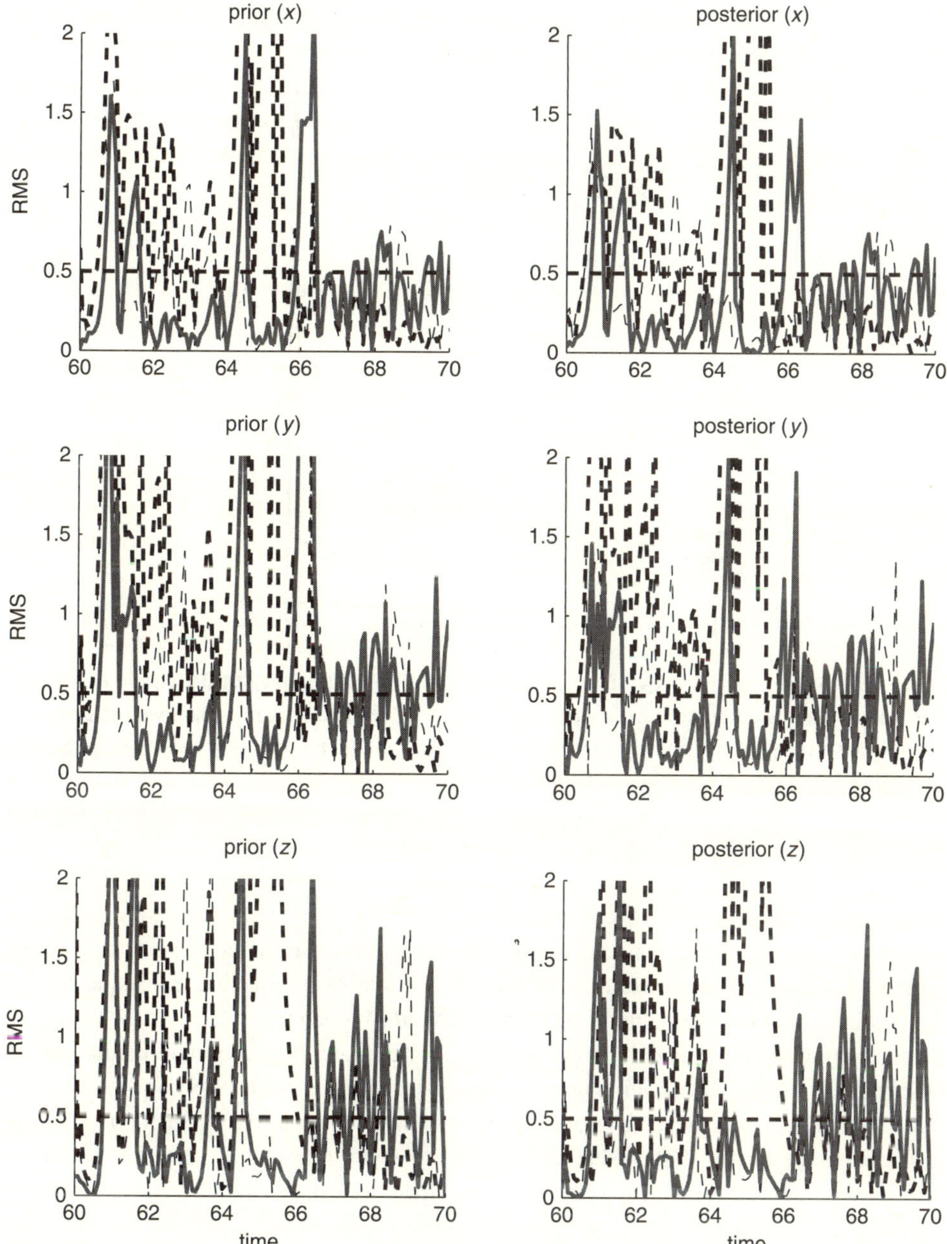

Figure 15.5 RMS errors as functions of time: $T_{\text{obs}} = 0.08$, $r^o = 0.25$. EAKF (thick dashes), MEPF (gray solid), and RHF (thin dashes). The first column denotes the RMS prior error and the second column denotes the posterior RMS error; each row denotes each variable of the L-63 model.

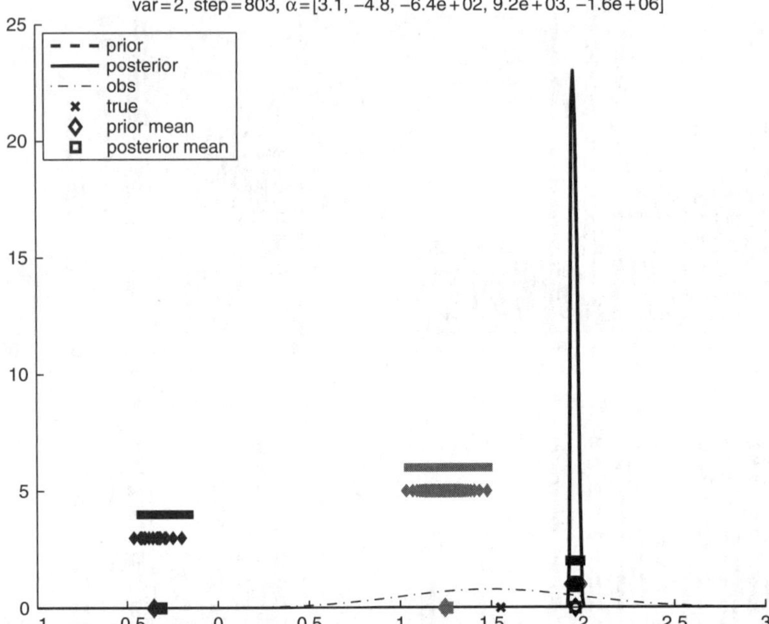

Figure 15.6 Regime where EAKF fails: MEPF continuous prior (dashes); MEPF continuous posterior (solid). They are both on top of each other. MEPF is denoted by white diamonds and squares (they are near the posterior density, close to 2), RHF (gray diamonds and squares with mean around 1.25), and EAKF (black diamond and squares with mean around −0.4). The true hidden state is denoted by '×' on the x-axis; the other diamonds and squares on the x-axis correspond to the mean prior and posterior of the corresponding schemes.

hidden true state (denoted by "×" on the x-axis) is in the tail of both the prior distribution and the likelihood function (much closer to the prior). As we discussed in Section 15.2, our remedy was to set the posterior density equal to the likelihood function, consequently the posterior particles (squares) are shifted further from the hidden true relative to the prior particles (diamonds). This scenario, that is, the hidden true state, is on the tail of the likelihood observation function due to large observation noise variance, is an ill-posed state estimation problem and will never be solved with the Bayesian framework for non-Gaussian distributions because the Bayesian setup is designed to approximate the distribution of solutions as opposed to estimate solutions. In contrast, we can use the mean and covariance of the Kalman-based (or Gaussian) filters for state estimation because these two statistics characterize a Gaussian distribution.

15.3.2 Regime where non-Gaussian filters (RHF and MEPF) are superior

Here, we consider longer observation time $T_{\text{obs}} = 0.25$ (beyond the shortest correlation time 0.16), small observation noise variance $r^o = 0.01$, and partial observations. In this

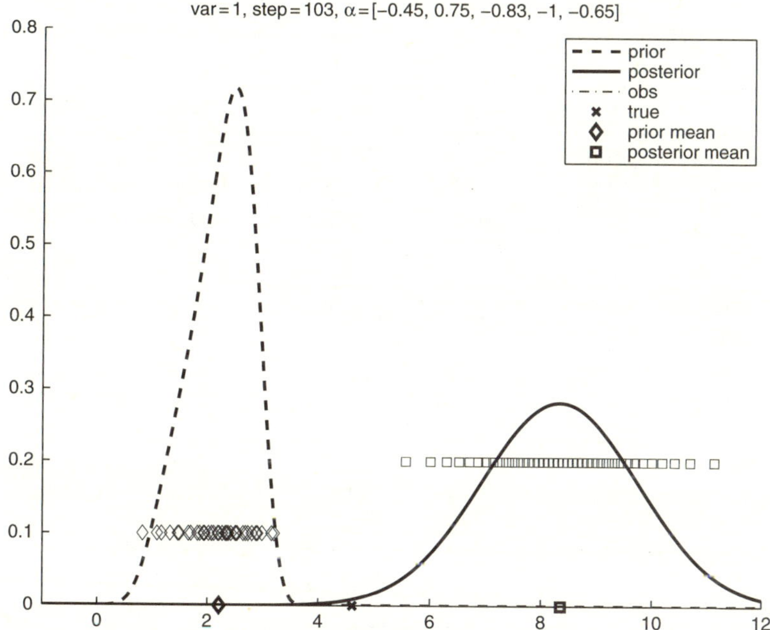

Figure 15.7 MEPF failure for the dynamic range A issue: MEPF continuous prior (dashes); MEPF continuous posterior (solid). MEPF prior and posterior particles are denoted by diamonds and squares, respectively.

section, we ignore the Gaussian filter EAKF and only consider the two non-Gaussian filters, RHF and EAKF, because EAKF diverges most of the time unless we extensively tune the filter parameters such as ensemble size, covariance inflation, or perhaps apply the sequential technique as we described in Chapter 11.

In Table 15.2, we show the average RMS errors over 10,000 assimilation cycles on each variable for filtering one and two observations. In this extremely difficult regime, the skill of MEPF supersedes RHF based on the mean RMS error as a measure. Most impressively, the accuracy of the filtered solutions only degrades slightly when we reduce the ensemble size to only 3 (see the prior and posterior RMS errors for each variable in Figs 15.8–15.9). For example, in Fig. 15.8, we don't see much degradation on the prior as well as on the posterior errors between using ensemble size 3 and 50 when we observe only variable y (see Fig. 15.8). In another case when we observe only the $\{y, z\}$ variables, notice that the errors in the unobserved variable x remain small even with only three ensemble members (see Fig. 15.9). In Figs 15.10–15.11, we show how each ensemble member tracks the true solution for a particular time domain with 10 ensemble members and partial observations of only y and z. Here, one can see at time 2495, at least some posterior particles of MEPF track the true signal whereas with RHF all of the particles deviate from the true signal.

Table 15.2 Average RMS analysis errors for 10,000 assimilation cycles with an ensemble size of 50, observation time $T_{\text{obs}} = 0.25$, and observation noise variance $r^o = 0.01$.

Obs x	MEPF	RHF	Obs $\{x, y\}$	MEPF	RHF
RMS x	0.10	0.30	RMS x	0.08	0.16
RMS y	9.16	8.92	RMS y	0.09	0.16
RMS z	8.47	8.42	RMS z	0.26	0.47
Obs y			Obs $\{y, z\}$		
RMS x	0.67	1.09	RMS x	0.59	1.10
RMS y	0.09	0.14	RMS y	0.14	0.18
RMS z	0.88	1.52	RMS z	0.09	0.16
Obs z			Obs $\{x, z\}$		
RMS x	8.00	8.07	RMS x	0.47	0.29
RMS y	9.11	9.09	RMS y	6.99	7.44
RMS z	0.23	0.27	RMS z	0.29	0.26

15.3.3 Nonlinear observations

In this section, we consider numerical simulations with nonlinear observation

$$v_m = \sum_{\ell=0}^{2} g_\ell y_m^\ell + \sigma_m^o \equiv g(y_m) + \sigma_m^o, \tag{15.22}$$

of only y-variables. In Table 15.3, we show the prior and posterior RMS errors for simulations on various cases where we vary the nonlinear observation operator coefficients and observation time for fixed ensemble size $K = 25$ and observation noise variance $r^o = 0.01$. From these results, we see that RHF has no filtering skill whereas the MEPF retains reasonable accuracy for weakly nonlinear observations with operator $g(y) = 0.8y + 0.1y^2$ and shorter observation time $T_{\text{obs}} = 0.15$, again, based on the RMS error as a performance measure. As the nonlinearity in the observation operator becomes stronger or the observation time becomes larger, the errors of both filters increase tremendously; however the errors of MEPF are still smaller than the natural variability of the model (7.91, 8.98, 8.6 for variables x, y and z, respectively).

We also consider an ill-posed filtering problem with quadratic observation operator

$$v_m = y_m^2 + \sigma_m^o. \tag{15.23}$$

This is an ill-posed problem because the inverse of y^2 is non-unique (except when $y = 0$) and consequently it is impossible to indicate which root is the hidden true y given observation v. Note that in order to implement MEPF in this ill-advised situation, we need an adjustment of the algorithm to determine the domain of interest for computing the appropriate normalization constant Z. In particular, we have to resample around one of these

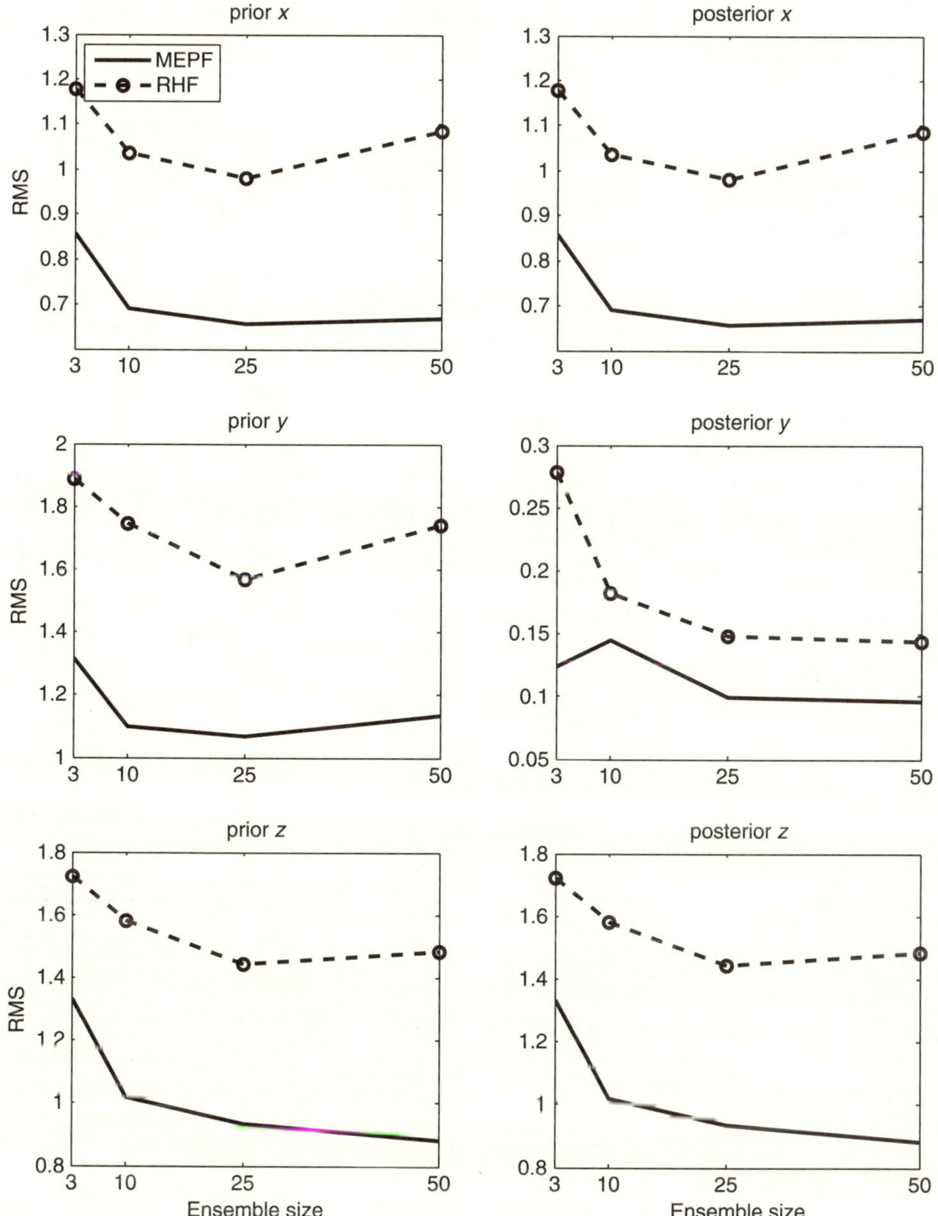

Figure 15.8 Partial observation $\{y\}$: Average RMS over 10,000 assimilation as a function of ensemble size for observation noise variance $r^o = 0.01$ and observation time $T_{\text{obs}} = 0.25$.

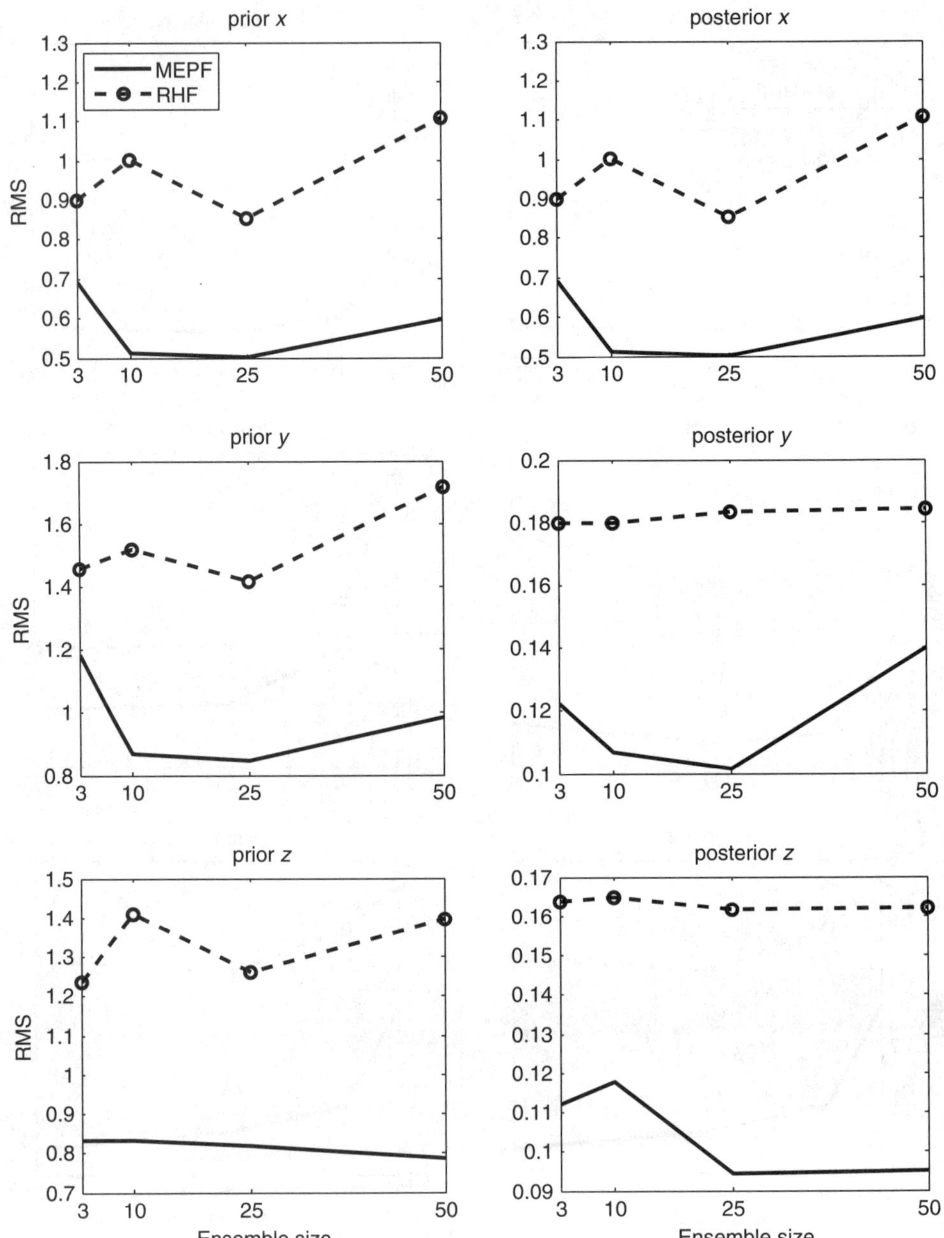

Figure 15.9 Partial observation $\{y, z\}$: Average RMS over 10,000 assimilation as a function of ensemble size for observation noise variance $r^o = 0.01$ and observation time $T_{\text{obs}} = 0.25$.

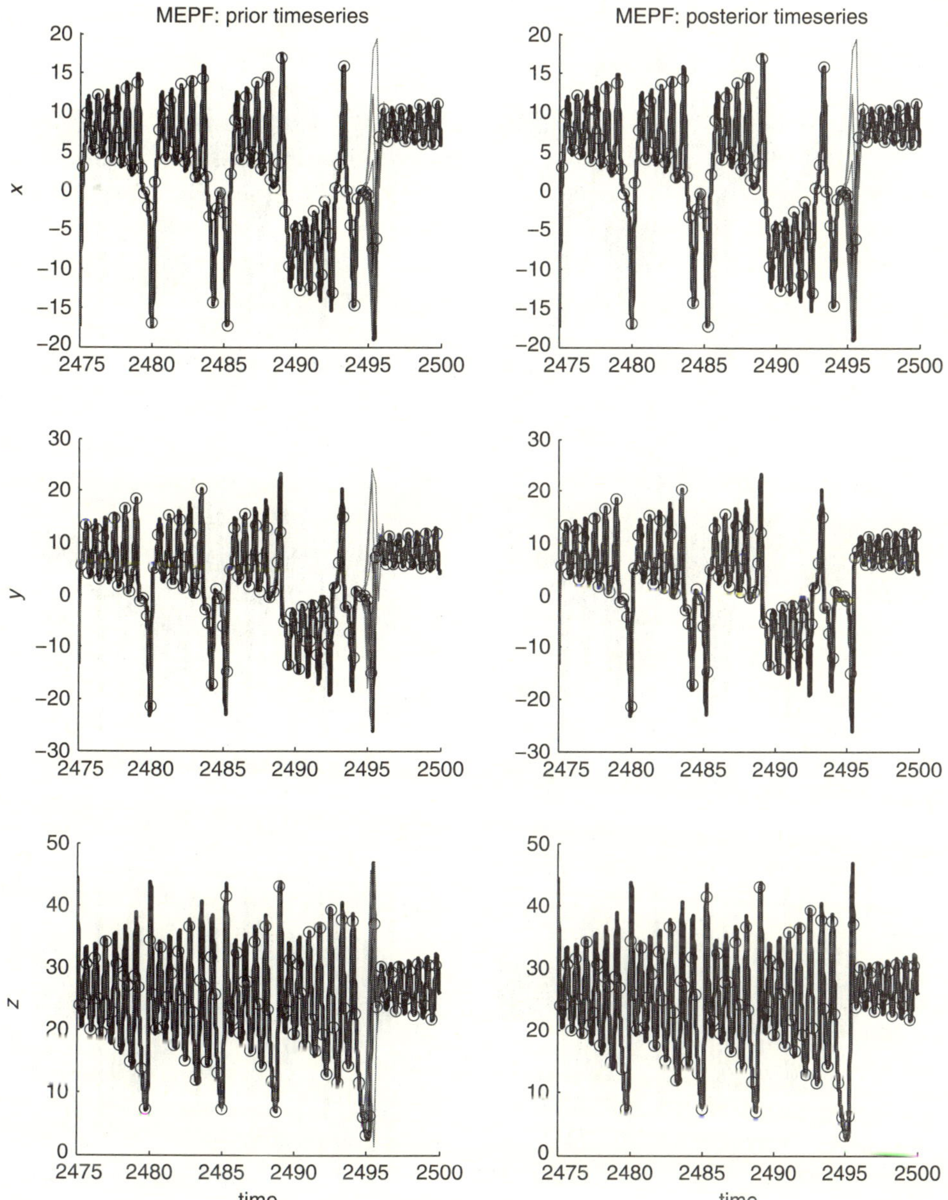

Figure 15.10 Time series with MEPF (partial observation $\{y, z\}$, observation time $T_{\text{obs}} = 0.25$, observation noise variance $r^o = 0.01$, ensemble size $K = 10$). particles (gray), true (black solid) and observation (circle).

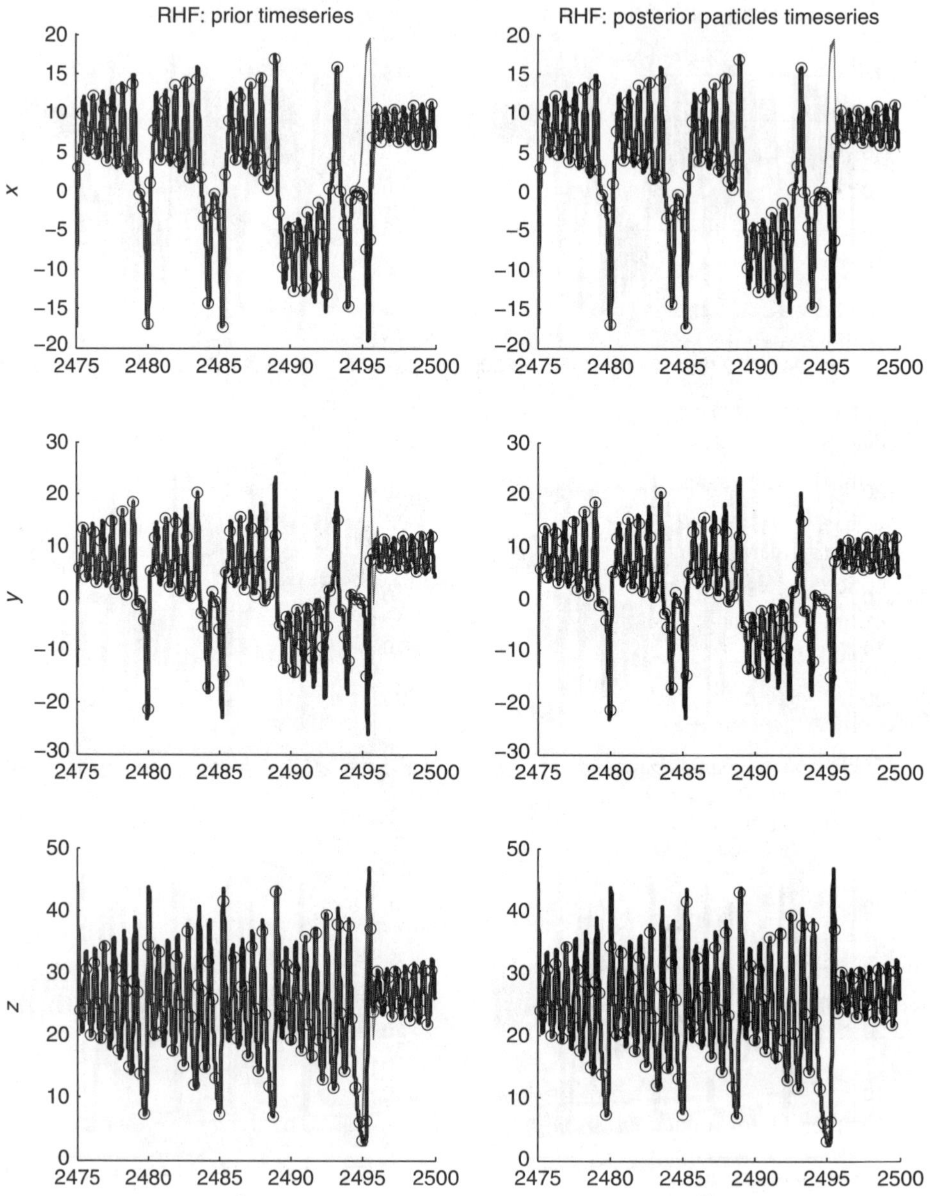

Figure 15.11 Time series with RHF (partial observation $\{y, z\}$, observation time $T_{\text{obs}} = 0.25$, observation noise variance $r^o = 0.01$, ensemble size $K = 10$): particles (gray), true (black solid) and observation (circle).

Table 15.3 Average RMS errors (over 10,000 cycles) of the observed variable y for simulations of the L-63 model, $r^o = 0.01$, $K = 25$.

$g(y)$	T_{obs}	Prior MEPF	Post MEPF	Prior RHF	Post RHF
$0.8y + 0.1y^2$	0.15	0.72	0.62	13.04	13.57
$0.8y + 0.1y^2$	0.25	2.98	2.18	10.95	13.07
$0.5y + 0.4y^2$	0.15	3.28	3.29	23.60	55.18

roots, otherwise we obtain two distinct clusters of particles (which are undesirable for state estimation). With the available information, the only sensible strategy is to pick the domain of interest to encompass one of the roots $\{\pm\sqrt{v}\}$ of the observation likelihood function that is the closest to the prior particle mean (which is an ad hoc choice here since one can perhaps also choose the median instead of the mean for example). Notice that if the noise σ_m^o is much smaller than y_m^2 in (15.23) such that v_m is negative with no real roots, we set $\sqrt{v_m} = 0$.

From a numerical simulation with $T_{\text{obs}} = 0.25$, $r^o = 0.01$, $K = 25$ (see Fig. 15.12), we notice that the posterior RMS error for the observed variable y is reasonably small until time 99.75 (or at assimilation cycle 399). The mechanism of the filter divergence is elucidated in Fig. 15.13. There, we find that the prior particles encompass $\sqrt{v_m}$ and our scheme automatically samples the posterior particles around this root but the true state is hidden near the other root, $-\sqrt{v_m}$, so filter divergence is unavoidable. This failure, again, is because we enforce a Bayesian approach to obtain a state estimate (which is ill-posed here) in addition to approximating the state distribution (which is a well-posed problem for any distribution).

15.4 Filter performance on the L-96 model

In this section, we show numerical results of filtering sparsely observed solutions of the 40-mode L-96 model with $F = 8$. This is the 40-mode turbulent dynamical system given by

$$\frac{du_j}{dt} = (u_{j+1} - u_{j-2})u_{j-1} - u_j + F, \quad j = 1, \ldots, 40, \tag{15.24}$$

see Chapter 11 for a detailed discussion on this model. In particular, we consider linear observations at 10 regularly spaced locations separated by $P = 4$ model grid points with observation frequency $T_{\text{obs}} = 0.25$ (this observation time is rather long, roughly corresponding to 28 hours according to Lorenz's doubling time argument but it is still shorter than the correlation time $T_{\text{corr}}/\sqrt{E_p} = 0.41$ where T_{corr} is recorded in Table 11.1 and E_p is the energy perturbation defined in (11.4)). These observations are relatively precise, i.e. they are corrupted by a spatially and temporally independent Gaussian noise with mean zero and small variance $r^o = 0.01$.

In our implementation, we apply the local least-squares (LLS) algorithm (see Chapter 11.5 or Anderson (2003)) to the MEPF that sequentially updates the model state using

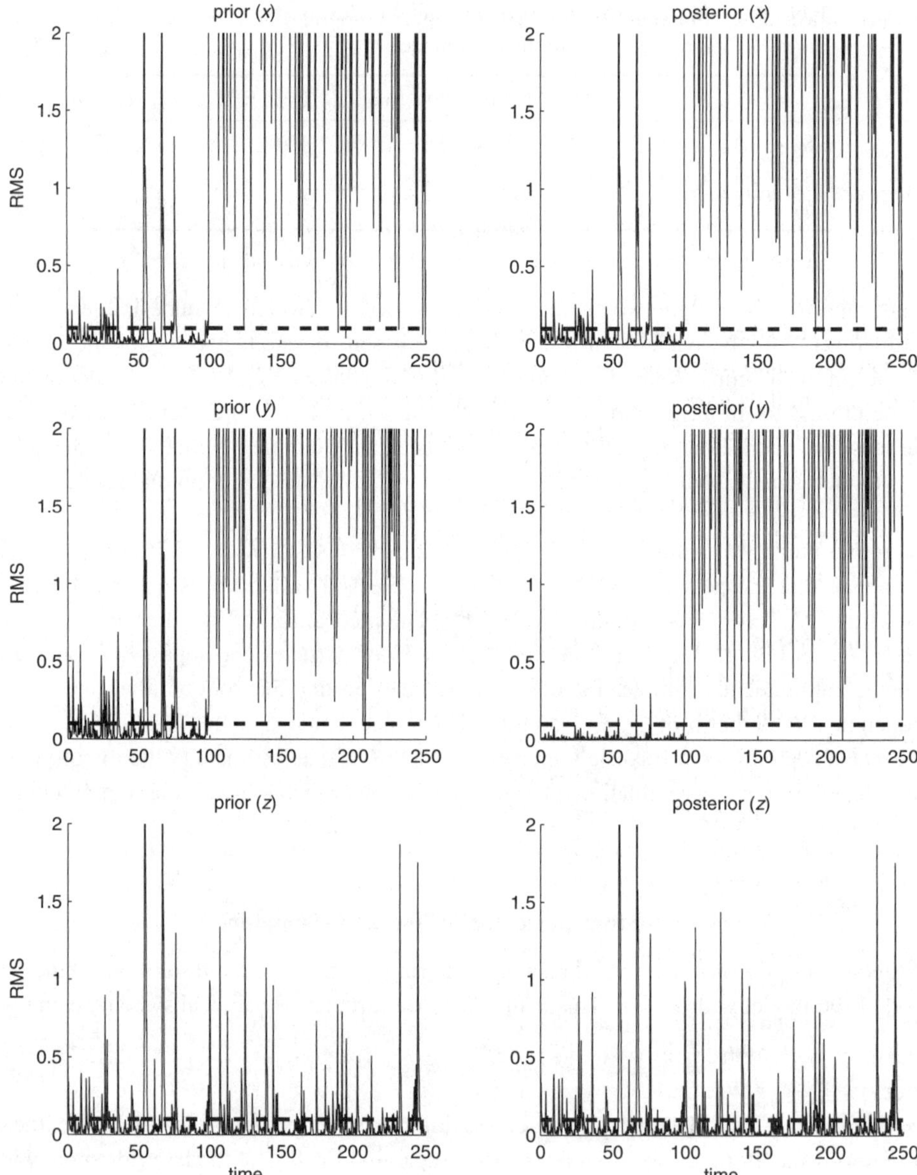

Figure 15.12 RMS errors as functions of time for simulation with quadratic observations $g(y) = y^2$, $T_{\mathrm{obs}} = 0.25$, $r^o = 0.01$, $K = 25$. The dash line denotes observation noise, $\sqrt{r^o}$.

one scalar observation at a time. This strategy follows the implementation of RHF in Anderson (2010). That is, given a scalar observation the analysis associated with this observation follows exactly the six steps discussed in Chapter 11.5 except that now we replace steps 4 and 5 with MEPF or RHF. In particular, the particle filter applies the Bayesian correction on each one-dimensional marginal observation space corresponding to the scalar

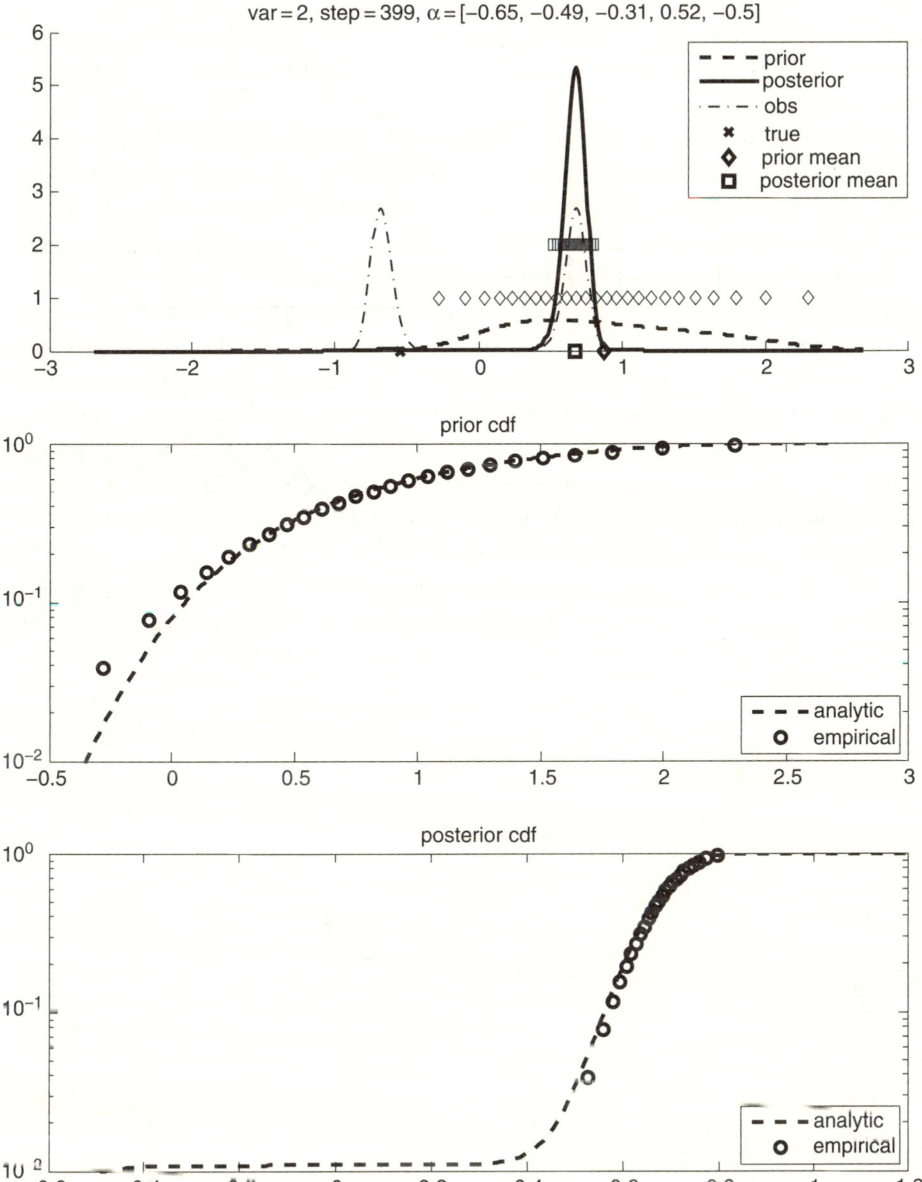

Figure 15.13 Filter divergence: All prior particles encompass the wrong peak. This filter divergence is not so surprising since the observation operator is not a one-to-one function.

observation and the posterior update is linearly interpolated (with local least squares) to its adjacent L grid points (see (11.7) in Chapter 11.5). Though this algorithm seems to work really well for RHF in various regimes in Anderson (2010), we found that it is not so robust in the difficult regime above.

In Fig. 15.14, we show the average RMS errors for MEPF, RHF and EAKF, as functions of time for simulation with ensemble size $K = 100$ and LLS applied with a local radius of $L = 3$ grid points. We have also exhaustively checked different ensemble sizes, observation networks, and also varied the local radius length L and we found that the characteristic solutions are not different from the case we present here so we don't report them. For the first few hundred iterations, MEPF performs as well as the other two algorithms but then we see a large burst of errors at time 100, both for RHF and MEPF. The EAKF sometimes fails too (e.g. in between periods of time 180–200). When particle filter divergence occurs (e.g. at time 102.5), the spatial snapshot in Fig. 15.15 suggests that the RHF ensemble collapses and the MEPF ensemble spreads out with large uncertainty especially at the unobserved locations. Notice also that all three schemes track the true signal accurately at the observed locations (see Fig. 15.16) but sometimes are inaccurate at unobserved grid points (see Fig. 15.17). This suggests that the LLS sometimes fails to interpolate the nonlinearly updated particles to their adjacent neighbors. When we use the Gaussian EAKF, this works fairly well because the LLS algorithm

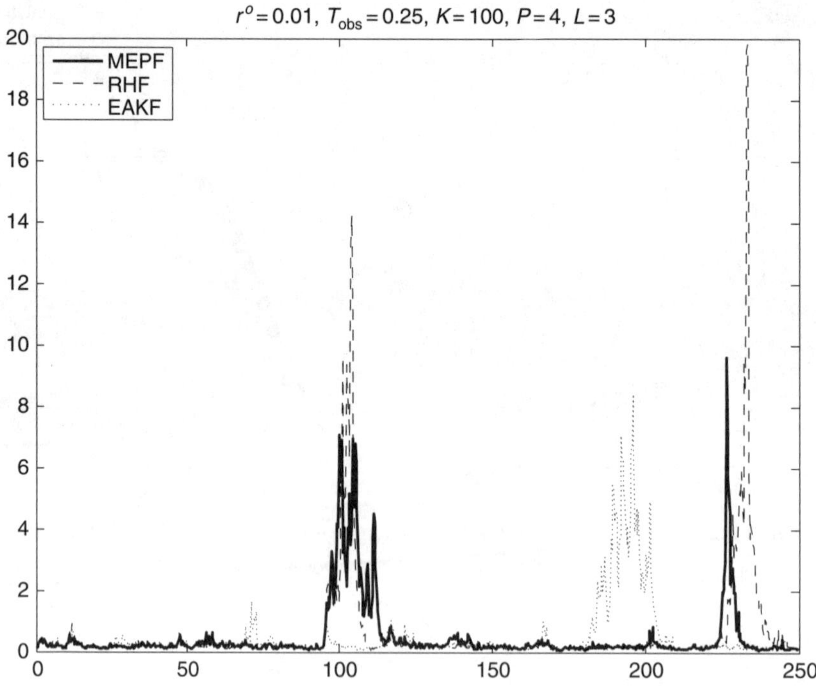

Figure 15.14 RMS errors as functions of time: $\Delta t = 0.25$, $r^o = 0.01$, $K = 100$, $P = 4$, $L = 3$.

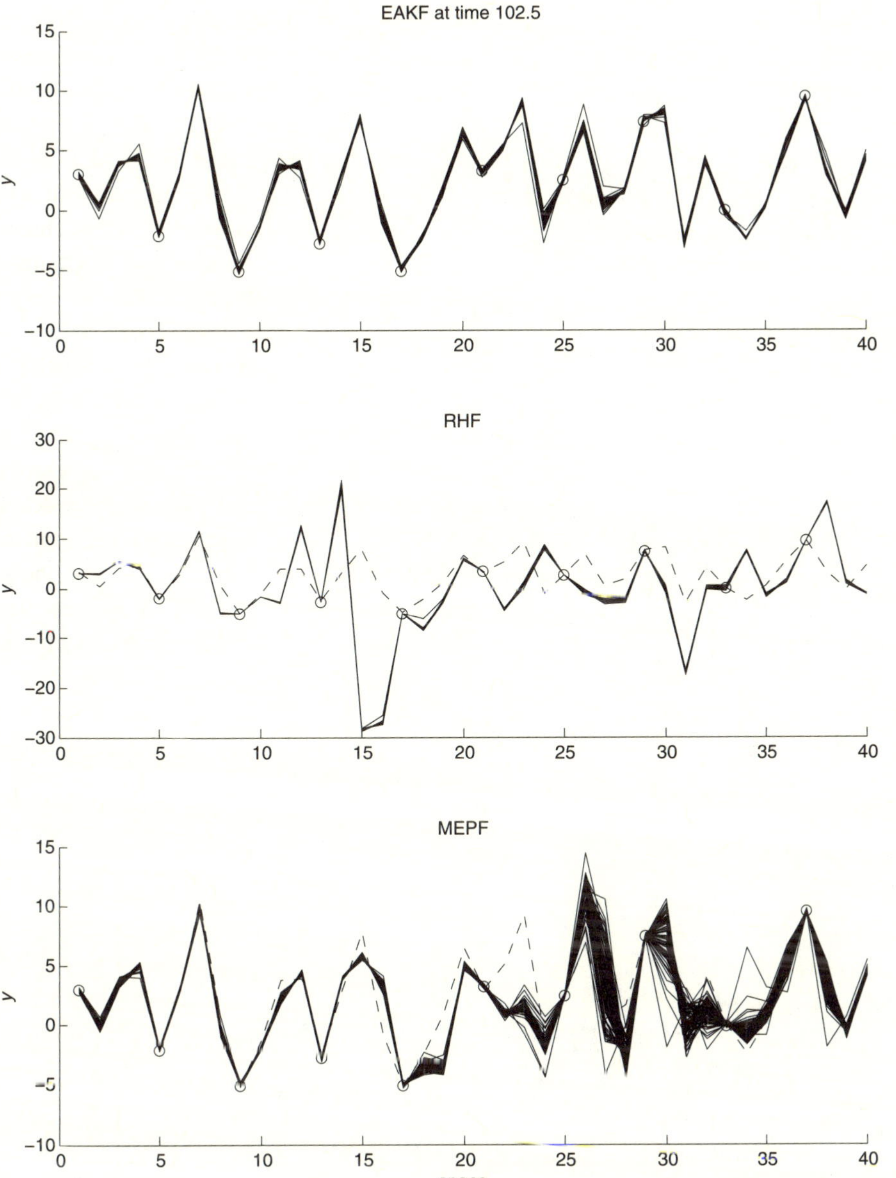

Figure 15.15 Snapshots at time 102.5 of the experiment in Fig. 15.14: true signal (dashes), observations (circle) and posterior ensemble solutions (solid line).

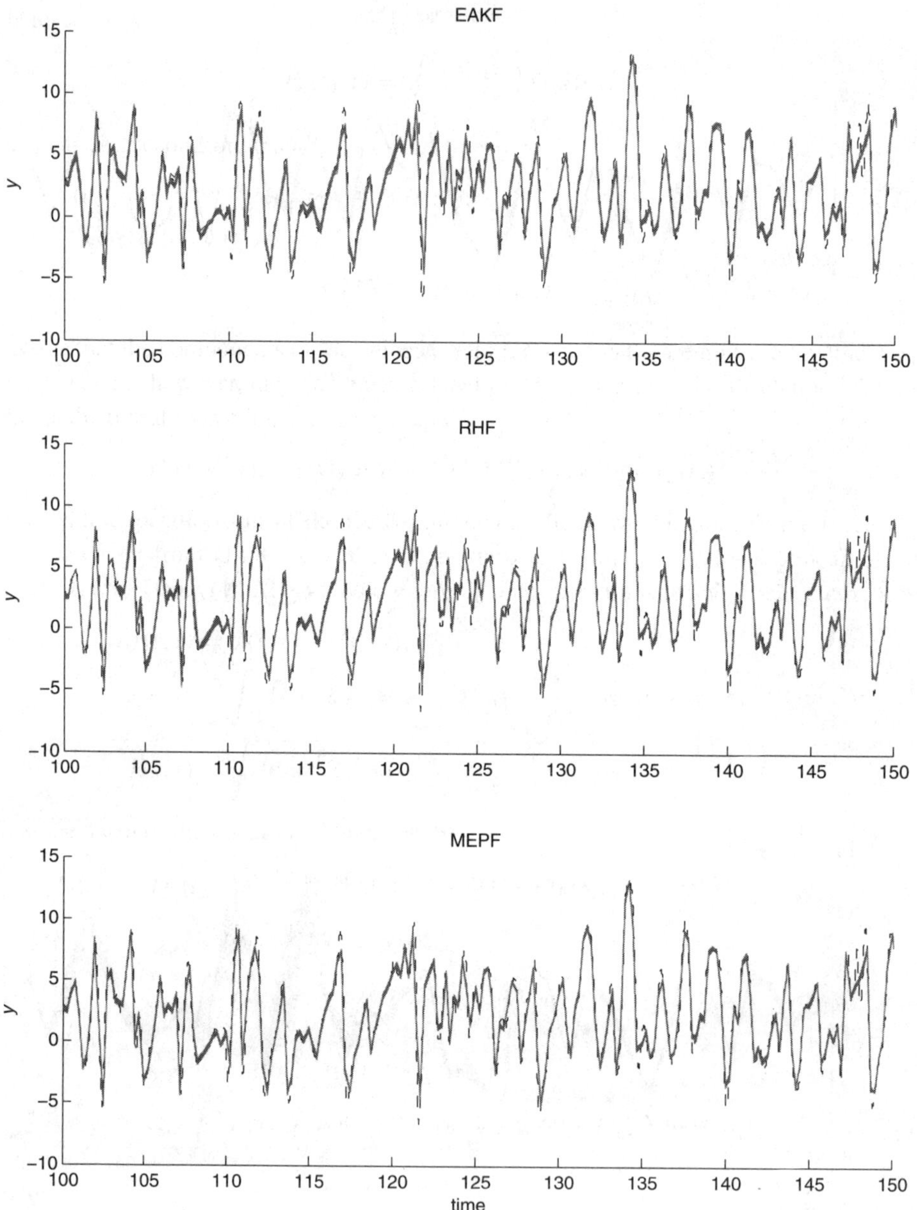

Figure 15.16 Time series of the observed grid point, x_{17}, of the experiment in Fig. 15.14: posterior ensemble solutions (gray solid line) and true signal (dashes).

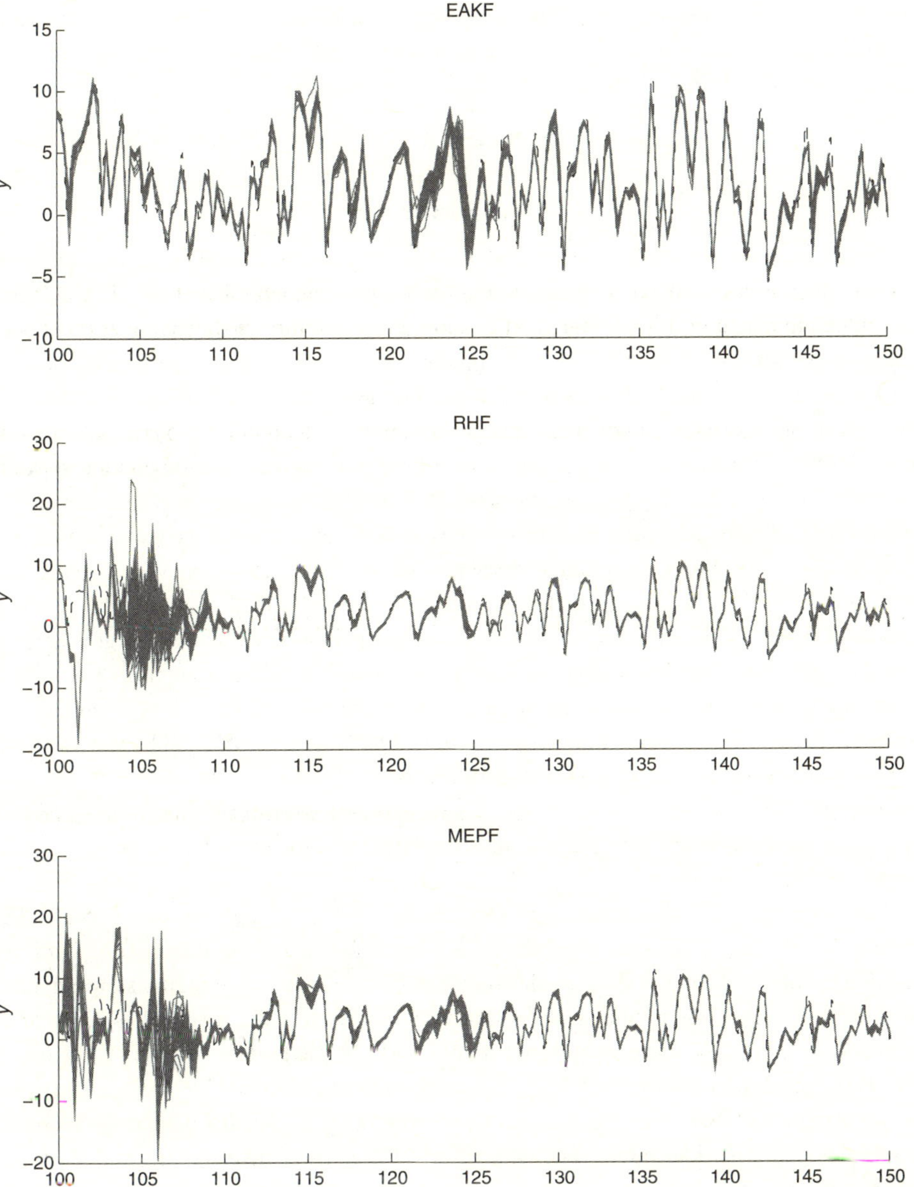

Figure 15.17 Time series of the unobserved field, x_{23}, of the experiment in Fig. 15.14: posterior ensemble solutions (grey solid line) and true signal (dashes).

was designed for optimality of a Gaussian filter (Anderson, 2003). The failure here corresponds to the difficulty in sampling dynamically balanced solutions between two marginal observation spaces and this issue is well-known in random sampling of high-dimensional systems.

15.5 Discussion

In this chapter, we provide an overview of what lies beyond the Gaussian-based Kalman filter and the nonlinear stochastic parametrization extended Kalman filter, NEKF and SPEKF developed in Chapters 12–14. Here, we explore the possibility of designing non-Gaussian (or particle) filtering strategies for high-dimensional systems. There are various different techniques for sampling high-dimensional variables as we discussed in Sections 15.1–15.2. One class of algorithms, known as sequential importance sampling, generates posterior particles by duplicating and eradicating some of the prior particles based on their posterior weights (Doucet *et al.*, 2008); another class of algorithms introduces various clever techniques to faithfully generate random samples in high-dimensional systems (Chorin and Krause, 2004; Del Moral, 1996; Del Moral and Jacod, 2001; Rossi and Vila, 2006; Chorin and Tu, 2009). The advantage of the former class over the latter class is that it always produces dynamically balanced solutions since prior solutions with large weights are kept. On the other hand, the latter class (assuming judicious sampling strategies) will not suffer from "an ensemble collapse" as often occurs in the former class on high-dimensional system (Bickel *et al.*, 2008; Bengtsson *et al.*, 2008; Snyder *et al.*, 2008; van Leeuwen, 2009). Moreover, the second class of algorithms is not restricted to stochastic models since no duplicate particles are used. All these long-standing known pros and cons make the design of robust and efficient particle filtering strategies for high-dimensional systems a very challenging problem.

The algorithm that we introduce in this chapter, the maximum entropy particle filter, belongs to the latter class of algorithms. We have shown its significant skill with very small ensemble size of 3 or 10 particles on the L-63 model in tough regimes of temporally sparse partial linear and nonlinear observations compared to other approaches. Despite this success, we report the strengths and weaknesses of our method and we do not claim that this is the method of choice for filtering high-dimensional systems. We advocate MEPF as an invaluable tool that elucidates various dynamical range issues that need to be accounted for in designing particle filters. The authors are still actively investigating different types of interpolators that can produce dynamically balanced solutions yet are able to interpolate the non-Gaussian estimates from MEPF at the observed grid points to the unobserved grid points (recall that the non-Gaussianity is lost when we interpolate with local least squares). We are also considering applying MEPF on different coordinate bases and in higher-dimensional marginal spaces (the latter of course will increase the computational complexity).

From our study in this chapter, we also see a lot of evidence that using the Bayesian approach that is designed to update the distribution of solutions may not be appropriate for state estimation with small ensemble size. The only reason why this approach is so successful with the Kalman-based strategy is because the Gaussian distribution is completely characterized by its first two moments and hence one can use the first moment, the mean, as the best estimate for the hidden true state and the second moment, the covariance, to quantify the uncertainty of the mean estimate. Commenting on our own work, this means that the average RMS error between the mean estimate and the hidden true state is not an appropriate quantifying measure for judging the filter performance. When the distribution of interest is non-Gaussian, the fair test is to compare the distribution of the true signal with the approximate posterior density at each observation time. This test, however, is an extremely difficult task for high-dimensional problems because we often do not know the distribution of the true signal (this involves solving the Fokker–Planck equation for high-dimensional problems).

References

Abramov, R. 2006. A practical computational framework for the multidimensional moment-constrained maximum entropy principle. *J. Comput. Phys.*, **211**(1), 198–209.

Abramov, R. V. 2007. An improved algorithm for the multidimensional moment-constrained maximum entropy problem. *J. Comput. Phys.*, **226**(1), 621–644.

Abramov, R. V. 2009. The multidimensional moment-constrained maximum entropy problem: A BFGS algorithm with constraint scaling. *J. Comput. Phys.*, **228**(1), 96–108.

Abramov, R. V. 2010. The multidimensional maximum entropy moment problem: A review on numerical methods. *Comm. Math. Sci.*, **8**(2), 377–392.

Abramov, R. and Majda, A. J. 2004. Quantifying uncertainty for non-Gaussian ensembles in complex systems. *SIAM J. Sci. Comput.*, **25**(2), 411–447.

Abramov, R. V. and Majda, A. J. 2007. Blended response algorithm for linear fluctuation–dissipation for complex nonlinear dynamical systems. *Nonlinearity*, **20**, 2793–2821.

Anderson, B. D. and Moore, J. B. 1979. *Optimal Filtering*. Prentice-Hall, Englewood Cliffs, NJ.

Anderson, J. L. 2001. An ensemble adjustment Kalman filter for data assimilation. *Monthly Weather Rev.*, **129**, 2884–2903.

Anderson, J. L. 2003. A local least squares framework for ensemble filtering. *Monthly Weather Rev.*, **131**(4), 634–642.

Anderson, J. L. 2007. An adaptive covariance inflation error correction algorithm for ensemble filters. *Tellus A*, **59**, 210–224.

Anderson, J. L. 2010. A non-Gaussian ensemble filter update for data assimilation. *Monthly Weather Rev.*, **138**(11), 4186–4198.

Anderson, J. L. and Anderson, S. L. 1999. A Monte Carlo implementation of the nonlinear filtering problem to produce ensemble assimilations and forecasts. *Monthly Weather Rev.*, **127**, 2741–2758.

Baek, S.-J., Hunt, B. R., Kalnay, E., Ott, E. and Szunyogh, I. 2006. Local ensemble Kalman filtering in the presence of model bias. *Tellus A*, **58**(3), 293–306.

Bain, A. and Crisan, D. 2009. *Fundamentals of Stochastic Filtering*. Springer, New York.

Bengtsson, T., Bickel, P. and Li, B. 2008. Curse of dimensionality revisited: Collapse of the particle filter in very large scale systems. In: Nolan, D. and Speed, T. (eds), *IMS Lecture Notes – Monograph Series in Probability and Statistics: Essays in Honor of David A. Freedman*, vol. 2. Institute of Mathematical Sciences, pp. 316–334.

Bensoussan, A. 2004. *Stochastic Control of Partially Observable Systems*. Cambridge University Press, Cambridge.

Berliner, L. M., Milliff, R. F. and Wikle, C. K. 2003. Bayesian hierarchical modeling of air–sea interaction. *J. Geophys. Res.*, **108**, 3104–3120.

Bickel, P., Li, B. and Bengtsson, T. 2008. Sharp failure rates for the bootstrap filter in high dimensions. In: *IMS Lecture Notes – Monograph Series: Essays in Honor of J. K. Gosh*, vol. 3. Institute of Mathematical Sciences, pp. 318–329.

Bishop, C. H., Etherton, B. and Majumdar, S. J. 2001. Adaptive sampling with the ensemble transform Kalman filter. Part I: The theoretical aspects. *Monthly Weather Rev.*, **129**, 420–436.

Bourlioux, A. and Majda, A. J. 2002. Elementary models with probability distribution function intermittency for passive scalars with a mean gradient. *Phys. Fluids*, **14**(2), 881–897.

Branicki, M., Gershgorin, B. and Majda, A. J. 2011. Filtering skill for turbulent signals for a suite of nonlinear and linear extended Kalman filters. *J. Comput. Phys.* doi:10.1016/j.jcp.2011.10.029.

Burgers, G., van Leeuwen, P. J. and Evensen, G. 1998. On the analysis scheme in the ensemble Kalman filter. *Monthly Weather Rev.*, **126**, 1719–1724.

Cane, M. A., Kaplan, A., Miller, R. N., Tang, B., Hackert, E. C. and Busalacchi, A. J. 1996. Mapping tropical Pacific sea level: Data assimilation via a reduced state space Kalman filter. *J. Geophys. Res.*, **101**, 22599–22617.

Castronovo, E., Harlim, J. and Majda, A. J. 2008. Mathematical test criteria for filtering complex systems: Plentiful observations. *J. Comput. Phys.*, **227**(7), 3678–3714.

Chorin, A. J. and Krause, P. 2004. Dimensional reduction for a Bayesian filter. *Proc. Natl Acad. Sci.*, **101**(42), 15013–15017.

Chorin, A. J. and Tu, X. 2009. Implicit sampling for particle filters. *Proc. Nat. Acad. Sci.*, **106**, 17249–17254.

Chui, C. and Chen, G. 1999. *Kalman Filtering: With Real-Time Applications*. Springer, New York.

Cohn, S. E. and Dee, D. P. 1988. Observability of discretized partial differential equations. *SIAM J. Numer. Anal.*, **25**(3), 586–617.

Cohn, S. E. and Parrish, S. F. 1991. The behavior of forecast error covariances for a Kalman filter in two dimensions. *Monthly Weather Rev.*, **119**, 1757–1785.

Constantin, P., Foias, C., Nicolaenko, B. and Temam, R. 1988. *Integral Manifolds and Inertial Manifolds for Dissipative Partial Differential Equations*. Applied Mathematical Sciences, vol. 70. Springer, Berlin.

Courtier, P., Andersson, E., Heckley, W., Pailleux, J., Vasiljevic, D., Hamrud, M., Hollingsworth, A., Rabier, F. and Fisher, M. 1998. The ECMWF implementation of three-dimensional variational assimilation (3D-Var). I: Formulation. *Quart. J. Roy. Meteorol. Soc.*, **124**, 1783–1807.

Daley, R. 1991. *Atmospheric Data Analysis*. Cambridge University Press, New York.

Dee, D. P. and Da Silva, A. M. 1998. Data assimilation in the presence of forecast bias. *Quart. J. Roy. Meteorol. Soc.*, **124**, 269–295.

Dee, D. P. and Todling, R. 2000. Data assimilation in the presence of forecast bias: the GEOS moisture analysis. *Monthly Weather Rev.*, **128**(9), 3268–3282.

Del Moral, P. 1996. Nonlinear filtering: Interacting particle solutions. *Markov Proce. Related Fields*, **2**(44), 555–580.

Del Moral, P. and Jacod, J. 2001. Interacting particle filtering with discrete observation. In: Doucet, A., de Freitas, N., and Gordon, N. (eds), *Sequential Monte Carlo Methods in Practice*, Springer, Berlin, pp. 43–75.

Delsole, T. 2004. Stochastic model of quasigeostrophic turbulence. *Surv. Geophys.*, **25**(2), 107–149.

Doucet, A., De Freitas, N. and Gordon, H. J. 2008. *Sequential Monte Carlo Methods in Practice*. Springer, Berlin.

Embid, P. F. and Majda, A. J. 1998. Low Froude number limiting dynamics for stably stratified flow with small or fixed Rossby Number. *Geophys. Astrophys. Fluid Dyn.*, **87**, 1–50.

Evensen, G. 1994. Sequential data assimilation with a nonlinear quasi-geostrophic model using Monte Carlo methods to forecast error statistics. *J. Geophys. Res.*, **99**, 10143–10162.

Evensen, G. 2003. The ensemble Kalman filter: theoretical formulation and practical implementation. *Ocean Dynam.*, **53**, 343–367.

Farrell, B. F. and Ioannou, P. J. 2001. State estimation using a reduced-order Kalman filter. *J. Atmos. Sci.*, **58**(23), 3666–3680.

Farrell, B. F. and Ioannou, P. J. 2005. Distributed forcing of forecast and assimilation error systems. *J. Atmos. Sci.*, **62**(2), 460–475.

Folland, G. B. 1999. *Real Analysis*. Wiley-Interscience, New York.

Franzke, C. and Majda, A. J. 2006. Low-order stochastic mode reduction for a prototype atmospheric GCM. *J. Atmos. Sci.*, **63**, 457–479.

Friedland, B. 1969. Treatment of bias in recursive filtering. *IEEE Trans. Automat. Contr.*, **AC-14**, 359–367.

Friedland, B. 1982. Estimating sudden changes of biases in linear dynamical systems. *IEEE Trans. Automat. Contr.*, **AC-27**, 237–240.

Gandin, L. S. 1965. Objective analysis of meteorological fields. *Gidrometerologicheskoe Izdatelstvo* (English translation by Israeli Program for Scientific Translations, Jerusalem, 1963).

Gardiner, C. W. 1997. *Handbook of Stochastic Methods for Physics, Chemistry, and the Natural Sciences*. Springer-Verlag, New York.

Gaspari, G. and Cohn, S. E. 1999. Construction of correlation functions in two and three dimensions. *Quart. J. Roy. Meteorol. Soc.*, **125**, 723–758.

Gershgorin, B. and Majda, A. J. 2008. A nonlinear test model for filtering slow–fast systems. *Comm. Math. Sci.*, **6**(3), 611–649.

Gershgorin, B. and Majda, A. J. 2010. Filtering a nonlinear slow–fast system with strong fast forcing. *Comm. Math. Sci.*, **8**(1), 67–92.

Gershgorin, B. and Majda, A. J. 2011. Filtering a statistically exactly solvable test model for turbulent tracers from partial observations. *J. Comput. Phys.*, **230**(4), 1602–1638.

Gershgorin, B., Harlim, J. and Majda, A. J. 2010a. Improving filtering and prediction of spatially extended turbulent systems with model errors through stochastic parameter estimation. *J. Comput. Phys.*, **229**(1), 32–57.

Gershgorin, B., Harlim, J. and Majda, A. J. 2010b. Test models for improving filtering with model errors through stochastic parameter estimation. *J. Comput. Phys.*, **229**(1), 1–31.

Ghil, M. and Malanotte-Rizolli, P. 1991. Data assimilation in meteorology and oceanography. *Adv. Geophys.*, **33**, 141–266.

Grote, M. J. and Majda, A. J. 2006. Stable time filtering of strongly unstable spatially extended systems. *Proce. Nat. Acad. Sci.*, **103**, 7548–7553.

Haken, H. 1975. Analogy between higher instabilities in fluids and lasers. *Phys. Lett. A*, **53**(1), 77–78.

Hamill, T. M., Whitaker, J. S. and Snyder, C. 2001. Distance-dependent filtering of background error covariance estimates in an ensemble Kalman filter. *Monthly Weather Rev.*, **129**, 2776–2790.

Harlim, J. 2006. *Errors in the initial conditions for numerical weather prediction: A study of error growth patterns and error reduction with ensemble filtering*. Ph.D. thesis, University of Maryland, College Park, MD.

Harlim, J. 2011. Interpolating irregularly spaced observations for filtering turbulent complex systems. *SIAM J. Sci. Comp.*, **33**(5), 2620–2640.

Harlim, J. and Hunt, B. R. 2007a. A non-Gaussian ensemble filter for assimilating infrequent noisy observations. *Tellus A*, **59**(2), 225–237.

Harlim, J. and Hunt, B. R. 2007b. Four-dimensional local ensemble transform Kalman filter: numerical experiments with a global circulation model. *Tellus A*, **59**(5), 731–748.

Harlim, J. and Majda, A. J. 2008a. Filtering nonlinear dynamical systems with linear stochastic models. *Nonlinearity*, **21**(6), 1281–1306.

Harlim, J. and Majda, A. J. 2008b. Mathematical strategies for filtering complex systems: Regularly spaced sparse observations. *J. Comput. Phys.*, **227**(10), 5304–5341.

Harlim, J. and Majda, A. J. 2010a. Catastrophic filter divergence in filtering nonlinear dissipative systems. *Comm. Math. Sci.*, **8**(1), 27–43.

Harlim, J. and Majda, A. J. 2010b. Filtering turbulent sparsely observed geophysical flows. *Monthly Weather Rev.*, **138**(4), 1050–1083.

Haven, K., Majda, A. J. and Abramov, R. 2005. Quantifying predictability through information theory: small sample estimation in a non-Gaussian framework. *J. Comput. Phys.*, **206**, 334–362.

Houtekamer, P. L. and Mitchell, H. L. 2001. A sequential ensemble Kalman filter for atmospheric data assimilation. *Monthly Weather Rev.*, **129**, 123–137.

Hunt, B. R., Kostelich, E. J. and Szunyogh, I. 2007. Efficient data assimilation for spatiotemporal chaos: a local ensemble transform Kalman filter. *Physica D*, **230**, 112–126.

Isaacson, E. and Keller, H. B. 1966. *Analysis of Numerical Methods*. John Wiley, New York.

Jazwinski, A. H. 1970. *Stochastic Processes and Filtering Theory*. Academic Press, San Diego, CA.

Junk, M. 2004. Maximum entropy moment problems and extended Euler equations. In: *Proceedings of IMA Workshop "Simulation of Transport in Transition Regimes"*, Springer, Berlin, pp. 189–198.

Kaipio, J. P. and Somersalo, E. 2005. *Statistical and Computational Inverse Problems*. Springer, New York.

Kallianpur, G. 1980. *Stochastic Filtering Theory*. Springer-Verlag, Berlin.

Kalman, R. E. and Bucy, R. 1961. New results in linear prediction and filtering theory. *Trans. AMSE J. Basic Eng.*, **83D**, 95–108.

Kalnay, E. 2003. *Atmospheric Modeling, Data Assimilation, and Predictability*. Cambridge University Press, Cambridge.

Kalnay, E., Li, H., Miyoshi, T., Yang, S.-C. and Ballabrera-Poy, J. 2007. 4D-Var or ensemble Kalman filter? *Tellus A*, **59A**, 758–773.

Kang, E. L. and Harlim, J. 2011. Filtering nonlinear spatio-temporal chaos with an autoregressive linear stochastic model. Submitted to *Physica D*.

Keating, S. R., Majda, A. J. and Smith, K. S. 2011. New methods for estimating poleward eddy heat transport using satellite altimetry. Submitted to *Monthly Weather Rev.*.

Keppenne, C. L. 2000. Data assimilation into a primitive-equation model with a parallel ensemble Kalman filter. *Monthly Weather Rev.*, **128**, 1971–1981.

Khouider, B. and Majda, A. J. 2006. A simple multicloud parametrization for convectively coupled tropical waves. Part I: Linear Analysis. *J. Atmos. Sci.*, **63**, 1308–1323.

Khouider, B. and Majda, A. J. 2007. A simple multicloud parametrization for convectively coupled tropical waves. Part II: Nonlinear simulations. *J. Atmos. Sci.*, **64**, 381–400.

Kleeman, R. and Majda, A. J. 2005. Predictabiliy in a model of geophysical turbulence. *J. Atmos. Sci.*, **62**, 2864–2879.

Lawler, G. 1995. *Introduction to Stochastic Processes*. Chapman & Hall/CRC, Bora Raton.

Lee, E. B. and Markus, L. 1967. *Foundations of Optimal Control Theory*. John Wiley, New York.

Lorenc, A. C. 1986. Analysis methods for numerical weather prediction. *Quart. J. Roy. Meteorol. Soc.*, **112**, 1177–1194.

Lorenc, A. C. 2003. The potential of the ensemble Kalman filter for NWP-a comparison with 4D-Var. *Quart. J. Roy. Meteorol. Soc.*, **129**, 3183–3203.

Lorenz, E. N. 1963. Deterministic nonperiodic flow. *J. Atmos. Sci.*, **20**, 130–141.

Lorenz, E. N. 1965. A study of the predictability of a 28-variable atmospheric model. *Tellus*, **17**, 321–333.

Lorenz, E. N. 1996. Predictability – a problem partly solved. In: *Proceedings on Predictability, ECMWF, 4–8 September 1995*, pp. 1–18.

Lorenz, E. N. and Emanuel, K. 1998. Optimal sites for supplementary weather observations: Simulations with a small model. *J. Atmos. Sci.*, **55**, 399–414.

Majda, A. J. 2000. Real world turbulence and modern applied mathematics. In: Arnold, V. I., Atiyah, M., Lax, P., and Mazur, B. (eds), *Mathematics: Frontiers and Perspectives 2000*. American Mathematical Society, Providence, RI, pp. 137–151.

Majda, A. J. 2003. *Introduction to PDEs and Waves for the Atmosphere and Ocean*. Courant Lecture Notes in Mathematics, vol. 9. American Mathematical Society, Providence, RI.

Majda, A. J. and Gershgorin, B. 2011. Elementary models for turbulent diffusion with complex physical features: eddy diffusivity, spectrum, and intermittency. *Phil. Trans. Roy. Soc., series B* (in press).

Majda, A. J. and Grote, M. J. 2007. Explicit off-line criteria for stable accurate time filtering of strongly unstable spatially extended systems. *Proc. Nat. Acad. Sci.*, **104**, 1124–1129.

Majda, A. J. and Kramer, P. 1999. Simplified models for turbulent diffusion: theory, numerical modeling, and physical phenomena. *Phys. Rep.*, **214**, 238–574.

Majda, A. J. and Timofeyev, I. 2002. Statistical mechanics for truncations of the Burgers–Hopf equation: A model for intrinsic behavior with scaling. *Milan J. Math.*, **70**, 39–96.

Majda, A. J. and Timofeyev, I. 2004. Low dimensional chaotic dynamics versus intrinsic stochastic chaos: A paradigm model. *Physica D*, **199**, 339–368.

Majda, A. J. and Wang, X. 2006. *Nonlinear Dynamics and Statistical Theories for Basic Geophysical Flows*. Cambridge University Press, Cambridge.

Majda, A. J., Abramov, R. V. and Grote, M. J. 2005. *Information Theory and Stochastics for Multiscale Nonlinear Systems*. CRM Monograph Series vol. 25, American Mathematical Society, Providence, RI.

Majda, A. J., Franzke, C. and Khouider, B. 2008. An applied mathematics perspective on stochastic modelling for climate. *Phil. Trans. A Math. Phys. Eng Sci.*, **366**(1875), 2429–2455.

Majda, A. J., Gershgorin, B. and Yuan, Y. 2009. Low-frequency climate response and fluctuation–dissipation theorems: Theory and practice. *J. Atmos. Sci.* **67**, 1186–1201.

Majda, A. J., Timofeyev, I. and Vanden-Eijnden, E. 1999. Models for stochastic climate prediction. *Proc. Natl. Acad. Sci.*, **96**, 15687–15691.

Majda, A. J., Timofeyev, I. and Vanden-Eijnden, E. 2001. A mathematical framework for stochastic climate models. *Comm. Pure Appl. Math.*, **54**, 891–974.

Majda, A. J., Timofeyev, I. and Vanden-Eijnden, E. 2003. Systematic strategies for stochastic mode reduction in climate. *J. Atmos. Sci.*, **60**, 1705–1722.

Majda, A. J., Timofeyev, I. and Vanden-Eijnden, E. 2006. Stochastic models for selected slow variables in large deterministic systems. *Nonlinearity*, **19**, 769–794.

Miller, R. N., Carter, E. F. and Blue, S. T. 1999. Data assimilation into nonlinear stochastic models. *Tellus A*, **51**(2), 167–194.

Mitchell, H. L., Houtekamer, P. L. and Pellerin, G. 2002. Ensemble size and model-error representation in an ensemble Kalman filter. *Monthly Weather Rev.*, **130**, 2791–2808.

Neelin, J. D., Lintner, B. R., Tian, B., Li, Q., Zhang, L., Patra, P. K., Chahine, M. T. and Stechmann, S. N. 2011. Long tails in deep columns of natural and anthropogenic tropospheric tracers. *Geophys. Res. Lett.*, **37**, L05804.

O'Kane, T. J. and Frederiksen, J. S. 2008. Comparison of statistical dynamical, square root and ensemble Kalman filters. *Entropy*, **10**(4), 684–721.

Oksendal, B. 2003. *Stochastic Differential Equations: An Introduction with Applications*. Universitext. Springer, Berlin.

Ott, E., Hunt, B. R., Szunyogh, I., Zimin, A. V., Kostelich, E. J., Corrazza, M., Kalnay, E. and Yorke, J. A. 2004. A local ensemble Kalman filter for atmospheric data assimilation. *Tellus A*, **56**, 415–428.

Parrish, D. F. and Derber, J. C. 1992. The National Meteorological Centers spectral statistical-interpolation analysis system. *Monthly Weather Rev.*, **120**(8), 1747–1763.

Pedlosky, J. 1979. *Geophysical Fluid Dynamics*. Springer, New York.

Pitcher, E. J. 1977. Application of stochastic dynamic prediction to real data. *J. Atmos. Sci.*, **34**(1), 3–21.

Richtmeyer, R. D. and Morton, K. W. 1967. *Difference Methods for Initial Value Problems*. John Wiley, New York.

Rossi, V. and Vila, J.-P. 2006. Nonlinear filtering in discrete time: A particle convolution approach. *Ann. Inst. Stat. Univ. Paris*, **3**, 71–102.

Salmon, R. 1998. *Lectures on Geophysical Fluid Dynamics*. Oxford University Press, Oxford.

Smith, S. K., Boccaletti, G., Henning, C. C., Marinov, I. N., Tam, C. Y., Held, I. M. and Vallis, G. K. 2002. Turbulent diffusion in the geostropic inverse cascade. *J. Fluid Mech.*, **469**, 13–48.

Snyder, C., Bengtsson, T., Bickel, P. and Anderson, J. L. 2008. Obstacles to high-dimensional particle filtering. *Monthly Weather Rev.* **136**, 4629–4640.

Solari, H. G., Natiello, M. A. and Mindlin, G. B. 1996. *Nonlinear Dynamics*. Institute of Physics, Bristol.

Sparrow, C. 1982. *The Lorenz Equations: Bifurcation, Chaos, and Strange Attractors*. Springer, New York.

Strikwerda, J. 2004. *Finite Difference Schemes and Partial Differential Equations*. (2nd ed.) SIAM, Philadelphia, PA.

Szunyogh, I., Kostelich, E. J., Gyarmati, G., Patil, D. J., Hunt, B. R., Kalnay, E., Ott, E. and Yorke, J. A. 2005. Assessing a local ensemble Kalman filter: perfect model experiments with the NCEP global model. *Tellus A*, **57**, 528–545.

Thomas, S. J., Hacker, J. P. and Anderson, J. L. 2007. A robust formulation of the ensemble Kalman filter. *Quart. J. Roy. Meteorol. Soc.*, **135**, 507–521.

Todling, R. and Ghil, M. 1994. Tracking atmospheric instabilities with the Kalman filter. Part I: Methodology and one-layer results. *Monthly Weather Rev.*, **122**(1), 183–204.

Tribbia, J. J. and Baumhefner, D. P. 1988. The reliability of improvements in deterministic short-range forecasts in the presence of initial state and modeling deficiencies. *Monthly Weather Rev.*, **116**, 2276–2288.

Vallis, G. K. 2006. *Atmospheric and Oceanic Fluid Dynamics: Fundamentals and Large-Scale Circulation.* Cambridge University Press, Cambridge.

van Leeuwen, P. J. 2009. Particle filtering in geosciences. *Monthly Weather Rev.*, **137**, 4089–4114.

Wang, X., Bishop, C. H. and Julier, S. J. 2004. Which is better, an ensemble of positive–negative pairs or a centered spherical simplex ensemble? *Monthly Weather Rev.*, **132**(7), 1590–1605.

Whitaker, J. S. and Hamill, T. M. 2002. Ensemble data assimilation without perturbed observations. *Monthly Weather Rev.*, **130**, 1913–1924.

Yang, S.-C., Baker, D., Cordes, K., Huff, M., Nagpal, F., Okereke, E., Villafane, J., Kalnay, E. and Duane, G. S. 2006. Data assimilation as synchronization of truth and model: experiments with the three-variable Lorenz system. *J. Atmos. Sci.*, **63**(9), 2340–2354.

Young, L.-S. 2002. What are SRB measures, and which dynamical systems have them? *J. Statist. Phys.*, **108**(5), 733–754.

Index